Long-Range
Control of Gene
Expression

Advances in Genetics, Volume 61

Serial Editors

Jeffery C. Hall
Waltham, Massachusetts

Jay C. Dunlap
Hanover, New Hampshire

Theodore Friedmann
La Jolla, California

Veronica van Heyningen
Edinburgh, United Kingdom

Long-Range Control of Gene Expression

Edited by
Veronica van Heyningen
MRC Human Genetics Unit
Western General Hospital
Edinburgh, United Kingdom

Robert E. Hill
MRC Human Genetics Unit
Western General Hospital
Edinburgh, United Kingdom

AMSTERDAM • BOSTON • HEIDELBERG • LONDON
NEW YORK • OXFORD • PARIS • SAN DIEGO
SAN FRANCISCO • SINGAPORE • SYDNEY • TOKYO
Academic Press is an imprint of Elsevier

Academic Press is an imprint of Elsevier
84 Theobald's Road, London WC1X 8RR, UK
Radarweg 29, PO Box 211, 1000 AE Amsterdam, The Netherlands
Linacre House, Jordan Hill, Oxford OX2 8DP, UK
30 Corporate Drive, Suite 400, Burlington, MA 01803, USA
525 B Street, Suite 1900, San Diego, CA 92101-4495, USA

First edition 2008

ISBN: 978-0-12-373881-3
ISSN: 0065-2660

For information on all Academic Press publications
visit our website at books.elsevier.com

Printed and bound in USA
08 09 10 11 12   10 9 8 7 6 5 4 3 2 1

Working together to grow
libraries in developing countries

www.elsevier.com | www.bookaid.org | www.sabre.org

ELSEVIER   BOOK AID International   Sabre Foundation

Contents

Contributors   xi
Preface   xiii

1 Chromatin Structure and the Regulation of Gene Expression: The Lessons of PEV in *Drosophila*   1
Jack R. Girton and Kristen M. Johansen

    I. Introduction: Position Effect in *Drosophila*   2
    II. Historical Background of the PEV Phenotype   5
    III. Types of PEV   8
    IV. Genome Organization and PEV   19
    V. Concluding Remarks   29
    References   31

2 Polycomb Group Proteins and Long-Range Gene Regulation   45
Julio Mateos-Langerak and Giacomo Cavalli

    I. Introduction   46
    II. Genetic and Biochemical Characterization of PcG Proteins   48
    III. PcG Mechanisms of Action   49
    IV. PcG Proteins and Long-Range Gene Silencing   53
    V. PcG and Very Long-Range Gene Silencing: "Teleregulation" of Gene Expression   57
    VI. Conclusions and Prospects   61
    References   62

3 Evolution of *Cis*-Regulatory Sequences in *Drosophila*   67
Pat Simpson and Savita Ayyar

    I. Introduction   68
    II. Developmental Homeostasis, Sequence Turnover, and Stabilizing Selection   72
    III. Enhancer Evolution and Loss or Gain of Traits   84

    IV. *Cis-Trans* Coevolution   91
    V. Evolution of New Regulatory Modules   95
    VI. Conclusions   98
    References   99

4 β-Globin Regulation and Long-Range Interactions   107

Robert-Jan Palstra, Wouter de Laat, and Frank Grosveld

    I. Introduction   108
    II. The β-Globin Locus   109
    III. Models of Long-Range Control of Gene Expression by Enhancers   115
    IV. Long-Range Activation by the β-Globin LCR   117
    V. Enhancement of Transcription by the β-Globin LCR: Rate-Limiting Steps   127
    VI. The Concept of an Active Chromatin Hub   131
    VII. Future Directions   132
    References   133

5 Long-Range Regulation of α-Globin Gene Expression   143

Douglas R. Higgs, Douglas Vernimmen, and Bill Wood

    I. Introduction   144
    II. The Normal Structure and Evolution of the α-Globin Cluster   146
    III. Functional Analysis of the α-Globin Regulatory Domain   147
    IV. Structure of the Upstream Regulatory Elements and the Promoters   152
    V. Transcription Factors Involved in Erythropoiesis   153
    VI. Cellular Resources for Studying the Key Stages of Hematopoiesis   157
    VII. Transcription Factor Binding to the Upstream Regulatory Elements   158
    VIII. Transcription Factor Binding to the Promoter   159
    IX. The Recruitment of RNA Polymerase and GTFs to the α-Globin Cluster   160
    X. What Role Do the Remote Regulatory Elements Play?   161
    XI. How Do the Upstream Elements Interact with the Promoter?   162

XII. Sequential Activation of the α-Globin Gene
   Cluster During Differentiation   165
XIII. Conclusions, Speculation, and Future Directions   166
   References   169

6   **Global Control Regions and Regulatory Landscapes**
   in Vertebrate Development and Evolution   175
   Francois Spitz and Denis Duboule

     I. Introduction   176
    II. Global Controls   178
   III. Co-Expression Chromosomal Territories,
       Regulatory Landscapes, and Global Control Regions   185
    IV. Mechanims of Underlying Global Regulation   188
     V. Co-Expression Chromosomal Territories, Regulatory
       Landscapes: Bystander Effects or Functional Operons?   192
    VI. Evolutionary Implications of Global Gene Control   195
   VII. Global Regulation, Chromosomal Architecture,
       and Genetic Disorders   197
  VIII. Concluding Remarks   197
       References   198

7   **Regulation of Imprinting in Clusters: Noncoding RNAs**
   Versus Insulators   207
   Le-Ben Wan and Marisa S. Bartolomei

     I. Introduction   208
    II. Insulator Model of Regulation   209
   III. The ncRNA Model of Regulation   212
    IV. *Dlk1/Gtl2* Imprinted Cluster: A Bit of Everything   215
     V. Conclusions   218
       References   219

8   **Genomic Imprinting and Imprinting Defects**
   in Humans   225
   Bernhard Horsthemke and Karin Buiting

     I. Introduction   226
    II. The Mechanisms of Genomic Imprinting   226
   III. Imprinting Defects   230
    IV. Conclusions   242
       References   243

9 **Epigenetic Gene Regulation in Cancer   247**
    Esteban Ballestar and Manel Esteller

    I. Introduction   248
    II. Cancer Cells Show A Disruption of DNA Methylation
        Patterns   250
    III. Disruption of the Histone Modification Profile
        in Cancer   253
    IV. Cascades of Epigenetic Deregulation in Cancer   256
    V. What Are the Mechanisms That Lead to
        Aberrant Methylation Patterns in Cancer?   258
    VI. Epigenetic Therapy for Cancer Treatment   261
        References   262

10 **Genomic Identification of Regulatory Elements by
    Evolutionary Sequence Comparison and Functional
    Analysis   269**
    Gabriela G. Loots

    I. Introduction   270
    II. Genomic Architecture of the Human Genome   271
    III. Computational Methods of Predicting
        Regulatory Elements   276
    IV. *In Vivo* Validation and Characterization of
        Transcriptional Regulatory Elements   282
    V. Conclusions   286
        References   287

11 **Regulatory Variation and Evolution: Implications for
    Disease   295**
    Emmanouil T. Dermitzakis

    I. Introduction   296
    II. Evolution and Variation of Noncoding DNA   297
    III. Natural Selection in Noncoding DNA   300
    IV. Gene Expression Studies   301
    V. Disease Implications   302
    VI. Conclusions   303
        References   304

12 Organization of Conserved Elements Near Key Developmental Regulators in Vertebrate Genomes 307

Adam Woolfe and Greg Elgar

    I. Introduction   308
    II. Gene-Regulatory Networks in Development   309
    III. Identification of Evolutionarily Constrained Sequences Using Phylogenetic Footprinting   310
    IV. Searches for Regulatory Elements Using Evolutionary Conservation   310
    V. *Takifugu Rubripes*: A Compact Model Genome   312
    VI. Identification of Enhancer Elements Through Fish-Mammal Comparisons   313
    VII. Fish-Mammal Conserved Noncoding Elements Are Associated with Vertebrate Development   315
    VIII. High-Resolution Analysis of the Organization of CNEs Around Key Developmental Regulators   317
    IX. General Genomic Environment Around CNEs   317
    X. CNEs Present in Transcripts   319
    XI. CNEs Located Within UTRs   321
    XII. CNEs Are Located Large Distances Away from Their Putative Target Gene   323
    XIII. Discussion   328
    References   333

13 Long-Range Gene Control and Genetic Disease 339

Dirk A. Kleinjan and Laura A. Lettice

    I. From Genetic Disease to Long-Range Gene Regulation   340
    II. Position Effect Revisited   342
    III. Loss of a Positive Regulator   345
    IV. TWIST, POU3F4, PITX2, SOX3, GLI3, and FOXP2   346
    V. The "Bystander" Effect   353
    VI. MAF, SDC2, TGFB2, REEP3, and PLP1   355
    VII. Two Position Effects—Different Outcomes   357
    VIII. Phenotypes Resulting from Position Effects on More than One Gene   362
    IX. Global Control Regions; HOXD, Gremlin, and Limb Malformations   363
    X. FOX Genes and Position Effects   365

XI. SOX9 and Campomelic Displasia   366
XII. Facioscapulohumeral Dystrophy   368
XIII. Aberrant Creation of an Illegitimate siRNA
        Target Site   370
XIV. Genetic Disease Due to Aberrant Gene Transcription
        Can Be Caused by Many Different Mechanisms   371
XV. Concluding Remarks   378
        References   379

Index   389

CONTRIBUTORS

Numbers in parentheses indicate the pages on which the authors' contributions begin.

Savita Ayyar (67) Department of Zoology, University of Cambridge, United Kingdom

Esteban Ballestar (247) Cancer Epigenetics Group, Molecular Pathology Programme, Spanish National Cancer Centre (CNIO), 28029 Madrid, Spain

Marisa S. Bartolomei (207) Department of Cell and Developmental Biology, University of Pennsylvania School of Medicine, Philadelphia, Pennsylvania 19104

Karin Buiting (225) Institut für Humangenetik, Universitätsklinikum Essen, Hufelandstrasse 55, 45122 Essen, Germany

Giacomo Cavalli (45) Chromatin and Cell Biology Lab, Institute of Human Genetics, CNRS, 141, rue de la Cardonille, 34396 Montpellier, France

Wouter de Laat (107) Department of Cell Biology and Genetics, Erasmus MC, 3000 CA Rotterdam, The Netherlands

Emmanouil T. Dermitzakis (295) Wellcome Trust Sanger Institute, Wellcome Trust Genome Campus, CB10 1SA Cambridge, United Kingdom

Denis Duboule (175) NCCR "Frontiers in Genetics" and Department of Zoology and Animal Biology, University of Geneva, 1211 Geneva, Switzerland
School of Life Sciences, Federal Polytechnic School, CH-1015 Lausanne, Switzerland

Greg Elgar (307) School of Biological and Chemical Sciences, Queen Mary, University of London, London E1 4NS, United Kingdom

Manel Esteller (247) Cancer Epigenetics Group, Molecular Pathology Programme, Spanish National Cancer Centre (CNIO), 28029 Madrid, Spain

Jack R. Girton (1) Department of Biochemistry, Biophysics, and Molecular Biology, Iowa State University, Ames, Iowa 50011

Frank Grosveld (107) Department of Cell Biology and Genetics, Erasmus MC, 3000 CA Rotterdam, The Netherlands

Douglas R. Higgs (143) MRC Molecular Haematology Unit, Weatherall Institute of Molecular Medicine, John Radcliffe Hospital, Headington, Oxford OX3 9DS, United Kingdom

Bernhard Horsthemke (225) Institut für Humangenetik, Universitätsklinikum Essen, Hufelandstrasse 55, 45122 Essen, Germany

Kristen M. Johansen (1) Department of Biochemistry, Biophysics, and Molecular Biology, Iowa State University, Ames, Iowa 50011

Dirk A. Kleinjan (339) MRC Human Genetics Unit, Western General Hospital, Edinburgh EH4 2XU, United Kingdom

Laura A. Lettice (339) MRC Human Genetics Unit, Western General Hospital, Edinburgh EH4 2XU, United Kingdom

Gabriela G. Loots (269) Biosciences and Biotechnology Division, Chemistry, Materials and Life Sciences Directorate, Lawrence Livermore National Laboratory, Livermore, California 94550

Julio Mateos-Langerak (45) Chromatin and Cell Biology Lab, Institute of Human Genetics, CNRS, 141, rue de la Cardonille, 34396 Montpellier, France

Robert-Jan Palstra (107) Department of Cell Biology and Genetics, Erasmus MC, 3000 CA Rotterdam, The Netherlands

Pat Simpson (67) Department of Zoology, University of Cambridge, United Kingdom

Francois Spitz (175) Developmental Biology Unit, EMBL, 69117 Heidelberg, Germany

Douglas Vernimmen (143) MRC Molecular Haematology Unit, Weatherall Institute of Molecular Medicine, John Radcliffe Hospital, Headington, Oxford OX3 9DS, United Kingdom

Le-Ben Wan (207) Department of Cell and Developmental Biology, University of Pennsylvania School of Medicine, Philadelphia, Pennsylvania 19104

Bill Wood (143) MRC Molecular Haematology Unit, Weatherall Institute of Molecular Medicine, John Radcliffe Hospital, Headington, Oxford OX3 9DS, United Kingdom

Adam Woolfe (307) School of Biological and Chemical Sciences, Queen Mary, University of London, London E1 4NS, United Kingdom

Preface

What constitutes long-range control is, by its nature, difficult to define. In Drosophila, *cis*-acting long-range elements located at a distance of a kilobase or so from the target gene's promoter are considered to be operating at a long distance. On the other hand, in mammals the classic long-range element is the locus control region (LCR) of the beta-globin locus, sitting 50kb from the globin gene cluster. We now know that *cis*-regulatory elements can lie at a distance of at least a megabase, but the absolute distance that regulators are able to span to control gene expression is unknown. The properties that dictate these regulatory limits are still open to investigation. But no matter the distance, the goal is the same; that is, to convey regulatory information across an intervening stretch of DNA to control target gene expression. Although some aspects of regulatory control have been known for three quarters of a century, the recent availability of multiple genomic sequences and high-throughput technologies has generated an emerging view of regulatory mechanisms. In this volume, we cover some of the most recent advances in this fast-moving area of genomic mechanisms involved in gene regulation.

Control of gene expression is the major key to regulating cellular function in developing and adult organisms. Transcriptional control may be imposed by different mechanisms, ranging from regional chromatin organization to very specific transcription factor binding at gene-specific sites. One of the earliest observed phenomena was position effect variegation in Drosophila, which is explored from the current vantage point here. Drosophila studies were frequently at the forefront in subsequent discoveries, such as the identification of more widely acting regulators like the polycomb group of proteins and some of the earliest analyses of *cis*-regulatory sequences. One advantage of Drosophila is the availability of multiple species sequences for comparison, so that divergence as well as conservation of homologous regulatory sites can be assessed at the bioinformatic and functional level. It has been possible to explore how alterations in regulatory elements mirror the evolutionary changes in detailed expression pattern and changing morphology.

Switching from Drosophila to mammals, and utilizing initial information from human disease variants, one of the best explored gene expression systems has been the beta-globin complex where the exquisitely controlled developmental switching from closely linked embryonic to fetal and adult gene copies has been studied in great detail. Many aspects of now more widely used technology for the study of expression control were developed for this system.

It is a salutary lesson then to see that the alpha-globin complex, which maps to a different chromosome, reveals significant differences, as well as some similarities in the regulatory mechanisms utilized to control expression of the other subunit for this crucial tetrameric protein where levels of expression of the two different subunits are very tightly coordinated. The four clustered Hox complexes in mammals constitute another classical set of multigene assemblies, which have been elegantly dissected to reveal additional layers of complexity.

Imprinting, or parent-of-origin-specific gene expression at a number of important growth regulatory loci, began to be explored systematically in mice a few decades ago. Here, methylation differences were identified and the role of noncoding RNAs and insulator elements was gradually uncovered, revealing a complex system of regulation that is re-set every generation. As with the globin genes, human disease-associated abnormalities have contributed significantly to our understanding of this esoteric level of gene expression control. Insights from some of the implicated genes suggested that altered methylation and other epigenetic changes may play a key role in the development of malignancies.

With the advent of multiple genome databases, sophisticated bioinformatics method have been developed for the identification of conserved regulatory elements across different evolutionary distances. Methodologies have also been developed to couple bioinformatic identification to functional analysis. A major emerging theme from this has been that regulatory changes play a key role in the evolution of different vertebrate species, much as has already been observed for Drosophila. Interestingly, the possibility of broader sequence comparison analysis has revealed that the existence of long-range control, sometimes over regions of more than a megabase, has helped to shape genome evolution significantly. The largest regulatory distances are associated with major developmental control genes which fulfill multiple roles during early development and often in later maintenance of gene function. Some of these gene domains are so large that they overlap with flanking gene territories so that neighboring genes have to maintain their links with each other through evolution, leading to regions of conserved synteny.

Not surprisingly, mutations in regulatory regions are increasingly shown to be associated with human disease. However, currently, we are only observing the tip of the iceberg. It is becoming clear that many common disease-association studies are identifying noncoding region variants as the underlying cause of these later onset disorders. It will be exciting, and potentially useful for disease management and treatment, to see what aspects of fine tuning are altered in different anomalies. Areas for future exploration will include the mechanisms through which physiological and environmental changes are translated into altered gene function through long-range regulatory networks.

We want to thank all the authors who delivered the 13 excellent reviews for this volume and our publishing editors for help and guidance in getting the book out.

Robert E. Hill and Veronica van Heyningen

1 Chromatin Structure and the Regulation of Gene Expression: The Lessons of PEV in *Drosophila*

Jack R. Girton and Kristen M. Johansen

Department of Biochemistry, Biophysics, and Molecular Biology,
Iowa State University, Ames, Iowa 50011

I. Introduction: Position Effect in *Drosophila*
II. Historical Background of the PEV Phenotype
   A. The PEV phenotype and heterochromatin
   B. Heterochromatin and euchromatin
III. Types of PEV
   A. Chromosomal rearrangement PEV
   B. Transposon insertion PEV
   C. Pairing-dependent dominant PEV: *Trans*-inactivation
IV. Genome Organization and PEV
   A. Chromatin structure
   B. Nuclear organization
V. Concluding Remarks
   Acknowledgments
   References

ABSTRACT

Position-effect variegation (PEV) was discovered in 1930 in a study of X-ray-induced chromosomal rearrangements. Rearrangements that place euchromatic genes adjacent to a region of centromeric heterochromatin give a variegated phenotype that results from the inactivation of genes by heterochromatin spreading from the breakpoint. PEV can also result from P element insertions that place euchromatic genes into heterochromatic regions and rearrangements

Advances in Genetics, Vol. 61
0065-2660/08 $35.00
DOI: 10.1016/S0065-2660(07)00001-6

that position euchromatic chromosomal regions into heterochromatic nuclear compartments. More than 75 years of studies of PEV have revealed that PEV is a complex phenomenon that results from fundamental differences in the structure and function of heterochromatin and euchromatin with respect to gene expression. Molecular analysis of PEV began with the discovery that PEV phenotypes are altered by suppressor and enhancer mutations of a large number of modifier genes whose products are structural components of heterochromatin, enzymes that modify heterochromatic proteins, or are nuclear structural components. Analysis of these gene products has led to our current understanding that formation of heterochromatin involves specific modifications of histones leading to the binding of particular sets of heterochromatic proteins, and that this process may be the mechanism for repressing gene expression in PEV. Other modifier genes produce products whose function is part of an active mechanism of generation of euchromatin that resists heterochromatization. Current studies of PEV are focusing on defining the complex patterns of modifier gene activity and the sequence of events that leads to the dynamic interplay between heterochromatin and euchromatin.   © 2008, Elsevier Inc.

I. INTRODUCTION: POSITION EFFECT IN *DROSOPHILA*

The discovery of position effects in *Drosophila* began with the efforts of many investigators to generate a coherent picture of the nature of genes and chromosomes and of the relationship between them. Between 1910 and 1930, many genes were discovered in *Drosophila* by the recovery of mutant alleles. Analysis of mutant phenotypes suggested that each gene had a specific function and linkage mapping indicated that each gene had a unique chromosomal location (its locus). Cytological studies of mitotic and salivary gland polytene chromosomes indicated that chromosomes contain two visibly different regions, heterochromatin and euchromatin. Heterochromatin consists of regions that remain condensed and densely stained during interphase, while euchromatin consists of regions that become diffused during interphase and form the visible, banded regions of the polytene chromosomes. Euchromatin contains high concentrations of genes and heterochromatin contains very few genes. Although many believed that differential regulation of gene expression was an underlying feature of eukaryotic development, these investigations did not provide information about how gene activity was regulated. Since chromosomal rearrangements, such as inversions or translocations, that changed gene location but did not eliminate genes often had no visible mutant phenotype, chromosome structure was not believed to have a role in regulating gene function. Chromosomes appeared to simply serve as vehicles for gene segregation during cell division. Two major findings challenged this view and launched an intensive study of the role of chromosome structure in gene regulation that continues today.

The first finding came in Sturtevant's study of the *Bar* eye mutation (Sturtevant, 1925). *Bar* is a dominant, sex-linked mutation that reduces the number of facets in the compound eye (normally about 800). The *Bar* mutation is a duplication of the *Bar* locus (*B*). Normal females have two copies of the locus, one on each X chromosome (*B/B*), but *Bar*-mutant females have four copies, two on each chromosome (BB/BB). Sturtevant described a new mutation (called *ultraBar* or *double-Bar*) that caused an even greater reduction in the number of facets. The *ultraBar* mutation has three copies of the *Bar* locus on one chromosome (*BBB*) and homozygous *ultraBar* females have a total of six copies (*BBB/BBB*). This suggested that increasing the number of copies of the *Bar* gene causes a decrease in eye size. However, Sturtevant noted that homozygous *Bar* females (*BB/BB*) have an average of 68×12 facets per eye, while females heterozygous for *ultraBar* (*BBB/B*) have an average of only 45×42 facets. Since both genotypes contain four copies of the *Bar* locus, this difference could not be due to differences in the number of copies of the locus. Sturtevant concluded that the arrangement of the copies of the *Bar* gene in the chromosome had an influence on *Bar* gene expression. He named this influence "position effect." This suggested that chromosomes are not passive collections of genes, but that they contain an internal structure that has a role in the regulation of gene expression.

The second finding came in 1930, when Muller reported the discovery of a series of X-ray-induced mutations affecting the *white* gene. Previously discovered *white* mutations reduced the amount of eye pigment, changing eye color from the normal uniform red to a uniform orange, yellow, or white (Lindsley and Zimm, 1992; Morgan *et al.*, 1925). Muller's new mutations showed a "mottled" phenotype, with each eye having some white (mutant) and some red (normal) regions (Fig. 1.1). The size, shape, and location of the white regions varied from eye to eye. The mutations causing these mottled phenotypes were chromosomal rearrangements that changed the location of the *white* gene in the chromosome (Muller, 1930). This rearrangement-induced mottled phenotype was named "variegated position effect" or "V-type" to distinguish it from the "stable position effect" or "S-type" position effect of Bar. Muller's discovery of the phenomenon of position-effect variegation (PEV) launched an intense study of the role of chromosome structure in gene expression that has continued for more than 75 years. This interest was partly because of the intriguing novel variegated phenotypes and partly because studies of PEV appeared likely to give important information about how gene expression was controlled by chromosomal organization.

PEV in *Drosophila* presents an interesting example of the historical progression of genetic analysis. During the past 75 years, the discovery of new investigative techniques has invariably led to a multitude of PEV studies using

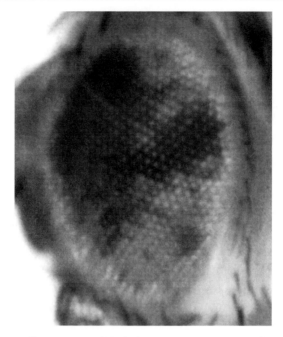

Figure 1.1. Position-effect variegation (PEV) of an inversion chromosome. The mottled pigmentation observed in the eye of an individual carrying the $In(1)w^{m4}$ inversion chromosome shows regions where the *white* gene has been silenced (white) and regions where the *white* gene is actively expressed (dark).

these techniques. The sheer number of studies and the volume of information accumulated about PEV present a formidable task for the reader. There have been a number of extensive reviews of the experimental literature about PEV (Farkas *et al.*, 2000; Huisinga *et al.*, 2006; Lewis, 1950; Schulze and Wallrath, 2007; Spofford, 1976; Spradling and Karpen, 1990; Weiler and Wakimoto, 1995; Zhimulev, 1998). It is not the purpose of this chapter to repeat these works. Our goal is to summarize what has been learned about certain key aspects of PEV that occur reliably and repeatedly and which are considered defining features of the PEV phenomenon. These reflect, in a manner that is not currently understood, fundamental characteristics of underlying molecular/biochemical processes of chromatin organization that affect gene regulation. We will attempt to connect these characteristics of PEV with the results of recent molecular studies of the formation and molecular structure of chromatin. These results reflect the current excitement in the field, as it appears that the long examination of PEV may be approaching the point of finally revealing the role that different forms of chromatin structure play in the regulation of gene expression, and how that regulation occurs. To cover such a large topic, we will have to be selective in the experiments we discuss and in the original publications we cite.

II. HISTORICAL BACKGROUND OF THE PEV PHENOTYPE

In 1930, Muller described a series of mutants induced by X-ray treatment. Muller originally discovered the first such mutation as a dominant *Notch* wing mutation. Dominant *Notch* mutations produce nicks or "notches" in the wings of heterozygous (*N*/+) flies. Muller knew *Notch* mutations sometimes resulted from deletions of portions of the X chromosome, so he generated flies heterozygous for the new *Notch* and a chromosome containing a recessive mutant allele of the nearby *white* locus (*N* (?)/+ *w*), expecting to see either a white eye phenotype, if a deletion removing both *Notch* and *white* was present, or a red eye phenotype if it were not. However, what he observed fit neither of these expectations. "To the great surprise of the writer the Notch-winged offspring of this cross had neither white nor normal red eyes, nor even eyes of any uniform intermediate colour. They had mottled eyes, and exhibited various grades and sizes of lighter and darker areas" (Muller, 1930). Further study revealed that the *Notch*-mutant phenotype was also "mottled" and that individuals showing a stronger mutant eye phenotype (lighter eyes) also had a stronger *Notch* phenotype (larger wing notches). This mutation was a translocation that had attached the distal tip of the X chromosome (the normal location of the *Notch* and *white* genes) to the third chromosome. Muller discovered four additional "white-mottled" mutations that did not involve mutations of *Notch* in additional irradiation experiments and all of these also were rearrangements (translocations or inversions) that moved the region of the X chromosome containing the *white* locus. Studying these mutations for more than 50 generations convinced Muller that the white mottled was a permanent mutant phenotype. He concluded that the change in the chromosomal arrangement of the genes on the X chromosome was responsible for the appearance of what he called the "eversporting displacement" phenotype, although he admitted to having no good explanation for how this worked. Muller's paper, with its striking color diagrams showing the white-mottled phenotype, made an enormous impression and stimulated a number of other investigators to begin studying the relationship between chromosomal rearrangements and "mottled" gene expression. Studies of several genes quickly demonstrated that the mottled phenotype was produced only by certain types of chromosomal rearrangements and that the PEV phenotypes given by all rearrangements had a number of consistent characteristics (Demerec, 1941; Dobzhansky, 1936; Lewis, 1950; Muller, 1932).

A. The PEV phenotype and heterochromatin

The PEV phenotypic effects Muller and other early investigators recovered were all produced by chromosomal rearrangements with one breakpoint close to the variegating gene and the other in a heterochromatic region that brought the affected gene into close proximity to heterochromatin. Examples of such PEV-inducing rearrangements are shown in Fig. 1.2. For example, the N^{264-52} X chromosome contains an inversion with one breakpoint in the centric heterochromatin and the other between *rst* and *w*. This places normal alleles of the *bi*, *ec*, *dm*, *fa*, and *rst* genes close to the centric heterochromatin. This inversion

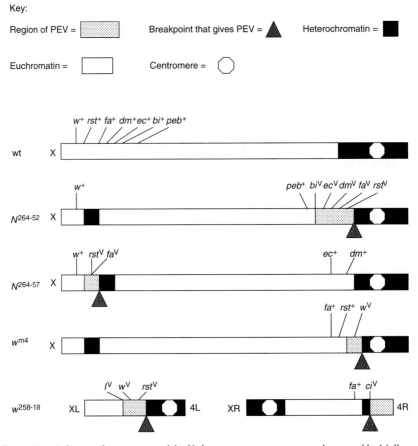

Figure 1.2. A diagram showing some of the X chromosome rearrangements discovered by Muller to give position-effect variegation (PEV). The relationship between the chromosomal breakpoint and several genes that are affected is shown. Adapted from Lewis (1950).

6 Girton and Johansen

generates a variegating loss-of-function phenotype for each of these genes. Females that are heterozygous for the N^{264-52} chromosome and a normal chromosome containing a recessive loss-of-function mutant allele of any of these genes have a variegating mutant phenotype for that gene. The variegation results from the alleles of the genes on the rearranged chromosome being repressed in some cells, and active in other cells. Females heterozygous for the rearranged chromosome and a chromosome with normal alleles of these genes have a normal phenotype, indicating that the PEV effect is recessive and does not affect the alleles on the normal chromosome. Other rearrangements, such as N^{264-57} or w^{m4}, have breakpoints in different locations and have different genes that show PEV but the effect is the same. PEV is produced in translocations, such as w^{258-18}, that move euchromatic genes on one chromosome adjacent to a heterochromatic region of a different chromosome, or in rearrangements, such as N^{264-57}, that move a region of heterochromatin into a euchromatic location (Fig. 1.2). The requirement for PEV is clearly that a euchromatic gene be moved adjacent to a heterochromatic breakpoint (Spofford, 1976; Zhimulev, 1998). This suggested that PEV phenotypes result from heterochromatic chromatin structure "spreading" from the breakpoint into the adjacent euchromatin and inactivating nearby euchromatic genes. The variegated phenotype was proposed to result from the length of spreading along the chromosome being variable.

B. Heterochromatin and euchromatin

Because heterochromatin plays a central role in any discussion of PEV, it is worthwhile to briefly discuss some key features of heterochromatin and euchromatin. Heitz (1928) discovered that portions of the chromosomes remain highly condensed and heavily stained throughout the cell cycle. He named these regions "heterochromatin" to distinguish them from the regions showing variable staining and condensation, which he called "euchromatin." The boundaries of heterochromatin as defined by such cytological analyses are not precise and may change in different tissues or with different analytical techniques. In *Drosophila*, heterochromatic regions are located near the centromeres (the pericentric heterochromatin), at the telomeres, interspersed throughout the fourth chromosome, throughout the Y chromosome, and at a number of defined "intercalary heterochromatic" sites scattered among the chromosome arms (Zhimulev, 1998; Zhimulev and Belyaeva, 2003). In most *Drosophila* somatic cells, 30–35% of the genome is heterochromatic. Heterochromatin makes up all of the Y chromosome, 40% of the X, 25% of chromosomes 2 and 3, and more than half of the fourth. These heterochromatic regions contain a high percentage of families of repetitive DNA sequences with specific highly and moderately repetitive sequences clustered at particular chromosomal locations. The arrangement of these clusters of sequences is conserved among populations of *Drosophila melanogaster* (Pimpinelli *et al.*, 1995; Weiler and Wakimoto, 1995). In *Drosophila*, the pericentric heterochromatin is divided into two visibly distinct categories, α-heterochromatin and β-heterochromatin. The α-heterochromatin is largely composed of highly repetitive satellite sequences and is located in and around the centromere. It coalesces into a highly condensed, uniform body in the center of the chromocenter in salivary gland nuclei. β-Heterochromatin is more diffuse and expands or "loops out" to form the bulk of the visible chromocenter. β-Chromatin more closely resembles euchromatin. It contains functioning genes interspersed with middle repeated sequences, and a significant percentage consists of sequences derived from transposons (Pimpinelli *et al.*, 1995). More than 40 functioning genes in heterochromatic regions have been identified by mutant phenotypes, including visible phenotypes, lethality, and sterility (Gatti and Pimpinelli, 1992; Weiler and Wakimoto, 1995). *Drosophila* chromosomes contain complex repeated sequences at the telomeres. These include copies of two retrotransposons (HeT-A and TART) at the end of the telomere that have the ability to repair or heal damaged telomeres (Pardue and DeBaryshe, 1999; Wallrath, 2000). Proximal to these sequences, *Drosophila* telomeres contain repeated sequences called telomere-associated sequences (Karpen and Spradling, 1992). These sequences differ in length and sequence from chromosome to chromosome (Walter *et al.*, 1995).

III. TYPES OF PEV

There are a variety of ways in which a mosaic phenotype can be produced in *Drosophila*, including somatic recombination, somatic chromosomal loss, and somatic P element transposition. In this chapter, we will consider as PEV only those variegated phenotypes that are produced by chromosomal rearrangements that share the common feature that they bring euchromatic genes into close association with heterochromatin without altering the gene itself. These can be grouped into three classes or types based on a number of characteristic features.

A. Chromosomal rearrangement PEV

The common feature of all chromosomal rearrangements that give PEV is that they either move a euchromatic gene into a pericentric region or move a block of heterochromatic sequences into a euchromatic region. Chromosomal rearrangements involving the X, the second, the third, and the fourth chromosomes give PEV phenotypes (Demerec, 1941; Kaufmann, 1946; Lewis, 1950; Weiler and Wakimoto, 1995; Zhimulev, 1998). The heterochromatic regions that can produce a PEV phenotype are the pericentric regions of all chromosomes, most of the fourth chromosome, and all of the Y chromosome. Several experiments in which the variegating allele is removed from the rearranged chromosome show the PEV effect is a repression of gene expression and not a permanent change in the allele. Dubinin and Sidorov (1935) used meiotic recombination to remove the variegated allele of the *hairy* gene from a translocation between the third and fourth chromosome. A double crossover in flies heterozygous for this translocation and normal chromosomes moved the variegating allele to the normal chromosome. In the normal chromosome, the allele gave a normal, *hairy*⁺ phenotype. When this same allele was reinserted into the translocated chromosome by another double crossover event, the allele again showed a variegating phenotype. Similar experiments with other variegating genes gave the same results (Lewis, 1950; Spofford, 1976). Inducing an additional chromosomal rearrangement has also been used to separate the variegating allele from the heterochromatin. Panshin (1938) used X-rays to induce a second translocation event that returned the portion of the X chromosome containing the variegating *white* allele in the w^{m11} translocation to a euchromatic chromosomal location where it again gave a normal phenotype. Several similar experiments demonstrated that variegating alleles of any gene separated from the heterochromatic breakpoint by another chromosomal rearrangement ceases to show the variegating phenotype (Demerec, 1941; Hinton, 1950; Judd, 1955; Lewis, 1950). Another type of evidence that PEV is not the result of a change in gene structure came from studies of gene products using alleles that produce proteins with detectable structural differences. In one such experiment, Bahn (1971) studied individuals heterozygous for T(1;2)OR32 carrying the *Amy2* allele and a normal chromosome carrying the *Amy4,6* allele using gel electrophoresis to detect which electrophoretic variant proteins were produced. In variegating individuals, the level of the Amy2 variant was reduced, and the level of reduction was greater when the strength of the PEV was increased, but the structure of the protein product was not altered. The level of the Amy4,6 enzyme produced by the allele on the normal chromosome was not altered, indicating that the PEV effect acted only on the allele in the rearranged chromosome. In other similar studies, PEV was also found to reduce enzyme levels from alleles of *sc*, *dor*, and *pn* (Gerasimova *et al.*, 1972).

In a study of PEV reversion, Tartof *et al.* (1984) induced a series of rearrangements of the w^{m4} chromosome. The w^{m4} chromosome has an inversion with one breakpoint about 25 kb from the *white* locus and the other in the centric heterochromatin and produces a white PEV phenotype. Tartof recovered three rearrangements that had a normal phenotype. Each had one breakpoint in the heterochromatin beyond the w^{m4} breakpoint and they all retained some heterochromatic sequences at the original breakpoint. This indicates that the breakpoint itself did not cause the PEV phenotype. Reuter *et al.* (1985) studied 51 reversions of the w^{m4} chromosome and found only 3 true revertants with no detectable PEV phenotype. The others showed a reduced PEV phenotype, and each of these retained blocks of heterochromatin at the original breakpoint site. These studies indicate that PEV results from the influence of a significant region of heterochromatic sequence adjacent to euchromatic genes, and that it does not simply result from a breakpoint, or from the gene being moved close to a centromere. They also suggest that the PEV effect is correlated with the amount of heterochromatic sequences and that there is a lower limit on the size of the block of heterochromatin that can cause PEV.

1. The spreading effect

The strength of PEV (the frequency with which a gene is inactivated) is inversely proportional to distance from the breakpoint (Lewis, 1950). For example, as diagrammed in Fig. 1.2, chromosomes containing the inversion N^{264-52} give a variegated phenotype for five genes (rst, fa, dm, ec, bi). The strength of the PEV effect on each was measured and genes further from the breakpoint showed less frequent or less strong mutant phenotypes, while those closer to the breakpoint showed more frequent or stronger mutant phenotypes (Demerec, 1940, 1941). This type of observation led to the hypothesis that PEV phenotypes resulted from a physical "spreading" of heterochromatin from the breakpoint into the euchromatic regions of the chromosome (Schultz, 1939). The distance between Notch and white is 330 kb, and in the original mutation of Muller both are frequently inactivated. The bi locus inactivated in the N^{264-52} chromosome is located at least 50 salivary gland chromosome bands away from the breakpoint, a molecular distance of more than 1mb of DNA. The PEV effect of $T(1;4)w^{m258-21}$ can extend over 67 bands (Hartman-Goldstein, 1967), that of $Dp(1;F)R$ over 86 bands (Spofford, 1976; Weiler and Wakimoto, 1995). These observations indicate that the spreading effect is able to cover significant chromosomal distances.

Two early studies suggested that the heterochromatic spreading effect in PEV is linear, that is, does not jump over genes. Demerec and Slizynska (1937) studied the PEV given by the w^{258-18} chromosome of roughest (rst) and white (w), two genes whose phenotypes can be scored in the eye. The gene order on the chromosome is: breakpoint, rst, w. An examination of the PEV phenotype showed the eye facets had one of three phenotypes: smooth and red (rst^+, w^+), rough and red (rst^V w^+), or rough and white (rst^V, w^V). They observed no facets with a smooth and white (rst^+, w^V) phenotype and concluded that the variegating effect passed in a linear manner from the breakpoint along the chromosome and always inactivated roughest before reaching white. Similar results in a study by Schultz (1941) of the PEV phenotype given by w^{258-21} on split and white reached the same conclusion. In a recent study, Talbert and Henikoff (2000) studied the PEV phenotypes given by X chromosome inversions that had the arrangement of breakpoint, w, and rst. They reported cases of w^+ cells in a rough (rst^V) region of the eye. Does this mean that inactivation has "skipped over" w to inactivate rst? These results must be interpreted with caution. Talbert and Henikoff (2000) themselves noted that the alleles of rst and w used in their studies are not necessarily affecting the same cells. The $white$ mutation is cell autonomous and its phenotype can be seen in each pigment cell in the eye. However, alleles of $roughest$ affect the shape of the facet by causing the inclusion of extra cells in the eye facet. Not all cells in the facet need to be rst^- to give a rough facet phenotype. PEV effects of w^{m4} and other rearrangements on $white$ give a fine-scale variegation and individual facets often contain both white and pigmented cells (Girton, personal observation). It is thus possible that a "rough" facet may be mosaic and that the pigmented cells are actually $w^+\ rst^+$, and the inactivation did not jump over w to inactivate rst. A definitive study of this question would need to use markers whose relationship to the breakpoint is known and whose expression can unambiguously be measured in the same cell.

2. All genes may show PEV

The first studies of PEV involved genes whose mutation produced a visible phenotype (such as w, fa, and rst). Later studies showed that PEV can be detected for a wide range of genes including those whose mutation produces a quantitative phenotype such as the number of bristles ($hairy$), the level of a detectable enzyme product (tryptophan pyrolase in v^+, Tobler $et\ al.$, 1971), enzymatic variants, (Amy and Pdg, Bahn, 1971), the amount of kynurenine production in fat body in $T(1;2)ras^v$ and $T(1;Y)y^+\ Yv^+$ (Rizki, 1961; Tobler $et\ al.$, 1971), or the rate of ribosomal RNA synthesis in female ovaries (sc^{S1} and sc^{L8}, Puckett and Snyder, 1973). Lewis (1950) listed 32 genes shown to variegate. Spofford (1976) concluded that all genes can show a PEV-variegated phenotype in the right rearrangement.

   Some genes are normally located in heterochromatic regions of chromosomes. Estimates of how many such genes exist in $Drosophila$ range from 40 to 100 or more (Schulze and Wallrath, 2007; Spofford, 1976; Weiler and Wakimoto, 1995). A question that was raised very early in the study of PEV was how a gene whose normal location is heterochromatic would respond in a rearrangement that moved it to a euchromatic location. The first such gene studied was the light gene (lt) studied by Schultz and Dobzhansky (1934). The lt maps to a region considered to be β heterochromatin in chromosome 2 (Hilliker, 1976; Schultz and Dobzhansky, 1934). Rearrangements that move lt adjacent to euchromatic regions give a variegated phenotype (Hessler, 1958; Schultz, 1936; Schultz and Dobzhansky, 1934). The conditions under which lt PEV inactivation occurs are the opposite of the conditions under which PEV inactivation of euchromatic genes occurs. At least five other heterochromatic genes normally located near lt show similar position effects when placed next to euchromatin (Wakimoto and Hearn, 1990; Weiler and Wakimoto, 1995). Other heterochromatic genes that show this pattern include *cubitus interruptus* (*ci*) (Dubinin and Sidorov, 1935), *peach* (*pe*) in *Drosophila virilis* (Baker, 1953), male fertility factors on the Y chromosome (Hackstein and Hochstenbach, 1995; Neuhaus, 1939), and the nucleolus organizer region (Baker, 1971). Not all euchromatic regions show the same efficiency in generating heterochromatic PEV. All regions of X chromosome euchromatin can cause heterochromatic gene variegation, but only the distal parts of the arms of chromosome 2 and 3. Breakpoints in the fourth do not induce PEV of heterochromatic genes, perhaps reflecting the fact that the fourth chromosome contains interspersed heterochromatic and euchromatic regions (Spofford, 1976). In two cases, heterochromatic genes showing variegation when removed from surrounding heterochromatin had this variegation eliminated when an additional rearrangement moved them back near a block of heterochromatin (Eberl *et al.*, 1993). This is one of the most intriguing observations of PEV, that genes normally located in hetrochromatin can be inactivated by being placed adjacent to euchromatin. Genes normally located in heterochromatic regions have some differences in molecular structure (Reugels *et al.*, 2000; Risinger *et al.*, 1997; Schulze and Wallrath, 2007; Schulze *et al.*, 2005). These observations suggest heterochromatic genes have evolved to not only tolerate being in a heterochromatic region but to actually require it for normal action.

3. Modifiers of the PEV phenotype

PEV induced by chromosomal rearrangements is strongly affected by *trans*-acting modifiers that increase the severity of the mutant phenotype (enhance variegation) or decrease the severity of the mutant phenotype (suppress variegation). Modifiers of PEV fall into two general classes, large-scale effects and single mutations. One large-scale effect is temperature. Rearing flies at higher temperature (29 °C) suppresses variegation (reduces the severity of the mutant phenotype) and rearing at lower temperature (16–18 °C) enhances variegation (increases the severity of the mutant phenotype). Gowen and Gay (1934) studied the phenotypic effects of temperature on PEV of w, spl, and ec caused by w^{m1}, w^{m2}, and w^{m3}. All of these genes showed more extreme mutant phenotypes at lower temperature. Kaufmann (1942) documented similar high- and low-temperature effects on PEV of rst^3. All other euchromatic genes tested showed similar results, indicating this is a general effect of PEV (Bahn, 1971; Benner, 1971; Schalet, 1969; Spofford, 1976; Wargent, 1972).

PEV phenotypes are modified by changes in the amount of heterochromatic sequences in the genome. This was first observed in studies in which adding an extra copy of the Y chromosome was found to suppress variegation and eliminating the Y chromosome enhanced variegation (Gowen and Gay, 1934). A PEV phenotype is more severely mutant in XO versus XY males or in XX versus XXY females (Lewis, 1950; Spofford, 1976). This effect is seen in genes producing visible phenotypes and those examined by measuring enzyme phenotypes (Bahn, 1971; Gerasimova et al., 1972). In extreme cases, a rearrangement may not show a variegation phenotype except when enhancing modifiers are present. One example studied by Gersh (1963) involved the inversion $In(1) rst^3$, which gives a white variegated phenotype in XO males but not in XY males. Changes in the Y chromosome affect all PEV-inducing rearrangements (Lindsley et al., 1960; Weiler and Wakimoto, 1995; Zhimulev, 1998). Changes in the amount of centric heterochromatin also modify PEV phenotypes in the same manner as the Y chromosome. The deletion of a significant portion of the centric heterochromatic region of the X enhanced PEV of the y, w, and ac loci (Panshin, 1938). The addition or subtraction of an entire X has about the same effect as the addition or subtraction of the Y chromosome (Grell, 1958; Hinton, 1949). Similar effects have been discovered for the centric heterochromatic regions of the autosomes. A second chromosome deficiency that removes heterochromatin from the right arm of the second chromosome, $Df(2R)M-S2^{10}$, enhances the PEV of several different genes in several different rearrangements (Morgan et al., 1941). Duplications of the heterochromatic region of the right arm of the second chromosome suppress PEV phenotypes (Grell, 1970; Hannah, 1951). The fact that centric heterochromatin and the Y chromosome have similar effects on all euchromatic genes showing PEV indicates that this is a general effect of heterochromatin, and not due to specific sequences on the Y.

4. Single gene Su(var) and E(var) mutations

In several of the early studies of PEV, the severity of the phenotypes given by a PEV-inducing rearrangement varied from strain to strain, even when the strains did not differ for any obvious modifying factor (Hinton, 1949). A rearrangement, such as $In(1)w^{m4}$, that gives a PEV phenotype of 50% white eye in one strain could give phenotypes of 25% white eye or 75% white eye in other strains. These "genetic background effects" were believed to be due to unmapped dominant, single gene suppressor or enhancer mutations located throughout the genome (Schultz, 1941; Spofford, 1967, 1976). The first such modifier mutation to be isolated and characterized was a dominant suppressor mutation, $Su(var)$, on the third chromosome (Spofford, 1967). Individuals with an $In(1)w^{m4}$ PEV-inducing genotype that also had the $Su(var)$ allele had almost completely red eyes, while siblings with the $Su(var)^+$ allele had light, variegated eyes, the standard PEV phenotype. The $Su(var)$ allele had similar effects on the PEV phenotypes of w, rst, and dm given by the $In(1)w^{m4}$, $Dp(1;3)N^{264-58}$, and $In(1)rst^3$ chromosomes

(Spofford, 1967, 1973, 1976). The discovery of the $Su(var)$ allele suggested that there might be a number of genes whose mutation could modify PEV phenotypes. This led to several experiments to recover suppressor and enhancer mutations (Dorn et al., 1993; Eissenberg et al., 1990, 1992; Grigliatti, 1991; Locke et al., 1988; Reuter and Wolff, 1981; Sinclair et al., 1983, 2000; Wustmann et al., 1989). A large number of dominant and recessive modifier alleles have been isolated, and it is estimated that there are between 50 and 150 loci that can mutate to give a PEV-modifying phenotype (Weiler and Wakimoto, 1995; Schulze and Wallrath, 2007). The effects of different modifier mutations appear to be similar; all suppressors reduce the severity of the PEV phenotype given by all rearrangements, and all enhancers increase the severity of the PEV phenotype given by all rearrangements (Fig. 1.3).

The PEV phenotype of genes that normally reside in heterochromatic regions shows a characteristic inverse response to modifiers of PEV. Temperature effects on the PEV of the heterochromatic gene lt are the reverse of those on euchromatic gene PEV, with higher temperatures producing a more mutant phenotype (Gersh, 1949). While extra Y chromosomes suppress PEV of euchromatic genes, they act as enhancers of variegation of lt (Baker and Reim, 1962; Schultz, 1936) and ci (Altorfer, 1967). Baker and Spofford (1959) studied the effects of 14 free Y chromosome fragments and reported that the same regions of

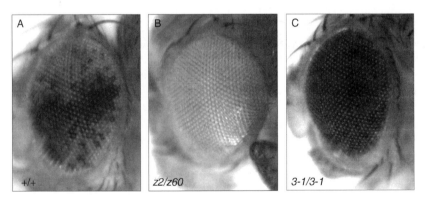

Figure 1.3. The effect of su(var) and e(var) mutations on PEV phenotypes. All individuals are homozygous for the pericentric P insertion line P118E-10 that gives a white PEV phenotype (Wallrath and Elgin, 1995). (A) The eye phenotype of an individual with no modifier mutation is variegated, with the white marker gene actively expressed in some cells but silenced in others. (B) An individual with a strong e(var) genotype, JIL-1^{z2}/JIL-1^{z60}, shows complete silencing of white gene expression. (C) An individual homozygous for a strong su(var) genotype, Su(var)3-1/Su(var)3-1, shows high levels of white gene expression. Photos: X. Bao.

the Y that suppress the w PEV phenotype given by $Dp(1;3)N^{264-58}$ enhanced the lt PEV phenotype given by $T(2;3)ltm^{100}$ (Baker and Reim, 1962). Single-gene mutations that suppress euchromatic gene PEV enhance the PEV phenotype of heterochromatic genes, and single-gene mutations that enhance euchromatic gene PEV suppress the PEV phenotype of heterochromatic genes (Schulze and Wallrath, 2007; Spofford, 1976; Weiler and Wakimoto, 1995). The reason why heterochromatic gene PEV and euchromatic gene PEV should show such consistent and opposite reactions is one of the most intriguing characteristics of PEV.

A paradigm in genetics is that the normal function of a gene can be deduced from the phenotype given by mutant alleles. The observation that mutations of modifier genes alter the PEV phenotype, and that heterochromatin is involved in generating PEV phenotypes, suggested that the normal functions of modifier genes are part of a genetic system that is essential for the establishment, maintenance, or function of chromatin structure. This is supported by several observations. The products of most modifier genes that have been studied to date are either structural components of heterochromatin (Baksa et al., 1993; Henikoff, 1979; James and Elgin, 1986; Locke et al., 1988; Reuter and Spierer, 1992; Reuter et al., 1982; Sinclair et al., 1983), enzymes that modify histones or nonhistone chromatin proteins (Ebert et al., 2006; Lerach et al., 2006; Mottes et al., 2000; Nishioka et al., 2002; Schotta et al., 2004), or nuclear architectural proteins (Bao et al., 2007a; Yamaguchi et al., 2001). For example, the product of the $Su(var)205$ gene is the HP-1 protein, a heterochromatic protein that is found in high concentrations in the chromocenter in salivary gland nuclei (Eissenberg et al., 1990, 1992; James and Elgin, 1986). The product of the $Su(var)3-7$ gene is a large zinc-finger protein that is mainly associated with the pericentric heterochromatin (Reuter et al., 1990). The JIL-1 protein is a tandem kinase that phosphorylates the serine 10 residue in the tail of histone 3 (Jin et al., 1999; Lerach et al., 2006; Zhang et al., 2003) and a dominant gain of function allele of JIL-1 is one of the strongest suppresors of PEV so far described (Ebert et al., 2004). The majority are haplo-abnormal, showing the dominant modifying effect when the gene is present in only single copy, and some (10%) have both haplo- and triplo-abnormal effects, modifying PEV when present in either one or three copies. For example, a deletion of Su(var)205 suppresses the $In(1)w^{m4}$ PEV phenotype, giving a pure red eye, and a duplication giving a genome three copies of the locus enhances the PEV phenotype, giving a nearly pure white eye (Schulze and Wallrath, 2007; Weiler and Wakimoto, 1995). This sensitivity to dosage suggests that a balance in the amount of the products of these genes is important for chromatin structure formation and function.

B. Transposon insertion PEV

Rearrangements that break and rejoin chromosomes are not the only method for placing a euchromatic gene adjacent to heterochromatin. In 1982, a genetically recombinant P element with an inserted cloned reporter gene (ry^+) was induced to transpose into a chromosomal site (Rubin and Spradling, 1982; Spradling and Rubin, 1982). The reporter gene in P element transformation is essentially a small piece of euchromatic sequence that is inserted into different regions of the genome. In an early transformation experiment (Hazelrigg et al., 1984), two insertions of a transposon containing as a reporter a wild-type copy of the *white* gene were recovered that gave a white variegated eye phenotype. There was no indication that the inserted w^+ gene was altered in the transposition process and it was concluded that the variegation was the result of a position effect. One of these variegating insertions was located in the basal region of chromosome 2L, a heterochromatic region that was known to induce PEV in chromosomal rearrangements. The second was inserted near the telomere of the right arm of chromosome 3. This finding that genes inserted in a heterochromatic region by P element transformation can show a variegated phenotype was soon confirmed by other investigators (Steller and Pirrotta, 1985). In a genome-wide screen, Wallrath and Elgin (1995) recovered insertions with a white variegated phenotype in all of the heterochromatic regions known to induce PEV in rearrangements (the pericentric regions of the X, the second, the third, throughout the fourth, and in the Y chromosome). Variegating insertions were also recovered in or near the telomeres of all chromosomes (Wallrath and Elgin, 1995). Ahmad and Golic (1996) also recovered variegating P insertions in euchromatic regions at a site of intercalary heterochromatin. These results suggested that there was more to learn about position effects than had been suspected before, because P insertions can give a variegated phenotype in locations that do not give a PEV phenotype in chromosomal rearrangements (Fig. 1.4).

   An important question about variegation produced by P element insertion is whether it is generated by the same mechanism as chromosomal rearrangement-induced PEV. One test of this is to determine the effect of known modifiers of PEV on the transposon insertion phenotype. Talbert et al. (1994) observed that the PEV phenotype of a transposon insertion in the telomeric region of 3R was not altered by modifiers of the PEV effect given by w^{m4}. Wallrath and Elgin (1995) showed the known PEV-modifying mutations, $Su(var)2-5^{02}$ and $Su-var(2)1^{01}$, suppressed the PEV phenotypes of all pericentric and fourth chromosome insertions, including insertions near the telomere of the fourth chromosome, but did not suppress the variegation produced by six different insertions located near the telomeres of 2R, 2L, or 3R. Comparing flies with XO and XXY genotypes showed that increasing the amount of heterochromatin suppressed the variegation phenotype of 13 inserts in the pericentric

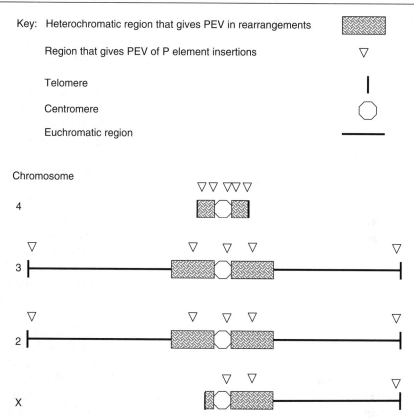

Figure 1.4. A diagram showing the chromosomal regions in which P element insertions showing a position-effect variegation (PEV) phenotype were recovered by Wallrath and Elgin (1995). Insertions were primarily recovered at centromeric and telomeric regions and distributed throughout the fourth chromosome.

heterochromatin or in the fourth chromosome, but had no effect on 7 insertions in or near the telomeres of 2R, 3L, or 3R (Wallrath and Elgin, 1995). A second test of PEV is to determine whether variegation is eliminated when the variegated allele is removed from its location. Ahmad and Golic (1996) demonstrated that if an inserted P element that is showing variegation in somatic cells is physically removed from the chromosome to an extrachromosomal circle using the flippase (FLP) recombinase system, the gene expression is restored. All of the insert lines they tested responded to modifiers of PEV (enhancement in XO males and suppression by the $Su(var)205^{502}$ mutation), but none of their insertions were located at a telomere. These results indicate that PEV generated by P element insertions in nontelomeric locations behaves like chromosomal rearrangement PEV, but that inserts in the telomeric regions have different properties. These results suggest that there may be differences in the mechanisms that produce telomeric insertion PEV phenotypes.

C. Pairing-dependent dominant PEV: *Trans*-inactivation

The first PEV-inducing rearrangements recovered gave mostly recessive phenotypes. For example, an individual heterozygous for the $In(1)w^{m4}$ chromosome and a normal chromosome $(In(1)w^{m4}/w^+)$ has a wild-type phenotype because the breakpoint in the $In(1)w^{m4}$ chromosome has no *trans*-inactivation effect on the w^+ allele in the normal chromosome. Chromosomal rearrangements that give a dominant PEV phenotype are rare. One well-studied example consists of rearrangements that affect the *brown (bw)* gene. Several inversions with one breakpoint in heterochromatin and another in the 59C6–59F3 interval of chromosome 2R have been recovered that give a dominant variegating brown eye PEV phenotype (Schultz, 1936; Slatis, 1955). Flies that are heterozygous for such a rearrangement and a normal chromosome containing a wild-type allele of the *brown* gene have a variegating brown eye color phenotype. This indicates that the breakpoint has a *trans*-inactivation effect on the bw^+ allele in the normal chromosome. A dominant *brown* allele (bw^D) was discovered that is not associated with an inversion but which carries an insertion of a block of 1.6 mb of centromeric heterochromatin in the *bw* locus (Hinton, 1940, 1942; Hinton and Goodsmith, 1950; Slatis, 1955). Flies with a bw^D/bw^+ genotype have a *bw* PEV phenotype. Removal of the heterochromatic block by further rearrangement eliminates the PEV effect (Hinton and Goodsmith, 1950). The dominant bw^D PEV phenotype is altered by suppressors of euchromatic gene PEV (Csink and Hennikoff, 1996). These findings confirmed that the dominant *trans*-inactivation effect of bw^D is due to the presence of adjacent heterochromatin, and that the effect is a PEV effect.

    Northern analysis indicates that bw^D/bw^+ individuals produce no detectable mRNA from either allele (Dreesen *et al.*, 1988; Henikoff and Dreesen, 1989). Thus, the dominant *bw* effect results from an inactivation of transcription of *bw* alleles in both chromosomes. The obvious question is how the insertion of a block of heterochromatin in one chromosome can affect the transcription of the allele on the other homologue. The key to this *trans*-inactivation effect is homologous chromosome pairing (Sass and Henikoff, 1999). When a fly has two *bw* variegating rearrangements with different breakpoints that disrupt somatic chromosome pairing, the PEV phenotype is also reduced (Henikoff and Dreesen, 1989). Confirmation that pairing is the key to *bw*-dominant *trans*-inactivation came when bw^D was combined with duplications containing extra copies of the bw^+ in other chromosomal locations, where they do not pair with the normal locus. In such cases, the inserted bw^+ is not inactivated by bw^D (Dreesen et al., 1991).

The idea that somatic pairing of chromosomes is important for normal function and might be a factor in PEV is an old one (Ephrussi and Sutton, 1944; Henikoff, 1994). How might pairing with a chromosome containing bw^D cause trans-inactivation of the bw^+ allele on the non-rearranged chromosome? The inserted heterochromatic block in bw^D contains multiple copies of the AAGAG sequence, a sequence that is one of the most abundant in the genome and which is found on each chromosome, and which is especially common in the pericentric region of chromosome 2 (Lohe et al., 1993). Tracking the nuclear location of the bw insert using fluorescent in situ hybridization (FISH) shows that in the interphase nucleus, this block becomes associated with a region of the nucleus where the 2R-centric heterochromatin is localized (Csink et al., 2002; Dernburg et al. 1996). The silencing of bw^D is thus correlated with the localization of the bw^D allele and the paired bw^+ allele to a heterochromatic region of the nucleus. This suggested that the dominant bw^D PEV effect is due to the bw^D allele dragging the paired bw^+ allele on the other homologue to a nuclear compartment where heterochromatic pairing results in gene inactivation. In a further investigation, the expressions of a series of P element insertions at different locations along the normal 2R chromosome were assayed when paired with the bw^D chromosome to determine the chromosomal extent of this trans-inactivation effect. An insertion 45 kb from the bw^D insertion was inactivated but inserts 280 and 460 kb from bw were not inactivated (Csink et al., 2002). When the bw^D insert was introgressed into Drosophila simulans, which has AAGAG sequences in the centric region of the X chromosome but not the second chromosome, the insert continued to associate with the centric region of the second chromosome, suggesting that the heterochromatic association effect is not sequence specific but is an association with the nearest large block of heterochromatin (Sage and Csink, 2003). Recent studies suggest that this heterochromatic-pairing brown-dominant PEV may not be unique, and that heterochromatic insertions in other euchromatic regions may show trans-association/inactivation (Sage and Csink, 2003; Thakar and Csink, 2005). This suggests that trans-inactivation by heterochromatic insertion may be an important mechanism of gene silencing.

IV. GENOME ORGANIZATION AND PEV

The study of PEV has been instrumental in advancing our appreciation of the role that chromatin structure and genome organization play on the regulation of gene expression. In particular, the ability to isolate modifiers of the PEV-silencing effect in genetic screens resulted in identification of a large number of chromosomal and nuclear architectural factors that are important for long-range control of gene expression (Table 1.1). Analysis of these factors has allowed us to develop a molecular understanding of the different processes contributing to the establishment or maintenance of active versus silent chromosomal domains and especially it has illuminated the critical role that heterochromatin plays in genome organization. In this section, we will review key aspects of chromatin structure and will examine some of the genes that affect PEV penetrance and the molecular mechanisms underlying the phenomenon. We will consider recent studies that underscore the link between heterochromatin formation and nuclear organization.

A. Chromatin structure

The primary packaging unit of chromatin is the nucleosome, composed of 147 bp of DNA wrapped around a histone octamer composed of two subunits each of the core histones H2A, H2B, H3, and H4 (reviewed in Luger, 2006). The histone globular domains are arranged in the interior of the nucleosome, while their unstructured tail domains project outwards where they are targets for a range of posttranslational modifications, including acetylation, phosphorylation, methylation, ADP-ribosylation, ubiquitination, sumoylation, and biotinylation. These modifications have been proposed to regulate genomic function by influencing chromatin structure both by altering the biophysical contacts between DNA and histones and by providing specific binding sites for different classes of chromatin-binding proteins (reviewed in Kouzarides, 2007). Thus, the modified histone tails provide a signaling platform that integrates output from various signal transduction pathways ultimately specifying the level of higher order chromatin folding (Cheung et al., 2000; Fischle et al., 2003; Turner et al., 1992).

Hyperacetylation of histones H3 and H4 and methylation of H3K4 are correlated with the establishment of transcriptionally active chromatin across a wide range of species from yeast to *Drosophila* to human (Bernstein et al., 2005; Pokholok et al., 2005; Schubeler et al., 2004). Acetylation neutralizes the basic charge of the lysine residue, which is thought to reduce histone tail interaction with the acidic phosphate residues of DNA, making it more accessible for transcription (Wade et al., 1997), and acetylation of the H4K16 residue, in particular, has been shown to physically impede the formation of higher order folded structures as well as block nucleosome sliding by the *Drosophila* ATP-utilizing chromatin assembly and remodeling factor (ACF) chromatin-remodeling complex *in vitro* (Shogren-Knaak et al., 2006). But besides changing the biophysical properties of chromatin fiber behavior, lysine acetylation also generates specific docking sites for distinct bromodomain proteins that have been found in a variety of different transcriptional activators and chromatin-remodeling complexes; binding to these sites can be influenced

Table 1.1. Nuclear PEV Modifiers of w^{m4}

Locus	Function	References
Histone-modifying enzymes/histone variants		
JIL-1	H3S10 kinase	Wang et al., 2001; Ebert et al., 2004; Bao et al., 2007a
Su(var)3–9	H3K9 methyltransferase	Reuter et al., 1986; Schotta et al., 2003b
G9a	H3K9 methyltransferase	Mis et al., 2006
HDAC1/RPD3	Histone deacetylase	De Rubertis et al., 1996; Mottus et al., 2000
Chameau	Histone acetyltransferase (HAT)	Grienenberger et al., 2002
Reptin	HAT complex	Qi et al., 2006
H2Av	H2A variant	Swaminathan et al., 2005
Chromosomal proteins		
Su(var)2–5/HP1	H3K9me-binding	James and Elgin, 1986; Eissenberg et al., 1990
HP2	AT Hook protein	Shaffer et al. 2002
D1	AT Hook protein	Aulner et al., 2002
Su(var)3–7	DNA binding (Zn finger)	Reuter et al., 1990; Cléard and Spierer, 2001
Modulo	DNA binding	Perrin et al., 1998; Büchner et al., 2000
trl/GAGA factor	DNA binding	Dorn et al., 1993; Farkas et al., 1994
Bonus	Nuclear receptor cofactor	Beckstead et al., 2005
mus209	Proliferating cell nuclear antigen (PCNA); DNA replication and repair	Henderson et al., 1994
Cramped	Pc-G gene; DNA replication	Yamamoto et al., 1997
Zeste	Transvection	Hazelrigg and Petersen, 1992
E(var)3–95E	E2F transcription factor; cell cycle	Seum et al., 1996
ORC2	Replication; complexes with HP1	Pak et al., 1997
HOAP	Complex with HP1/ORC; HMG-like	Shareef et al., 2001
DRE4/spt16	Facilitates chromatin transcription (FACT) complex	Nakayama et al., 2007
Nuclear and chromosome architecture		
Lamin Dm0	Nuclear lamina/architecture	Bao et al., 2007b
Su(var)2–10/PIAS	Nuclear organization	Reuter and Wolff, 1981; Hari et al., 2001
Piwi	RNAi	Pal-Bhadra et al., 2004

(Continues)

Table 1.1. (*Continued*)

Locus	Function	References
Homeless/Spn-E	RNAi	Pal-Bhadra *et al.*, 2004
Aubergine	RNAi	Pal-Bhadra *et al.*, 2004
mod/mdg4 [E(var) 3–93D)]	Boundary element	Dorn *et al.*, 1993; Gerasimova *et al.*, 1995
BEAF-32	Boundary element	Zhao *et al.*, 1995; Gilbert *et al.*, 2006
Lip/Rm62	Helicase	Csink *et al.*, 1994
dCAP-G	Condensin subunit	Dej *et al.*, 2004
gluon/SMC4	Condensin subunit	Cobbe *et al.*, 2006
wapl	Chromosome adhesion?	Verni *et al.*, 2000

by neighboring posttranslational modifications on the histone tail, thus providing a combinatorial complexity to the signaling process (Dyson *et al.*, 2001; Yang, 2004). Although histone acetylation is generally associated with active chromatin regions, acetylation of H4K12 is found in heterochromatic regions, suggesting that acetylation at particular sites may mediate specific effects on chromatin function (Turner *et al.*, 1992).

Whereas acetylation is, with few exceptions, associated with euchromatin, different methylated histone modifications have been found to play an important role in differentiating active (euchromatic) chromosomal domains from silenced (heterochromatic) domains (reviewed in Lachner *et al.*, 2003). In general, the "activating" modifications consist of methylated H3K4, H3K36, and H3K79, while the "silencing" modifications are composed of methylated H3K9, H3K27, and H4K20. Thus, identification of the enzymes affecting these modifications and elucidating the mechanisms underlying their targeting promised to provide significant insight in the specification of chromatin domains. Given that the early genetic and cytogenetic experiments indicated that PEV occurs when heterochromatin spreads into juxtaposed euchromatic sequences, it was anticipated that many of the modifiers of position effect would fall into the categories of histone-modifying and histone-binding proteins (Schotta *et al.*, 2003a).

1. Methylation and the spread of heterochromatin

Three Su(*var*) gene products that were of early interest due to their heterochromatic localization, their "haplo-suppressor/triplo-enhancer" behavior, and their physical and genetic interactions were SU(VAR)2–5 (also known as SU(VAR) 205, heterochromatin protein 1, or HP1) (Eissenberg *et al.*, 1990; James and Elgin, 1986; James *et al.*, 1989), SU(VAR)3–7 (Cléard and Spierer, 2001; Cléard *et al.*, 1997; Reuter *et al.*, 1990), and SU(VAR)3–9, the latter of which is one of the strongest modifiers of PEV known, showing dominance over nearly all other enhancers of variegation tested (Reuter *et al.*, 1986; Tschiersch *et al.*, 1994; Wustmann *et al.*, 1989). Interestingly, one of the only two exceptions was ptn^D, which is a gain of function allele of *Su(var)3–9* that causes ectopic localization of the protein to euchromatic sites (Kuhfittig *et al.*, 2001). A critical advance in the field of heterochromatin biology came from the findings that (1) the mammalian orthologs of SU(VAR)3–9 encode a methyltransferase that specifically methylates H3K9 (Rea *et al.*, 2000), an activity that is conserved in *Drosophila* SU(VAR)3–9 (Czermin *et al.*, 2001; Schotta *et al.*, 2002) and (2) the chromodomain of HP1 binds to the methylated K9 of histone H3 (Bannister *et al.*, 2001; Lachner *et al.*, 2001). The silencing activity of HP1 was disrupted when its ability to bind methylated H3K9 was abolished by mutation (Jacobs *et al.*, 2001). Because the H3K9 methyltransferase SU(VAR)3–9 associates both with the H3K9 deacetylase HDAC1 (Czermin *et al.*, 2001) and with HP1 (Schotta *et al.*, 2002; Yamamoto and Sonoda, 2003), a spreading model for heterochromatin involving these components was proposed in which deacetylation of H3K9 clears the residue for methylation by SU(VAR)3–9 to enable HP1 binding, which, in turn, recruits additional SU(VAR)3–9 to methylate the adjacent nucleosome that provides another HP1-binding site in a self-propagating process (Grewal and Elgin, 2002; Schotta *et al.*, 2003b).

2. Initiation of heterochromatin formation

In recent years, a number of studies have indicated that RNAi-mediated silencing pathways can initiate the formation of heterochromatin (reviewed in Grewal and Rice, 2004; Matzke and Birchler, 2005; Zaratiegui *et al.*, 2007). This pathway was first identified in *S. pombe* where mutations in components of the RNAi pathway were found to disrupt centric heterochromatin silencing (Verdel *et al.*, 2004). Transcripts from repetitive elements in the centromeric region are processed into siRNAs that are incorporated into a RITS (RNAi-induced transcriptional silencing) complex that recognizes and binds homologous regions to initiate gene silencing mediated via H3K9 methylation (Verdel and Moazed, 2005; Verdel *et al.*, 2004). *Drosophila* centromeres are also composed of short satellite- and transposon-fragment repeats (Sun *et al.*, 1997) that are actively transcribed (Lakhotia and Jacob, 1974) and transgenes inserted into these regions show PEV (Cryderman *et al.*, 1998; Wallrath and Elgin, 1995; Wallrath *et al.*, 1996). Mutations in genes encoding the RNAi pathway Argonaute homologues *piwi*, *aubergine*, and the helicase *homeless* disrupt HP1 localization and suppress transgene silencing, indicating a similar mechanism for RNAi-mediated heterochromatin assembly operates in *Drosophila* as well (Pal-Bhadra *et al.*, 2004).

Mutants in *Argo2*, another RNAi argonaut family member, showed defects in centromeric heterochromatin assembly, including abnormal HP1 localization and defects in H3K9 methylation (Deshpande *et al.*, 2005).

3. Terminating heterochromatic spreading

That PEV is, by definition, a *variegated* phenotype indicates that chromosomal silencing can spread different lengths in different cells, and the question of what terminates the spreading is not well understood. Because of the "haplo-supressor/ triplo-enhancer" behavior of certain key components of heterochromatin, some models have postulated that heterochromatic-promoting factors are present in limiting amounts and spreading continues until components are exhausted (Locke *et al.*, 1988; Zuckerkandl, 1974). Alternatively, a "boundary model" was proposed in which discrete sites promoted initiation and termination of heterochromatin (Tartof *et al.*, 1984), but discrete sites conferring these activities could not be identified. Recently, the examination of PEV in *JIL-1* mutants in *Drosophila* has revealed a role for H3S10 phosphorylation in antagonizing heterochromatic spreading that suggests both models may apply in different genomic contexts (Bao *et al.*, 2007b; Ebert *et al.*, 2004; Lerach *et al.*, 2006; Zhang *et al.*, 2006). The JIL-1 histone H3S10 tandem kinase localizes specifically to euchromatic interband regions of polytene chromosomes (Jin *et al.*, 1999; Wang *et al.*, 2001) and loss of JIL-1 results in an array of chromosomal defects, including disruption of banded regions that normally do not contain JIL-1 (Deng *et al.*, 2005). Alterations in band morphology in these mutants is accompanied by extensive spreading of the major heterochromatin markers H3K9me2 and HP1 to ectopic locations on the chromosome arms with the most pronounced increase on the X chromosomes (Zhang *et al.*, 2006). As would be expected, if *JIL-1* is required to "mark" euchromatic regions and terminate heterochromatic spreading, when a *white* reporter transgene is inserted into pericentric heterochromatin *JIL-1* loss-of-function alleles act as enhancers of variegation (i.e., the *white* gene is silenced; Bao *et al.*, 2007b). However, these very same *JIL-1* loss-of-function alleles act as *suppressors* of variegation for the w^{m4} inversion chromosome (Lerach *et al.*, 2006). In this case, the redistribution of the major heterochromatin markers to ectopic locations on the chromosome arms in *JIL-1* loss-of-function mutants decreases the concentration of these components at the centromere, and the increased expression of the *white* gene reflects the reduced extent of pericentromeric silencing (Lerach *et al.*, 2006) (Fig. 1.5). A gain-of-function *JIL-1* allele *JIL-1*$^{Su(var)3-1}$, however, is able to mark the euchromatin and prevent silencing in both instances (Bao *et al.*, 2007b; Ebert *et al.*, 2004). JIL-1$^{Su(var)3-1[3]}$ is a C-terminal truncation that retains its ability to phosphorylate H3S10 but is mislocalized, showing a broad distribution on polytene chromosomes

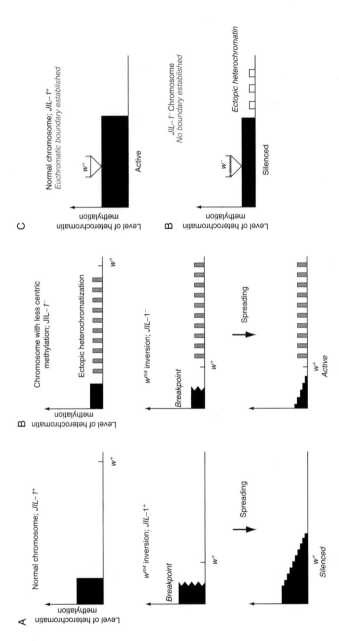

Figure 1.5. Three diagrams illustrating position-effect variegation (PEV) inactivation of genes by heterochromatic spreading. (A) In a genome with normal amounts of JIL-1 protein, heterochromatic spreading from the w^{m4} inversion breakpoint often reaches the *white* gene; the gene is heterochromatized and is silenced. (B) In a genotype with significantly reduced JIL-1 levels, ectopic heterochromatization is widespread, resulting in lower levels of heterochromatic factors at the centromere. Therefore, heterochromatic spreading does not extend as far into the euchromatin and usually does not reach the *white* gene, leaving the gene euchromatic and active. (C) If a P element inserts a copy of the *white* gene into heterochromatin in a genotype where JIL-1 is present, JIL-1 often establishes a euchromatic chromatin structure at the *white* gene, the gene is protected from becoming heterochromatized, and the gene remains active. If the same insert is in a genome where JIL-1 is not present, the *white*

(Ebert *et al.*, 2004; Zhang *et al.*, 2006). Ectopic H3S10ph appears to restrict the formation of heterochromatin, both at the w^{m4} inversion breakpoint and in the pericentric transgenes, resulting in a dominant suppression of variegation (Bao *et al.*, 2007b; Ebert *et al.*, 2004). Recently, H3S10 phosphorylation has been shown to displace HP1 binding to methylated H3K9 during mitosis (Fischle *et al.*, 2005; Hirota *et al.*, 2005). Likewise, JIL-1 kinase phosphorylation of H3S10 may displace or prevent HP1 binding, thereby terminating heterochromatic spreading at that site in order to establish a euchromatic domain (Johansen and Johansen, 2006). Consistent with this model, the $JIL\text{-}1^{Su(var)3\text{-}1}$ C-terminal truncation alleles are dominant over the triplo-enhancer effects of *Su(var)3–9* and *HP1* (Ebert *et al.*, 2004).

A second mechanism that operates to restrict heterochromatic spreading is the incorporation of the histone variant H3.3 into actively transcribed regions (Ahmad and Henikoff, 2002; Schwartz and Ahmad, 2005). Genome-wide profiling of histone H3.3 replacement patterns revealed that H3.3 is also enriched far upstream and downstream of active genes, but it was suggested this may reflect intergenic transcription that is now known to be widespread (Mito *et al.*, 2005). Recently, GAGA factor was found to interact with FACT (*f*acilitates *c*hromatin *t*ranscription) complex to direct H3.3 replacement (Nakayama *et al.*, 2007). GAGA factor had originally been identified as the enhancer of variegation E(var)3–93D (Dorn *et al.*, 1993) and in this recent study, mutation in either GAGA factor or in DRE4/spt16, a subunit of FACT, enhanced w^{m4} PEV (Nakayama *et al.*, 2007). Furthermore, the GAGA factor–FACT complex occupies a site where H3K4 methylation peaks and H3K9 methylation dips, suggesting that the remodeling activity of GAGA–FACT serves a barrier function against heterochromatic spreading and silencing (Nakayama *et al.*, 2007). The absence of such a barrier may result in expanded heterochromatic silencing, thus titrating heterochromatic factors away from the pericentromeric region resulting in an e(var) phenotype in the GAGA mutant.

B. Nuclear organization

The phenomenon of bw^D PEV in which chromosomal pairing brings a functional *bw* allele into a "silencing compartment" underscores the impact of nuclear organization on gene expression (Csink and Hennikoff, 1996; Csink *et al.*, 2002;

gene is not protected. The establishment of ectopic heterochromatic sites throughout the chromosomes lowers the level of heterochromatic factors in the pericentric region. However, the basal level of pericentric heterochromatin, maintained by the RNAi system that generates heterochromatin, is sufficient to spread into the inserted *white* gene, the gene becomes heterochromatized, and is silenced.

Dernburg *et al.*, 1996). The functional consequences of long-range nuclear reorganization on gene expression were examined by FISH for three known variegating genes located on different chromosomes and in each case a strong correlation was observed between silencing and association with centromeric satellite sequences (Harmon and Sedat, 2005). Thus, changes in nuclear chromosomal organization have clear consequences on gene expression. What establishes the organization of the nucleus, however, is still not well understood (Jackson, 2003; Lanctôt *et al.*, 2007; Misteli, 2007).

Lamins are the best characterized nuclear architectural components to date and their role in chromosomal organization and gene expression is increasingly appreciated (reviewed in Hutchison, 2002; Goldman *et al.*, 2005). Association of genes with the nuclear lamina is generally linked to gene silencing (Andrulis *et al.*, 1998; Taddei and Gasser, 2004), although in some cases interaction with nuclear pore complex components is associated with transcriptional activation (Casolari *et al.*, 2004; Ishii *et al.*, 2002; Mendjan *et al.*, 2006). In a recent study, Pickersgill *et al.* (2006) identified $\sim$500 genes that interact with lamin Dm_0 (the *Drosophila* B-type lamin) and found that they are transcriptionally repressed. Lamin-bound genes tend to be clustered in the genome, developmentally coregulated, have long intergenic domains, and lack active histone marks, suggesting that these regions likely comprise compacted chromatin (Pickersgill *et al.*, 2006). Consistent with lamins playing a role in epigenetic regulation of gene expression, a lamin Dm_0 C-terminal truncation mutant Lam^{Ari3} modified the PEV of w^{m4} (Bao *et al.*, 2007a). However, this allele shows an *enhancer* of variegation effect indicating *increased* silencing when the C-terminus is deleted, indicating that the role of lamin B in organizing chromatin domains is complex. Although interaction with lamin B is associated with gene silencing (Pickersgill *et al.*, 2006), the PEV studies suggest that lamin B's C-terminus may play a role in reversing silencing effects (Bao *et al.*, 2007a).

Interaction between nuclear lamins and chromatin is thought to regulate higher order chromatin organization and a direct physical interaction between lamin Dm_0 and histone H2A has been reported (Mattout *et al.*, 2006; Rzepecki *et al.*, 1998). In addition, different chromatin-organizing proteins have been found to interact with nuclear lamins and nuclear envelope proteins and the nature of these activities may facilitate the complex integration of various signaling pathways to influence genome organization and gene regulation (Gruenbaum *et al.*, 2005). Signaling to the chromatin may be mediated at several different levels besides physical association to the lamina. For example, the nuclear envelope protein LAP2β recruits a histone deacetyltransferase to the periphery, thus triggering silencing histone modifications in this domain (Somech *et al.*, 2005).

The nuclear lamina has also been found to bind the ubiquitin ligase dTopors (Topoisomerase-I-interacting protein), which functions in concert with proteins that bind to a chromatin regulatory domain known as an insulator or boundary element (Capelson and Corces, 2005; Wei et al., 2005; West et al., 2002). Insulators block enhancer interactions (Geyer and Corces, 1992; Kellum and Schedl, 1992) and block the spreading of one chromatin state into another (Sun and Elgin, 1999). The gypsy insulator sequence was originally identified as a gypsy retrotransposon insertion that blocked enhancer function at the yellow locus by introducing a Su(Hw) (suppressor of Hairy wing)-binding site (Geyer and Corces, 1992; Parkhurst et al., 1988; Roseman et al., 1993). The Su(Hw) protein binds to BTB/POZ-domain proteins (Broad complex Tramtrack, Bricabrac/Poxvirus and Zinc finger), CP190 (Pai et al., 2004) and mod/mdg4 (Gerasimova et al., 1995; Ghosh et al., 2001) to form a complex with dTopors at the nuclear periphery that is thought to isolate regions of DNA in "insulator bodies" (Capelson and Corces, 2004; Gerasimova et al., 2000). The presence of gypsy insulator sequences at endogenous sites in the genome suggests that this may provide a means to compartmentalize the genome and prevent heterochromatic spreading into active euchromatic regions (Parnell et al., 2006; Ramos et al., 2006). Consistent with a role in limiting heterochromatic spreading, mod/mdg4 was originally identified as the enhancer of variegation $E(var)3-93D$ (Dorn et al., 1993).

Because there are significantly more chromosomal-binding sites for Su(Hw) and mod/mdg4 than genomic consensus Su(Hw) DNA-binding sites (Parnell et al., 2006; Ramos et al., 2006), it is not known whether the PEV modifier effects observed in the mod/mdg4 mutant reflect activities related directly to gypsy insulator function, activities at other as-yet uncharacterized chromosomal sites, or is the indirect consequence of the nuclear reorganization that has been found to occur when gypsy insulator components are disrupted (Gerasimova et al., 2000; Pai et al., 2004). One plausible scenario is that if the mod/mdg4 mutation compromises the insulator's barrier function to heterochromatic spreading, the resultant increased euchromatic distribution of heterochromatic factors reduces their levels at the pericentromeric regions resulting in decreased spreading at the w^{m4} inversion and an E(var) phenotype, much as occurs for JIL-1 (Bao et al., 2007b). Likewise, proper insulator function also depends on the presence of functional RNAi-mediated pathways, perhaps to ensure the generation of heterochromatin and thus the appropriate sequestration of heterochromatic components at centromeric regions. When Argonaute RNAi pathway genes are mutated, nuclear architecture is disrupted and insulator activity is decreased (Lei and Corces, 2006). Furthermore, CP190 interacts with Rm62/Lip (Lei and Corces, 2006), a DEAD-box helicase that has also been identified as a modifier of PEV and is required for dsRNA-mediated silencing, heterochromatin formation, and transposon silencing (Csink et al., 1994; Ishizuka et al., 2002). These results underscore the relationship between genome organization/chromatin pack-aging of euchromatic sites and the levels of heterochromatic factors that are available to spread at pericentromeric sites (i.e., the inversion breakpoint).

Another nuclear domain besides the periphery that has been associated with silencing is the nucleolus (Lawrence and Pikaard, 2004; Lewis and Pikaard, 2001; Wang *et al.*, 2005; Yasufzi *et al.*, 2004) and chromosomal inversions that place the *white* gene adjacent to the nucleolar organizer region (NOR) exhibit PEV consistent with different levels of gene silencing (Spofford and DeSalle, 1991). Recently, it was found that heterochromatin-mediated silencing pathways including the RNAi pathway, HP1, and H3K9 methylation regulate nucleolar organization and structural integrity (Peng and Karpen, 2006). In most of the RNAi pathway mutants examined as well as in *Su(var)2–5/HP1* and *Su(var)3–9* mutants, heterochromatic repeated DNAs became dispersed and multiple, ectopic nucleoli were formed. In addition, significant amounts of extrachromosomal circles (ecc) of rDNA accumulated (Peng and Karpen, 2006). It was proposed that loss of heterochromatization in the NOR correlates with increased recombination or repair rates in this repetitive sequence region with subsequent accumulation of ecc rDNAs when errors occurred (Peng and Karpen, 2006). In support of this idea, mutation in *Ligase4*, a DNA repair enzyme, suppressed ecc generation (Peng and Karpen, 2006). A connection between DNA synthesis/repair and heterochromatin behavior has been inferred by a number of studies showing that mutations in different factors involved in these processes act as PEV modifiers (Table 1.1). Another mutation that disrupts nucleolar and chromosomal organization in *Drosophila*, *Su(var)2–10/dPIAS* (Hari *et al.*, 2001), also was originally identified as a modifier of PEV (Reuter and Wolff, 1981). Interestingly, a correlation between deregulation of chromatin-silencing and disruption of nuclear organization has also been observed in yeast (Teixeira *et al.*, 2002), suggesting that the link between heterochromatin behavior and nuclear organization is a general feature of eukaryotes.

V. CONCLUDING REMARKS

Studies of the control of gene regulation that began with the discovery of PEV and the chromosomal conditions that cause it have brought us a long way toward understanding the role of chromatin structure in the regulation of gene expression. Studies of chromosome rearrangements and P element insertions demonstrated that a change in chromatin structure from euchromatic to hetero-chromatic leads to gene silencing and that this silencing effect has specific characteristics that are amenable to analysis. The identification of mutations that modify the PEV phenotype identified a number of factors that are part of or regulate the formation or maintenance of heterochromatin and euchromatin. The findings of these studies indicate that control of gene expression is a complex process involving regulatory mechanisms that are integrated at multiple hierarchical levels ranging from the primary DNA sequence to the chromatin packaging of DNA and higher order chromosomal folding to the three-dimensional spatial organization of chromosomes within the nucleus (reviewed in van Driel *et al.*, 2003; Misteli, 2007). A key component of this process may be the modification of histone N-terminal tails to "mark" a region of chromatin for the activation of a number of specific downstream responses (Fischle *et al.*, 2003; Kouzarides, 2007; Strahl and Allis, 2000; Turner, 2002) such as the establishment of transcriptionally active chromatin (Brownell and Allis, 1996; Csordas, 1990; Turner, 2000) or higher level folding associated with silencing (Turner, 2002; Zhang and Reinberg, 2001). However, regulation of gene expression at the level of the nucleosome goes beyond the modification of histone tails. Genes that are actively transcribed are often subject to nucleosomal remodeling by chromatin-remodeling complexes (Becker and Horz, 2002; Workman and Kingston, 1998) and characterized by nucleosome-free or "DNase-hypersensitive" sites, whereas silenced or heterochromatic regions are characterized by tightly packed, ordered nucleosomal arrays (Elgin, 1988; Wallrath and Elgin, 1995). Besides the local remodeling of a small number of nucleosomes in the promoter region, larger scale nucleosomal remodeling is also implicated in the unfolding of large chromatin domains (Dietzel *et al.*, 2004; Neely and Workman, 2002; Peterson, 2003; Tumbar *et al.*, 1999). Current models suggest that within the nucleus, there exist regions of condensed silent chromatin interspersed with regions of decondensed, active chromatin. A clear example of this is found within the band-interband pattern observed in *Drosophila* larval polytene chromosomes (Labrador and Corces, 2002). The challenge is to identify the molecules and molecular mechanisms that determine how this organization is established and maintained and the signal transduction events that regulate this process.

A constructive way to think about heterochromatin is in quantitative terms. The *Drosophila* genome contains a finite number of repeated sequences that interact with the RNAi system to utilize a fixed quantity of heterochromatic factors that can bind to and heterochromatize a set length of chromatin. The concentration of these factors is normally high in the pericentric chromosomal regions, which serve as a reservoir for heterochromatic components. The amount of chromatin that is heterochromatic during interphase is a dynamic balance between the number of heterochromatic factors and the number of chromatin sites available for binding. Anything that increases the number of sites or decreases the number of factors will decrease the pericentric concentration and anything that decreases the number of sites or increases the number of factors will increase the pericentric concentration. Consider three types of

Su(var) phenotypes. (1) The addition of Y chromosomes or heterochromatic regions increases the sequences competing to bind heterochromatic factors. This increased competition lowers the concentration of heterochromatic factors in the pericentric heterochromatin. (2) Mutations that reduce the dose of genes that code for structural components of heterochromatin (e.g., HP1) also reduce the quantity of heterochromatic factors in pericentric regions. (3) JIL-1 proteins establish euchromatic regions that are resistant to heterochromatic factor binding. In loss-of-function JIL-1 mutants, ectopic-binding sites for heterochromatic factors are established throughout the genome that compete for heterochromatic factors. This reduces the concentration of heterochromatic factors in pericentric regions. In all three cases, the lower level of heterochromatic factors in pericentric regions means that heterochromatin cannot spread as far into euchromatin from a breakpoint and adjacent genes are not inactivated as often, resulting in a Su(var) phenotype. E(var) phenotypes result from mutations that increase the concentration of heterochromatic factors in the pericentric regions, leading to increased lengths of spreading from breakpoints.

The details of a number of the most interesting aspects of PEV remain to be explained. How does the regulation of genes normally residing in heterochromatic regions differ from that of genes residing in euchromatic regions? How does the PEV of genes inserted into telomeres by P element transposition differ from that of those inserted into pericentric regions? What controls nuclear compartmentalization and how does this normally act to control gene regulation? These and other questions are currently being actively investigated in a continuation of an examination that began more than 75 years ago with the observation of a novel color of the eyes of a fly.

Acknowledgments

We thank members of the lab for critical reading of the manuscript and for helpful comments. We thank Drs. Lori Wallrath and Diane Cryderman (University of Iowa) for helpful discussion and for providing the 118E-10 P element insertion line. The authors' work on JIL-1 is supported by NIH grant GM62916.

References

Ahmad, K., and Golic, K. G. (1996). Somatic reversion of chromosomal position effects in *Drosophila melanogaster*. *Genetics* **144**, 657–670.
Ahmad, K., and Henikoff, S. (2002). The histone variant H3.3 marks active chromatin by replication-independent nucleosome assembly. *Mol. Cell* **9**, 1191–1200.
Altorfer, N. (1967). Effect of the absence of the Y chromosome on the expression of the cubitus interuptus phenotype in various translocations in *Drosophila melanogaster*. *Genetics* **55**, 755–767.
Andrulis, E. D., Neiman, A. M., Zappulla, D. C., and Sternglanz, R. (1998). Perinuclear localization of chromatin facilitates transcriptional silencing. *Nature* **394**, 592–595.

Aulner, N., Monod, C., Mandicourt, G., Julien, D., Cuvier, O., Sall, A., Janssen, S., Laemmli, U. K., and Käs, E. (2002). The AT-hook protein D1 is essential for *Drosophila melanogaster* development and is implicated in position-effect variegation. *Mol. Cell. Biol.* **22,** 1218–1232.

Bahn, E. (1971). Position effect variegation for an isoamylase in *Drosophila melanogaster*. *Hereditas* **67,** 79–82.

Baker, W. K. (1953). V-type position effects on a gene normally found in heterochromatin. *Genetics* **38,** 328–344.

Baker, W. K. (1971). Evidence for position effect suppression of the ribosomal RNA cistrons in *Drosophila melanogaster*. *Proc. Natl. Acad. Sci. USA* **68,** 2472–2476.

Baker, W. K., and Reim, A. (1962). The dichotomous action of Y chromosomes on the expression of position-effect variegation. *Genetics* **47,** 1399–1407.

Baker, W. K., and Spofford, J. B. (1959). Heterochromatic control of position-effect variegation in *Drosophila*. *Univ. Texas Publ.* **5914,** 135–154.

Baksa, K., Morawietz, H., Dombradi, V., Axton, M., Taubert, H., Szabo, G., Torol, I., Udvardy, A., Gurkovics, B., Szoor, B., Gilover, D., Reuter, G., et al. (1993). Mutations in the protein phosphatase 1 gene at 87B can differentially affect suppression of position-effect variegation and mitosis in *Drosophila melanogaster*. *Genetics* **135,** 117–125.

Bannister, A. J., Zegerman, P., Partridge, J. F., Miska, E. A., Thomas, J. O., Allshire, R. C., and Kouzarides, T. (2001). Selective recognition of methylated lysine 9 on histone H3 by the HP1 chromo domain. *Nature* **410,** 120–124.

Bao, X., Girton, J., Johansen, J., and Johansen, K. M. (2007a). The lamin Dm0 allele *Ari3* acts as an enhancer of position effect variegation of the wm4 allele in *Drosophila*. *Genetica* **129,** 339–342.

Bao, X., Deng, H., Johansen, J., Girton, J., and Johansen, K. M. (2007b). Loss-of-function alleles of the JIL-1 histone H3S10 kinase enhance position-effect-variegation at pericentric sites in *Drosophila* heterochromatin. *Genetics* **176,** 1355–1358.

Becker, P. B., and Horz, W. (2002). ATP-dependent nucleosome remodeling. *Annu. Rev. Biochem.* **71,** 247–273.

Beckstead, R. B., Ner, S. S., Hales, K. G., Grigliatti, T. A., Baker, B. S., and Bellen, H. J. (2005). Bonus, a Drosophila TIF1 homolog, is a chromatin-associated protein that acts as a modifier of position-effect variegation. *Genetics* **169**(2), 783–794.

Benner, D. B. (1971). Some evidence against the presence of suppressors of variegation on the Y chromosome. *Dros. Inform. Serv.* **47,** 72.

Bernstein, B. E., Kamal, M., Lindblad-Toh, K., Bekiranov, S., Bailey, D. K., Huebert, D. J., McMahon, E. K., Karlsson, E. K., Kulbokas, E. J. 3rd., Gingeras, T. R., Schreiber, S. L., and Lander, E. S. (2005). Genomic maps and comparative analysis of histone modifications in human and mouse. *Cell* **120,** 169–181.

Brownell, J. E., and Allis, C. D. (1996). Special HATs for special occasions: Linking histone acetylation to chromatin assembly and gene activation. *Curr. Opin. Genet. Dev.* **6,** 176–184.

Büchner, K., Roth, P., Schotta, G., Krauss, V., Saumweber, H., Reuter, G., and Dorn, R. (2000). Genetic and molecular complexity of the position effect variegation modifier mod(mdg4) in *Drosophila*. *Genetics* **155,** 141–157.

Capelson, M., and Corces, V. G. (2004). Boundary elements and nuclear organization. *Biol. Cell* **96,** 617–629.

Capelson, M., and Corces, V. G. (2005). The ubiquitin ligase dTopors directs the nuclear organization of a chromatin insulator. *Mol. Cell* **20,** 105–116.

Casolari, J. M., Brown, C. R., Komili, S., West, J., Hieronymus, H., and Silver, P. A. (2004). Genome-wide localization of the nuclear transport machinery couples transcriptional status and nuclear organization. *Cell* **117,** 427–439.

Cheung, P., Allis, C. D., and Sassone-Corsi, P. (2000). Signaling to chromatin through histone modifications. *Cell* **103,** 263–271.

Cléard, F., and Spierer, P. (2001). Position-effect variegation in *Drosophila:* The modifier *Su(var)3–7* is a modular DNA-binding protein. *EMBO Rep.* **21,** 1095–1100.

Cléard, F., Delattre, M., and Spierer, P. (1997). SU(VAR)3–7: A *Drosophila* heterochromatin-associated protein and companion of HP1 in the genomic silencing of position-effect variegation. *EMBO J.* **16,** 5280–5288.

Cobbe, N., Savvidou, E., and Heck, M. M. S. (2006). Diverse mitotic and interphase functions of condensins in *Drosophila. Genetics* **172,** 991–1008.

Cryderman, D. E., Cuaycong, M. H., Elgin, S. C., and Wallrath, L. L. (1998). Characterization of sequences associated with position-effect variegation at pericentric sites in *Drosophila* heterochromatin. *Chromosoma* **107,** 277–285.

Csink, A. K., and Hennikoff, S. (1996). Genetic modification of heterochromatic association and nuclear organization in *Drosophila. Nature* **381,** 529–531.

Csink, A. K., Linsk, R., and Birchler, J. A. (1994). The *Lighten up (Lip)* gene of *Drosophila melanogaster,* a modifier of retroelement expression, position effect variegation and *white* locus insertion alleles. *Genetics* **138,** 153–163.

Csink, A. K., Bounoutas, A., Griffith, M. L., Sabl, J. F., and Sage, B. T. (2002). Differential gene silencing by trans-heterochromatin in *Drosophila melanogaster. Genetics* **160,** 257–269.

Csordas, A. (1990). On the biological role of histone acetylation. *Biochem. J.* **265,** 23–38.

Czermin, B., Schotta, G., Hulsmann, B. B., Brehm, A., Becker, P. B., Reuter, G., and Imhof, A. (2001). Physical and functional association of SU(VAR)3–9 and HDAC1 in *Drosophila. EMBO Rep.* **2,** 915–919.

Dej, K. J., Ahn, C., and Orr-Weaver, T. L. (2004). Mutations in the *Drosophila* condensin subunit dCAP-G: defining the role of condensin for chromosome condensation in mitosis and gene expression in interphase. *Genetics* **168,** 895–906.

Demerec, M. (1940). Genetic behavior of euchromatic segments inserted into heterochromatin. *Genetics* **25,** 618–627.

Demerec, M. (1941). The nature of changes in the *white–Notch* region of the X chromosome of *Drosophila melanogaster. Proc. 7th Int. Genet. Congr.* 99–103.

Demerec, M., and Slizynska, H. (1937). Mottled *white*[258–18] of *Drosophila melanogaster. Genetics* **22,** 641–649.

Deng, H., Zhang, W., Bao, X., Martin, J. N., Girton, J., Johansen, J., and Johansen, K. M. (2005). The JIL-1 kinase regulates the structure of *Drosophila* polytene chromosomes. *Chromosoma* **114,** 173–182.

Dernburg, A. F., Broman, K. W., Fung, J. C., Marshall, W. F., Philips, J., Agard, D. A., and Sedat, J. W. (1996). Perturbation of nuclear architecture by long-distance chromosome interactions. *Cell* **85,** 745–759.

De Rubertis, F., Kadosh, D., Henchoz, S., Pauli, D., Reuter, G., Struhl, K., and Spierer, P. (1996). The histone deacetylase RPD3 counteracts genomic silencing in *Drosophila* and yeast. *Nature* **384,** 589–591.

Deshpande, G., Calhoun, G., and Schedl, P. (2005). *Drosophila* argonaute-2 is required early in embryogenesis for the assembly of centric/centromeric heterochromatin, nuclear division, nuclear migration, and germ cell formation. *Genes Dev.* **19,** 1680–1685.

Dietzel, S., Zolghadr, K., Hepperger, C., and Belmont, A. S. (2004). Differential large-scale chromatin compaction and intranuclear positioning of transcribed versus non-transcribed transgene arrays containing β-globin regulatory sequences. *J. Cell Sci.* **117,** 4603–4614.

Dobzhansky, T. (1936). Position effects on genes. *Biol. Rev.* **11,** 364–384.

Dorn, R., Szidonya, J., Korge, G., Sehnert, M., Taubert, H., Archoukieh, E., Tschiersch, B., Morawietz, H., Wustmann, G., Hoffmann, G., and Reuter, G. (1993). P transposon-induced dominant enhancer mutations of position-effect variegation in *Drosophila melanogaster*. *Genetics* **133,** 279–290.

Dreesen, T. D., Johnson, D. H., and Henikoff, S. (1988). The brown protein of *Drosophila melanogaster* is similar to the white protein and to components of active transport complex. *Mol. Cell. Biol.* **8,** 5206–5215.

Dreesen, T. D., Henikoff, S., and Loughney, K. (1991). A pairing-sensitive element that mediates trans-inactivation is associated with *Drosophila* brown gene. *Genes Dev.* **5,** 331–340.

Dubinin, N. P., and Sidorov, B. N. (1935). Position effect of the gene *hairy*. *Biol. Zh.* **4,** 555–568.

Dyson, M. H., Rose, S., and Mahadevan, L. C. (2001). Acetyllysine-binding and function of bromodomain-containing proteins in chromatin. *Front. Biosci.* **6,** D853–D865.

Eberl, D. F., Duyf, B. J., and Hilliker, A. J. (1993). The role of heterochromatin in the expression of a heterochromatic gene, the *rolled* locus of *Drosophila melanogaster*. *Genetics* **134,** 277–292.

Ebert, A., Schotta, G., Lein, S., Kubicek, S., Krauss, V., Jenuwein, T., and Reuter, G. (2004). Su(var) genes regulate the balance between euchromatin and heterochromatin in *Drosophila*. *Genes Dev.* **18,** 2973–2983.

Ebert, A., Lein, S., Schotta, G., and Reuter, G. (2006). Histone modification and the control of heterochromatic gene silencing in *Drosophila*. *Chromosome Res.* **14,** 377–392.

Eissenberg, J. C., James, T. C., Foster-Hartnett, D. M., Ngan, V., and Elgin, S. C. R. (1990). Mutation in a heterochromatin-specific protein is associated with suppression of position effect variegation in *Drosophila melanogaster*. *Proc. Natl. Acad. Sci. USA* **87,** 9923–9927.

Eissenberg, J. C., Morris, G. D., Reuter, G., and Hartnett, T. (1992). The heterochromatin-associated protein HP-1 is an essential protein in *Drosophila* with dosage-dependent effects on position effect variegation. *Genetics* **131,** 345–352.

Elgin, S. C. (1988). The formation and function of DNase I hypersensitive sites in the process of gene activation. *J. Biol. Chem.* **263,** 19259–19262.

Ephrussi, B., and Sutton, E. (1944). A reconsideration of the mechanism of position effect. *Proc. Natl. Acad. Sci. USA* **30,** 183–197.

Farkas, G., Gausz, J., Galloni, M., Reuter, G., Gyurkovics, H., and Karch, F. (1994). The *Trithorax-like* gene encodes the *Drosophila* GAGA factor. *Nature* **371,** 806–808.

Farkas, G., Leibovitch, B. A., and Elgin, S. C. R. (2000). Chromatin organization and transcriptional control of gene expression in *Drosophila*. *Gene* **253,** 117–136.

Fischle, W., Wang, Y., and Allis, C. D. (2003). Regulation of HP1-chromatin binding by histone H3 methylation and phosphorylation. *Nature* **425,** 475–479.

Fischle, W., Tseng, B. S., Dormann, H. L., Ueberheide, M. B., Garcia, B. A., Shabanowitz, J., Hunt, D. F., Funabiki, H., and Allis, C. D. (2005). Regulation of HP1-chromatin binding by histone H3 methylation and phosphorylation. *Nature* **438,** 1116–1122.

Gatti, M., and Pimpinelli, S. (1992). Functional elements in *Drosophila melanogaster* heterochromatin. *Annu. Rev. Genet.* **26,** 239–275.

Gerasimova, T. I., Gvozdev, V. A., and Birstein, V. (1972). Position effect variegation of *Pgl* locus determining 6-phospogluconate dehydrogenase in *Drosophila melanogaster*. *Dros. Inform. Serv.* **48,** 81.

Gerasimova, T. I., Gdula, D. A., Gerasimov, D. V., Simonova, O., and Corces, V. G. (1995). A *Drosophila* protein that imparts directionality on a chromatin insulator is an enhancer of position-effect variegation. *Cell* **82,** 587–597.

Gerasimova, T. I., Byrd, K., and Corces, V. G. (2000). A chromatin insulator determines the nuclear localization of DNA. *Mol. Cell* **6,** 1025–1035.

Gersh, E. S. (1949). Influence of temperature on the expression of position effects in the *scute-8* stock of *Drosophila melanogaster* and its relation to heterochromatization. *Genetics* **34,** 701–707.

Gersh, E. S. (1963). Variegation at the *white* locus in In(1)rst. *Dros. Inform. Serv.* **37,** 81.

Geyer, P. K., and Corces, V. G. (1992). DNA position-specific repression of transcription by a *Drosophila* zinc finger protein. *Genes Dev.* **6**, 1865–1873.

Ghosh, D., Gerasimova, T. I., and Corces, V. G. (2001). Interactions between the Su(Hw) and Mod(mdg4) proteins required for gypsy insulator function. *EMBO J.* **20**, 2518–2527.

Gilbert, M. K., Tan, Y. Y., and Hart, C. M. (2006). The *Drosophila* boundary element-associated factors BEAF-32A and BEAF-32B affect chromatin structure. *Genetics* **173**, 1365–1375.

Goldman, R. D., Goldman, A. E., and Shumaker, D. K. (2005). Nuclear lamins: Building blocks of nuclear structure and function. *Novartis Found. Symp.* **264**, 3–16.

Gowen, J. M., and Gay, E. H. (1934). Chromosome constitution and behavior in ever-sporting and mottling in *Drosophila melanogaster*. *Genetics* **19**, 189–126.

Grell, R. F. (1958). The effect of X chromosome loss on variegation. *Dros. Inform. Serv.* **32**, 124.

Grell, R. F. (1970). The time of initiation of segregational pairing between nonhomologues in *Drosophila melanogaster*. A reexamination of w^{m4}. *Genetics* **64**, 337–365.

Grewal, S. I., and Elgin, S. C. (2002). Heterochromatin: New possibilities for the inheritance of structure. *Curr. Opin. Genet. Dev.* **12**, 178–187.

Grewal, S. I., and Rice, J. C. (2004). Regulation of heterochromatin by histone methylation and small RNAs. *Curr. Opin. Cell Biol.* **16**, 230–238.

Grienenberger, A., Miotto, B., Sagnier, T., Cavalli, G., Schrmake, V., Geli, V., Mariol, M.-C., Berenger, Y., Graba, Y., and Pradel, J. (2002). The MYST domain acetyltransferase Chameau functions in epigenetic mechanisms of transcriptional repression. *Curr. Biol.* **12**, 762–766.

Grigliatti, T. (1991). Position-effect variegation—an assay for nonhistone chromosomal proteins and chromatin assembly and modifying factors. *Methods Cell Biol.* **35**, 588–625.

Gruenbaum, Y., Margalit, A., Goldman, R. D., Shumaker, D. K., and Wilson, K. L. (2005). The nuclear lamina comes of age. *Nat. Rev. Mol. Cell Biol.* **6**, 21–31.

Hackstein, J. H. P., and Hochstenbach, R. (1995). The elusive fertility genes of *Drosophila*: The ultimate haven for selfish genetic elements. *Trends Genet.* **11**, 195–200.

Hannah, A. H. (1951). Localization and function of heterochromatin in *Drosophila melanogaster*. *Adv. Genet.* **113**, 191–203.

Hari, K. L., Cook, D. R., and Karpen, G. H. (2001). The *Drosophila* Su(var)2-10 locus regulates chromosome structure and function and encodes a member of the PIAS family. *Genes Dev.* **15**, 1334–1348.

Harmon, B., and Sedat, J. (2005). Cell-by-cell dissection of gene expression and chromosomal interactions reveals consequences of nuclear reorganization. *PLoS Biol.* **3**, e67.

Hartman-Goldstein, I. J. (1967). On the relationship between heterochromatin and variegation in *Drosophila*, with special references to temperature sensitive periods. *Genet. Res.* **10**, 143–159.

Hazelrigg, T., and Petersen, S. (1992). An unusual genomic position effect on *Drosophila white* gene expression: pairing dependence, interactions with zeste, and molecular analysis of revertants. *Genetics* **130**, 125–138.

Hazelrigg, T., Levis, R., and Rubin, G. M. (1984). Transformation of *white* locus DNA in *Drosophila*: Dosage compensation, *zeste* interaction and position effects. *Cell* **36**, 469–481.

Heitz, E. (1928). Das Heterochromatin der Moose. *Jahrb Wiss Botanik* **69**, 762–818.

Henderson, D. S., Banga, S. S., Grigliatti, T. A., and Boyd, J. B. (1994). Mutagen sensitivity and suppression of position-effect variegation result from mutations in mus209, the *Drosophila* gene encoding PCNA. *EMBO J.* **13**, 1450–1459.

Henikoff, S. (1979). Position effect and variegation enhancers in an autosomal region of *Drosophila melanogaster*. *Genetics* **93**, 105–115.

Henikoff, S. (1994). A reconsideration of the mechanism of position effect. *Genetics* **138**, 1–5.

Henikoff, S., and Dreesen, T. D. (1989). Trans-inactivation of the *Drosophila brown* gene: Evidence for transcriptional repression and somatic pairing dependence. *Proc. Nat. Acad. Sci. USA* **86**, 6704–6708.

Hessler, A. Y. (1958). V-type position effect at the *light* locus in *Drosophila melanogaster. Genetics* **43,** 395–403.

Hilliker, A. J. (1976). Genetic analysis of the centromeric heterochromatin of chromosome 2 of *Drosophila melanogaster*: Deficiency mapping of EMS-induced lethal complementation groups. *Genetics* **83,** 765–782.

Hinton, T. (1940). Report on *D. melanogaster* new mutants. *Dros. Inform. Serv.* **13,** 49.

Hinton, T. (1942). A comparative study of certain heterochromatic regions in the mitotic and salivary gland chromosomes of *Drosophila melanogaster. Genetics* **27,** 119–127.

Hinton, T. (1949). The role of heterochromatin in position effect. *Proc. 8th Int. Congr. Genet.* 595–596.

Hinton, T. (1950). A correlation of phenotypic changes and chromosomal rearrangements at the two ends of an inversion. *Genetics* **35,** 188–205.

Hinton, T., and Goodsmith, W. (1950). An analysis of phenotypic reversions at the *brown* locus in *Drosophila. J. Exp. Zool.* **114,** 103–114.

Hirota, T., Lipp, J. J., Toh, B. H., and Peters, J. M. (2005). Histone H3 serine 10 phosphorylation by Aurora B causes HP1 dissociation from heterochromatin. *Nature* **438,** 1176–1180.

Huisinga, K. L., Brower-Toland, B., and Elgin, S. C. R. (2006). The contradictory definitions of heterochromatin: Transcription and silencing. *Chromosoma* **115,** 110–122.

Hutchison, C. J. (2002). Lamins: Building blocks or regulators of gene expression? *Nat. Rev. Mol. Cell Biol.* **3,** 848–858.

Ishii, K., Arib, G., Lin, C., Van Houwe, G., and Laemmli, U. K. (2002). Chromatin boundaries in budding yeast: The nuclear pore connection. *Cell* **109,** 551–562.

Ishizuka, A., Siomi, M. C., and Siomi, H. A. (2002). A *Drosophila* fragile X protein interacts with components of RNAi and ribosomal proteins. *Genes Dev.* **16,** 2497–2508.

Jackson, D. A. (2003). The principles of nuclear structure. *Chromosome Res.* **11,** 387–401.

Jacobs, S. A., Taverna, S. D., Zhang, Y., Briggs, S. D., Li, J., Eissenberg, J. C., Allis, C. D., and Khorasanizadeh, S. (2001). Specificity of the HP1 chromo domain for the methylated N-terminus of histone H3. *EMBO J.* **20,** 5232–5241.

James, T. C., and Elgin, S. C. R. (1986). Identification of a nonhistone chromosomal protein associated with heterochromatin in *Drosophila melanogaster* and its gene. *Mol. Cell Biol.* **6,** 3862–3872.

James, T. C., Eissenberg, J. C., Craig, C., Dietrich, V., Hobson, A., and Elgin, S. C. (1989). Distribution patterns of HP1, a heterochromatin-associated nonhistone chromosomal protein of *Drosophila. Eur. J. Cell Biol.* **50,** 170–180.

Jin, Y., Wang, Y., Walker, D. L., Dong, H., Conley, C., Johansen, J., and Johansen, K. M. (1999). JIL-1: A novel chromosomal tandem kinase implicated in transcriptional regulation in *Drosophila. Mol. Cell* **4,** 129–135.

Johansen, K. M., and Johansen, J. (2006). Regulation of chromatin structure by histone H3S10 phosphorylation. *Chromosome Res.* **14,** 393–404.

Judd, B. (1955). Direct proof of a variegated-type position effect at the *white* locus in *Drosophila melanogaster. Genetics* **40,** 739–744.

Karpen, G. H., and Spradling, A. C. (1992). Analysis of subtelocentric heterochromatin in the *Drosophila* minichromosome Dp1187 by single P element insertional mutagenesis. *Genetics* **132,** 737–753.

Kaufmann, B. P. (1942). Reversion from *roughest* to wild type in *Drosophila melanogaster. Genetics* **27,** 537–549.

Kaufmann, B. P. (1946). Organization of the chromosome. I. Break distribution and chromosome recombination in *Drosophila melanogaster. J. Exp. Zool.* **102,** 293–320.

Kellum, R., and Schedl, P. (1992). A group of scs elements function as domain boundaries in an enhancer-blocking assay. *Mol. Cell Biol.* **12,** 2424–2431.

Kouzarides, T. (2007). Chromatin modifications and their function. *Cell* **128,** 693–705.

Kuhfittig, S., Szabad, J., Schotta, G., Hoffmann, J., Mathe, E., and Reuter, G. (2001). *pitkin^D*, a novel gain-of-function enhancer of position-effect variegation, affects chromatin regulation during oogenesis and early embryogenesis in *Drosophila.* *Genetics* **157,** 1227–1244.

Labrador, M., and Corces, V. G. (2002). Setting the boundaries of chromatin domains and nuclear organization. *Cell* **111,** 151–154.

Lachner, M., O'Carroll, D., Rea, S., Mechtler, K., and Jenuwein, T. (2001). Methylation of histone H3 lysine 9 creates a binding site for HP1 proteins. *Nature* **410,** 116–120.

Lachner, M., O'Sullivan, R. J., and Jenuwein, T. (2003). An epigenetic road map for histone lysine methylation. *J. Cell Sci.* **116,** 2117–2124.

Lakhotia, S. C., and Jacob, J. (1974). EM autoradiographic studies on polytene nuclei of *Drosophila melanogaster.* *Exp. Cell Res.* **86,** 253–263.

Lanctôt, C., Cheutin, T., Cremer, M., Cavalli, G., and Cremer, T. (2007). Dynamic genome architecture in the nuclear space: Regulation of gene expression in three dimensions. *Nat. Rev. Genet.* **8,** 104–115.

Lawrence, R. J., and Pikaard, C. S. (2004). Chromatin turn ons and turn offs of ribosomal RNA genes. *Cell Cycle* **3,** 880–883.

Lei, E. P., and Corces, V. G. (2006). RNA interference machinery influences the nuclear organization of a chromatin insulator. *Nat. Genet.* **38,** 936–941.

Lerach, S., Zhang, W., Bao, X., Deng, H., Girton, J., Johansen, J., and Johansen, K. M. (2006). Loss-of-function alleles of the JIL-1 kinase are strong suppressors of position effect variegation of the w^{m4} allele in *Drosophila.* *Genetics* **173,** 2403–2406.

Lewis, E. B. (1950). The phenomenon of position effect. *Adv. Genet.* **3,** 73–115.

Lewis, M. S., and Pikaard, C. S. (2001). Restricted chromosomal silencing in nucleolar dominance. *Proc. Natl. Acad. Sci. USA* **98,** 14536–14540.

Lindsley, D. L., and Zimm, G. G. (1992). "The Genome of *Drosophila Melanogaster.*" Academic Press, San Diego.

Lindsley, D. L., Edington, C., and von Halle, E. S. (1960). Sex-linked recessive lethals in *Drosophila* whose expression is suppressed by the Y chromosome. *Genetics* **45,** 1649–1670.

Locke, J., Kotarski, M. A., and Tartof, K. D. (1988). Dosage dependent modifiers of position effect variegation in *Drosophila* and a mass action model that explains their effects. *Genetics* **120,** 181–198.

Lohe, A. R., Hilliker, A. J., and Roberts, P. A. (1993). Mapping simple repeated DNA sequences in heterochromatin of *Drosophila melanogaster.* *Genetics* **134,** 1149–1174.

Luger, K. (2006). Dynamic nucleosomes. *Chromosome Res.* **14,** 5–16.

Mattout, A., Goldberg, M., Tzur, Y., Margalit, A., and Gruenbaum, Y. (2006). Specific and conserved sequences in *D. melanogaster* and *C. elegans* lamins and histone H2A mediate the attachment of lamins to chromosomes. *J. Cell. Sci.* **120,** 77–85.

Matzke, M. A., and Birchler, J. A. (2005). RNAi-mediated pathways in the nucleus. *Nat. Rev. Genet.* **6,** 24–35.

Mendjan, S., Taipale, M., Kind, J., Holz, H., Gebhardt, P., Schelder, M., Vermeulen, M., Buscaino, K., Duncan, K., Mueller, J., Wilm, M., Stunnenberg, H. G., *et al.* (2006). Nuclear pore components are involved in the transcriptional regulation of dosage compensation in *Drosophila.* *Mol. Cell* **21,** 811–823.

Mis, J., Ner, S. S., and Grigliatti, T. A. (2006). Identification of three histone methyltransferases in Drosophila: dG9a is a suppressor of PEV and is required for gene silencing. *Mol. Genet. Genomics* **275,** 513–526.

Misteli, T. (2007). Beyond the sequence: Cellular organization of genome function. *Cell* **128**, 787–800.

Mito, Y., Henikoff, J. G., and Henikoff, S. (2005). Genome-scale profiling of histone H3.3 replacement patterns. *Nat. Genet.* **37**, 1090–1097.

Morgan, T. H., Bridges, C. B., and Sturtevant, A. H. (1925). The genetics of *Drosophila melanogaster*. *Biblphia Genet.* **2**, 262.

Morgan, T. H., Schultz, J., and Curry, V. (1941). Investigation on the constitution of the germinal material in relation to heredity. *Carnegie Inst. Yearbook* **40**, 282–287.

Mottus, R., Sobels, R. E., and Grigliatti, T. A. (2000). Mutational analysis of a histone deacetylase in *Drosophila melanogaster*: Missense mutations suppress gene silencing associated with position effect variegation. *Genetics* **154**, 657–668.

Muller, H. J. (1930). Types of visible variations induced by X-rays in *Drosophila*. *J. Genet.* **22**, 299–334.

Muller, H. J. (1932). Further studies on the nature and causes of gene mutations. *Proc. Sixth Int. Congr. Genet.* **1**, 213–255.

Nakayama, T., Nishioka, K., Dong, Y.-X., Shimojima, T., and Hirose, S. (2007). *Drosophila* GAGA factor directs histone H3.3 replacement that prevents the heterochromatin spreading. *Genes Dev.* **21**, 552–561.

Neely, K. E., and Workman, J. L. (2002). Histone acetylation and chromatin remodeling: Which comes first? *Mol. Genet. Metab.* **76**, 1–5.

Neuhaus, M. J. (1939). Genetic study of the Y chromosome in *Drosophila melanogster*. *J. Genet.* **37**, 229–254.

Nishioka, K., Rice, J. C., Sarma, K., Erdjument-Bromage, H., Werner, J., Wang, Y., Sergei Chuikov, P., Valenzuela, P., Tempst, P., Steward, R., Lis, J. T., Allis, C. D., *et al.* (2002). PR-set7 is a nucleosome-specific methyltransferase that modifies lysine 20 of histone H4 and is associated with silent chromatin. *Mol. Cell* **9**, 1201–1213.

Pai, C. Y., Lei, E. P., Ghosh, D., and Corces, V. G. (2004). The centrosomal protein CP190 is a component of the gypsy chromatin insulator. *Mol. Cell* **16**, 737–748.

Pak, D. T., Pflumm, M., Chesnokov, I., Huang, D. W., Kellum, R., Marr, J., Romanowski, P., and Botchan, M. R. (1997). Association of the origin recognition complex with heterochromatin and HP1 in higher eukaryotes. *Cell* **91**, 311–323.

Pal-Bhadra, M., Leibovitch, B. A., Gandhi, S. G., Rao, M., Bhadra, U., Birchler, J. A., and Elgin, S. C. (2004). Heterochromatic silencing and HP1 localization in *Drosophila* are dependent on the RNAi machinery. *Science* **303**, 669–672.

Panshin, I. B. (1938). Cytogenic nature of position effect of genes *white* (mottled) and *cubitus interuptus*. *Biol. Zh.* **7**, 837–868.

Pardue, M. L., and DeBaryshe, P. G. (1999). Telomeres and telomerase: More than the end of the line. *Chromosoma* **108**, 73–82.

Parkhurst, S. M., Harrison, D. A., Remington, M. P., Spana, C., Kelley, R. L., Coyne, R. S., and Corces, V. G. (1988). The *Drosophila su(hw)* gene, which controls the phenotypic effect of the *gypsy* transposable element, encodes a putative DNA-binding protein. *Genes Dev.* **2**, 1205–1215.

Parnell, T. J., Kuhn, E. J., Gilmore, B. L., Helou, C., Wold, M. S., and Geyer, P. K. (2006). Identification of genomic sites that bind the *Drosophila* suppressor of Hairy-wing insulator protein. *Mol. Cell Biol.* **26**, 5983–5993.

Peng, J. C., and Karpen, G. H. (2006). H3K9 methylation and RNA interference regulate nucleolar organization and repeated DNA stability. *Nat. Cell Biol.* **9**, 25–35.

Perrin, L., Demakova, O., Fanti, L., Kallenbach, S., Saingery, S., Macl'ceva, N. I., Pimpinelli, S., Zhimulev, I., and Pradel, J. (1998). Dynamics of the sub-nuclear distribution of Modulo and the regulation of position-effect variegation. *J. Cell Sci.* **111**, 2753–2761.

Peterson, C. L. (2003). Transcriptional activation: Getting a grip on condensed chromatin. *Curr. Biol.* **13**, R195–R197.

Pickersgill, H., Kalverda, B., de Wit, E., Talhout, W., Fornerod, M., and van Steensel, B. (2006). Characterization of the *Drosophila melanogaster* genome at the nuclear lamina. *Nat. Genet.* **38**, 1005–1014.

Pimpinelli, S., Berloco, M., Fanti, L., Dimitri, P., Bonaccorsi, S., Marchetti, E., Caizzi, R., Aggese, C. C., and Gatti, M. (1995). Transposable elements are stable structural components of *Drosophila melanogaster* heterochromatin. *Proc. Natl. Acad. Sci. USA* **92**, 3804–3808.

Pokholok, D. K., Harbison, C. T., Levine, S., Cole, M., Hannett, N. M., Lee, T. I., Bell, G. W., Walker, P. A., Rolfe, P. A., Herbolsheimer, E., Zeitlinger, J., Lewitter, F., *et al.* (2005). Genome-wide map of nucleosome acetylation and methylation in yeast. *Cell* **122**, 517–527.

Puckett, L. D., and Snyder, L. A. (1973). Analysis of ribosomal RNA synthesis in two X chromosome inversions in *Drosophila* melanogaster. *Genetics* **74**, S221.

Qi, D., Jin, H., Lilja, T., and Mannervik, M. (2006). *Drosophila* Reptin and other TIP60 complex components promote generation of silent chromatin. *Genetics* **174**, 241–251.

Ramos, E., Ghosh, D., Baxter, E., and Corces, V. G. (2006). Genomic organization of *gypsy* chromatin insulators in *Drosophila melanogaster. Genetics* **172**, 2337–2349.

Rea, S., Eisenhaber, F., O'Carroll, D., Strahl, B. D., Sun, Z. W., Schmid, M., Opravil, S., Mechtler, C. P., Ponting, C. P., Allis, C. D., and Jenuwein, T. (2000). Regulation of chromatin structure by site-specific histone H3 methyltransferases. *Nature* **406**, 593–599.

Reugels, A. M., Kurek, R., Lammermann, U., and Bunemann, H. (2000). Mega-introns in the dynein gene DhDhc7(Y) on the heterochromatic Y chromosome give rise to the giant thread loops in primary spermatocytes of *Drosophila hydei. Genetics* **154**, 759–769.

Reuter, G., and Spierer, P. (1992). Position effect variegation and chromatin proteins. *Bioessays* **14**, 605–612.

Reuter, G., and Wolff, I. (1981). Isolation of dominant suppressor mutations for position-effect variegation in *Drosophila melanogaster. Mol. Gen. Genet.* **182**, 516–519.

Reuter, G., Werner, W., and Hoffmann, H. J. (1982). Mutants affecting position-effect heterochromatization in *Drosophila melanogaster. Chromosoma* **85**, 539–551.

Reuter, G., Wolff, I., and Friede, B. (1985). Functional properties of the heterochromatic sequences inducing w^{m4} position-effect variegation in *Drosophila melanogaster. Chromosoma* **93**, 132–139.

Reuter, G., Dorn, R., Wustmann, G., Friede, B., and Rauh, G. (1986). Third chromosome suppressor of position-effect variegation in *Drosophila melanogaster. Mol. Gen. Genet.* **202**, 481–487.

Reuter, G., Giarre, M., Farah, J., Gausz, J., Spierer, A., and Spierer, P. (1990). Dependence of position-effect variegation in *Drosophila* on the dose of a gene encoding an unusual zinc-finger protein. *Nature* **344**, 219–223.

Risinger, C., Deiteher, D. L., Lundell, I., Schwartz, T. L., and Larhammar, D. (1997). Complex gene organization of synaptic protein SNAP-25 in *Drosophila melanogaster. Gene* **194**, 169–177.

Rizki, T. M. (1961). Intracellular localization of kynurenine in the fat body of *Drosophila. J. Biophys. Biochem. Cytol.* **9**, 567–572.

Roseman, R. R., Pirrotta, V., and Geyer, P. K. (1993). The su(Hw) protein insulates expression of the *Drosophila melanogaster white* gene from chromosomal position-effects. *EMBO J.* **12**, 435–442.

Rubin, G. M., and Spradling, A. C. (1982). Genetic transformation of *Drosophila* with transposable element vectors. *Science* **218**, 348–353.

Rzepecki, R., Bogachev, S. S., Kokoza, E., Stuurman, N., and Fisher, P. A. (1998). In vivo association of lamins with nucleic acids in *Drosophila melanogaster. J. Cell Sci.* **111**, 121–129.

Sage, B. T., and Csink, A. K. (2003). Heterochromatic self-association, a determinant of nuclear organization, does not require sequence homology in *Drosophila. Genetics* **165**, 1183–1193.

Sass, G. L., and Henikoff, S. (1999). Pairing-dependent mislocalization of a *Drosophila brown* gene reporter to a heterochromatic environment. *Genetics* **152**, 596–604.

Schalet, A. (1969). Three Y suppressed phenotypes associated with X chromosomes carrying *In(1)B*^{M1}. *Dros. Inf. Serv.* **44**, 87.

Schotta, G., Ebert, A., Krauss, V., Fischer, A., Hoffmann, J., Rea, S., Jenuwein, T., Dorn, R., and Reuter, G. (2002). Central role of *Drosophila* SU(VAR)3–9 in histone H3-K9 methylation and heterochromatic gene silencing. *EMBO J.* **21**, 1121–1131.

Schotta, G., Ebert, A., Dorn, R., and Reuter, G. (2003a). Position effect variegation and the genetic dissection of chromatin regulation in *Drosophila*. *Semin. Cell Dev. Biol.* **14**, 67–75.

Schotta, G., Ebert, A., and Reuter, G. (2003b). SU(VAR)3–9 is a conserved key function in heterochromatic gene silencing. *Genetica* **117**, 149–158.

Schotta, G., Lachner, M., Sarma, K., Ebert, A., Sengupta, R., Reuter, G., Reinberg, D., and Jenuwein, T. (2004). A silencing pathway to induce H3-K9 and H4-K20 trimethylation at constitutive heterochromatin. *Genes Dev.* **18**, 1251–1262.

Schubeler, D., MacAlpine, D. M., Scalzo, D., Wirbelauer, C., Kooperberg, C., van Leeuwen, F., Gottschling, D. E., O'Neill, L. P., Turner, B. M., Delrow, J., Bell, S. P., and Groudine, M. (2004). The histone modification pattern of active genes revealed through genome-wide chromatin analysis of a higher eukaryote. *Genes Dev.* **18**, 1263–1271.

Schultz, J. (1936). Variegation in *Drosophila* and the inert chromosome regions. *Proc. Natl. Acad. Sci. USA* **22**, 27–33.

Schultz, J. (1939). The function of heterochromatin. *Proc. Natl. Acad. Sci. USA* **22**, 27–33.

Schultz, J. (1941). The function of heterochromatin. *Proc. 7th Intern. Congr. Genet.* **25**, 7–262.

Schultz, J., and Dobzhansky, T. (1934). The relation of a dominant eye color in *Drosophila melanogaster* to the associated chromosome rearrangement. *Genetics* **19**, 344–364.

Schulze, R. S., and Wallrath, L. L. (2007). Gene regulation by chromatin structure: Paradigms established in *Drosophila melanogaster*. *Annu. Rev. Entomol.* **52**, 171–192.

Schulze, S. R., Sinclair, D. A., Fitzpatrick, K. A., and Honda, B. M. (2005). A genetic and molecular characterization of two proximal heterochromatic genes on chromosome 3 of *Drosophila melanogaster*. *Genetics* **169**, 2165–2177.

Schwartz, B. E., and Ahmad, K. (2005). Transcriptional activation triggers deposition and removal of the histone variant H3.3. *Genes Dev.* **19**, 804–814.

Seum, C., Spierer, A., Pauli, D., Szidonya, J., Reuter, G., and Spierer, P. (1996). Position-effect variegation in *Drosophila* depends on the dose of the gene encoding the E2F transcriptional activator and cell cycle regulator. *Development* **122**, 1949–1956.

Shaffer, C. D., Stephens, G. E., Thompson, B. A., Funches, L., Bernat, J. A., Craig, C. A., and Elgin, S. C. R. (2002). Heterochromatin protein 2 (HP2), a partner of HP1 in *Drosophila* heterochromatin. *Proc. Natl. Acad. Sci. USA* **99**, 14332–14337.

Shareef, M. M., King, C., Damaj, M., Badagu, R., Huang, D. W., and Kellum, R. (2001). *Drosophila* heterochromatin protein 1 (HP1)/Origin recognition complex (ORC) protein is associated with HP1 and ORC and functions in heterochromatin-induced silencing. *Mol. Biol. Cell* **12**, 1671–1685.

Shogren-Knaak, M., Ishii, H., Sun, J.-M., Pazin, M. J., Davie, J. R., and Peterson, C. L. (2006). Histone H4-K16 acetylation controls chromatin structure and protein interactions. *Science* **311**, 844–847.

Sinclair, D. A. R., Mottus, R. C., and Grigliatti, T. A. (1983). Genes which suppress position-effect variegation in *Drosophila melanogaster* are clustered. *Mol. Gen. Genet.* **191**, 326–333.

Sinclair, D. A. R., Schulze, S., Silva, E., Fitzpatrick, K. A., and Honda, B. M. (2000). Essential genes in autosomal heterochromatin of *Drosophila melanogaster*. *Genetica* **109**, 9–18.

Slatis, H. M. (1955). Position effects at the *brown* locus in *Drosophila melanogaster*. *Genetics* **40**, 5–23.

Somech, R., Shaklai, S., Geller, O., Amariglio, N., Simon, A. J., Rechavi, G., and Gal-Yam, E. N. (2005). The nuclear-envelope protein and transcriptional repressor LAP2β interacts with HDAC3 at the nuclear periphery, and induces histone H4 deacetylation. *J. Cell Sci.* **118**, 4017–4025.

Spofford, J. B. (1967). Single-locus modification of position-effect variegation in *Drosophila melanogaster*. I. White variegation. *Genetics* **57**, 751–766.

Spofford, J. B. (1973). Variegation for *dm* in *Dp(1:3)*[N264-58]. *Dros. Inf. Serv.* **50**, 98.

Spofford, J. B. (1976). Position-effect variegation in *Drosophila*. *In* "The Genetics and Biology of *Drosophila*" (M. Ashburner and E. Novitski, eds.), Vol. 1c, pp. 955–1018. Academic Press, London.

Spofford, J. B., and DeSalle, R. (1991). Nucleolus organizer-suppressed position-effect variegation in *Drosophila melanogaster*. *Genet. Res.* **57**, 245–255.

Spradling, A. C., and Karpen, G. H. (1990). Sixty years of mystery. *Genetics* **126**, 779–784.

Spradling, A. C., and Rubin, G. M. (1982). Transposition of cloned P elements into *Drosophila* germ line chromosomes. *Science* **218**, 341–347.

Steller, H., and Pirrotta, V. (1985). Expression of the *Drosophila white* gene under the control of the hsp-70 heat shock promoter. *EMBO J.* **4**, 3765–3772.

Strahl, B. D., and Allis, C. D. (2000). The language of covalent histone modifications. *Nature* **403**, 1–45.

Sturtevant, A. H. (1925). The effect of unequal crossing over at the *Bar* locus in *Drosophila*. *Genetics* **10**, 117–147.

Sun, F. L., and Elgin, S. C. (1999). Putting boundaries on silence. *Cell* **99**, 459–462.

Sun, X., Wahlstrom, J., and Karpen, G. (1997). Molecular structure of a functional *Drosophila* centromere. *Cell* **91**, 1007–1019.

Swaminathan, J., Baxter, E. M., and Corces, V. G. (2005). The role of histone H2Av variant replacement and histone H4 acetylation in the establishment of *Drosophila* heterochromatin. *Genes Dev.* **19**, 65–76.

Taddei, A., and Gasser, S. M. (2004). Multiple pathways for telomere tethering: Functional implications of subnuclear position for heterochromatin formation. *Biochim. Biophys. Acta* **1677**, 120–128.

Talbert, P. B., and Henikoff, S. (2000). A reexamination of spreading of position-effect variegation in the *white-roughest* region of *Drosophila melanogaster*. *Genetics* **154**, 259–272.

Talbert, P. B., LeCiel, C. D. S., and Hennikoff, S. (1994). Modifications of the *Drosophila* heterochromatic mutation *brown Dominant* by linkage alterations. *Genetics* **136**, 559–571.

Tartof, K. D., Hobbs, C., and Jones, M. (1984). A structural basis for variegating position effects. *Cell* **37**, 869–878.

Teixeira, M. T., Dujon, B., and Fabre, E. (2002). Genome-wide nuclear morphology screen identifies novel genes involved in nuclear architecture and gene-silencing in *Saccharomyces cerevisiae*. *J. Mol. Biol.* **321**, 551–561.

Thakar, R., and Csink, A. K. (2005). Changing chromatin dynamics and nuclear organization during differentiation in *Drosophila* larval tissue. *J. Cell Sci.* **118**, 951–960.

Tobler, J., Bowman, J. T., and Simmons, J. B. (1971). Gene modulation in *Drosophila*: Dosage compensation and relocated v$^+$ genes. *Biochem. Genet.* **5**, 11–117.

Tschiersch, B., Hofmann, A., Krauss, V., Dorn, R., Korge, G., and Reuter, G. (1994). The protein encoded by the *Drosophila* position-effect variegation suppressor gene *Su(var)3–9* combines domains of antagonistic regulators of homeotic gene complexes. *EMBO J.* **13**, 3822–3831.

Tumbar, T., Sudlow, G., and Belmont, A. S. (1999). Large-scale chromatin unfolding and remodeling induced by VP16 acidic activation domain. *J. Cell Biol.* **145**, 1341–1354.

Turner, B. M. (2000). Histone acetylation and an epigenetic code. *Bioessays* **22**, 836–845.

Turner, B. M. (2002). Cellular memory and the histone code. *Cell* **111**, 285–291.

Turner, B. M., Birley, A. J., and Lavender, J. (1992). Histone H4 isoforms acetylated at specific lysine residues define individual chromosomes and chromatin domains in *Drosophila* polytene nuclei. *Cell* **69**, 375–384.

van Driel, R., Fransz, P. F., and Verschure, P. J. (2003). The eukaryotic genome: A system regulated at different hierarchical levels. *J. Cell Sci.* **116**, 4067–4075.

Verdel, A., and Moazed, D. (2005). RNAi-directed assembly of heterochromatin in fission yeast. *FEBS Lett.* **579**, 5872–5878.

Verdel, A., Jia, S., Gerber, S., Sugiyama, T., Gygi, S., Grewal, S. I., and Moazed, D. (2004). RNAi-mediated targeting of heterochromatin by the RITS complex. *Science* **303**, 672–676.

Verni, F., Gandhi, R., Goldberg, M. L., and Gatti, M. (2000). Genetic and molecular analysis of *wings apart-like* (*wapl*), a gene controlling heterochromatin organization in *Drosophila melanogaster*. *Genetics* **154**, 1693–1710.

Wade, P. A., Pruss, D., and Wolffe, A. P. (1997). Histone acetylation: Chromatin in action. *Trends Biochem. Sci.* **22**, 128–132.

Wakimoto, B., and Hearn, M. G. (1990). The effect of chromosome rearrangements on the expression of heterochromatic genes in chromosome 2L of *Drosophila melanogaster*. *Genetics* **125**, 141–154.

Wallrath, L. L. (2000). *Drosophila* telomeric transgenes provide insights on mechanisms of gene silencing. *Genetica* **109**, 25–33.

Wallrath, L. L., and Elgin, S. C. R. (1995). Position effect variegation in *Drosophila* is associated with an altered chromatin structure. *Genes Dev.* **9**, 1263–1277.

Wallrath, L. L., Guntur, V. P., Rosman, L. E., and Elgin, S. C. (1996). DNA representation of variegating heterochromatic P-element inserts in diploid and polytene tissues of *Drosophila melanogaster*. *Chromosoma* **104**, 519–527.

Walter, M. F., Jang, C., Kasravi, B., Donath, J., Mechler, J. M., and Biessmann, H. (1995). DNA organization and polymorphism of a wild-type *Drosophila* telomere region. *Chromosoma* **104**, 229–241.

Wang, Y., Zhang, W., Jin, Y., Johansen, J., and Johansen, K. M. (2001). The JIL-1 tandem kinase mediates histone H3 phosphorylation and is required for maintenance of chromatin structure in *Drosophila*. *Cell* **105**, 433–443.

Wang, L., Haeusler, R. A., Good, P. D., Thompson, M., Nagar, S., and Engelke, D. R. (2005). Silencing near tRNA genes requires nucleolar localization. *J. Biol. Chem.* **280**, 8637–8639.

Wargent, J. M. (1972). Position-effect variegation in the *Revolute of Bridges* strain of *Drosophila melanogaster*. *Dros. Inform. Serv.* **49**, 50–51.

Wei, G. H., Liu, E. P., and Liang, C. C. (2005). Chromatin domain boundaries: Insulators and beyond. *Cell Res.* **15**, 292–300.

Weiler, K. S., and Wakimoto, B. T. (1995). Heterochromatin and gene expression in *Drosophila*. *Annu. Rev. Genet.* **29**, 577–605.

West, A. G., Gaszner, M., and Felsenfeld, G. (2002). Insulators: Many functions, many mechanisms. *Genes Dev.* **16**, 271–288.

Workman, J. L., and Kingston, R. E. (1998). Alteration of nucleosome structure as a mechanism of transcriptional regulation. *Annu. Rev. Biochem.* **67**, 545–579.

Wustmann, G., Szidonya, J., Taubert, H., and Reuter, G. (1989). The genetics of position-effect modifying loci in *Drosophila melanogaster*. *Mol. Gen. Genet.* **217**, 520–527.

Yamaguchi, M., Yoshida, H., Hirose, F., Inoue, Y. H., Hayashi, Y., Yamagishi, M., Nishi, Y., Tamai, K., Sakaguchi, K., and Matsukage, A. (2001). Ectopic expression of BEAF32A in the *Drosophila* eye imaginal disc inhibits differentiation of photoreceptor cells and induces apoptosis. *Chromosoma* **110**, 313–321.

Yamamoto, K., and Sonoda, M. (2003). Self-interaction of heterochromatin protein 1 is required for direct binding to histone methyltransferase, SUV39H1. *Biochem. Biophys. Res. Commun.* **301,** 287–292.

Yamamoto, Y., Girard, F., Bello, B., Affolter, M., and Gehring, W. J. (1997). The cramped gene of *Drosophila* is a member of the Polycomb-group, and interacts with *mus209*, the gene encoding Proliferating Cell Nuclear Antigen. *Development* **124,** 3385–3394.

Yang, X. J. (2004). Lysine acetylation and the bromodomain: A new partnership for signaling. *Bioessays* **26,** 1076–1087.

Yasufzi, T. M., Tagami, H., Nakatani, Y., and Felsenfeld, G. (2004). CTCF tethers an insulator to subnuclear sites, suggesting shared insulator mechanisms across species. *Mol. Cell* **13,** 291–298.

Zaratiegui, M., Irvine, D. V., and Martienssen, R. A. (2007). Noncoding RNAs and gene silencing. *Cell* **128,** 763–776.

Zhang, Y., and Reinberg, D. (2001). Transcription regulation by histone methylation: Interplay between different covalent modifications of the core histone tails. *Genes Dev.* **15,** 2343–2360.

Zhang, W., Jin, Y., Ji, Y., Girton, J., Johansen, J., and Johansen, K. M. (2003). Genetic and phenotypic analysis of alleles of the *Drosophila* chromosomal JIL-1 kinase reveals a functional requirement at multiple developmental stages. *Genetics* **165,** 1341–1354.

Zhang, W., Deng, H., Bao, X., Lerach, S., Girton, J., Johansen, J., and Johansen, K. M. (2006). The JIL-1 histone H3S10 kinase regulates dimethyl histone H3K9 modifications and heterochromatic spreading in *Drosophila*. *Development* **133,** 229–235.

Zhao, K., Hart, C. M., and Laemmli, U. K. (1995). Visualization of chromosomal domains with boundary element-associated factor BEAF-32. *Cell* **81,** 879–889.

Zhimulev, I. F. (1998). Polytene chromosomes, heterochromatin, and position effect variegation. *Adv. Genet.* **37,** 1–566.

Zhimulev, I. F., and Belyaeva, E. S. (2003). Intercalary heterochromatin and genetic silencing. *BioEssays* **25,** 1040–1051.

Zuckerkandl, E. (1974). A possible role of "inert" heterochromatin in cell differentiation. *Biochemie* **56,** 937–954.

2

Polycomb Group Proteins and Long-Range Gene Regulation

Julio Mateos-Langerak and Giacomo Cavalli

Chromatin and Cell Biology Lab, Institute of Human Genetics, CNRS, 141, rue de la Cardonille, 34396 Montpellier, France

I. Introduction
II. Genetic and Biochemical Characterization of PcG Proteins
III. PcG Mechanisms of Action
IV. PcG Proteins and Long-Range Gene Silencing
V. PcG and Very Long-Range Gene Silencing: "Teleregulation" of Gene Expression
VI. Conclusions and Prospects
References

ABSTRACT

Genome regulation takes place at different hierarchically interconnected levels: the DNA sequence level, the chromatin level, and the three-dimensional (3D) organization of the nucleus. Polycomb group (PcG) proteins are silencers that regulate transcription at all these three levels. They are targeted to specific sequences in the genome, contributing to maintain cellular identity. Recent research reveals that PcG proteins may be important actors at the level of the nuclear 3D structure. Here, we discuss our current knowledge of how PcG proteins regulate transcription across the three mentioned levels, and in particular their possible role in regulation of remote genes. We suggest the possibility that PcG proteins establish 3D networks of chromatin contacts as a mechanism to orchestrate gene expression. © 2008, Elsevier Inc.

Advances in Genetics, Vol. 61

Copyright 2008, Elsevier Inc. All rights reserved.

0065-2660/08 $35.00

DOI: 10.1016/S0065-2660(07)00002-8

I. INTRODUCTION

Chromatin is a very special biological structure formed by the DNA, histone and nonhistone proteins, and RNA molecules. These components are directly or indirectly associated with the DNA molecule, forming large supramolecular complexes. All these elements interact to maintain and to use in an appropriate way the genetic and epigenetic information. What makes chromatin so unique is the extraordinary combination of mechanisms that manage the huge amount of information stored in the genome. It is clear that regulation of the genome is crucial for a proper functioning of the organism. This regulation requires a strict and complex control that involves many different actors—DNA, histones and other proteins, and RNAs—and a complex choreography where spatial and temporal aspects are crucial. We are just beginning to understand this complexity.

We can discriminate between three hierarchically interconnected levels at which genome control takes place (van Driel *et al.*, 2003). The first level of regulatory mechanisms is found on the DNA molecule. Specific DNA sequences mediate the targeting of nuclear factors that regulate a multitude of processes in the genome. Good examples of this are the promoters, sequences responsible for the binding of transcription initiation machineries at the right positions on the genome. Multicellular organisms are constituted by cells with different identities. These cellular identities involve different gene expression profiles, yet all these cells share identical genomes. Therefore, the regulatory elements found in the DNA nucleotide sequence cannot fully explain the different cellular identities found in multicellular organisms. These inheritable identities that cannot be explained by differences in the DNA sequence are referred to as *epigenetic* and are, at least partially, explained by the chromatin composition. This second regulatory level includes marks such as methylation of the DNA molecule, covalent posttranslational modifications of the core histones, histone variants, and other chromatin-associated proteins. Contrasting with the DNA sequence level which, excluding mutation and recombination events, is stable throughout the life of the organism, the chromatin level may be regulated by enzymatic activities acting as a "switch" that can change its functional state (Jenuwein, 2001; Jenuwein and Allis, 2001; van Driel *et al.*, 2003). On the other hand, chromatin states may sometimes be maintained through the cell cycle, conferring inheritance of specific cell identities, which is necessary, for example, for organism development (Dejardin and Cavalli, 2005; Rando, 2007; Santoro and De Lucia, 2005).

A third regulatory level that has been poorly studied, when compared to the other two, is the three-dimensional (3D) organization of chromatin in the nucleus. DNA molecules contained in the nucleus of, for example, human cells have a total length of 2 m if they are released from all other chromatin components. This has to be fitted into a nucleus of a diameter of $\sim$10 μm. This formidable process is comparable to fitting a string of 2 km into a marble of 1 cm in diameter. To achieve this high degree of compaction, the DNA molecule is organized together with the histones and other chromatin components into higher order 3D structures. In the past decades, evidence has emerged that this spatial organization of chromatin, and also other nuclear components, has a role in regulating the function of the genome (Cremer *et al.*, 2004; Lanctot *et al.*, 2007; Misteli, 2007; van Driel *et al.*, 2003).

Polycomb group (PcG) proteins are regulators of the genome that act at these three levels, constituting a key example of how they are functionally interconnected (Fig. 2.1). PcG proteins were first identified in *Drosophila melanogaster* as mutants deregulating *Hox* gene expression patterns. *Hox* genes acquire a specific expression pattern during the early development of the fly. This pattern is maintained through development thanks to the action of two antagonistic groups of proteins, the PcG and the trithorax group (trxG). According to this view, PcG proteins would be responsible for the maintenance of the silenced state of genes through cell division, while trxG proteins would maintain the active state of genes (reviewed in Schuettengruber *et al.*, 2007; Schwartz and

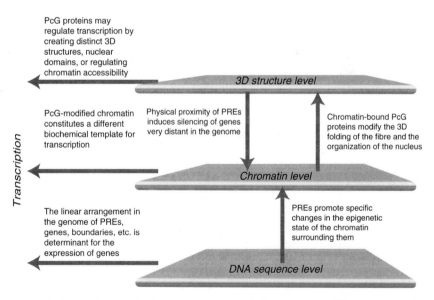

Figure 2.1. Polycomb group (PcG) proteins act as transcription regulators at different levels. In this simplified scheme, we represent the different levels at which transcription regulation takes place. Arrows represent how PcG proteins contribute to regulate global genome transcription by connecting these three levels with each other and with gene transcription.

Pirrotta, 2007; Sparmann and van Lohuizen, 2006). In this chapter, we will review our current knowledge of the role of the PcG proteins in regulating genome expression, focusing on the two higher levels, that is, the chromatin and 3D structure. Comprehensive reviews of the biochemical characterization of PcG proteins can be found in the literature (Calonje and Sung, 2006; Grimaud et al., 2006b; Levine et al., 2004; Schuettengruber et al., 2007; Schwartz and Pirrotta, 2007). Much of the data available come from the model system *Drosophila melanogaster*. Nevertheless, we will refer to the corresponding knowledge in vertebrates and plants when available.

II. GENETIC AND BIOCHEMICAL CHARACTERIZATION OF PcG PROTEINS

The PcG is a diverse group of proteins that form at least three different complexes: Polycomb repressive complex 1 and 2 (PRC1 and PRC2) and pleiohomeotic (Pho) repressive complex (PhoRC) (Grimaud et al., 2006b; Schuettengruber et al., 2007; Schwartz and Pirrotta, 2007). A list of the proteins pertaining to any of the complexes, the respective human and mouse homologues, and some of the known functions can be found in Table 2.1. We will use here the *D. melanogaster* nomenclature, as most of the data relevant for this chapter are proceeding from the fruit fly. The function of PcG proteins in vertebrates and plants has been discussed in other reviews (Calonje and Sung, 2006; Delaval and Feil, 2004; Heard, 2005; Otte and Kwaks, 2003; Sparmann and van Lohuizen, 2006). The biochemical functions attributed to some of the components of PcG complexes are key to understanding how they can maintain silencing through cell division. The core of the proteins contained in PRC1 contains two identified functions, namely, a methyl-lysine-binding function on the chromodomain of (polycomb) PC (Bannister et al., 2001; Lachner et al., 2001; Paro and Hogness, 1991) and an E3 ubiquitin ligase activity on RING (Wang et al., 2004a). Meanwhile, the PRC2 component E(Z) has a Su(var) 3–9, E(z) and Trithorax (SET) domain known to tri-methylate lysine 27 of the histone H3 (H3K27) and, with a lower efficiency, the lysine 9 of the same histone (H3K9) (Czermin et al., 2002). Unlike the two other complexes, PhoRC can bind specific DNA sequences through the Zing-finger domain of Pho (Brown et al., 1998; Klymenko et al., 2006), and is involved in recruitment of PRC2 and PRC1 (Mohd-Sarip et al., 2005; Mohd-Sarip et al., 2006; Wang et al., 2004b). It is important to note that although PcG proteins form specific protein complexes *in vitro*, we do not know the real composition of these in cells (Otte and Kwaks, 2003). It is likely that their configuration is dynamic, and that a subset of the PcG components can participate in different complexes or function as individual entities.

Table 2.1. PcG Components in *Drosophila* and Their Human Orthologues

Complex	*Drosophila melanogaster*	Human	Domains	Function
PRC1	Pc	HPC1, HPC2, HPC3	Chromodomain	Methyl-lysine binding
	Ph	HPH1, HPH2, HPH3	Zinc-finger SPM domain	
	Sce	RING1A, RING1B	RING-finger domain	Ubiquitin ligase
	Psc	BMI1, PCGF2, ZNF134	RING-finger domain	
	Scm	SCML1	Zinc-finger SPM domain	
PRC2	Esc	EED	WD repeats	
	E(z)	EZH1, EZH2	SET domain	Histone methyl-transferase
	Su(z)12	SUZ12	Zinc-finger domain	
	Caf1	RBAP48, RBAP46	WD repeats	Histone binding
	Pcl	PHF1	PHD-finger domain	
PhoRC	Pho	YY1	Zinc-finger domain	Sequence-specific DNA binding
	Phol	YY1		DNA binding
	DSfmbt	SFMBT1, SFMBT2	MBT-repeat	Methyl-lysine binding

Proteins are classified according to the polycomb repressive complex (PRC) they belong to. Protein domains and identified functions are also indicated.

III. PcG MECHANISMS OF ACTION

PcG proteins, in order to silence genes, bind to chromatin at specific loci called polycomb response elements (PREs) (Dejardin and Cavalli, 2004; reviewed in Muller and Kassis, 2006; Schwartz and Pirrotta, 2007). What makes a PRE a PcG-binding site? There are several known factors that contribute to PcG targeting to chromatin at the DNA sequence level and at the chromatin level. Nevertheless, the precise combination of requirements is unknown. Certainly, there must be specific DNA sequences necessary for PcG binding to chromatin, as PcG proteins are bound to specific loci in the genome (Dejardin and Cavalli, 2004; Lee *et al.*, 2006; Negre *et al.*, 2006; Schwartz *et al.*, 2006; Tolhuis *et al.*, 2006). Some of these sequences have been identified in *D. melanogaster* (Dejardin and Cavalli, 2004; Negre *et al.*, 2006; Schwartz *et al.*, 2006; Tolhuis *et al.*, 2006), but a sequence or pattern of sequences common to a majority of the PREs is yet to be found and the presence of a PRE cannot yet be predicted by just looking at the DNA sequence (Ringrose and Paro, 2007; Ringrose *et al.*, 2003) . However, recent experiments using DamID (Tolhuis *et al.*, 2006) and chromatin immunoprecipitation followed by hybridization on a microarray (ChIP on chip) (Lee et al., 2006; Negre et al., 2006; Schwartz et al., 2006) have localized at a genome-wide scale a set of PcG-binding sites that most likely correspond to PREs. These techniques provide a relatively high resolution when compared with the traditional immunostaining of polytene chromosomes. Comparison of these data with genome sequence promises important advances in our understanding of the elements of the sequence that determine PREs. Remarkably, only two of the components of the PcG family, namely, Pho and Pho-like, have strong affinity for a specific DNA sequence (Brown et al.,1998, 2003; Mohd-Sarip et al., 2005). Other chromatin-binding factors have been suggested to cooperatively help the targeting of PcG to PREs. Identified PREs frequently contain consensus sequences for GAF (Strutt et al., 1997), Zeste (Dejardin and Cavalli, 2004; Dellino et al., 2002; Ringrose et al., 2003), and Dsp1 (Dejardin et al., 2005) proteins. This suggested that these factors could be required for the targeting of PcG proteins to PREs. Recent genome-wide analysis of PcG components and GAF-binding sites disclosed the fact that there is not a one-to-one relationship between PcG and GAF targeting to chromatin (Negre et al., 2006). These findings discard GAF as an absolute requirement for the PcG–PRE interaction. They do, however, suggest the potential existence of different subclasses of PREs. Other unidentified DNA-binding proteins could certainly contribute to drive the formation of a functional PRE (Schwartz and Pirrotta, 2007).

At the chromatin level, there is abundant evidence relating histone posttranslational modifications with PcG-mediated silencing, supposedly conferring the means for the inheritability and stability of a particular chromatin "on/off state" (Muller and Kassis, 2006; Schuettengruber et al., 2007; Schwartz and Pirrotta, 2007). Histone methyltransferase and methyl-lysine-binding activities have been reported for some PcG members (Bannister et al., 2001; Czermin et al., 2002; Lachner et al., 2001; Paro and Hogness, 1991). In addition, H3K27me3 and H3K9me3 strongly colocalize with PcG proteins at PREs in ChIP experiments (Lee et al., 2006; Papp and Muller, 2006; Schwartz et al., 2006).

PcG protein binding to chromatin is not limited to PREs, but extends to the flanking chromatin forming relatively large PcG domains (PcDs) (Comet et al., 2006; Negre et al., 2006; Schwartz et al., 2006; Tolhuis et al., 2006). These domains extend for many kilobases along the chromatin fiber reaching sizes up to a few hundreds of kilobases. The series of events that directs the binding and consequent spreading of PcG proteins along the chromatin is not yet clear. However, a plausible scenario is a sequential process initiated with the binding of the PhoRC and/or PRC2 alone, or in cooperation with other chromatin factors, to specific DNA sequences at the PRE. In a second step, PcG proteins may spread the H3K27me3 mark through a self-maintaining mechanism involving the methyl-lysine binding and histone methyltransferase activities of PC and E(Z), respectively (Muller and Kassis, 2006; Schwartz and Pirrotta, 2007; Wang et al., 2004b). PREs are known to be depleted, at least partially, of nucleosomes. This fact points to DNA sequences as the triggering factor for PcG targeting to PREs, although it cannot be discarded that histone modifications are an important factor in the early establishment of the functional PRE–PcG interaction. As stated above, it is plausible that this binding is a cooperative process where specific DNA sequences and other unidentified binding factors are required for proper targeting.

How do PcG proteins inhibit transcription? Although, again, the precise mechanism is unknown, there is strong evidence that the inhibition of transcription by PcG proteins is related to changes in the structure of the chromatin fiber. In vitro studies show that PRC1 can inhibit the ATP-dependent remodeling of nucleosomes associated to transcription (Francis et al., 2001). Also, the addition of core components of PRC1 to nucleosomal arrays in vitro is able to rearrange the "beads on a string" fiber into a more compact structure (Francis et al., 2004), suggesting inaccessibility for transcription factors as a mechanism of silencing. Other mechanisms could involve direct inhibition of transcriptional initiation by RNA polymerase II (Dellino et al., 2004), and the H2AK119 ubiquitin ligase activity found for RING, a component of the PRC1, as this histone modification has been linked to gene silencing (de Napoles et al., 2004; Wang et al., 2004a). Moreover, a recent publication by Vire et al. (2006) shows an interesting link between PcG and DNA methylation, another epigenetic mark extensively linked to gene silencing. They show that EZH2, a mammalian homologue of the fly E(Z) protein, can recruit DNA methyltransferases (DNMTs), and the methylation activity itself, to at least some of the EZH2-repressed genes. It is important to note that these different mechanisms are not necessarily mutually exclusive. It is also likely that the diversity of these silencing mechanisms serves, at least in part, to regulate different subsets of target genes during different stages of development or in different cell types.

PcG proteins can mediate specific silencing of genes that are far away from PREs in the genome (Fig. 2.2). This phenomenon is called "long-range" gene silencing. In contrast to "short-range" regulation, where control depends on sequences next to the gene, typically several hundreds of base pairs away from the start site. Long-range regulating sequences are located many kilobases away or even in other chromosomes. We will now focus on this long-range silencing. To do so, we would first like to differentiate between what we will call long-range silencing and *very* long-range silencing. Under long-range silencing, we will treat the PRE–gene interactions that take place through the chromatin fiber. This silencing is, therefore, dependent on the spreading of chromatin marks along the DNA fiber and is necessarily restricted to a single PcD. In contrast, in

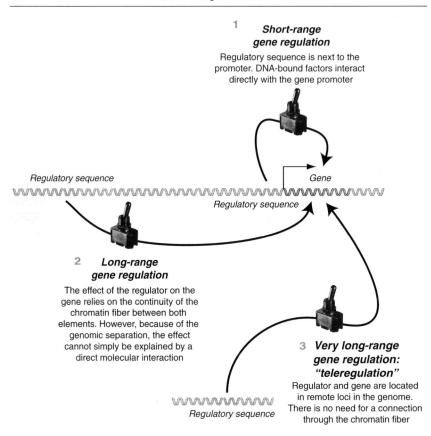

1 *Short-range*
gene regulation

Regulatory sequence is next to the
promoter. DNA-bound factors interact
directly with the gene promoter

Regulatory sequence

Gene

Regulatory sequence

2 *Long-range*
gene regulation

The effect of the regulator on the
gene relies on the continuity of the
chromatin fiber between both
elements. However, because of the
genomic separation, the effect
cannot simply be explained by a
direct molecular interaction

3 *Very long-range*
gene regulation:
"teleregulation"

Regulator and gene are located
in remote loci in the genome.
There is no need for a connection
through the chromatin fiber

Regulatory sequence

Figure 2.2. Regulatory sequences are located at various distances from the gene. Regulatory
sequences may interact with genes in various ways depending on their relative position
in the linear structure of the genome. 1: When the regulator element is located close to
the promoter, in the range of a few hundred base pairs, direct molecular interactions
between factors bound to the regulatory sequence and the promoter may drive the
regulation. 2: Other regulatory sequences, like is the case of polycomb responsive
elements (PREs), are located many kilobase pairs away from the regulated genes.
This long-range regulation may not be easily explained with direct molecular contacts.
The effect on gene transcription relies, however, in the fact that the chromatin
fiber connects the regulatory sequence and the gene. This invokes mechanisms that,
somehow (see Fig. 2.3), use the chromatin fiber as a linker structure. 3: In other cases,
including PRE-mediated silencing, the regulatory sequence is located many megabases
away from the affected gene, or even in different chromosomes. In these very long-range
interactions, physical contacts have been detected between both elements. The
mechanisms responsible for these contacts are still a mystery.

very long-range interaction, the PRE silences genes located somewhere else in
the genome, in the same or in a different chromosome, but necessarily outside

the PcD surrounding that particular PRE. This differentiation between long- and very long-range silencing holds a certain parallelism with the concepts *in cis* and *in trans*. We will, nevertheless, avoid these terms, as they lead to confusion when interactions are between loci within one chromosome, semantically (technically) *in cis* but, mechanistically speaking, probably similar to an interaction *in trans*.

IV. PcG PROTEINS AND LONG-RANGE GENE SILENCING

Regardless of the exact mechanism in charge of the initial targeting of PcG proteins to PREs, there is an apparent spreading of H3K27me3 in the surrounding region. Polycomb (PC) and, to a lesser extent, polyhomeotic (PH), two core components of the PRC1, are also capable of spreading in regions flanking PREs (Comet *et al.*, 2006; Negre *et al.*, 2006; Schwartz *et al.*, 2006). PcG-dependent gene silencing spreads along with PcG proteins on the chromatin fiber, establishing repressive domains in the order of tens of kilobases in the chromatin up- and downstream of PREs (Bantignies *et al.*, 2003; Comet *et al.*, 2006). There is so far no evidence that elements in the DNA sequence will, directly or indirectly, target PC through the whole domain. On the contrary, deletion of the PRE leads to the reactivation of gene promoters under its control, as well as loss of PC and PH (Busturia *et al.*, 1997; Comet *et al.*, 2006). This has been shown, however, only for transgenic PREs. A well-studied case of this long-range silencing is found in the bithorax complex (BX-C) of *Drosophila*. The BX-C is a cluster of three *Hox* genes (*Ubx*, *abd-A*, and *Abd-B*) spanning about 300 kbp in the 3R chromosome arm of the fly genome. This cluster constitutes a large PcD, containing many PREs, from which six have been identified. As an extreme example, one of these PREs, the Mcp element, controls the expression of *Abdominal-B* (*Abd-B*), placed as far as 60 kbp away (Busturia *et al.*, 1997; Lewis, 1978). Although 60 kbp is an extreme example of the reach of PREs control over gene expression, it is well in the range of the genomic distances where other elements, such as enhancers, control gene expression.

But how do these epigenetic marks and the associated gene silencing extend into the flanking chromatin fiber? Some models have been proposed to interpret this phenomenon (Muller and Kassis, 2006; Schuettengruber *et al.*, 2007; Schwartz and Pirrotta, 2007; Wang *et al.*, 2004b). Independently of the precise PRCs involved, we can classify these models into two: the "chromatin-walking" and the "chromatin-probing" models. These models start from the assumption of a number of PRE-targeted PcG complexes with H3K27 methyltransferase activity. The first possible explanation is that the PRE-bound factors "walk" into the neighboring chromatin fiber, modifying chromatin marks as they do so. In a first possible scenario (Fig. 2.3A), PRC2, PRC1, and/or PhoRC would be steadily bound to the PRE. In a first step, PRC1 would bind, through the

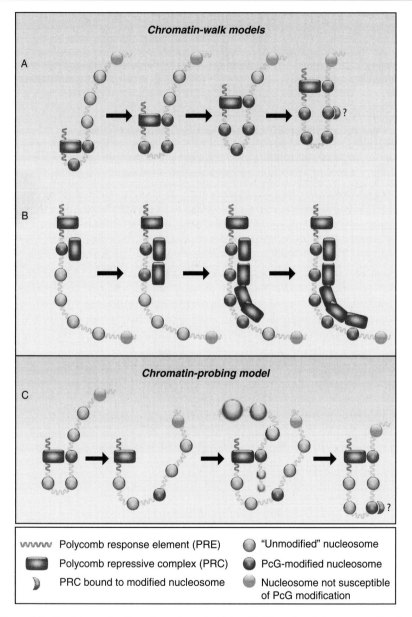

Figure 2.3. Possible mechanisms for polycomb group (PcG)-mediated long-range silencing. A polycomb repressive complex (PRC) binds a polycomb responsive elements (PRE) through the DNA-binding domain of pleiohomeotic (Pho) alone or together with other DNA-binding factors. This PRC can covalently modify histone tails and create a silent chromatin state. This illustration schematically represents the two possible

chromodomain of PC, the H3K27me3 from neighboring nucleosomes, bringing them into the proximity of the PRE. In a second step, PRC2 would tri-methylate contiguous mono- or di-methylated H3K27. Alternatively, the PRE could be bridged to the hypomethylated nucleosomes via the PhoRC, which selectively binds mono- or di-methylated H3K9 and H4K20. These *de novo* tri-methylated H3K27 domains would form a binding substrate for PC and maybe other PcG components. In a second scenario (Fig. 2.3B), *de novo* methylated histones, and not the PRE, would be the support for the PC-containing complexes. These complexes would then methylate the next nucleosomes, allowing the progression of the silencing state along the chromatin fiber.

The chromatin-probing model (Fig. 2.3C) assumes that the PRE-bound factors randomly contact, and consequently methylate, other nucleosomes. Given the fact that chromatin is a flexible fiber, these contacts will be significantly more frequent with nucleosomes that are proximal in the linear structure of the genome, resulting in the formation of transient loops of chromatin. Consequently, the flanking chromatin will be more efficiently methylated the closer it is to the PRE. According to this model, a complex bringing hypomethylated nucleosomes into the proximity of the PRE would not be necessary, as it is the proximity in the genome that drives the random contacts. Indeed, it is known that the average physical distance between two loci in the same chromatin fiber is dependent on the genomic distance separating them (Sachs *et al.*, 1995; Yokota *et al.*, 1995). An elegant example of a similar process is the DamID technique (van Steensel *et al.*, 2001). A DNA methyltransferase (Dam) fused to GAL4 is artificially targeted to a known locus containing a UAS array (GAL4 binds to UAS sequences). After analysis, DNA methylation is quantitatively more frequent in the proximity of the UAS site. In this case, a "walk" along the

models for the spreading of the PRE-mediated silencing into the chromatin flanking the PRE. In the chromatin-walk models, the PRC spreads its silencing activity by following the chromatin fiber. A. In the upper model, the PRC remains bound to the PRE and is able to bind unmodified nucleosomes, tri-methylate H3K27 and/or H3K9, and subsequently "step" to the next unmodified nucleosome. B. In the lower model, the PRC is bound to the tri-methylated nucleosome (not to the PRE) and it recruits a new PRC to the next unmodified nucleosome. In principle, a continuous binding of the PRC(s) to the PRE or the already tri-methylated H3K27 and/or H3K9 is not necessary in this scenario. However, these PRCs could remain "coating" the silenced chromatin. In both cases, other chromatin states could counteract or inhibit the spreading of the polycomb domain. C. The chromatin-probing is a more stochastic model. PRE-bound PRC may again tri-methylate H3K27 and/or H3K9. However, in contrast with the chromatin-walk model, this methylation is not spreading along the fiber but to the nucleosomes that happen to interact with the enzymatic pocket of the PRC. Given the fact that chromatin is a flexible fiber, the frequency of these interactions will be significantly correlated to the proximity in the linear dimension of the genome. In all cases, other PRCs or individual PcG proteins may bind to the modified nucleosomes acting as transcriptional repressors.

chromatin fiber cannot explain the spreading of the Dam enzymatic activity along the chromatin. Similarly, as explained above, the spreading of H3K27me3 would not require a chromatin-walk to be explained.

We consider it important to differentiate between all these possibilities, not only for didactic purposes but also for the implications they have for understanding the molecular mechanisms for PcG-mediated silencing. It is also important to note here that these models are not mutually exclusive per se. It is perfectly possible that PRE-bound PRCs probe the surrounding chromatin transiently and, then, "walk" locally modifying relatively short stretches of fiber. Another important factor to take into account is the existence of counteracting forces. Whereas PcG proteins mark chromatin and prevent transcription, there are other factors that promote the activation of chromatin or impede the effect of PcG proteins. Indeed, trxG proteins, targeted at DNA elements frequently overlapping with PREs, are in charge of the maintenance of the active state of genes (Papp and Muller, 2006; Schuettengruber et al., 2007). Strong activation of transcription itself has also been shown to displace PcG proteins from PREs (Schmitt et al., 2005). Ultimately, transcription is the result of an intricate and precisely balanced network of silencing and activating factors.

A recent finding from a study analyzing the relation between PREs and elements called chromatin insulators provides strong support for the chromatin-probing model at least in some cases. Chromatin insulators are elements that are suggested to segregate the eukaryotic chromosomes in functionally independent domains (Valenzuela and Kamakaka, 2006). The spreading of the PcG proteins and the associated silencing along the fiber can be blocked by the action of an insulator (Comet et al., 2006; Sigrist and Pirrotta, 1997). Surprisingly, however, the spreading of PcG-mediated silencing was shown to be bypassed when two insulators are present between the PRE and the reporter gene (Comet et al., 2006). Importantly, the chromatin region between the two bypassed insulators is neither silenced nor coated with PC and PH, two PRC1 members. These findings thus demonstrate the existence of chromatin elements that can isolate specific sequences from the action of PcG proteins. Although the molecular mechanism of insulator action is still unknown, these data emphasize the importance of linking the regulatory elements on the linear dimension of the chromatin with its 3D folding.

As we lack knowledge concerning the distribution of chromatin marks such as histone modifications at the single-cell level, there is little direct evidence for either of the two models. Current data on the biochemical composition of chromatin at specific loci are based on ChIP or DamID (Negre et al., 2006; Schwartz et al., 2006; Tolhuis et al., 2006). These techniques collect information on large ensembles of cells, obviating any information of the distribution of chromatin components at the single-cell level. Microscopy-based techniques, in contrast, do provide this type of information. However, the optical resolution of current microscopes limits very much the reach of these approaches. Immuno-staining of polytene chromosomes in *Drosophila* can, for example, show the distribution of PcG proteins in the linear genome with a resolution of about 50 kb in most cases, while the resolution with ChIP on chip data can reach 500 bp (Qi *et al.*, 2006), and possibly a higher sensitivity. Then again, polytene chromosomes are present only in the salivary glands. A viable alternative in the future might be the combination of high-resolution microscopy with "chromatin combing," where the chromatin is stretched and, thereby, allows higher resolutions on the single-molecule basis (Blower *et al.*, 2002).

V. PcG AND VERY LONG-RANGE GENE SILENCING: "TELEREGULATION" OF GENE EXPRESSION

In contrast to long-range interactions, where silencing takes place within the range of the PcD, very long-range silencing takes place between different PcDs, across large regions in the order of several megabases, or even between different chromosomes. The first evidence for very long-range interactions came from the discovery of a phenomenon that was termed transvection. In transvection, two alleles, located in homologous chromosomes, will influence each other when they are in the vicinity because of homologous pairing (Duncan, 2002). In *Drosophila* somatic cells, homologous chromosomes are paired most of the time. This cir-cumstance assembles all sequences in close physical contact with their homo-logues. During transvection, regulatory sequences, such as enhancers, present in one of the alleles can activate gene expression of the allele present on the homologous chromosome. When the pairing is disrupted by, for example, chromosomal rearrangements, the physical as well as the genetic interactions are lost. A similar phenomenon associated with PREs and PcG action was later defined as "pairing-sensitive silencing" (PSS) (Kassis, 2002). In transgenic assays, the mini-*white* reporter gene is very useful to assess gene expression by simply looking at the fly's eye color, as the intensity of pigmentation is roughly propor-tional to the gene transcription. Therefore, the mini-*white* transgene gives more intense pigmentation in homozygosis than in heterozygosis, as there is double the amount of transcripts. The insertion of a PRE next to the mini-*white* gene induces a partial silencing of the gene in heterozygous flies. Interestingly, however, when the transgene PRE-mini-*white* is homozygous, it does not always show the higher pigmentation expected from two copies of the gene. On the contrary, the eyes are less pigmented than in the heterozygote. The implication of PcG proteins in PSS is confirmed by the fact that mutation of some PcG components suppresses PSS. However, this phenomenon is observed only in a fraction of the transgenic lines, suggesting that the chromatin environment surrounding the inserted transgene influences its function. Moreover, PSS is much less frequent when the two copies of the transgene are located in different chromosomes, suggesting that the molecular contact between the two copies is important for enhancement of silencing.

The existence of PSS demonstrates that the repressive effect of PREs is not limited to genes in the same DNA molecule. On the contrary, they can exert their influence into the neighboring chromatin, whether it pertains to the same fiber or not. It seems logical to look back into the two models proposed for the action of PRE on genes in the same chromatin fiber, that is, the chromatin-probing and chromatin-walking models. Can these models explain PSS?

Certainly, the chromatin-probing model, as described above, could explain the phenomena observed during PSS. A PRE, and the protein complexes associated with it, could probe for chromatin in its vicinity. Given that paired chromosomes are relatively close to each other, there is no *a priori* reason to think that PREs would have a preference for any of the two chromosomes. In contrast, in the chromatin-walking model, the PRE-associated complexes would walk along the chromatin fiber and not "jump" to the other fiber. There is, therefore, strong evidence that the silencing activity of PREs and the PRE-coupled factors do probe for the chromatin in the surrounding space of the nucleus, consistent with the fact that chromatin has a certain degree of mobility *in vivo*, in the range of several hundred nanometers (Abney *et al.*, 1997; Lanctot *et al.*, 2007; Marshall *et al.*, 1997).

Although transvection phenomena are thought to depend on the somatic pairing of homologous chromosomes, there are some cases in which chromosomal interactions can take place between remote loci (Bantignies *et al.*, 2003; Grimaud *et al.*, 2006a; Muller *et al.*, 1999; Vazquez *et al.*, 2006). One example is Mcp, a PRE-containing element from the BX-C. Mcp functions in transgenic assays like other PREs which silence mini-*white* in a pairing-sensitive manner; that is, Mcp next to mini-*white* will have a stronger silencing effect in homozygosis. However, when the two copies of the Mcp-mini-*white* are in different sites of the genome, one can sometimes observe a stronger silencing of mini-*white* when compared with the heterozygous parental lines. These genetic interactions may be detected when the two copies are located many megabases away from each other, or even in different chromosomes (Muller *et al.*, 1999). This clearly suggests a physical interaction between the two alleles that is not driven by pairing of homologous chromosomes. Recently, this point has been confirmed by Vazquez and coworkers (Vazquez *et al.*, 2006), who showed *in vivo* that physical interactions are established between Mcp-mini-*white*-containing transgenes. These interactions are seen in a high percentage (more than 90%) of the cells and they are stable through time frames in the range of hours. A similar phenomenon seems to involve the *Fab-7* PRE-containing element, though the frequency of contacts is lower in this case. Moreover, the contacts between remote *Fab-7* transgenes are correlated with an enhanced PcG-mediated gene silencing (Bantignies *et al.*, 2003). Apart from these interactions between remote transgenic PREs, contacts have been reported for endogenous gene clusters controlled by PREs, namely, Antp-C and the BX-C (Grimaud *et al.*, 2006a). These contacts, like in the transgenic experiments, are correlated with the expression of the genes in these clusters. However, a causal relation between the contacts and the silencing has not yet been established. In contrast with the PSS, where the rearrangements disrupt the contacts and a cause–effect relation can be postulated, a technique to physically interfere with the contacts between sequences in nonhomologous chromosomal regions is not available.

Contacts between remote loci have also been found in other cases, including other species and involving both silencing and activation. Moreover, these chromosomal contacts have been proposed to affect gene regulation in those cases as well (Lanctot *et al.*, 2007). Therefore, a working hypothesis that is currently being studied postulates that remote regulatory elements can establish very long-range regulatory contacts with a subset of the genomic loci. We propose to define the regulation of genes by these elements located in remote positions as "teleregulation" in order to distinguish it from regulation by distant elements in the same chromosomal domains as the gene promoters they regulate.

Recent genome-wide analysis of PcG binding in *Drosophila* has identified PcDs by ChIP on chip and DamID. These PcDs are regions of the chromatin bound to PC and trimethylated on H3K27 (Lee *et al.*, 2006; Negre *et al.*, 2006; Schwartz *et al.*, 2006; Tolhuis *et al.*, 2006). These data reveal that, inside PcDs, PREs can be distinguished by a local higher binding of PH and PC. In the fly genome, there are ~200 distinct PcDs, each of them containing one or more PREs (Schwartz *et al.*, 2006; Tolhuis *et al.*, 2006). However, microscopy studies show that, in the interphase diploid nucleus, PC and H3K27me3 accumulate in less than 40 PC bodies, a much lower number of bodies when compared with the total number of detected PcDs in the linear sequence. Moreover, PC bodies change in number and size during development (Buchenau *et al.*, 1998; Ficz *et al.*, 2005). This clearly suggests that PcDs come together in PC bodies and that this clustering is relevant to regulate gene expression during development. Therefore, in the case of PcDs, teleregulation might reflect clustering of PC bodies, whereby PREs in a certain PcD may affect gene promoters contained in another PcD.

In mammals, it has been observed that the *Hoxb* gene cluster, a large group of PcG-regulated genes, moves outside the chromosome territory upon activation (Chambeyron *et al.*, 2005). This rearrangement of the cluster is, moreover, correlated with the decondensation of the chromatin in the cluster itself (Morey *et al.*, 2007). This suggests not only an interesting correlation between the 3D structure of chromatin and PcG-mediated gene regulation but also the possibility of contacting other PcG-binding loci in the genome.

Similarly to the previous cases, it is relatively easy to imagine that factors bound to one PRE would probe, and chemically modify, the surrounding nucleosomes. But, in this case there is a major complication. How can we explain the contacts of PREs and genes when they are located very far in the linear dimension of the genome? While, during PSS, the pairing of homologous chromosomes would bring alleles close to each other, we do not know what drives the interactions between PREs that are extremely far in the genome. The relatively high percentages of interactions observed under the microscope, and the time-lapses in which they are maintained, indicate that these interactions are specific and regulated. What are the molecular mechanisms that drive and stabilize the contacts? This is still an open question for which we can only speculate. There are, however, some clues that might guide our speculations, and our experiments.

Recently, an intriguing link has been found between the machinery of the RNA interference pathway, PcG-mediated silencing, and the associated gene contacts (Grimaud et al., 2006a). It has also been suggested that sequence homology might be involved in driving specific interactions between distant PREs (Bantignies et al., 2003). Indeed, the high colocalization rates and the genetic studies with transgenic elements support this possibility. However, although sequence homology might play a role in the maintenance of the interactions, it is unlikely that it moves PREs closer to each other. To date, evidence for mechanisms that could bring specific sequences together remains highly speculative. Is it possible that these sequences roam randomly in the nucleus until they find each other at a distance sufficient to establish a direct and specific molecular interaction based on sequence? Currently, we do not have an answer to this question. The probability for two single molecules to find each other, given the dynamic restrictions of chromatin and the size of the nucleus, seems very low. In vivo microscopy studies on chromatin mobility show that although chromatin can freely diffuse through the fly nucleus, this diffusion is confined to subregions of ~ 0.6 μm in diameter (Marshall et al., 1997). These subregions account for only a fraction ($\approx 1/6$) of the nuclear volume, significantly restricting the possibilities to establish contacts. In mammalian cells, the fraction of the nucleus "explored" by chromatin is relatively much smaller (Abney et al., 1997; Chubb et al., 2002). However, there might be mechanisms that facilitate these "random encounters," for example, by restricting the roaming to specific nuclear domains, for example, the nuclear periphery. These possibilities need further analysis using biophysical models.

PcG-regulated genes are certainly not an isolated example of physical contacts between remote DNA sequences. Fundamental processes, such as transcription and replication, take place in localized domains in the nucleus. These so-called transcription and replication factories are thought to assemble the transcription and replication of a number of different loci of the genome (reviewed in Misteli, 2007). 3C (for Chromatin Conformation Capture) is a technique where interactions between specific loci can be detected by cross-linking chromatin *in situ*. After a digestion and a religation, DNA fragments that were originally next to each other may be specifically amplified and detected by quantitative polymerase chain reaction (qPCR). Recent studies using 4C, a recent development of 3C where the same principle is applied on a genome-wide scale, have revealed many more inter- and intrachromosomal interactions. Although the 4C technique is still in development, these preliminary studies suggest that very long-range contacts constitute a more general mechanism to regulate gene expression (Simonis *et al.*, 2006; Zhao *et al.*, 2006).

VI. CONCLUSIONS AND PROSPECTS

There remain many open questions about the molecular mechanisms that drive the PcG-mediated 3D rearrangements of the chromatin in the nucleus. Future research should learn to combine data obtained using "average-cell" approaches, like ChIP, DamID, 3C and 4C, and "single-cell" approaches, mainly microscopy. A further challenge for the future is to advance in our understanding of the physical principles that rule the nuclear organization.

The view of a large network of interacting genes all through the genome is in apparent contradiction with our knowledge of the compartmentalization of the chromatin fiber. Chromosomes in the interphase nucleus occupy a discrete region of the nucleus, the chromosome territory (Bolzer *et al.*, 2005). Although this is a matter of discussion at present, the amount of intermingling between the different chromosome territories is remarkably limited, and it can be estimated, depending on the technique used, ranging from 0 to 19% of the nuclear volume (Branco and Pombo, 2006, 2007; Lanctot *et al.*, 2007; Visser *et al.*, 2000). Not only the chromosomes are separated in territories, but also the subchromosomal fragments do not significantly intermingle (Dietzel *et al.*, 1998; Goetze *et al.*, 2007). Moreover, the mobility of chromatin in the nucleus is restricted to only a small fraction of the nuclear volume (Abney *et al.*, 1997; Marshall *et al.*, 1997). How do genes contact each other under such spatial constraints? Answer to this question has been often based more on the intuition of researchers than in scientific models. This question in particular and, in general, understanding how 3D chromatin structure contributes to fundamental functioning of the genome demand contributions from biophysics. In recent years, biophysical models have been developed to explain the highly dynamic properties of single molecules in the nucleus. However, we remarkably lack satisfactory models to describe the physical properties of the chromatin fiber. Only a handful of models have been proposed to date (Munkel *et al.*, 1999; Sachs *et al.*, 1995; van den Engh *et al.*, 1992; Yokota *et al.*, 1995). What is more, these models have, for good reasons, a very limited explanatory potential. One of these reasons is the highly heterogeneous and dynamic nature of chromatin. Obviously, chromatin constitutes a continuous fiber with certain physical properties, such as density, thickness, flexibility, and so on. However, as certain as these properties exist, they are heterogeneously distributed along the chromosome. Different regions of the genome have different structures related to their gene density, transcription rate, repetitive sequences, and so on. (Goetze et al., 2007; Yokota et al., 1997). Furthermore, these properties are not constant in time but rather change depending on the functional state of the chromatin. It thus remains difficult to develop models that include all of these changing parameters. One of the options to model these heterogeneous fibers is the use of block copolymer physics, where different physical properties are attributed to different blocks of a modeled polymer. Another limiting factor to make progress here is the amount of computing time required to produce simulations of polymer dynamics. As in a polymer fiber, the behavior of one of the segments will influence the rest of the fiber. Monte-Carlo simulations result in extremely costly experiments in terms of computer time (computing times increase exponentially with the length of the simulated fiber). One alternative here is to use analytical models that, rather than attempting to simulate the real scenario, dissect the physical principles via mathematical analysis.

The possibility that PcG proteins establish 3D networks of chromatin contacts as a mechanism to orchestrate gene expression is certainly intriguing. Despite our lack of knowledge, there are sufficient indications to believe that it is an important factor controlling the function of the genome. Although much of the facts relating PcG proteins, the 3D structure of the chromatin, and the genome function are collected in the model *Drosophila* system, there is no reason to believe that the means in mammalian cells are different. On the contrary, our current knowledge of the organization of the mammalian nucleus tells us that it is an extremely structured organelle (Lanctot et al., 2007; Misteli, 2007). It is important to be aware of the significance of these facts, and the drastic consequences that it might have in our understanding of the regulation of the genome.

References

Abney, J. R., Cutler, B., Fillbach, M. L., Axelrod, D., and Scalettar, B. A. (1997). Chromatin dynamics in interphase nuclei and its implications for nuclear structure. *J. Cell Biol.* **137,** 1459–1468.
Bannister, A. J., Zegerman, P., Partridge, J. F., Miska, E. A., Thomas, J. O., Allshire, R. C., and Kouzarides, T. (2001). Selective recognition of methylated lysine 9 on histone H3 by the HP1 chromo domain. *Nature* **410,** 120–124.
Bantignies, F., Grimaud, C., Lavrov, S., Gabut, M., and Cavalli, G. (2003). Inheritance of Polycomb-dependent chromosomal interactions in *Drosophila*. *Genes Dev.* **17,** 2406–2420.

Blower, M. D., Sullivan, B. A., and Karpen, G. H. (2002). Conserved organization of centromeric chromatin in flies and humans. *Dev. Cell* **2**, 319–330.

Bolzer, A., Kreth, G., Solovei, I., Koehler, D., Saracoglu, K., Fauth, C., Muller, S., Eils, R., Cremer, C., Speicher, M. R., and Cremer, T. (2005). Three-dimensional maps of all chromosomes in human male fibroblast nuclei and prometaphase rosettes. *PLoS Biol.* **3**, e157.

Branco, M. R., and Pombo, A. (2006). Intermingling of chromosome territories in interphase suggests role in translocations and transcription-dependent associations. *PLoS Biol.* **4**, e138.

Branco, M. R., and Pombo, A. (2007). Chromosome organization: New facts, new models. *Trends Cell Biol.* **17**, 127–134.

Brown, J. L., Mucci, D., Whiteley, M., Dirksen, M. L., and Kassis, J. A. (1998). The *Drosophila* Polycomb group gene pleiohomeotic encodes a DNA binding protein with homology to the transcription factor YY1. *Mol. Cell* **1**, 1057–1064.

Brown, J. L., Fritsch, C., Mueller, J., and Kassis, J. A. (2003). The Drosophila pho-like gene encodes a YY1-related DNA binding protein that is redundant with pleiohomeotic in homeotic gene silencing. *Development* **130**, 285–294.

Buchenau, P., Hodgson, J., Strutt, H., and Arndt-Jovin, D. J. (1998). The distribution of polycomb-group proteins during cell division and development in *Drosophila* embryos: Impact on models for silencing. *J. Cell Biol.* **141**, 469–481.

Busturia, A., Wightman, C. D., and Sakonju, S. (1997). A silencer is required for maintenance of transcriptional repression throughout *Drosophila* development. *Development* **124**, 4343–4350.

Calonje, M., and Sung, Z. R. (2006). Complexity beneath the silence. *Curr. Opin. Plant Biol.* **9**, 530–537.

Chambeyron, S., Da Silva, N. R., Lawson, K. A., and Bickmore, W. A. (2005). Nuclear re-organisation of the Hoxb complex during mouse embryonic development. *Development* **132**, 2215–2223.

Chubb, J. R., Boyle, S., Perry, P., and Bickmore, W. A. (2002). Chromatin motion is constrained by association with nuclear compartments in human cells. *Curr. Biol.* **12**, 439–445.

Comet, I., Savitskaya, E., Schuettengruber, B., Negre, N., Lavrov, S., Parshikov, A., Juge, F., Gracheva, E., Georgiev, P., and Cavalli, G. (2006). PRE-mediated bypass of two Su(Hw) insulators targets PcG proteins to a downstream promoter. *Dev. Cell* **11**, 117–124.

Cremer, T., Kupper, K., Dietzel, S., and Fakan, S. (2004). Higher order chromatin architecture in the cell nucleus: On the way from structure to function. *Biol. Cell.* **96**, 555–567.

Czermin, B., Melfi, R., McCabe, D., Seitz, V., Imhof, A., and Pirrotta, V. (2002). *Drosophila* enhancer of Zeste/ESC complexes have a histone H3 methyltransferase activity that marks chromosomal Polycomb sites. *Cell* **111**, 185–196.

de Napoles, M., Mermoud, J. E., Wakao, R., Tang, Y. A., Endoh, M., Appanah, R., Nesterova, T. B., Silva, J., Otte, A. P., Vidal, M., Koseki, H., and Brockdorff, N. (2004). Polycomb group proteins Ring1A/B link ubiquitylation of histone H2A to heritable gene silencing and X inactivation. *Dev. Cell* **7**, 663–676.

Dejardin, J., and Cavalli, G. (2004). Chromatin inheritance upon Zeste-mediated Brahma recruitment at a minimal cellular memory module. *EMBO J.* **23**, 857–868.

Dejardin, J., and Cavalli, G. (2005). Epigenetic inheritance of chromatin states mediated by Polycomb and trithorax group proteins in *Drosophila*. *Prog. Mol. Subcell. Biol.* **38**, 31–63.

Dejardin, J., Rappailles, A., Cuvier, O., Grimaud, C., Decoville, M., Locker, D., and Cavalli, G. (2005). Recruitment of *Drosophila* Polycomb group proteins to chromatin by DSP1. *Nature* **434**, 533–538.

Delaval, K., and Feil, R. (2004). Epigenetic regulation of mammalian genomic imprinting. *Curr. Opin. Genet. Dev.* **14**, 188–195.

Dellino, G. I., Schwartz, Y. B., Farkas, G., McCabe, D., Elgin, S. C., and Pirrotta, V. (2004). Polycomb silencing blocks transcription initiation. *Mol. Cell* **13**, 887–893.

Dellino, G. I., Tatout, C., and Pirrotta, V. (2002). Extensive conservation of sequences and chromatin structures in the bxd polycomb response element among *Drosophilid* species. *Int. J. Dev. Biol.* **46,** 133–141.

Dietzel, S., Jauch, A., Kienle, D., Qu, G., Holtgreve-Grez, H., Eils, R., Munkel, C., Bittner, M., Meltzer, P. S., Trent, . P. S, and Cremer, T. (1998). Separate and variably shaped chromosome arm domains are disclosed by chromosome arm painting in human cell nuclei. *Chromosome Res.* **6,** 25–33.

Duncan, I. W. (2002). Transvection effects in *Drosophila*. *Annu. Rev. Genet.* **36,** 521–556.

Ficz, G., Heintzmann, R., and Arndt-Jovin, D. J. (2005). Polycomb group protein complexes exchange rapidly in living *Drosophila*. *Development* **132,** 3963–3976.

Francis, N. J., Saurin, A. J., Shao, Z., and Kingston, R. E. (2001). Reconstitution of a functional core polycomb repressive complex. *Mol. Cell* **8,** 545–556.

Francis, N. J., Kingston, R. E., and Woodcock, C. L. (2004). Chromatin compaction by a polycomb group protein complex. *Science* **306,** 1574–1577.

Goetze, S., Mateos-Langerak, J., Gierman, H. J., de Leeuw, W., Giromus, O., Indemans, M. H., Koster, J., Ondrej, V., Versteeg, R., and van Driel, R. (2007). The three-dimensional structure of human interphase chromosomes is related to the transcriptome map. *Mol. Cell. Biol.* **27,** 4475–4487.

Grimaud, C., Bantignies, F., Pal-Bhadra, M., Ghana, P., Bhadra, U., and Cavalli, G. (2006a). RNAi components are required for nuclear clustering of Polycomb group response elements. *Cell* **124,** 957–971.

Grimaud, C., Negre, N., and Cavalli, G. (2006b). From genetics to epigenetics: The tale of Polycomb group and trithorax group genes. *Chromosome Res.* **14,** 363–375.

Heard, E. (2005). Delving into the diversity of facultative heterochromatin: The epigenetics of the inactive X chromosome. *Curr. Opin. Genet. Dev.* **15,** 482–489.

Jenuwein, T. (2001). Re-SET-ting heterochromatin by histone methyltransferases. *Trends Cell. Biol.* **11,** 266–273.

Jenuwein, T., and Allis, C. D. (2001). Translating the histone code. *Science* **293,** 1074–1080.

Kassis, J. A. (2002). Pairing-sensitive silencing, polycomb group response elements, and transposon homing in *Drosophila*. *Adv. Genet.* **46,** 421–438.

Klymenko, T., Papp, B., Fischle, W., Kocher, T., Schelder, M., Fritsch, C., Wild, B., Wilm, M., and Muller, J. (2006). A Polycomb group protein complex with sequence-specific DNA-binding and selective methyl-lysine-binding activities. *Genes Dev.* **20,** 1110–1122.

Lachner, M., O'Carroll, D., Rea, S., Mechtler, K., and Jenuwein, T. (2001). Methylation of histone H3 lysine 9 creates a binding site for HP1 proteins. *Nature* **410,** 116–120.

Lanctot, C., Cheutin, T., Cremer, M., Cavalli, G., and Cremer, T. (2007). Dynamic genome architecture in the nuclear space: Regulation of gene expression in three dimensions. *Nat. Rev. Genet.* **8,** 104–115.

Lee, T. I., Jenner, R. G., Boyer, L. A., Guenther, M. G., Levine, S. S., Kumar, R. M., Chevalier, B., Johnstone, S. E., Cole, M. F., Isono, K., Koseki, H., Fuchikami, T., *et al.* (2006). Control of developmental regulators by Polycomb in human embryonic stem cells. *Cell* **125,** 301–313.

Levine, S. S., King, I. F., and Kingston, R. E. (2004). Division of labor in polycomb group repression. *Trends Biochem. Sci.* **29,** 478–485.

Lewis, E. B. (1978). A gene complex controlling segmentation in *Drosophila*. *Nature* **276,** 565–570.

Marshall, W. F., Straight, A., Marko, J. F., Swedlow, J., Dernburg, A., Belmont, A., Murray, A. W., Agard, D. A., and Sedat, J. W. (1997). Interphase chromosomes undergo constrained diffusional motion in living cells. *Curr. Biol.* **7,** 930–939.

Misteli, T. (2007). Beyond the sequence: Cellular organization of genome function. *Cell* **128,** 787–800.

Mohd-Sarip, A., Cleard, F., Mishra, R. K., Karch, F., and Verrijzer, C. P. (2005). Synergistic recognition of an epigenetic DNA element by Pleiohomeotic and a Polycomb core complex. *Genes Dev.* **19**, 1755–1760.

Mohd-Sarip, A., van der Knaap, J. A., Wyman, C., Kanaar, R., Schedl, P., and Verrijzer, C. P. (2006). Architecture of a polycomb nucleoprotein complex. *Mol. Cell* **24**, 91–100.

Morey, C., Da Silva, N. R., Perry, P., and Bickmore, W. A. (2007). Nuclear reorganisation and chromatin decondensation are conserved, but distinct, mechanisms linked to Hox gene activation. *Development* **134**, 909–919.

Muller, J., and Kassis, J. A. (2006). Polycomb response elements and targeting of Polycomb group proteins in *Drosophila. Curr. Opin. Genet. Dev.* **16**, 476–484.

Muller, M., Hagstrom, K., Gyurkovics, H., Pirrotta, V., and Schedl, P. (1999). The mcp element from the *Drosophila* melanogaster bithorax complex mediates long-distance regulatory interactions. *Genetics* **153**, 1333–1356.

Munkel, C., Eils, R., Dietzel, S., Zink, D., Mehring, C., Wedemann, G., Cremer, T., and Langowski, J. (1999). Compartmentalization of interphase chromosomes observed in simulation and experiment. *J. Mol. Biol.* **285**, 1053–1065.

Negre, N., Hennetin, J., Sun, L. V., Lavrov, S., Bellis, M., White, K. P., and Cavalli, G. (2006). Chromosomal distribution of PcG proteins during *Drosophila* development. *PLoS Biol.* **4**, e170.

Otte, A. P., and Kwaks, T. H. (2003). Gene repression by Polycomb group protein complexes: A distinct complex for every occasion? *Curr. Opin. Genet. Dev.* **13**, 448–454.

Papp, B., and Muller, J. (2006). Histone trimethylation and the maintenance of transcriptional ON and OFF states by trxG and PcG proteins. *Genes Dev.* **20**, 2041–2054.

Paro, R., and Hogness, D. S. (1991). The Polycomb protein shares a homologous domain with a heterochromatin-associated protein of *Drosophila. Proc. Natl. Acad. Sci. USA* **88**, 263–267.

Qi, Y., Rolfe, A., MacIsaac, K. D., Gerber, G. K., Pokholok, D., Zeitlinger, J., Danford, T., Dowell, R. D., Fraenkel, R. D., Jaakkola, T. S., Young, R. A., and Gifford, D. K. (2006). High-resolution computational models of genome binding events. *Nat. Biotechnol.* **24**, 963–970.

Rando, O. J. (2007). Global patterns of histone modifications. *Curr. Opin. Genet. Dev.* **17**, 94–99.

Ringrose, L., and Paro, R. (2007). Polycomb/Trithorax response elements and epigenetic memory of cell identity. *Development* **134**, 223–232.

Ringrose, L., Rehmsmeier, M., Dura, J. M., and Paro, R. (2003). Genome-wide prediction of Polycomb/Trithorax response elements in *Drosophila* melanogaster. *Dev. Cell.* **5**, 759–771.

Sachs, R. K., van den Engh, G., Trask, B., Yokota, H., and Hearst, J. E. (1995). A random-walk/giant-loop model for interphase chromosomes. *Proc. Natl. Acad. Sci. USA* **92**, 2710–2714.

Santoro, R., and De Lucia, F. (2005). Many players, one goal: How chromatin states are inherited during cell division. *Biochem. Cell. Biol.* **83**, 332–343.

Schmitt, S., Prestel, M., and Paro, R. (2005). Intergenic transcription through a polycomb group response element counteracts silencing. *Genes Dev.* **19**, 697–708.

Schuettengruber, B., Chourrout, D., Vervoort, M., Leblanc, B., and Cavalli, G. (2007). Genome regulation by polycomb and trithorax proteins. *Cell* **128**, 735–745.

Schwartz, Y. B., and Pirrotta, V. (2007). Polycomb silencing mechanisms and the management of genomic programmes. *Nat. Rev. Genet.* **8**, 9–22.

Schwartz, Y. B., Kahn, T. G., Nix, D. A., Li, X. Y., Bourgon, R., Biggin, M., and Pirrotta, V. (2006). Genome-wide analysis of Polycomb targets in *Drosophila* melanogaster. *Nat. Genet.* **38**, 700–705.

Sigrist, C. J., and Pirrotta, V. (1997). Chromatin insulator elements block the silencing of a target gene by the *Drosophila* polycomb response element (PRE) but allow trans interactions between PREs on different chromosomes. *Genetics* **147**, 209–221.

Simonis, M., Klous, P., Splinter, E., Moshkin, Y., Willemsen, R., de Wit, E., van Steensel, B., and de Laat, W. (2006). Nuclear organization of active and inactive chromatin domains uncovered by chromosome conformation capture-on-chip (4C). *Nat. Genet.* **38**, 1348–1354.

Sparmann, A., and van Lohuizen, M. (2006). Polycomb silencers control cell fate, development and cancer. *Nat. Rev. Cancer* **6**, 846–856.

Strutt, H., Cavalli, G., and Paro, R. (1997). Co-localization of Polycomb protein and GAGA factor on regulatory elements responsible for the maintenance of homeotic gene expression. *EMBO J.* **16**, 3621–3632.

Tolhuis, B., de Wit, E., Muijrers, I., Teunissen, H., Talhout, W., van Steensel, B., and van Lohuizen, M. (2006). Genome-wide profiling of PRC1 and PRC2 Polycomb chromatin binding in *Drosophila* melanogaster. *Nat. Genet.* **38**, 694–699.

Valenzuela, L., and Kamakaka, R. T. (2006). Chromatin insulators. *Annu. Rev. Genet.* **40**, 107–138.

van den Engh, G., Sachs, R., and Trask, B. J. (1992). Estimating genomic distance from DNA sequence location in cell nuclei by a random walk model. *Science* **257**, 1410–1412.

van Driel, R., Fransz, P. F., and Verschure, P. J. (2003). The eukaryotic genome: A system regulated at different hierarchical levels. *J. Cell Sci.* **116**, 4067–4075.

van Steensel, B., Delrow, J., and Henikoff, S. (2001). Chromatin profiling using targeted DNA adenine methyltransferase. *Nat. Genet.* **27**, 304–308.

Vazquez, J., Muller, M., Pirrotta, V., and Sedat, J. W. (2006). The Mcp element mediates stable long-range chromosome-chromosome interactions in *Drosophila*. *Mol. Biol. Cell* **17**, 2158–2165.

Vire, E., Brenner, C., Deplus, R., Blanchon, L., Fraga, M., Didelot, C., Morey, L., Van Eynde, A., Bernard, D., Vanderwinden, . D., Bollen, M., Esteller, M., *et al.* (2006). The Polycomb group protein EZH2 directly controls DNA methylation. *Nature* **439**, 871–874.

Visser, A. E., Jaunin, F., Fakan, S., and Aten, J. A. (2000). High resolution analysis of interphase chromosome domains. *J. Cell Sci.* **113**(Pt. 14), 2585–2593.

Wang, H., Wang, L., Erdjument-Bromage, H., Vidal, M., Tempst, P., Jones, R. S., and Zhang, Y. (2004a). Role of histone H2A ubiquitination in Polycomb silencing. *Nature* **431**, 873–878.

Wang, L., Brown, J. L., Cao, R., Zhang, Y., Kassis, J. A., and Jones, R. S. (2004b). Hierarchical recruitment of polycomb group silencing complexes. *Mol. Cell* **14**, 637–646.

Yokota, H., van den Engh, G., Hearst, J. E., Sachs, R. K., and Trask, B. J. (1995). Evidence for the organization of chromatin in megabase pair-sized loops arranged along a random walk path in the human G0/G1 interphase nucleus. *J. Cell Biol.* **130**, 1239–1249.

Yokota, H., Singer, M. J., van den Engh, G. J., and Trask, B. J. (1997). Regional differences in the compaction of chromatin in human G0/G1 interphase nuclei. *Chromosome Res.* **5**, 157–166.

Zhao, Z., Tavoosidana, G., Sjolinder, M., Gondor, A., Mariano, P., Wang, S., Kanduri, C., Lezcano, M., Sandhu, K. S., Singh, U., Pant, V., Tiwari, V., *et al.* (2006). Circular chromosome conformation capture (4C) uncovers extensive networks of epigenetically regulated intra- and interchromosomal interactions. *Nat. Genet.* **38**, 1341–1347.

3

Evolution of *Cis*-Regulatory Sequences in *Drosophila*

Pat Simpson and Savita Ayyar

Department of Zoology, University of Cambridge, United Kingdom

I. Introduction
   A. Evolution of promoter sequences
   B. Modularity
   C. Experimental approaches to promoter evolution
II. Developmental Homeostasis, Sequence Turnover,
   and Stabilizing Selection
   A. The P2 promoter of *hunchback*
   B. The stripe 2 enhancer of *even-skipped*
   C. The dorsocentral enhancer of *scute*
   D. Conclusions
III. Enhancer Evolution and Loss or Gain of Traits
   A. Evolution of *yellow* and variation in pigment patterns
   B. Evolution of *scute* and variation in bristle patterns
   C. Conclusions
IV. *Cis-Trans* Coevolution
   A. The interaction between Bicoid and the *hunchback* P2 promoter
   B. Conclusions
V. Evolution of New Regulatory Modules
VI. Conclusions
   References

Advances in Genetics, Vol. 61
Copyright 2008, Elsevier Inc. All rights reserved.

0065-2660/08 $35.00
DOI: 10.1016/S0065-2660(07)00003-X

ABSTRACT

Altered expression of genes during development is one mechanism that might underlie morphological diversity in animals. Comparison has shown that differences in gene expression often correlate with differences in morphology between species. However, many of these examples involve slowly evolving traits between widely diverged taxa, making investigation of how such changes came about all but impossible. Changes in expression of a specific gene can be due to changes in the activity of *trans*-acting regulatory factors or to evolution of the *cis*-acting sequences of the gene itself. A number of studies indicate that *cis*-regulatory regions can undergo significant sequence turnover even when their function is maintained. In other cases, however, regulatory regions of considerable sequence similarity mediate a different gene expression pattern. These observations make it difficult to predict a change in transcriptional output from an examination of the sequence alone. Here we review recent observations on the evolution of *cis*-regulatory sequences between *Drosophila* species and some other species of dipteran flies. We look at specific cases that have been investigated in detail, where the function of the regulatory element is either maintained or has evolved to mediate a different transcriptional pattern of gene activity. The examples chosen illustrate the necessity to identify the interactions between the proteins that bind the element, and to verify binding sites through *in vivo* assays. Although the number of such studies is still small, they suggest that changes in gene regulation might be an important factor in the evolution of animal morphology.   © 2008, Elsevier Inc.

I. INTRODUCTION

Evolutionary biology seeks to identify and understand how changes at the molecular level mediate differences in phenotype. Much of the morphological diversity seen in animals can be traced back to the activity of genes regulating development. There are a number of ways in which the function of genes can evolve. These include changes in the activity of micro-RNAs (Chen and Rajewsky, 2007), gene duplication (Force *et al.*, 1999; Holland, 1999; Ohno, 1970), evolution of protein specificity, and changes in the expression patterns of genes. There are some good examples of the evolution of protein sequences that are associated with phenotypic change (e.g., Galant and Carroll, 2002; Hoekstra *et al.*, 2006). However, coding sequences are quite constrained because of the triplet code and the need to maintain protein function. In addition to this we now know that there are only a finite number of protein motifs and that most animals employ the same conserved set of proteins with which to regulate embryogenesis (Carroll *et al.*, 2001). Differences in protein specificity are unlikely to account for all of the diversity seen. Changes in gene expression can have immediate and sometimes profound effects on the phenotype that would be visible to selection. Such effects can be quantitative, spatial, or temporal. They can affect many aspects of the phenotype of an organism. Many correlations have been made between differences in spatial expression of a gene and anatomical differences between species. Mutation altering the regulation of a single gene has been shown to be responsible for phenotypic change in a number of cases (Belting *et al.*, 1998; Colosimo *et al.*, 2005; Wang and Chamberlin, 2004). Therefore, differences in *cis*-regulatory sequences might be important for phenotypic diversity.

A. Evolution of promoter sequences

Promoter sequences might evolve more readily than coding sequences. In general terms, there are fewer constraints on their sequence. Gene expression is a function of the composition and configuration of transcription factor binding sites within a promoter. These are, therefore, potential targets for natural selection on gene expression. However, regulatory elements do not have universal structural features, and their organization does not show the regularity typical of coding sequences (Arnosti, 2003; Ludwig, 2002; Wittkopp, 2006; Wray *et al.*, 2003). Binding sites for transcription factors are often only a few base pairs long and are therefore present at a high frequency in genomes and many apparent sites are not actually *bona fide*. True binding sites are often clustered into modules but may not necessarily be present in great numbers and are not usually spaced with any degree of precision. In some instances, there may be functional constraints resulting from the binding of dimers or protein complexes but often considerable variation in sequence is tolerated. The predictions of regulatory modules by sequence examination and phylogenetic comparison are improving all the time (see Chapter 10 by Gabriela Loots), but detailed studies on model organisms show that identification of true binding sites needs to be determined both biochemically and through *in vivo* functional assays. There are very few cases where binding site topology and interactions between the binding proteins are known in sufficient detail to allow identification of possible changes in phenotypic outcome from examination of sequence alone (Erives and Levine, 2004; Markstein *et al.*, 2002; Senger *et al.*, 2004).

Promoter sequences evolve at very different rates. Mechanisms such as slippage, unequal recombination, and point mutation all contribute to sequence turnover within regulatory elements (Hancock *et al.*, 1999; MacArthur and Brookfield, 2004; Stone and Wray, 2001). Long-term conservation is not a feature so far observed in Diptera although it has been described in a number of vertebrate genes (Forest *et al.*, 2007; Pennacchio *et al.*, 2006; Plessy *et al.*, 2005; Richards *et al.*, 2005; Wittkopp, 2006). Even when function is conserved, sequences can be very divergent between related species (Balhoff and Wray, 2005; Ludwig *et al.*, 1998). Many of the gene networks acting in early development are quite conserved (Davidson, 2006; Levine and Davidson, 2005; Olivieri and Davidson, 2007). This suggests that gene expression patterns are constrained. The question of developmental constraints has been much debated (Alberch, 1982; Maynard-Smith *et al.*, 1985). It seems that some traits in development are so well canalized that deviation from them would be selected against although others can be overcome by selection (Beldade *et al.*, 2002). One of the reasons for the conservation of early gene networks may be that in early development few spatial coordinates exist as yet, the developmental histories (in terms of gene expression) of individual cells are short, and little redundancy of developmental pathways has been established. Therefore, the expression patterns of some genes need to be maintained in spite of the continued turnover of the regulatory sequences driving them. So enhancers have to evolve in such a way as to maintain functional output. A prevailing view is that they are subject to stabilizing selection whereby selection for compensatory mutations counterbalances those that lead to small changes in output but that are sufficiently neutral to be tolerated (Ludwig, 2002; Wray *et al.*, 2003).

While divergent promoters can mediate similar transcriptional outputs, it is also the case that similar promoters can mediate quite different outputs. Exactly how enhancers evolve to mediate different patterns of gene expression is still unknown (Stern, 2000; Wray *et al.*, 2003). The process is assumed to occur by the loss of existing transcription factor binding sites or the accretion of new ones that arise by chance mutation (Costas *et al.*, 2003; Guss *et al.*, 2001; Hancock *et al.*, 1999; MacArthur and Brookfield, 2004; Stone and Wray, 2001). Some rare cases have been described where loss of binding sites or gain of new ones leads to alterations in gene expression that correlate with changes in morphology (Jeong *et al.*, 2006; Prud'homme *et al.*, 2006). Sequence turnover in promoter regions is presumably responsible for such events. In a very few cases a change in gene expression has been shown to result from differences in only a few nucleotides.

Many genes are pleiotropic and are expressed at multiple times and places. Pleiotropic genes often have modular promoters in which binding sites for regulatory proteins are organized into independent modules within a promoter. The transcriptional machinery itself assembles on the basal promoter and individual modules may be a long way from this. Modules therefore have to loop up to the correct promoter and there are mechanisms to ensure this. These include selective tethering, basal promoter selectivity, and the presence of insulator sequences. Little is known about how evolution of these mechanisms could alter gene expression patterns and phenotype. The fact that protein–protein interactions are necessary for enhancer action to take place via DNA looping means that the process is relatively insensitive to position and orientation of binding sites within enhancers and the gene will often tolerate changes in position of regulatory sequences relative to the transcription unit.

B. Modularity

Within a complex promoter, each individual module acts independently to direct gene expression at a specific time and/or place. The modules provide specificity of gene activity due to the different transcription factors that bind them. Arnone and Davidson (1997) define a *cis*-regulatory module as the smallest sequence fragment that will drive reporter gene expression in a manner consistent with that of the native gene *in situ*. While the "minimal regulatory element" is a useful definition for experimental analysis, it should be noted that sequences outside the minimal fragment may contribute to expression and have evolved with it. Genes with complex expression profiles often have modular promoters. Even genes expressed early in development can have complex expression patterns in spite of the fact that the process of pattern formation is at an early stage and there are fewer spatial coordinates to refer to. Although many genes are expressed at several times and places, regulatory genes are often the most complex although modularity is not restricted to these. Changes in expression patterns of transcriptional regulators can affect expression of many downstream genes, whereas those in structural genes have limited effects on other genes. Transcriptional regulators interact within gene networks and structural genes are usually at the terminal node (Wray *et al.*, 2003). *Cis–trans* effects are important and there can be coevolution of regulators with their target-binding sites.

A modular organization of gene promoters allows the loss of specific functions of a gene and also addition of new functions through loss of modules, changes to existing modules, or the addition of new ones. This flexibility is an important feature for the evolution of gene function since evolution of a single module allows alteration of only one function of the gene leaving the others intact. Such evolutionary tinkering may underlie gradual changes in gene activity (Duboule and Wilkins, 1998; Jacob, 1977). The modularity of eukaryotic promoters may also contribute to retention of duplicate genes. Mutation in regulatory modules can lead to a sharing out of ancestral functions between the two gene copies, a process called subfunctionalization for which there are a number of examples (Lynch and Force, 2000).

C. Experimental approaches to promoter evolution

Comparisons of gene expression patterns between species have revealed correlations between different spatiotemporal expression domains of regulatory genes and morphological differences. However, many of these have involved slowly evolving traits between distantly related species. On the contrary, direct genetic analyses have been restricted to intraspecific variation and interspecific differences between closely related species that interbreed. Recently, some authors have sought to bring these two approaches together through analysis of rapidly evolving traits between closely related species with small differences in morphology (Carroll, 2005). For this it is necessary to identify a tractable trait, the development of which is well documented in one of the species (Stern, 2000). Ideally, the *trans*-regulators are known as well as the genetic makeup of the gene whose expression is evolving. There is the need to verify the true binding sites by demonstrating binding and by altering those sites and examining the consequences in transgenic animals. Although this is becoming possible in a number of organisms, rapid technological advances have made *Drosophila* one of the animals of choice for such studies. As promoters are modular, the activity of individual modules can be examined in transgenic animals by means of reporter genes. These constructs comprise the regulatory module coupled to a basal promoter that drives expression of an easily visible reporter gene such as that encoding green fluorescent protein. Transformation is routine in *Drosophila melanogaster* but is now possible in other dipteran species (Christophides *et al.*, 2000; Spradling and Rubin, 1982; Wittkopp *et al.*, 2002b). A good phylogenetic framework is also required in order to be able to infer the direction of phenotypic change, and a number of studies on the phylogeny of *Drosophila* species have been carried out.

In this review we focus on a few well-characterized *cis*-regulatory modules in *Drosophila* and other species of Diptera, whose functions have either been maintained over long periods of evolutionary time or have undergone changes that have contributed to phenotypic evolution. This is because at present only a handful of cases have examined the functional consequences of naturally occurring variation in promoter sequence. We will discuss the maintenance of promoter function in the face of continual sequence turnover, sequence change associated with phenotypic diversity, coevolution of *trans*-regulatory factors and their target sequences, and briefly consider the evolution of modular promoters.

II. DEVELOPMENTAL HOMEOSTASIS, SEQUENCE TURNOVER, AND STABILIZING SELECTION

There are still only a small number of well-characterized *cis*-regulatory modules of *Drosophila* that have been subjected to evolutionary studies. Three examples will be discussed in this section to illustrate common themes that are emerging from examination of modules whose function is constrained and has remained unchanged. All three display a high turnover of transcription factor binding sites such that the overall topology is quite divergent between species. It appears that function can be conserved in spite of rapid sequence divergence. The redundancy of binding sites together with changes in binding site topology suggests that multiple mutations can contribute to variation in gene expression. An enhancer can therefore be viewed as a quantitative character (Ludwig, 2002). The model most frequently evoked for enhancer evolution is that of stabilizing selection

(Ludwig, 2002). Mutational changes with only mild effects might be nearly neutral and might reach fixation by genetic drift (Kimura, 1981, 1983; Ohta and Tachida, 1990). Functionally, compensatory mutations would then be selected for. Such coevolution of binding sites means that binding site turnover does not disrupt enhancer function.

A. The P2 promoter of *hunchback*

Hunchback (*hb*) encodes a transcription factor that functions during segmentation in *Drosophila* and is essential for development of the gnathal and thoracic segments (Lehmann and Nusslein-Volhard, 1987; Tautz *et al.*, 1987). Zygotic expression of *hb* in the anterior embryo depends on the P2 promoter, an enhancer containing binding sites for the Bicoid (Bcd) homeodomain protein. In the *D. melanogaster hb* P2 enhancer, there are seven Bcd-binding sites (Driever *et al.*, 1989; Ma *et al.*, 1999; Struhl *et al.*, 1989) four of which are conserved in *D. virilis* (Lukowitz *et al.*, 1994; Fig. 3.1). It is assumed that binding sites situated at similar positions, with a similar spacing and orientation relative to each other, are likely to be homologous (from a common ancestor). The Bcd-binding sites between the two species show a similar pattern of clustering. The *hb* P2 promoters of three species distantly related to *Drosophila*—*Musca domestica*, *Calliphora vicina*, and *Lucilia sericata*—have 10, 9, and 7 Bcd-binding sites, respectively, spread over a larger distance (Bonneton *et al.*, 1997; McGregor *et al.*, 2001). The promoters of these three species, however, cannot be aligned with those of the *Drosophila* species in spite of the fact that they mediate a conserved spatial expression pattern. Detailed analysis indicates that the sequences are mainly composed of simple repetitive motifs indicative of slippage-like mechanisms of turnover (Hancock *et al.*, 1999). An intraspecific comparison between 10 strains of *M. domestica* revealed that the sequence of all 10 binding sites is conserved despite indel polymorphism in other regions of the promoter (McGregor *et al.*, 2001). The *hb* P2 promoters of *M. domestica*, *L. sericata*, and *C. vicina*, however, extensively share short repetitive motifs, suggesting local turnover of a small number of shared motifs. The degree of similarity reflects the evolutionary divergence between species: *L. sericata* and *C. vicina* are phylogenetically more closely related to each other than to *M. domestica* (McGregor *et al.*, 2001; Fig. 3.2). They also share a cluster of binding sites further upstream that do not have a homeodomain canonical core sequence (McGregor *et al.*, 2001). This is a rare example where the details of binding site specificity of the regulatory protein, Bcd, have been well examined. Binding to the consensus sequence is mainly determined by the lysine residue at homeodomain position 50. However, Bcd also binds a range of sequences that differ from the consensus, a fact attributable to the presence of an arginine at homeodomain position 54 [for review, see McGregor (2005)]. This is important for the correct expression of *hb* as well as for other genes (Ma *et al.*, 1999; Zhao *et al.*, 2000) and will be discussed below.

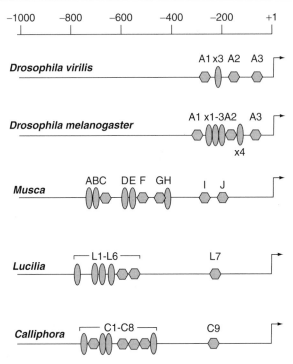

Figure 3.1. Conservation of the *hunchback* promoter in higher Diptera. A comparison of the number and distribution of Bicoid-binding sites in the P2 promoter of the gene *hunchback* of five species of Diptera. Hexagons represent strong binding sites, ovals weak binding sites. The *hb*P2 of *Drosophila melanogaster* has seven Bcd-binding sites, four of which are conserved in *D. virilis* and have been shown by footprinting analysis to mediate strong binding. They are spread over 280 base pairs. In contrast the *hb*P2 of *Musca domestica* has 10 binding sites spread over 700 base pairs, that of *Lucilia sericata* has 7, and that of *Calliphora vicina* 9, spread over 560 and 504 base pairs, respectively. Arrows indicate transcription start sites. Bar at the top represents distance in base pairs 5' from the transcription start site. Reproduced from McGregor *et al* (2001).

There has therefore been significant restructuring of the *hb* P2 promoters between species with respect to the spacing, number, and orientation of Bcd-binding sites. The binding sites are clustered and the pattern of clustering is shared between *L. sericata*, *C. vicina*, and *M. domestica* on the one hand and between the *Drosophila* species on the other hand.

The *tailless* (*tll*) promoter has been similarly restructured between species and cannot be aligned between *M. domestica* and *D. melanogaster* (Wratten *et al.*, 2006; Fig. 3.3). Thus even though, unlike the *hb* P2 element which is exclusively regulated by Bcd, this element has multiple inputs from Bcd, Dorsal, Tramtrack, and the Torso-signaling pathway (Chen *et al.*, 2002; Liaw and

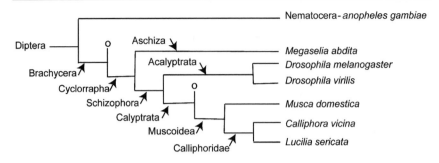

Figure 3.2. Simplified phylogenetic tree of the Diptera. Species mentioned in the text are shown in italics.

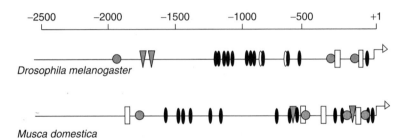

Figure 3.3. Conservation of the *tailless* promoter between *Drosophila melanogaster* and *Musca domestica*. Binding sites for Bicoid (ovals), Dorsal (triangles), and Tramtrack 69 (rectangles) are shown, together with the response element for Torso signaling (circles). Numbering is in base pairs with respect to the transcription start site (arrows). Reproduced from Wratten *et al* (2006).

Lengyel, 1992; Liaw *et al.*, 1995), it is not constrained and evolves as rapidly as the *hb* P2 promoter. Similar to *hb* P2, binding sites for Bcd in the *tll* promoter are more widely spaced in M. *domestica* than in D. *melanogaster*, indicating that site spacing requirements differ. The M. *domestica tll* enhancer has been shown to drive a conserved expression pattern in transgenic D. *melanogaster* with only minor differences (Wratten *et al.*, 2006).

B. The stripe 2 enhancer of *even-skipped*

The *even-skipped* (*eve*) gene is part of a regulatory network governing segmentation that is well conserved in cyclorraphous flies (species that pupate inside a modified larval skin, the puparium) (Sommer and Tautz, 1991). It is a primary pair rule gene that is expressed in the early embryo in seven stripes (Frasch and Levine, 1987). It has a modular promoter and expression in different stripes results from the activity of independent regulatory sequences (Fig. 3.4A). The most intensively studied of these is the *eve* stripe two enhancer, *eve* S2E. It is activated by Bcd and Hb and repressed at the anterior and posterior borders by the products of the gap genes *Giant* (*Gt*) and *Kruppel* (*Kr*), respectively (Arnosti *et al.*, 1996; Small *et al.*, 1992; Stanojevic *et al.*, 1991). A minimal stripe element of 480 bp has been delimited that correctly drives reporter gene expression (Goto *et al.*, 1989; Harding *et al.*, 1989; Small *et al.*, 1992). All four regulatory proteins bind *eve* S2E with high affinity, and cooperative binding between repressors and activators is important to sharpen the boundaries of stripe expression. There are five binding sites for Bcd, six for Kr, three for Gt, and three for Hb. Despite this redundancy they are all probably important for function (Arnosti *et al.*, 1996; Small *et al.*, 1992). They are not equivalent, however, and are of variable affinity. A comparison between *D. melanogaster, D. erecta, D. simulans, D. yakuba, D. pseudoobsura,* and *D. piticornis*

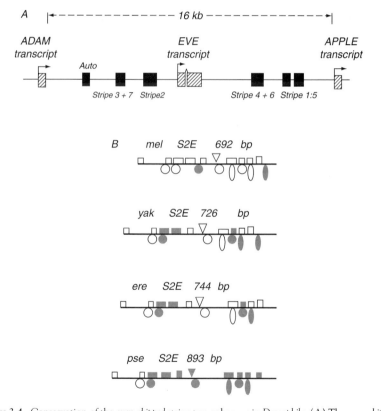

Figure 3.4. Conservation of the *even-skipped* stripe two enhancer in *Drosophila*. (A) The *even-skipped* gene has a modular promoter containing five regulatory elements regulating expression in seven stripes in the embryo. (B) Comparison of the *even-skipped* stripe 2 elements of four species showing binding sites for Bicoid (circles), Hunchback (ovals), Kruppel (squares), Giant (rectangles), and Sloppy paired (triangles). Open symbols represent sites that are conserved with *D. melanogaster*. Reproduced from Ludwig *et al* (2005).

revealed that of the 17 binding sites in *D. melanogaster*, only three are completely conserved in all species (Kreitman and Ludwig, 1996; Ludwig *et al.*, 1998; Ludwig *et al.*, 2005; Fig. 3.4B). Some sites present changes but only of one or two base pairs, indicating that they are probably functionally constrained. Indeed most of these changes occurred only once in the phylogeny of these species (Ludwig *et al.*, 1998). There has also been turnover of the sites themselves. For example, a novel Bcd-binding site has appeared in the *D. melanogaster* lineage and a novel Hb site is present only in *D. melanogaster* and *D. erecta* (Kreitman and Ludwig, 1996; Ludwig *et al.*, 2005). There is also an evolutionary pattern of length changes for *eve* S2E; that is, within the *melanogaster* subgroup 33 out of 34 indels are deletions suggesting that there has been selection for decreased length (Kreitman and Ludwig, 1996). Nucleotide polymorphism within binding sites was also observed between different populations of *D. melanogaster* and of *D. simulans*, as were length variants (Ludwig and Kreitman, 1995). Thus, as with the *hb* and *tll* promoters, there has been significant turnover in enhancer length, and binding site number, affinity, and spacing within the genus *Drosophila*.

Reporter genes carrying the *eve* S2E from *D. melanogaster*, *D. yakuba*, *D. erecta*, and *D. pseudoobscura* were assayed for their activity in *D. melanogaster* hosts. They were found to faithfully reproduce expression that coincides perfectly with native *eve* expression at syncytial blastoderm stages (Ludwig *et al.*, 1998). The evolutionary differences in the *eve* S2Es from these species have therefore not impaired their ability to direct gene expression at the correct location of stripe 2. Therefore, in terms of spatial expression there has been no species-specific coevolution with the morphogens to which they are responding. However, the levels of reporter gene expression driven by the heterologous *eve* S2Es were noticeably lower. Rescue experiments revealed that, indeed, the output of these enhancers has evolved and many cannot drive expression to high enough levels to rescue *D. melanogaster* hosts devoid of native *eve* expression. Ludwig *et al.* (2005) expressed transgenes with the S2Es from *D. melanogaster*, *D. yakuba*, *D. erecta*, and *D. pseudoobscura*, together with the *eve* coding region, in *D. melanogaster* hosts in which the endogenous *eve* S2E had been deleted. Surprisingly, while the *eve* S2E from *D. pseudoobscura* was able to rescue the mutant, that of *D. erecta* was not able to do so (Fig. 3.5). This is unexpected given the much closer evolutionary divergence of *D. erecta* from *D. melanogaster*. The *eve* S2E of *D. yakuba* is functional, but two copies are required for complete rescue. Therefore, the rescue does not correlate with either phylogenetic relationship or sequence divergence. The *eve* S2E of *D. pseudoobscura* is 25% longer than that of *D. melanogaster*, is missing two binding sites, and has only three conserved sites. The newly acquired Bcd3 site in *D. melanogaster*, missing from *D. pseudoobscura*, is essential for function, and so the authors argue that an ancestral *eve* S2E without this site would be unlikely to rescue in this assay. They suggest that the sensitivity of the enhancer has oscillated over evolutionary time presumably as a

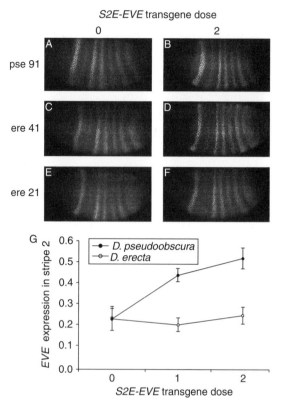

Figure 3.5. Orthologous *even-skipped* stripe 2 elements contribute differentially to the abundance of *eve* protein. (A–F) Fluorescence-labeled antibody staining of Eve in embryos with no (0) or two (2) copies of a transgene bearing the S2E elements of *D. pseudoobscura* (pse) and *D. erecta* (ere) and the *eve* coding sequence. A dose effect is seen only with *D. pseudoobscura*. (G) The relative quantities of Eve protein are significant in the transgenic embryos. Reproduced from Ludwig *et al* (2005).

response to evolutionary shifts of Bcd and /or Hb activity. The rescuing ability of the *D. pseudoobscura eve* S2E may therefore be a result of functional convergence between the enhancers of *D. pseudoobscura* and *D. melanogaster*. In turn the failure of the *D. erecta eve* S2E to rescue may reflect a lower sensitivity to activation. These experiments indicate that functional differences can arise very rapidly in evolutionary terms and that, furthermore, one cannot assume that closely related species will display greater functional conservation in enhancer structure than more distant ones.

The *eve* S2Es of *D. pseudoobscura* and *D. melanogaster* are biologically indistinguishable. They both drive correct spatiotemporal gene expression to the appropriate levels. In spite of significant changes in transcription factor binding sites in the enhancers the trait remains unchanged. This suggests that early expression of *eve* in stripe two is under stabilizing selection (Ludwig *et al.*, 2005). Evidence for stabilizing selection was obtained through investigation into the activities of chimaeric enhancers in which the 5′ and 3′ halves of the native *D. pseudoobscura* and *D. melanogaster* S2Es were exchanged (Ludwig *et al.*, 2000; Fig. 3.6). In contrast to the intact native enhancers, reporter gene expression driven by the chimaeric enhancers is altered: a posterior shift of expression or an expansion of both anterior and posterior borders was observed.

C. The dorsocentral enhancer of *scute*

scute (*sc*) is a regulatory gene encoding a basic helix-loop-helix transcription factor required for the development of neural cells (Alonso and Cabrera, 1988; Bertrand *et al.*, 2002; Ghysen and Dambly-Chaudiere, 1988; Villares and Cabrera, 1987). It has a modular promoter and is sited within a complex of four genes containing both shared and nonshared regulatory sequences (Gomez-Skarmeta *et al.*, 1995; Modolell and Campuzano, 1998; Ruiz-Gomez and Modolell, 1987; Fig. 3.7) (see below). Evolution of expression has mainly been examined in the context of later developmental stages when the positional information and spatial coordinates are well established. So each module may not require multiple inputs as in the case of many early genes. *scute* is part of a regulatory gene network. On the thorax of the fly it is expressed in small clusters of cells at the site of each of the future sensory bristles (Cubas *et al.*, 1991; Skeath and Carroll, 1991). One module, the dorsocentral enhancer (DCE), that regulates expression in only a few cells of the wing imaginal disc for the development of the two dorsocentral bristles has been well described in *D. melanogaster* (Garcia-Garcia *et al.*, 1999). Loss of expression at this site results in a loss of the two bristles. Activation is mediated by the GATA factor Pannier (Pnr) and numerous GATA binding sites are present in the enhancer (Garcia-Garcia *et al.*, 1999; Ramain *et al.*, 1993, 2000). Although the protein binds at least seven of these *in vitro*, they each mediate quantitatively different transcriptional outputs (Garcia-Garcia *et al.*, 1999).

Comparisons of the *sc* DCE between different drosophilids show little sequence conservation. Sequence alignments reveal that the enhancers of *D. melanogaster*, *D. virilis*, *D. eugracilis*, and *D. quadrilineata* are greatly variable in size and have undergone considerable turnover (Marcellini and Simpson, 2006; Fig. 3.8). Only the extremities display significant levels of similarity; the central region is poorly conserved. The enhancers from *D. melanogaster* (1.5 kb) and *D. eugracilis* (2 kb) are more similar to each other than to the others, consistent in this case with their closer phylogenetic relationship. The enhancers from *D. virilis* and *D. quadrilineata* share a relatively large size (4.1 and 3.3 kb, respectively) and a conserved stretch (of unknown function) of about 300 nucleotides that is absent from the *D. melanogaster* and *D. eugracilis* sequences. Putative-binding sites for Pnr

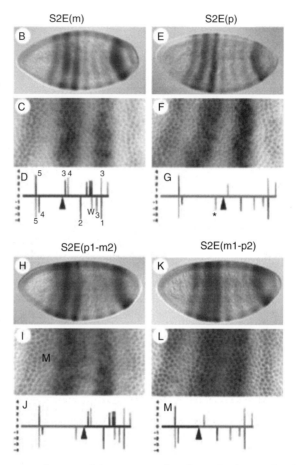

Figure 3.6. Expression of native and chimaeric *even-skipped* stripe 2 enhancers from *D. melanogaster*
and *D. pseudoobscura*. (B–F) Expression of *lacZ* mRNA driven by reporter genes comprising the natural enhancers from *D. melanogaster* (S2E(m)) and *D. pseudoobscura* (S2E
(p)). Stripes 3 and 7 are also labelled for comparison. (D) and (G) summarize the
enhancer sequences with their different binding sites. (H and I) Expression of *lacZ*
mRNA driven by reporter genes comprising the chimaeric enhancers (S2E(p1-m2))
and (S2E(m1-p2)). The composite enhancer structure is summarized in J and M.
Reproduced from Ludwig *et al* (2000).

are present in all species but have only been verified for *D. melanogaster*. Mutation
of a specific Pnr binding site severely reduces activity of the *D. melanogaster sc* DCE
(Garcia-Garcia *et al.*, 1999). This site is embedded within a stretch of 16 nucleotides that is perfectly conserved between the four species. Two other neighboring
GATA sequences can be recognized as homologous between all species.

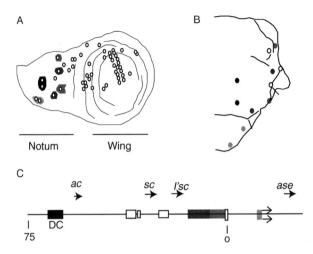

Figure 3.7. The *achaete-scute* complex bears many *cis*-regulatory elements driving expression in different proneural clusters. (A) The positions of the proneural clusters of *achaete-scute* expression in the imaginal disc. (B) The positions of the bristles on the notum arising from each cluster. Each proneural cluster gives rise to one or two bristle precursors on the notum. (C) The *achaete-scute* complex showing some of the independently acting *cis*-regulatory elements that mediate expression of *achaete* and *scute* in different proneural clusters. The dorsocentral regulatory element and its corresponding proneural cluster and bristles are shown in black. Adapted from Gomez-Skarmeta *et al.* (1995).

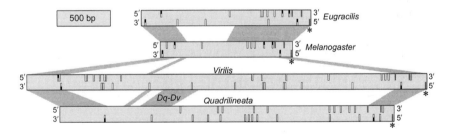

Figure 3.8. Comparison of the dorsocentral enhancer from four species of *Drosophila*. The sequence organization of the enhancers of *D. eugracilis*, *D. melanogaster*, *D. virilis*, and *D. quadrilineata* is shown. The scale is shown at the top-left. Only the dark regions connecting adjacent enhancers are alignable. Small vertical rectangles symbolize all the GATA sites found in the forward (top) or reverse (bottom) strand. The asterisk marks a Pannier binding site essential for normal activity of the *D. melanogaster* enhancer. Reproduced from Marcellini and Simpson (2006).

Conservation overall, however, is low and the number, spacing, and orientation of the remaining putative Pnr binding sites are extremely variable. Alone of these four species *D. quadrilineata* displays four dorsocentral bristles and the *D. quadrilineata* enhancer mediates reporter gene expression in a significantly altered domain in transgenic *D. melanogaster* (see below). This difference could not be predicted from examination of the sequences.

  D. virilis and *D. melanogaster* are diverged from one another by 40–60 Myr. The DCE from *D. virilis* drives expression of a reporter gene in *D. melanogaster* in a similar region of the disc to that of the native enhancer but with detectable differences in the extent of spatial expression (Marcellini and Simpson, 2006; Fig. 3.9A). In spite of this, when coupled to *sc* coding sequences and expressed in a host animal devoid of endogenous *sc* expression, it is able to rescue the two bristles with a remarkable degree of precision (Fig. 3.9B). This is because of other inputs to bristle patterning that are independent of the spatial activation of *sc* at this site (Rodriguez *et al.*, 1990).

  It is interesting to contrast evolution of the DCE of *sc* with that of the S2E of *eve*. The spatiotemporal expression driven by orthologous *eve* S2Es is conserved but the levels are sufficiently variable such that not all enhancers can rescue (Ludwig *et al.*, 2005). The *D. virilis sc* DCE does not mediate completely conserved spatial expression but is nevertheless able to rescue the phenotype, indicating that the levels of expression are adequate. Expression of *eve* is undoubtedly constrained by the fact that it is expressed in the early embryo and its expression is crucial for the process of segmentation. The embryo is composed of fewer cells at this early stage; each stripe is only three to four cells wide and stripe two must be equidistant from stripes one and three. In addition, fewer spatial coordinates have been established. Bristle patterning by comparison takes place at a much later stage in development when positional information is more defined. There are many cells in the dorsocentral cluster of *sc* expression and significant redundancy in the patterning processes, so that other mechanisms participate in the positioning of the two bristle precursors (Huang *et al.*, 1995; Rodriguez *et al.*, 1990). So bristle patterning is very well canalized in *Drosophila*, and there are developmental constraints to prevent deviations from the pattern (Usui *et al.*, 2004). Bristle patterns are often stereotypical and species-specific among cyclorraphous Diptera, and many patterns remain stable for long periods of evolutionary time (Grimaldi, 1987; McAlpine, 1981). This suggests that deviations from the pattern will probably be selected against. It is possible to select for additional bristles in *D. melanogaster* but selection leads to predictable patterns due to certain developmental constraints [see Usui *et al.* (2004) and references therein]. The *eve* S2E may be constrained for spatiotemporal expression but less so for gene expression levels (Ludwig *et al.*, 2005). The loss and gain of binding sites and their polymorphism may allow for continuous adjustment of activity levels in response to changes in the *trans*-acting environment.

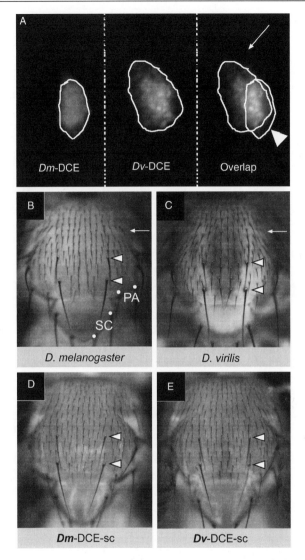

Figure 3.9. (A) The dorsocentral enhancers from *D. melanogaster* and *D. virilis* drive reporter gene expression in overlapping domains in transgenic *D. melanogaster* flies. The domain of expression mediated by the enhancer of *D. virilis* is larger and displaced medially when compared with the endogenous enhancer of *D. melanogaster*. (B–E) Thoraces of both species bear two dorsocentral bristles on each hemithorax. A transgene bearing either the *D. melanogaster* or the *D. virilis* enhancer together with Scute rescues two bristles in *D. melanogaster* hosts. Reproduced from Marcellini and Simpson (2006).

D. Conclusions

These three examples, *hbP2*, *eveS2E*, and *scDCE*, illustrate features common to many *cis*-regulatory elements of *Drosophila*. There is redundancy with multiple copies of binding sites for regulatory proteins. The number of sites and their spacing and orientation differs between species. The enhancers undergo significant turnover, suggesting a loss and gain of binding sites (Hancock *et al.*, 1999). Costas *et al.* (2003) examined verified binding sites for early regulators and promoters and found that 37.5% of those in D. *melanogaster* are absent from D. *virilis*. Assuming a 40-Myr split between these species, they estimate that 0.94% of sites are lost or gained per million years. Twenty-three percent of the sites are absent from D. *pseudoobscura*, a more closely related species, so the turnover is consistent with a molecular clock (Costas *et al.*, 2003). This is twice the turnover estimated between mouse and man, although the mammalian study was not confined to early regulatory genes (Dermitzakis and Clark, 2002). There is a general higher divergence of insect sequences when compared to vertebrate sequences (Zdobnov *et al.*, 2002). Evolutionary changes in enhancer structure between species are likely to be due to slippage-like processes and unequal crossing-over (Hancock *et al.*, 1999), but new binding sites probably also arise by point mutations (MacArthur and Brookfield, 2004; Stone and Wray, 2001). The extent of turnover is such that, sometimes, even between closely related species sequences cannot be aligned and orthologous transcription factor binding sites cannot be identified. In spite of this, in many cases the function of the enhancer as a unit is maintained and mediates a conserved gene expression pattern.

Investigation of the S2E of *eve* provides good evidence for the model of stabilizing selection operating on *cis*-regulatory modules. Compensatory evolution, whereby a pair of mutations, each of which is deleterious, confer fitness when in combination could also apply to enhancer evolution (Kimura, 1985; Ludwig, 2002). Theoretical predictions suggest that this is more likely in large populations, which could explain the higher turnover rates in *Drosophila* enhancers when compared to vertebrate ones (Carter and Wagner, 2002).

III. ENHANCER EVOLUTION AND LOSS OR GAIN OF TRAITS

How do enhancers evolve to mediate different patterns of gene expression? In this section we review examples from the *yellow* (*y*) and *sc* genes of regulatory elements that have undergone changes between species that are associated with different phenotypic outcomes. There is evidence for association between a loss of binding sites and anatomical difference, and for the acquisition of new expression domains by preexisting enhancers.

A. Evolution of *yellow* and variation in pigment patterns

yellow is an example of a gene with a modular promoter that does not have a regulatory function and is in addition expressed late in development. It plays a role in pigmentation and since pigmentation is spatially varied in different body regions this gene has acquired a complex expression pattern. In *D. melanogaster* it has been shown that patterns of pigmentation are attributable to the restricted transcription of two enzymes encoded by the *y* and *ebony* (*e*) genes (Wittkopp et al., 2002a, 2003). *yellow* is required for black melanin (dopa melanin) and its expression correlates with black pigmentation. Ebony converts dopamine to N-β-alanyl dopamine synthetase (NBAD), which is oxidized to produce tan pigment. Although *e* is not spatially regulated, the expression levels vary. Both genes are expressed very late in development at the time of deposition of the cuticle (True et al., 1999).

Some *Drosophila* species display dark pigmentation on the 5th and 6th abdominal segments (A5 and A6) of male flies. The sexual dimorphism results from regulatory interactions between *Abdominal-B* (*Abd-B*), *bric a brac* (*bab*), and *doublesex* (*dsx*) (Kopp et al., 2000). Abd-B represses *bab* in segments A5 and A6 in males but not in females where repression is over-ridden by a female-specific isoform of *dsx*, Dsx-F. In addition, it has recently been shown that Abd-B directly activates *y* through a specific regulatory element, the "body" element (Jeong et al., 2006). Phylogenetic analysis indicates that this interaction was present in a common ancestor of the *melanogaster* species group, but has been lost in some species. Inactivation of a key Abd-B binding site through two base substitutions appears to underlie the loss of pigmentation in the *D. kikkawai* lineage: the orthologous "body" element of this species is unable to mediate expression of a reporter gene in transgenic *D. melanogaster* flies (Jeong et al., 2006). Indeed experimental mutation of the two Abd-B binding sites in the "body" element of *D. melanogaster* also results in a loss of male-specific expression of reporter genes. This illustrates how a trait might be lost through changes in a binding site for a *trans*-regulatory factor.

Many species of Drosophilidae display intricate species-specific patterns of pigmented wings. A dark spot of pigment, preceded by the expression of *y*, is found on the anterior distal wing of *D. biarmipes* and *D. elegans*, both members of the *melanogaster* group of drosophilids (Fig. 3.10). Phylogenetic analysis indicates that the common ancestor of this group was unspotted, so the spot is a derived character; it was secondarily lost in *D. melanogaster* (Prud'homme et al., 2006). A specific, orthologous, 675bp regulatory module from *D. biarmipes*, the "spot" enhancer, recapitulates most of the expression of *y* in the spot region: when coupled to a reporter gene it mediates expression in a similar spot region in transgenic *D. melanogaster* (Gompel et al., 2005; Prud'homme et al., 2006). The "spot" enhancers of *D. biarmipes* and *D. elegans* have therefore evolved to

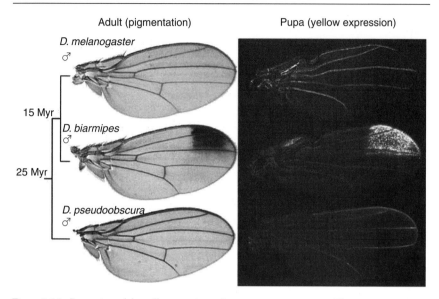

Figure 3.10. Expression of the yellow protein prefigures wing pigmentation. The pigmentation spot of *D. biarmipes* is a derived trait that evolved among species of the *D. melanogaster* group. Distribution of yellow prefigures pigmentation in the wing veins and spot. Reproduced from Gompel *et al.* (2005).

make use of proteins of a conserved wing regulatory network present in *D. melanogaster*, a species devoid of a pigment spot, and have evolved to acquire binding sites for these. The "spot" element contains sequences for activation and repression. That of *D. biarmipes* contains two binding sites for the repressor Engrailed, not present in the orthologous *cis*-regulatory region of the *D. melanogaster* y gene. Loss of these causes the expression of y to expand into the posterior compartment (Gompel *et al.*, 2005). A comparison of the "spot" element of *D. elegans* with the orthologous region of the closely related *D. gunungcola*, a species with no spot, allowed Prud'homme *et al.* (2006) to pinpoint a sequence divergence of only 10 nucleotides. Removal of these nucleotides from the spot element of *D. elegans* renders the sequence inactive; their addition to the orthologous region of *D. gunungcola* confers the ability to direct expression in the region of the pigment spot when assayed in *D. melanogaster* hosts. So one of the, as yet unknown, transcription factors required for activation binds this sequence, indicating that loss of the spot would not require substantial sequence change.

The authors determined that the "spot" sequence is embedded in a larger region that drives expression over most of the wing. An orthologous region of the y gene of *D. pseudoobscura*, an outgroup species devoid of a spot, drives ubiquitous wing expression (Gompel *et al.*, 2005). They suggest that the "spot" sequence arose from a preexisting element responsible for expression in the wing that, by subfunctionalization, gave rise to independent wing and spot elements.

The same authors examined spot formation in *D. tristis*, a species in the *obscura* group that has independently gained a similar spot of pigment on the wing. *yellow* is also expressed in the spot region in this species, but, remarkably, expression is regulated by a different *cis*-regulatory element situated in the intron, known as "intron spot" (Fig. 3.11). The fact that the enhancers do not display sequence similarity is a further indication that the two sequences have originated independently (Prud'homme *et al.*, 2006). The "intron-spot" element of *D. tristis* might have evolved through the co-option of an ancestral element for vein expression. Therefore, more than one molecular path has evolved to direct expression of *y* in the spot region but both cases involve evolution of an ancestral regulatory element that mediated expression in the wing.

B. Evolution of *scute* and variation in bristle patterns

The large sensory bristles on the notum are arranged into species-specific patterns and often occupy stereotypical locations particularly in cyclorrhaphous flies (McAlpine, 1981; Simpson *et al.*, 1999). In those species examined so far there is a correlation between the spatial expression of *sc* and the location of bristles. *scute* is expressed in discrete domains on the thorax, either in stripes or in small clusters of cells called proneural clusters, at the sites of formation of the future bristle precursors (Fig. 3.7; Cubas *et al.*, 1991; Pistillo *et al.*, 2002; Skeath and Carroll, 1991; Wülbeck and Simpson, 2000). Evolution of the regulation of *sc* with reference to different bristle patterns has been examined within Drosophilidae.

Most of the nearly 4000 species of Drosophilidae bear two dorsocentral bristles (Bachli, 1998; Wheeler, 1981). However, *D. quadrilineata*, a species belonging to the *immigrans* subgroup, displays four dorsocentral bristles (Fig. 3.12A). The four bristles form a row and the additional bristles are located in an anterior position. The anterior bristles represent a trait newly acquired in *D. quadrilineata*, since it is thought that the ancestor to the Drosophilidae had only posterior dorsocentral bristles (Bachli, 1998; McAlpine, 1981; Wheeler, 1981). The dorsocentral bristles of *D. melanogaster* arise from a small cluster of *sc* expressing cells, and those of *D. quadrilineata* from an elongated streak (Cubas *et al.*, 1991; Marcellini and Simpson, 2006; Skeath and Carroll, 1991). Within *D. melanogaster*, a transgene bearing the *D. melanogaster* dorsocentral enhancer of *sc* (*sc* DCE, described above) together with the coding sequences of *sc* is able to rescue the two dorsocentral bristles in an animal devoid of endogenous *sc* expression (Culi and Modolell, 1998). An orthologous *sc* DCE from *D. quadrilineata* coupled with *sc* is able to mediate the formation of a row of

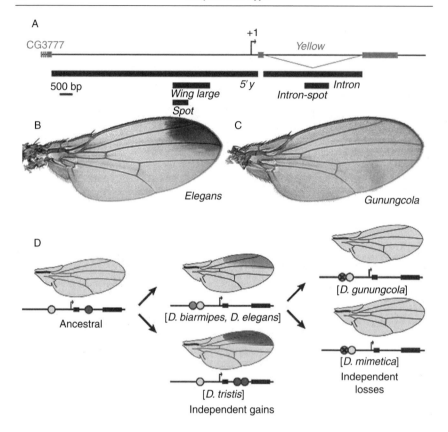

Figure 3.11. Regulation of *yellow* expression and the wing spot of pigmentation. (A) The promoter of *yellow* showing the position of two distinct regulatory elements driving expression in the distal, anterior wing of different species, labelled "spot" and "intron spot". (B) The "spot" element drives expression in *D. elegans*. (C) The orthologous region in *D. gunungcola* has lost this activity. (D) The "intron-spot" element drives *yellow* expression in the distal, anterior wing of *D. tristis*. Phylogenetic analysis indicates that *yellow* spot expression patterns have evolved twice independently in two different lineages through the cooption of ancestral regulatory elements, symbolized by circles. Expression has also been lost independently in different lineages by the repeated inactivation of the "spot" element. Reproduced from Prud'homme *et al.* (2006).

four dorsocentral bristles in *D. melanogaster*, instead of the two usually present in this species (Marcellini and Simpson, 2006; Fig. 3.12B). It is not yet known which region(s) of the *sc* DCEs is responsible for this difference.

The domain of expression mediated by the *D. quadrilineata sc* DCE in transgenic *D. melanogaster* extends further in an anterior direction and it is this modified domain of expression that allows the formation of additional anterior

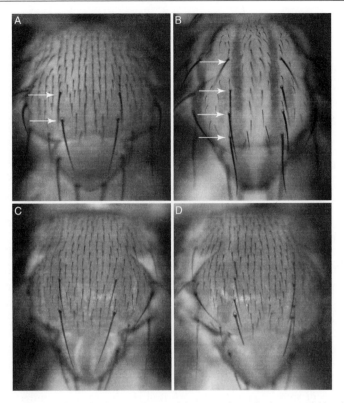

Figure 3.12. The dorsocentral enhancer of *D. quadrilineata* mediates the formation of four bristles in transgenic *D. melanogaster*. (A) Thorax of *D. melanogaster* showing two dorsocentral bristles (arrows). (B) Thorax of *D. quadrilineata* with four dorsocentral bristles. (C–D) Thoraces of transgenic *D. melanogaster* flies bearing the dorsocentral regulatory elements of *D. melanogaster* (C) or *D. quadrilineata* (D) coupled with *scute* coding sequences. The enhancer of *D. quadrilineata* mediates the development of four bristles.

dorsocentral bristles. However, the upstream regulator Pnr has a conserved expression domain and is part of a regulatory network that is little changed between species as diverged as *Calliphora* and *Drosophila* (Pistillo *et al.*, 2002; Richardson and Simpson, 2005). The activity of the *D. quadrilineata* sequence is modulated in a similar fashion to that of *D. melanogaster* in different *pnr* mutant backgrounds: expression is expanded in gain of function mutants and diminished in loss of function ones. This suggests that, in spite of significant sequence turnover, the divergent enhancers require Pnr and are efficiently repressed by U-shaped, a repressor of Pnr (Cubadda *et al.*, 1997; Haenlin *et al.*, 1997; Marcellini and Simpson, 2006). The *D. quadrilineata* enhancer therefore responds differently to the conserved *trans*-regulatory landscape of D. *melanogaster*, indicating that the morphological difference is largely due to evolution of these *cis*-regulatory sequences. The D. *quadrilineata sc* DCE does not perfectly recapitulate the D. *quadrilineata* bristle pattern when expressed in D. *melanogaster*: the anterior-most bristles are not located on the prescutum, the anterior-most region of the notum (Fig. 3.12B). Some changes in *trans*-acting factors or other regulatory regions of *sc* may, therefore, also have taken place.

C. Conclusions

The prepatterns or *trans*-regulatory landscapes to which y and *sc* respond are made earlier in development and remain unchanged between species. They are deeply conserved regulatory networks that govern most aspects of the morphology of the wing or notum, not merely the bristles or the pigment. The networks involve conserved selector genes such as *engrailed* or *pannier* (Blair, 2003; Calleja *et al.*, 2000; Lawrence and Struhl, 1996; Mann and Morata, 2000). This means the regulatory gene network is present in species that are devoid of wing spots or anterior dorsocentral bristles. Altering the response to a conserved regulatory landscape allows slight, localized modifications of the final pigment or bristle pattern. The *trans*-regulatory proteins of D. *melanogaster* recognize the regulatory sequences from the other *Drosophila* species so the prepattern of D. *melanogaster*, itself devoid of spots and anterior dorsocentral bristles, carries the information allowing development of these traits. This explains why independently derived wing pigment spots or anterior dorsocentral bristles are very similar in pattern. It suggests that if a gene acquires a new binding site for a regulator, this is likely to be a regulator that is already expressed at that time and place and gene expression is likely to change in accordance with the existing regulatory landscape (Gompel *et al.*, 2005).

Evolution of expression of both y and *sc* appears to have involved regulatory sequences that were already present. Therefore, tinkering with these sequences, presumably in small incremental steps, has allowed the introduction of a new expression domain associated with a new trait. The fact that evolution of different, but nevertheless pre-existing, regulatory elements of the same gene underlies convergent evolution of y expression in D. *tristis* and D. *elegans* suggests that a new regulatory activity is likely to evolve within the context of a regulatory element that already bears information relative to the expression of that gene (Prud'homme *et al.*, 2006). These examples also serve to illustrate the fact that it is almost impossible at the present time to predict phenotypic outcome from comparison of the enhancer sequences alone. Enhancers undergo significant sequence divergence in the absence of any phenotypic change.

In contrast, evolution of y expression in the region of the pigment spot in wings of different *Drosophila* species suggests that a phenotypic difference might result from change in a small number of base pairs.

IV. *CIS-TRANS* COEVOLUTION

Interspecific differences in gene expression can result from divergence of *cis*-regulatory sequences or of *trans*-regulatory transcription factors. In addition, these two components, the transcription factor and its target binding sequences, might evolve together. Molecular coevolution involves the selection of compensatory changes that restore the functional interaction between two components (Carter and Wagner, 2002; Castillo-Davis *et al.*, 2004; Dover, 2000; Dover and Flavell, 1984; Kimura, 1985). Coevolution between a transcription factor and its target sequences has been demonstrated in fungi, nematodes, and sea urchins (Castillo-Davis *et al.*, 2004; Gasch *et al.*, 2004; Romano and Wray, 2003). It can lead to regulatory incompatibility between species that hybridize. An elegant study by Wittkopp *et al.* (2004) examined hybrids between *D. simulans* and *D. melanogaster* to determine whether species-specific differences in gene expression are due to changes in *cis* or in *trans*. The authors looked at species-specific allelic expression of 29 (mostly structural) genes, and measured the abundance of transcripts in the parent species and the hybrid. Since the *trans*-regulatory background is the same in the hybrid, unequal abundance of transcripts of two alleles of a specific gene indicates that the changes are in *cis*. Differences between the parental lines not due to changes in *cis* are then assigned to *trans* divergence. The analysis revealed that the regulation of expression of 97% of the genes has evolved in *cis*, half of these in *cis* alone; 57% were also affected in *trans*. This is consistent with coevolution. It also suggests that differences in gene expression between the two species are not caused by a small number of changes in *trans*-regulatory factors with widespread effects, but rather by a large number of *cis*-acting changes (Wittkopp *et al.*, 2004). Hybrids between species often display novel gene expression patterns and indeed 13 genes with *cis-trans* coevolution are misexpressed in the hybrid. This is likely to be a consequence of coevolution between *cis*- and *trans*-regulatory elements within species (Landry *et al.*, 2005).

A. The interaction between Bicoid and the *hunchback* P2 promoter

This is an example of the interaction between a specific transcription factor and its target genes for which there is a suggestion of regulatory incompatibility between *D. melanogaster* and other cyclorrhaphous flies. The interaction between Bicoid and its target promoters has been intensively investigated in the context of morphogen function in *D. melanogaster*. Bicoid acts in anterior–posterior patterning of the embryo and is expressed in an anterior to posterior concentration gradient. Different target genes are activated at different concentrations along the anterior–posterior axis of the embryo and this implies different responses from target regulatory sequences (Fig. 3.13). The differences are to be found in the number, affinity, and spacing of Bcd-binding sites. Most of these studies are confined to *D. melanogaster*, but they may provide pointers to the nature of species-specific differences that could lead to regulatory incompatibilities between species and so will be briefly discussed here.

A simple threshold model would predict that enhancers with low affinity binding sites would respond only to high Bcd concentrations restricting gene expression to the anterior-most region of the embryo, while those with high affinity sites would allow expression to extend to more posterior regions. Two Bcd targets are *hb* and *orthodenticle* (*otd*) (Driever and Nusslein-Volhard, 1989; Finkelstein and Perrimon, 1990; Struhl *et al.*, 1989). Consistent with a threshold model, the *otd* promoter, responsible for gene activation in only the anterior region, has lower affinity sites than that of *hb* whose expression extends more posteriorly (Gao and Finkelstein, 1998; Ochoa-Espinosa *et al.*, 2005).

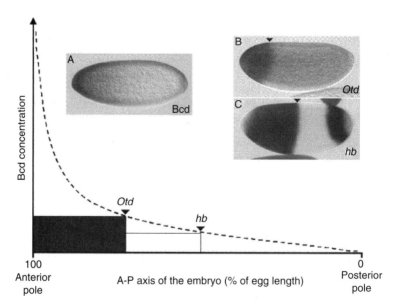

Figure 3.13. The Bicoid morphogen and concentration-dependent target gene activation. (A) The Bicoid gradient with the highest concentration at the anterior end of the embryo that is to the left. This is indicated by dotted lines on the graph. (B–C) Expression of *orthodenticle* (B) and *hunchback* (C) revealed by *in situ* hybridization. Arrowheads show the posterior border of expression. Reproduced from Crauk and Dostatni (2005).

The threshold model appears incomplete, however. In a study of 21 regulatory modules responding to Bcd no correlation was found between Bcd binding strength and anterior–posterior expression of reporter genes (Ochoa-Espinosa *et al.*, 2005). Instead, Ochoa-Espinosa *et al.* (2005) suggest that a combinatorial mechanism of gene activation is more likely. Bicoid alone regulates only five of the modules and all of these drive expression in anterior-most regions where Bcd concentrations are high. Four modules respond to both Bcd and Hb (this includes *hb* itself) and they mediate more posterior expression. Bicoid and Hb have been shown to exhibit synergistic activities (Simpson-Brose *et al.*, 1994). Six modules have Bcd and Kruppel sites, while six have Bcd, Kruppel, and Hb sites and drive expression even more posteriorly (Ochoa-Espinosa *et al.*, 2005).

Binding sites for Bcd exhibit different topologies in the responding modules. For example, those of *otd* are spread over a much greater genomic region than those of *hb* P2 (Ochoa-Espinosa *et al.*, 2005). A significant body of knowledge has been gleaned about Bcd function that illustrates just how important spacing of binding sites may be to enhancer function in some cases. Two properties of Bcd affect the way in which it interacts with target sequences [reviewed by McGregor (2005)]. First, Bcd binds cooperatively (Ma *et al.*, 1996). Therefore, the distance between sites in Bcd target promoters determines cooperative binding between Bcd molecules and modulates transcriptional output (Hanes *et al.*, 1994; Ma *et al.*, 1996). Binding to sites of strong affinity can induce binding to adjacent weaker sites and furthermore mutation of specific amino acids can abolish this cooperative binding (Burz and Hanes, 2001; Burz *et al.*, 1998; Lebrecht *et al.*, 2005; Yuan *et al.*, 1996). This observation may explain why weak transcription factor binding sites may be retained for significant evolutionary periods. Second, a self-inhibition domain in the N-terminal region of Bcd has recently been discovered (Fu *et al.*, 2003; Zhao *et al.*, 2002, 2003). This is masked when Bcd binds to closely spaced sites. However, when Bcd is bound to widely spaced sites the inhibition domain can recruit the Sin3 corepressor, which dramatically reduces transcriptional activity. For example, the promoters of both *knirps* (*kni*) (a Bcd target gene expressed quite posteriorly) and *hb*P2 each have six Bcd-binding sites, but their topology differs. The sites are closely spaced and arranged symmetrically in *kni* (Fu *et al.*, 2003; Fig. 3.14). The *bcd* protein prefers sites such as those of *kni* that are arranged tail to tail and separated by 7–15 base pairs or head to head and separated by 3 base pairs (Yuan *et al.*, 1999). This allows cooperative binding to the *kni* promoter preventing self-inhibition and recruitment of Sin3 (Fu *et al.*, 2003). In contrast, the binding sites in *hb* are dispersed and arranged in tandem, as a result of which the N-terminal of Bcd plays an inhibitory role (Fu *et al.*, 2003; Fig. 3.14). Reporter genes bearing the *kni* promoter therefore respond to lower concentrations of Bcd than do those bearing the *hb*P2 promoter.

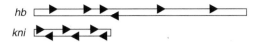

Figure 3.14. The distribution of Bicoid binding sites in the promoters of *hunchback* and *knirps*. Schematic diagram of the 205 bp *hunchbackP2* enhancer element and the 64 bp *knirps* enhancer element. Arrowheads represent Bicoid binding sites. These are close together in the *knirps* element but spaced apart in the *hunchback* element. Reproduced from Fu *et al.* (2003).

Cooperative binding and the self-inhibition properties of Bcd, together with the appropriate arrangement of binding sites in the target promoters, allow the correct activation of different genes at precise spatial locations along the embryo of *D. melanogaster*. This suggests that changes in the activity of Bcd and/ or in the spacing of Bcd-binding sites in specific target promoters might coevolve, eventually resulting in incompatibilities between species. This possibility has been examined for several species of Cyclorrapha where it has been proposed that deleterious mutations in Bcd-binding sites in target promoters are the cause of selective pressure leading to coevolution of Bcd and its target sequences (McGregor *et al.*, 2001; Shaw *et al.*, 2002). Bicoid is found only in cyclorraphous dipterans and its function is conserved throughout this taxon (Sommer and Tautz, 1991; Stauber *et al.*, 1999). *M. domestica* bcd can rescue, albeit inefficiently, *D. melanogaster* bcd mutants (Shaw *et al.*, 2001, 2002; Stauber *et al.*, 2000). The divergence in *hb* promoters between different species of Cyclorrapha was described above. It has been accompanied by changes in Bcd itself, which might reflect a coevolutionary response to changes in promoter sequence. *bicoid* is a new gene that arose at the base of the Cyclorrapha by duplication from an ancestral *Hox3* gene (Stauber *et al.*, 1999, 2002). The *bcd* protein has evolved rapidly due to the relaxed constraints on a new gene and possibly the reduced selection operating on maternal effect genes (Barker *et al.*, 2005). In the homeodomain alone, there are differences in 5 out of 60 amino acids between *D. melanogaster* and *M. domestica* and in 18 out of 60 amino acids between *D. melanogaster* and *Megaselia abdita*, a basal cyclorraphan, a significant difference (Fig. 3.2) (Shaw *et al.*, 2001; Stauber *et al.*, 1999). Indeed, while the anterior cytoplasm of *M. domestica* and a related species *L. sericata* can rescue anterior structures in *Drosophila* bcd mutants, that of *M. abdita* cannot (Schröder and Sander, 1993). As a result of changes in the homeodomain Bcd recognizes a wide range of sequences that vary from the consensus, see review by McGregor (2005).

    There are hints that coevolution between Bcd and *hb* P2 may have occurred in the 100 Myr separating *M. domestica* and *D. melanogaster*. *M. domestica* Bcd binds its own *hb* P2 promoter more efficiently than the *hb* P2 promoter of *D. melanogaster* and *vice versa* (Shaw *et al.*, 2002). Furthermore, transcriptional assays using reporter genes revealed that while *D. melanogaster* Bcd activates transcription from the *D. melanogaster hb* P2 better than from the *M. domestica hb* P2, Bcd from both *Drosophila* and *M. domestica* activate transcription from the *M. domestica hb* P2 to the same extent (Shaw *et al.*, 2002). The *M. domestica* promoter is therefore more sensitive to Bcd-dependent activation than that of *D. melanogaster*. Differences have also been reported for the *tll* promoter: *M. domestica* Bcd binds *M. domestica tll* more efficiently than *D. melanogaster tll*, but surprisingly *D. melanogaster* Bcd binds *M. domestica tll* with higher affinity than it does *D. melanogaster tll* (Shaw *et al.*, 2002; Wratten *et al.*, 2006). So possibly, the *M. domestica hb* and *tll* promoters have evolved a more sensitive configuration of binding sites. The amino acid sequence of the self-inhibition domain of *M. domestica* and *D. melanogaster* Bcd is identical (McGregor, 2005). However, the Bcd-binding sites in both the *M. domestica tll* and *hb* promoters are more widely spaced when compared to those in *D. melanogaster* (Figs. 3.1 and 3.3); therefore, the site-spacing requirements for Bcd may differ between these species. The egg of *M. domestica* is twice the size that of *D. melanogaster*. Therefore, activation of *hb* to the same extent (about 55% egg length) as in the *D. melanogaster* embryos would require either increased levels of Bcd or greater sensitivity of target promoters to respond to a shallower gradient, discussed in the work of McGregor (2005).

B. Conclusions

The study of Bcd targets in *Drosophila* shows just how important the spacing and orientation of binding sites can be for some transcription factors. In this case transcriptional output is significantly modulated by binding site architecture because of specific properties of the *bcd* protein. Cooperativity between Bcd molecules means that closely positioned sites may be no further apart than the size of the protein they bind. This also illustrates the apparent conservation of some sites of weak affinity that are adjacent to sites of strong affinity.

V. EVOLUTION OF NEW REGULATORY MODULES

How do new *cis*-regulatory elements arise? The examples discussed above appear to have involved evolutionary modifications of preexisting modules. The changes involve the loss of binding sites and the creation of new ones by slippage mechanisms or point mutation (Hancock *et al.*, 1999; MacArthur and Brookfield, 2004; Stone and Wray, 2001). Binding sites for transcription factors that did not ancestrally regulate a specific module can in theory arise by chance and lead to new gene expression patterns. It has been argued that small modifications of preexisting regulatory elements would entail fewer steps than the *de novo* generation of new ones (Prud'homme *et al.*, 2006). Whether a *cis*-regulatory module can evolve from DNA with no prior regulatory activity, however, is still an open question.

New regulatory elements could more easily arise through gene duplication. Ohno (1970) proposed that gene duplication could lead to phenotypic diversity because it enables one copy to acquire new functions while the other performs the original function. For developmentally important genes, after duplication, it is inferred that both descendants often have different roles from the original. This could be due to acquisition of additional roles by each copy and/or to the partitioning of the ancestral functions between them (Force *et al.*, 1999; Holland, 1992). This is especially relevant to regulatory genes in view of their pleiotropic functions and modular promoters. Duplication would allow evolution of gene expression by freeing regulatory modules that are no longer required for ancestral functions (Lynch and Force, 2000). For example, the homoeotic (HOM/Hox) genes are thought to have arisen through tandem gene duplication of a small number of ancestral genes (Ferrier and Holland, 2001). A complex of 10 Hox genes is found in most arthropods (Hughes and Kaufman, 2002). The genes are expressed in specific domains in an anterior to posterior order (Hughes and Kaufman, 2002). In *D. melanogaster* the expression of each HOM/Hox gene is a result of its own specific *cis*-regulatory information. Thus, these regulatory elements are likely to have diverged from the ancestral one(s). This is illustrated by the fact that although in many species some of the genes remain clustered, this is not a requirement for function and clustering differs between different *Drosophila* species indicating splits between different genes of the ancestral cluster (Negre and Ruiz, 2007).

An example of gene duplication and increasing complexity of expression patterns may be found in the genes of the *achaete* (*ac*)-*sc* complex (AS-C) within Diptera. *D. melanogaster*, a member of the suborder Brachycera, bears four genes at the AS-C resulting from three independent duplication events, two of which have occurred within the Diptera (Skaer *et al.*, 2002). In contrast, *Anopheles gambiae*, a mosquito belonging to the suborder Nematocera (Fig. 3.2), has only two (Skaer *et al.*, 2002). Sequence comparison suggests that AgASH (*A. gambiae Ac-Sc-Homolog*) is orthologous to the three proneural genes *ac*, *sc*, and *lethal of scute* of *D. melanogaster* (Skaer *et al.*, 2002; Wülbeck and Simpson, 2002). The AS-C of *D. melanogaster* comprises a number of discrete *cis*-regulatory sequences scattered over nearly 100 kb (Gomez-Skarmeta *et al.*, 1995; Modolell and Campuzano, 1998; Fig. 3.7). Those regulating *ac* and *sc* expression in the wing disc are shared (Gomez-Skarmeta *et al.*, 1995). They respond to local positional cues, conveyed by transcriptional activators such as Pnr (Garcia-Garcia *et al.*, 1999). Pannier is expressed on the medial half of the notum and activates *ac–sc* in small subdomains within its own broad domain of expression, an effect mediated by the enhancer elements (Garcia-Garcia *et al.*, 1999).

Expression of *pnr* is also restricted to the medial notum in A. *gambiae*, but in this species *AgASH* is transcribed in all *Agpnr*-expressing cells on the notum (Wülbeck and Simpson, 2002). This is consistent with a conserved role for AgPnr in the regulation of *AgASH* in A. *gambiae*, but it suggests that AgPnr may activate *AgASH* in a simple straightforward fashion in every cell in which it is expressed. The nature of the regulatory interactions between these two genes appears to have diverged. The observations suggest that the regulatory region of the *AS-C* of A. *gambiae* may have a less complex organization than that of *Drosophila*, with few, if any, discrete *cis*-regulatory elements. If so, this suggests that the *cis*-regulatory sequences mediating spatial expression of *ac-sc* have been acquired after the separation of the Nematocera and the Cyclorrapha ∼250 Myr ago. This hypothesis suggests that the acquisition of a modular promoter coincided with the duplication events that led to an increase in number of copies of the *AS-C* genes. Notably, the increased complexity of the spatial expression of *ac-sc* of those cyclorraphous species studied to date correlates with the emergence of stereotypical bristle patterns not characteristic of the Nematocera.

A suggestion as to how the organization of regulatory modules at the *eve* locus may evolve can be made on the basis of comparisons of the mode of segmentation and the expression of genes regulating segmentation. The gap genes are activated by maternal gene products such as Bcd and products of the gap genes in turn activate pair rule genes like *eve*. The cascade of activity of these genes differs between different species of insects and flies. In the *Drosophila* embryo all of the segments are generated simultaneously before gastrulation, via a mechanism that relies on free diffusion of transcription factors in a syncytium (so-called long germ band embryos). In other insects and arthropods, however, many segments are added sequentially from a posterior growth zone in a cellularized environment (so-called short germ band embryos). It has been suggested that the transition from short to long germ bands might have occurred gradually, and that the striped expression of pair rule genes such as *eve* gradually came under the control of gap genes as the number of segments patterned simultaneously in the anterior increased and the number patterned sequentially in the posterior correspondingly decreased (Peel, 2004). Under this hypothesis the *cis*-regulatory modules for expression in each stripe would have arisen sequentially one by one.

Observations of the expression patterns of gap and pair rule genes in the basal dipteran A. *gambiae* suggest that the number of *cis*-acting elements regulating *eve* expression might differ from that in D. *melanogaster* (Goltzev et al., 2004). Five *cis*-regulatory modules mediate expression of *eve* in seven stripes in D. *melanogaster*. Two regulate stripes 3/7 and 4/6, respectively (Clyde et al., 2003). In these cases the modules are activated by ubiquitous factors and then the stripes are carved out by repression at either ends by the products of the gap genes Hb and Kni. Knirps establishes the posterior border of stripe 3 and the anterior border of stripe 7, while Hb establishes the anterior border of stripe 3 and the posterior border of stripe 7. This is made possible by differential affinities of the regulatory modules to Kni and Hb and by the opposing gradients of Hb and Kni that repress one another. (This process relies on a second, posterior domain of expression of *hb*, dependent on a different promoter to P2.) In *A. gambiae*, *hb* is not expressed in a posterior domain, thus excluding the use of a strategy involving cross repression between Hb and Kni (Goltzev *et al.*, 2004). Instead, different combinations of gap repressors appear to define the borders of each stripe, suggesting that they are each regulated by a separate module. The authors argue that an individual module for each stripe of expression might represent the ancestral situation.

VI. CONCLUSIONS

It is still the case that relatively few regulatory modules have been subjected to detailed investigation. Nevertheless, a few common features are emerging from those that have. All cases examined display significant sequence turnover even when function is conserved. This is in contrast to some vertebrate genes where significant sequence conservation is seen between species in regulatory regions. The differences may relate to the greater population sizes and generation times of flies (Carter and Wagner, 2002). Not all apparent binding sites are equal and some appear not to be used. Turnover in flies is so great that even when the true binding sites have been identified, it is often difficult to trace those that are homologous between species. There is considerable variation in the binding site architecture of regulatory modules but the study of Bcd and its target promoters indicates that the arrangement of sites can be very important in some cases. There is just as much sequence turnover in modules that have evolved to mediate a different transcriptional output. This makes it difficult to make predictions about gene expression based on sequence alone.

   Comparative genomic approaches using phylogenetic footprinting (sequence conservation between species) and/or statistical methods to identify clusters of known transcription factor binding sites will enable detection of regulatory modules between species (see Chapter 10 by Gabriela Loots). However, there is still the need to examine more cases in as much detail as *eve* SR2 for example. Furthermore, in order to determine the possible importance of a nucleotide change it remains necessary to mutate the site and examine the consequences *in vivo* in transgenic animals. This is unfortunately time-consuming. In addition, some of the examples described above, such as evolution of expression of *y* and *sc* between different *Drosophila* species, indicate that some changes are likely to have occurred in *trans* as well as in *cis*. Examination of the activity of heterologous regulatory elements in reporter genes has mainly been restricted to assays in *D. melanogaster*. If the properties of the *trans*-acting factors have diverged between species, then activity of heterologous elements might be expected to differ in *D. melanogaster* when compared to the species from which they are derived. This illustrates the importance of performing reciprocal tests between species to look at *trans*-regulation (Christophides *et al.*, 2000; Romano and Wray, 2003; Wittkopp *et al.*, 2002b).

It has been shown that altered gene expression can be attributed to a difference of a few base pairs in a single binding site, for example, loss of expression of *y* mediated by the "body" element in the *D. kikkawai* (Jeong *et al.*, 2006). There are few examples of this so far and it is not known whether this will be a frequent occurrence. It is equally possible that, in other cases, a difference in gene expression could be attributable to an additive effect of changes to a number of binding sites resulting in overall reduced binding affinity for example. Quantitative effects in gene expression could also in principle be associated with morphological differences. Finally coevolution of regulatory proteins and their target binding sequences might be a factor underlying regulatory incompatibility between species. In-depth investigation of other *cis*-regulatory elements will reveal to what extent evolution of regulatory sequences contributes to morphological diversity.

References

Alberch, P. (1982). Developmental constraints in evolutionary processes. In "Evolution and Development" (J. T. Bonner, ed.), pp. 313–332. Springer-Verlag, Berlin.

Alonso, M. C., and Cabrera, C. V. (1988). The *achaete-scute* gene complex of *Drosophila melanogaster* comprises four homologous genes. *EMBO J.* **7**, 2585–2591.

Arnone, M. I., and Davidson, E. H. (1997). The hardwiring of development: Organization and function of genomic regulatory systems. *Development* **124**, 1851–1864.

Arnosti, D. N. (2003). Analysis and function of transcriptional regulatory elements: Insights from *Drosophila*. *Annu. Rev. Entomol.* **48**, 579–602.

Arnosti, D. N., Barolo, S., Levine, M., and Small, S. (1996). The *eve* stripe 2 enhancer employs multiple modes of transcriptional synergy. *Development* **122**, 205–214.

Bachli, G. (1998). Drosophilidae. In "Manual of Palaearctic Diptera" (L. Papp and B. Darvas, eds.), pp. 503–513. Science Herald, Budapest.

Balhoff, J. P., and Wray, G. A. (2005). Evolutionary analysis of the well characterized *endo16* promoter reveals substantial variation within functional sites. *Proc. Natl. Acad. Sci. USA* **102**, 8591–8596.

Barker, M. S., Demuth, J. P., and Wade, M. J. (2005). Maternal expression relaxes constraint on innovation of the anterior determinant, bicoid. *PLoS Genet.* **1**, 57.

Beldade, P., Koops, K., and Brakefield, P. M. (2002). Developmental constraints versus flexibility in morphological evolution. *Nature* **416**, 844–847.

Belting, H. G., Shashikant, C. S., and Ruddle, F. H. (1998). Modification of expression and cis-regulation of Hoxc8 in the evolution of diverged axial morphology. *Proc. Natl. Acad. Sci. USA* **95**, 2355–2360.

Bertrand, N., Castro, D. S., and Guillemot, F. (2002). Proneural genes and the specification of neural cell types. *Nat. Rev. Neurosci.* **3,** 517–530.

Blair, S. S. (2003). Lineage compartments in *Drosophila. Curr. Biol.* **13,** R548–R551.

Bonneton, F., Shaw, P. J., Fazakerley, C., Shi, M., and Dover, G. A. (1997). Comparison of bicoid-dependent regulation of *hunchback* between *Musca domestica* and *Drosophila melanogaster. Mech. Dev.* **66,** 143–156.

Burz, D. S., and Hanes, S. D. (2001). Isolation of mutations that disrupt cooperative DNA binding by the *Drosophila* bicoid protein. *J. Mol. Biol.* **305,** 219–230.

Burz, D. S., Rivera-Pomar, R., Jackle, H., and Hanes, S. D. (1998). Cooperative DNA-binding by Bicoid provides a mechanism for threshold-dependent gene activation in the *Drosophila* embryo. *EMBO J.* **17,** 5998–6009.

Calleja, M., Herranz, H., Estella, C., Casal, J., Lawrence, P., Simpson, P., and Morata, G. (2000). Generation of medial and lateral dorsal body domains by the *pannier* gene of *Drosophila. Development* **127,** 3971–3980.

Carroll, S. B. (2005). Evolution at two levels: On genes and form. *PLoS Biol.* **3,** e245.

Carroll, S. B., Grenier, J. K., and Weatherbee, S. D. (2001). From DNA to diversity. "Molecular Genetics and the Evolution of Animal Design" Blackwell Science.

Carter, A., and Wagner, G. (2002). Evolution of functionally conserved enhancers can be accelerated in large populations: A population-genetic model. *Proc. R. Soc. Lond. Ser. B.* **269,** 953–960.

Castillo-Davis, C. I., Kondrashov, F. A., Hartl, D. L., and Kulathinal, R. J. (2004). The functional genomic distribution of protein divergence in two animal phyla: Coevolution, genomic conflict, and constraint. *Genome Res.* **14,** 802–811.

Chen, K., and Rajewsky, N. (2007). The evolution of gene regulation by transcription factors and microRNAs. *Nat. Rev. Genet.* **8,** 93–103.

Chen, Y., Chiang, C., Weng, L., Lengyel, J., and Liaw, G. (2002). Tramtrack69 is required for the early repression of *tailless* expression. *Mech. Dev.* **116,** 75–83.

Christophides, G. K., Livadaras, I., Savakis, C., and Komitopoulou, K. (2000). Two medfly promoters that have originated by recent gene duplication drive distinct sex, tissue, and temporal expression patterns. *Genetics* **156,** 173.

Clyde, D. E., Corado, M. S., Wu, X., Pare, A., Papatsenko, D., and Small, S. (2003). A self-organizing system of repressor gradients establishes segmental complexity in *Drosophila. Nature* **426,** 849–853.

Colosimo, P. F., Hosemann, K. E., Balabhadra, S., Villareal, G. J., Dickson, M., Grimwood, J., Schmutz, J., Muyers, R. M., Schluter, D., and Kingsley, D. M. (2005). Widespread parallel evolution in sticklebacks by repeated fixation of Ectodysplasin alleles. *Science* **307,** 1928–1933.

Costas, J., Casares, F., and Vieira, J. (2003). Turnover of binding sites for transcription factors involved in early *Drosophila* development. *Gene* **310,** 215–220.

Crauk, O., and Dostatni, N. (2005). Bicoid determines sharp and precise target gene expression in the Drosophila embryo. *Curr. Biol.* **15,** 1888–1898.

Cubadda, Y., Heitzler, P., Ray, R. P., Bourouis, M., Ramain, P., Gelbart, W., Simpson, P., and Haenlin, M. (1997). *u-shaped* encodes a zinc finger protein that regulates the proneural genes *achaete* and *scute* during the formation of bristles in *Drosophila. Genes Dev.* **11,** 3083–3095.

Cubas, P., de Celis, J. F., Campuzano, S., and Modolell, J. (1991). Proneural clusters of *achaete-scute* expression and the generation of sensory organs in the *Drosophila* imaginal wing disc. *Genes Dev.* **5,** 996–1008.

Culi, J., and Modolell, J. (1998). Proneural gene self-stimulation in neural precursors: An essential mechanism for sense organ development that is regulated by Notch signaling. *Genes Dev.* **12,** 2036–2047.

Davidson, E. H. (2006). "The Regulatory Genome: Gene Regulatory Networks in Development and Evolution." Academic Press, San Diego, CA.

Dermitzakis, E. T., and Clark, A. G. (2002). Evolution of transcription factor binding sites in Mammalian gene regulatory regions: Conservation and turnover. *Mol. Biol. Evol.* **19,** 1114–1121.

Dover, G. (2000). How genomic and developmental dynamics affect evolutionary processes. *Bioessays* **22,** 1153–1159.

Dover, G. A., and Flavell, R. B. (1984). Molecular coevolution: DNA divergence and the maintenance of function. *Cell* **38,** 622–623.

Driever, W., and Nusslein-Volhard, C. (1989). The bicoid protein is a positive regulator of *hunchback* transcription in the early *Drosophila* embryo. *Nature* **337,** 138–143.

Driever, W., Thoma, G., and Nusslein-Volhard, C. (1989). Determination of spatial domains of zygotic gene expression in the *Drosophila* embryo by the affinity of binding sites for the bicoid morphogen. *Nature* **340,** 363–367.

Duboule, D., and Wilkins, A. S. (1998). The evolution of 'bricolage.' *Trends Genet.* **14,** 54–59.

Erives, A., and Levine, M. (2004). Coordinate enhancers share common organizational features in the *Drosophila* genome. *Proc. Natl. Acad. Sci. USA* **101,** 3851–3856.

Ferrier, D. E., and Holland, P. W. (2001). Ancient origin of the Hox gene cluster. *Nat. Rev. Genet.* **2,** 33–38.

Finkelstein, R., and Perrimon, N. (1990). The *orthodenticle* gene is regulated by *bicoid* and *torso* and specifies *Drosophila* head development. *Nature* **346,** 485–488.

Force, A., Lynch, M., Pickett, F. B., Amores, A., Yan, Y. L., and Postlethwait, J. (1999). Preservation of duplicate genes by complementary, degenerate mutations. *Genetics* **151,** 1531–1545.

Forest, D., Nishikawa, R., Kobayashi, H., Parton, A., Bayne, C. J., and Barnes, D. W. (2007). RNA expression in a cartilaginous fish cell line reveals ancient 3' noncoding regions highly conserved in vertebrates. *Proc. Natl. Acad. Sci. USA* **104,** 1224–1229.

Frasch, M., and Levine, M. (1987). Complementary patterns of *even-skipped* and *fushi tarazu* expression involve their differential regulation by a common set of segmentation genes in *Drosophila*. *Genes Dev.* **1,** 981–995.

Fu, D., Zhao, C., and Ma, J. (2003). Enhancer sequences influence the role of the amino-terminal domain of bicoid in transcription. *Mol. Cell. Biol.* **23,** 4439–4448.

Galant, R., and Carroll, S. B. (2002). Evolution of a transcriptional repression domain in an insect Hox protein. *Nature* **415,** 910–913.

Gao, Q., and Finkelstein, R. (1998). Targeting gene expression to the head: The *Drosophila orthodenticle* gene is a direct target of the Bicoid morphogen. *Development* **125,** 4185–4193.

Garcia-Garcia, M. J., Ramain, P., Simpson, P., and Modolell, J. (1999). Different contributions of *pannier* and *wingless* to the patterning of the dorsal mesothorax of *Drosophila*. *Development* **126,** 3523–3532.

Gasch, A. P., Moses, A. M., Chiang, D. Y., Fraser, H. B., Berardini, M., and Eisen, M. B. (2004). Conservation and evolution of cis-regulatory systems in ascomycete fungi. *PLoS Biol.* **2,** e398.

Ghysen, A., and Dambly-Chaudiere, C. (1988). From DNA to form: The *achaete-scute* complex. *Genes Dev.* **2,** 495–501.

Goltzev, Y., Hsiong, W., Lanzaro, G., and Levine, M. (2004). Different combinations of gap repressors for common stripes in *Anopheles* and *Drosophila* embryos. *Dev. Biol.* **275,** 435–446.

Gomez-Skarmeta, J. L., Rodriguez, I., Martinez, C., Culi, J., Ferres-Marco, D., Beamonte, D., and Modolell, J. (1995). Cis-regulation of *achaete* and *scute*: Shared enhancer-like elements drive their coexpression in proneural clusters of the imaginal discs. *Genes Dev.* **9,** 1869–1882.

Gompel, N., Wittkopp, P. J., Prud'homme, B., Kassner, V. A., and Carroll, S. B. (2005). Chance caught on the wing: Cis-regulatory evolution and the origin of pigment patterns in *Drosophila*. *Nature* **433,** 481–487.

Goto, T., Macdonald, P. M., and Maniatis, T. (1989). Early and late periodic patterns of *even-skipped* expression are controlled by distinct regulatory elements that respond to different spatial cues. *Cell* **57,** 413–422.

Grimaldi, D. (1987). Amber fossil Drosophilidae (Diptera), with particular reference to the Hispaniola taxa. *Am. Mus. Novit.* **2880,** 1–23.

Guss, K. A., Nelson, C. E., Hudson, A., Kraus, M. E., and Carroll, S. B. (2001). Control of a genetic regulatory network by a selector gene. *Science* **292,** 1164–1167.

Haenlin, M., Cubadda, Y., Blondeau, F., Heitzler, P., Lutz, Y., Simpson, P., and Ramain, P. (1997). Transcriptional activity of *pannier* is regulated negatively by heterodimerization of the GATA DNA-binding domain with a cofactor encoded by the *u-shaped* gene of *Drosophila. Genes Dev.* **11,** 3096–3108.

Hancock, J. M., Shaw, P. J., Bonneton, F., and Dover, G. A. (1999). High sequence turnover in the regulatory regions of the developmental gene *hunchback* in insects. *Mol. Biol. Evol.* **16,** 253–265.

Hanes, S. D., Riddihough, G., Ish-Horowicz, D., and Brent, R. (1994). Specific DNA recognition and intersite spacing are critical for action of the bicoid morphogen. *Mol. Cell. Biol.* **14,** 3364–3375.

Harding, K., Hoey, T., Warrior, R., and Levine, M. (1989). Autoregulatory and gap response elements of the *even-skipped* promoter of *Drosophila. EMBO J.* **8,** 1205–1212.

Hoekstra, H. E., Hirschmann, R. J., Bundey, R. A., Insel, P. A., and Crossland, J. P. (2006). A single amino acid mutation contributes to adaptive beach mouse color pattern. *Science* **313,** 101–104.

Holland, P. (1992). Homeobox genes in vertebrate evolution. *BioEssays* **14,** 267–273.

Holland, P. (1999). Gene duplication: Past, present, and future. *Semin. Cell Dev. Biol.* **10,** 541–547.

Huang, F., van Helden, J., Dambly-Chaudiere, C., and Ghysen, A. (1995). Contribution of the gene *extramacrochaetae* to the precise positioning of bristles in *Drosophila. Roux's. Arch. Dev. Biol.* **204,** 336–343.

Hughes, C., and Kaufman, T. (2002). Hox genes and the evolution of the arthropod body plan. *Evol. Dev.* **4,** 459–499.

Jacob, F. (1977). Evolution and tinkering. *Science* **196,** 1161–1166.

Jeong, S., Rokas, A., and Carroll, S. B. (2006). Regulation of body pigmentation by the abdominal-B hox protein and its gain and loss in *Drosophila* evolution. *Cell* **125,** 1387–1399.

Kimura, M. (1981). Possibility of extensive neutral evolution under stabilizing selection with special reference to non-random usage of synonymous substitutions. *Proc. Natl. Acad. Sci. USA* **78,** 5773–5777.

Kimura, M. (1983). "The Neutral Theory of Molecular Evolution." Cambridge University Press, Cambridge.

Kimura, M. (1985). The role of compensatory neutral mutations in molecular evolution. *J. Genet.* **64,** 7–19.

Kopp, A., Duncan, I., Godt, D., and Carroll, S. B. (2000). Genetic control and evolution of sexually dimorphic characters in *Drosophila. Nature* **408,** 553–559.

Kreitman, M., and Ludwig, M. (1996). Tempo and mode of *even-skipped* stripe 2 enhancer evolution in *Drosophila. Sem. Cell Dev. Biol.* **7,** 583–592.

Landry, C. R., Wittkopp, P. J., Taubes, C. H., Ranz, J. M., Clark, A. G., and Hartl, D. L. (2005). Compensatory cis-trans evolution and the dysregulation of gene expression in interspecific hybrids of *Drosophila. Genetics* **171,** 1813–1822.

Lawrence, P. A., and Struhl, G. (1996). Morphogens, compartments, and pattern: Lessons from *Drosophila. Cell* **85,** 951–961.

Lebrecht, D., Foehr, M., Smith, E., Lopes, F. J., Vanario-Alonso, C. E., Reinitz, J., Burz, D. S., and Hanes, S. D. (2005). Bicoid cooperative DNA binding is critical for embryonic patterning in *Drosophila. Proc. Natl. Acad. Sci. USA* **102,** 13176–13181.

Lehmann, R., and Nusslein-Volhard, C. (1987). *hunchback,* a gene required for segmentation of an anterior and posterior region of the *Drosophila* embryo. *Dev. Biol.* **119,** 402–417.

Levine, M., and Davidson, E. H. (2005). Gene regulatory networks for development. *Proc. Natl. Acad. Sci. USA* **102,** 4936–4942.

Liaw, G., and Lengyel, J. (1992). Control of *tailless* expression by *bicoid, dorsal* synergistically interacting terminal system regulatory elements. *Mech. Dev.* **40,** 47–61.

Liaw, G., Rudolf, K. M., Huang, J. D., Dubnicoff, T., Courey, A. J., and Lengyel, J. (1995). The *torso* response element binds GAGA and NTF-1/Elf-1, and regulates *tailless* by relief of repression. *Genes Dev.* **9,** 3163–3176.

Ludwig, M. (2002). Functional evolution of noncoding DNA. *Curr. Opin. Genet. Dev.* **12,** 634–639.

Ludwig, M. Z., and Kreitman, M. (1995). Evolutionary dynamics of the enhancer region of *even-skipped* in *Drosophila*. *Mol. Biol. Evol.* **12,** 1002–1011.

Ludwig, M. Z., Patel, N. H., and Kreitman, M. (1998). Functional analysis of *eve* stripe 2 enhancer evolution in *Drosophila*: Rules governing conservation and change. *Development* **125,** 949–958.

Ludwig, M. Z., Bergman, C., Patel, N. H., and Kreitman, M. (2000). Evidence for stabilizing selection in a eukaryotic enhancer element. *Nature* **403,** 564–567.

Ludwig, M. Z., Palsson, A., Alekseeva, E., Bergman, C. M., Nathan, J., and Kreitman, M. (2005). Functional evolution of a cis-regulatory module. *PLoS Biol.* **3,** e93.

Lukowitz, W., Schroder, C., Glaser, G., Hulskamp, M., and Tautz, D. (1994). Regulatory and coding regions of the segmentation gene hunchback are functionally conserved between *Drosophila virilis* and *Drosophila melanogaster*. *Mech. Dev.* **45,** 105–115.

Lynch, M., and Force, A. (2000). The probability if duplicate gene preservation by subfunctionalization. *Genetics* **154,** 459–473.

Ma, X., Yuan, D., Diepold, K., Scarborough, T., and Ma, J. (1996). The *Drosophila* morphogenetic protein Bicoid binds DNA cooperatively. *Development* **122,** 1195–1206.

Ma, X., Yuan, D., Scarborough, T., and Ma, J. (1999). Contributions to gene activation by multiple functions of Bicoid. *Biochem. J.* **338**(Pt. 2), 447–455.

MacArthur, S., and Brookfield, J. F. (2004). Expected rates and modes of evolution of enhancer sequences. *Mol. Biol. Evol.* **21,** 1064–1073.

Mann, R. S., and Morata, G. (2000). The developmental and molecular biology of genes that subdivide the body of *Drosophila*. *Annu. Rev. Cell Dev. Biol.* **16,** 243–271.

Marcellini, S., and Simpson, P. (2006). Two or four bristles: Functional evolution of an enhancer of scute in drosophilidae. *PLoS Biol.* **4,** e386.

Markstein, M., Markstein, P., Markstein, V., and Levine, M. S. (2002). Genome-wide analysis of clustered Dorsal binding sites identifies putative target genes in the *Drosophila* embryo. *Proc. Natl. Acad. Sci. USA* **99,** 763–768.

Maynard-Smith, J., Burian, R., Kauffman, S., Alberch, P., Campbell, J., Goodwin, B., Lande, R., Raup, D., and Wolpert, L. (1985). Developmental constraints and evolution. *Q. Rev. Biol.* **60,** 265–287.

McAlpine, J. F. (1981). *In* "Manual of Nearctic Diptera" (J. F. McAlpine, ed.). Research Branch Agriculture, Canada.

McGregor, A. P. (2005). How to get ahead: The origin, evolution and function of bicoid. *Bioessays* **27,** 904–913.

McGregor, A. P., Shaw, P. J., Hancock, J. M., Bopp, D., Hediger, M., Wratten, N. S., and Dover, G. A. (2001). Rapid restructuring of bicoid-dependent *hunchback* promoters within and between Dipteran species: Implications for molecular coevolution. *Evol. Dev.* **3,** 397–407.

Modolell, J., and Campuzano, S. (1998). The *achaete-scute* complex as an integrating device. *Int. J. Dev. Biol.* **42,** 275–282.

Negre, B., and Ruiz, A. (2007). HOM-C evolution in *Drosophila*: Is there a need for Hox gene clustering? *Trends. Genet.* **23,** 55–59.

Ochoa-Espinosa, A., Yucel, G., Kaplan, L., Pare, A., Pura, N., Oberstein, A., Papatsenko, D., and Small, S. (2005). The role of binding site cluster strength in Bicoid-dependent patterning in *Drosophila*. *Proc. Natl. Acad. Sci. USA* **102,** 4960–4965.

Ohno, S. (1970). "Evolution by Gene Duplication." Springer Verlag, Heidelberg.

Ohta, T., and Tachida, H. (1990). Theoretical study of near neutrality. I. Heterozygosity and rate of mutant substitution. *Genetics* **126**, 219–229.

Olivieri, P., and Davidson, E. H. (2007). Built to run, not fail. *Science* **315**, 1510–1511.

Peel, A. (2004). The evolution of arthropod segmentation mechanisms. *BioEssays* **26**, 1108–1116.

Pennacchio, L. A., Ahituv, N., Moses, A. M., Prabhakar, S., Nobrega, M. A., Shoukry, M., Minovitsky, S., Dubchak, I., Holt, A., Lewis, K. D., Plajzer-Frick, I., Akiyama, J., et al. (2006). *In vivo* enhancer analysis of human conserved non-coding sequences. *Nature* **444**, 499–502.

Pistillo, D., Skaer, N., and Simpson, P. (2002). *scute* expression in *Calliphora vicina* reveals an ancestral pattern of longitudinal stripes on the thorax of higher Diptera. *Development* **129**, 563–572.

Plessy, C., Dickmeis, T., Chalmel, F., and Strahle, U. (2005). Enhancer sequence conservation between vertebrates is favoured in developmental regulator genes. *Trends. Genet.* **21**, 207–210.

Prud'homme, B., Gompel, N., Antonis, R., Kassner, V. A., Thomas, M. W., Shu-Dan, Y., John, R. T., and Carroll, S. B. (2006). Repeated morphological evolution through cis-regulatory changes in a pleiotropic gene. *Nature* **440**, 1050–1053.

Ramain, P., Heitzler, P., Haenlin, M., and Simpson, P. (1993). *pannier*, a negative regulator of *achaete* and *scute* in *Drosophila*, encodes a zinc finger protein with homology to the vertebrate transcription factor GATA-1. *Development* **119**, 1277–1291.

Ramain, P., Khechumian, R., Khechumian, K., Arbogast, N., Ackermann, C., and Heitzler, P. (2000). Interactions between chip and the *achaete/scute-daughterless* heterodimers are required for pannier-driven proneural patterning. *Mol. Cell.* **6**, 781–790.

Richards, S., Liu, Y., Bettencourt, B. R., Hradecky, P., Letovsky, S., and Nielson, R. (2005). Comparative genome sequencing of *Drosophila pseudoobscura*: Chromosomal, gene and cis-element evolution. *Genome Res.* **15**, 1–18.

Richardson, J., and Simpson, P. (2005). A conserved trans-regulatory landscape for *scute* expression on the notum of cyclorraphous Diptera. *Dev. Genes Evol.* **216**, 29–38.

Rodriguez, I., Hernandez, R., Modolell, J., and Ruiz-Gomez, M. (1990). Competence to develop sensory organs is temporally and spatially regulated in *Drosophila* epidermal primordia. *EMBO J.* **9**, 3583–3592.

Romano, L. A., and Wray, G. A. (2003). Conservation of *Endo16* expression in sea urchins despite evolutionary divergence in both *cis* and *trans*-acting components of transcriptional regulation. *Development* **130**, 4187–4199.

Ruiz-Gomez, M., and Modolell, J. (1987). Deletion analysis of the *achaete-scute* locus of *Drosophila melanogaster*. *Genes Dev.* **1**, 1238–1246.

Schröder, R., and Sander, R. (1993). A comparison of transplantable Bicoid activity and partial Bicoid homeobox sequences in several *Drosophila* and blowfly species. *Roux's Arch. Dev. Biol.* **203**, 34–43.

Senger, K., Armstrong, G. W., Rowell, W. J., Kwan, J. M., Markstein, M., and Levine, M. (2004). Immunity regulatory DNAs share common organizational features in *Drosophila*. *Mol. Cell.* **13**, 19–32.

Shaw, P. J., Salameh, A., McGregor, A. P., Bala, S., and Dover, G. A. (2001). Divergent structure and function of the *bicoid* gene in Muscoidea fly species. *Evol. Dev.* **3**, 251–262.

Shaw, P. J., Wratten, N. S., McGregor, A. P., and Dover, G. A. (2002). Coevolution in bicoid-dependent promoters and the inception of regulatory incompatibilities among species of higher Diptera. *Evol. Dev.* **4**, 265–277.

Simpson, P., Woehl, R., and Usui, K. (1999). The development and evolution of bristle patterns in Diptera. *Development* **126**, 1349–1364.

Simpson-Brose, M., Treisman, J., and Desplan, C. (1994). Synergy between the Hunchback and Bicoid morphogens is required for anterior patterning in *Drosophila*. *Cell* **78**, 855–865.

Skaer, N., Pistillo, D., Gibert, J.-M., Lio, P., Wulbeck, C., and Simpson, P. (2002). Gene duplication at the *achaete-scute* complex and morphological complexity of the peripheral nervous system in Diptera. *Trends Genet.* **18,** 399–405.

Skeath, J. B., and Carroll, S. B. (1991). Regulation of *achaete-scute* gene expression and sensory organ pattern formation in the *Drosophila* wing. *Genes Dev.* **5,** 984–995.

Small, S., Blair, A., and Levine, M. (1992). Regulation of *even-skipped* stripe 2 in the *Drosophila* embryo. *EMBO J.* **11,** 4047–4057.

Sommer, R., and Tautz, D. (1991). Segmentation gene expression in the housefly Musca *domestica*. *Development* **113,** 419–430.

Spradling, A. C., and Rubin, G. M. (1982). Transposition of cloned P elements into *Drosophila* germ line chromosomes. *Science* **218,** 341–347.

Stanojevic, D., Small, S., and Levine, M. (1991). Regulation of a segmentation stripe by overlapping activators and repressors in the *Drosophila* embryo. *Science* **254,** 1385–1387.

Stauber, M., Jackle, H., and Schmidt-Ott, U. (1999). The anterior determinant bicoid of *Drosophila* is a derived Hox class 3 gene. *Proc. Natl. Acad. Sci. USA* **96,** 3786–3789.

Stauber, M., Taubert, H., and Schmidt-Ott, U. (2000). Function of *bicoid* and *hunchback* homologs in the basal cyclorrhaphan fly *Megaselia* (Phoridae). *Proc. Natl. Acad. Sci. USA* **97,** 10844–10849.

Stauber, M., Prell, A., and Schmidt-Ott, U. (2002). A single *Hox3* gene with composite *bicoid* and *zerknullt* expression characteristics in non-Cyclorrhaphan flies. *Proc. Natl. Acad. Sci. USA* **99,** 274–279.

Stern, D. L. (2000). Evolutionary developmental biology and the problem of variation. *Evol. Int. J. Org. Evol.* **54,** 1079–1091.

Stone, J. R., and Wray, G. A. (2001). Rapid evolution of cis-regulatory sequences via local point mutations. *Mol. Biol. Evol.* **18,** 1764–1770.

Struhl, G., Struhl, K., and Macdonald, P. M. (1989). The gradient morphogen bicoid is a concentration-dependent transcriptional activator. *Cell* **57,** 1259–1273.

Tautz, D., Lehmann, R., Schnurch, H., Schuh, R., Seifert, E., Kienlin, A., Jones, K., and Jackle, H. (1987). Finger protein of novel structure encoded by *hunchback*, a second member of the gap class of *Drosophila* segmentation genes. *Nature* **327,** 383–389.

True, J. R., Edwards, K. A., Yamamoto, D., and Carroll, S. B. (1999). *Drosophila* wing melanin patterns form by vein-dependent elaboration of enzymatic prepatterns. *Curr. Biol.* **9,** 1382–1391.

Usui, K., Pistillo, D., and Simpson, P. (2004). Mutual exclusion of sensory bristles and tendons on the notum of dipteran flies. *Curr. Biol.* **14,** 1047–1055.

Villares, R., and Cabrera, C. V. (1987). The *achaete-scute* gene complex of *D. melanogaster*: Conserved domains in a subset of genes required for neurogenesis and their homology to *myc. Cell* **50,** 415–424.

Wang, X., and Chamberlin, H. M. (2004). Evolutionary innovation of the excretory system in *Caenorhabditis elegans*. *Nat. Genet.* **36,** 231–232.

Wheeler, M. (1981). Drosophilidae. *In* "Manual of Nearctic Diptera" (J. McAlpine, ed.), pp. 1011–1018. Research Branch Agriculture Canada, Ottawa.

Wittkopp, P. J. (2006). Evolution of cis-regulatory sequence and function in Diptera. *Heredity* **97,** 139–147.

Wittkopp, P. J., True, J. R., and Carroll, S. B. (2002a). Reciprocal functions of the *Drosophila yellow* and *ebony* proteins in the development and evolution of pigment patterns. *Development* **129,** 1849–1858.

Wittkopp, P. J., Vaccaro, K., and Carroll, S. B. (2002b). Evolution of *yellow* gene regulation and pigmentation in *Drosophila*. *Curr. Biol.* **12,** 1547–1556.

Wittkopp, P. J., Carroll, S. B., and Kopp, A. (2003). Evolution in black and white: Genetic control of pigment patterns in *Drosophila*. *Trends. Genet.* **19,** 495–504.

Wittkopp, P. J., Haerum, B. K., and Clark, A. G. (2004). Evolutionary changes in cis and trans gene regulation. *Nature* **430,** 85–88.

Wratten, N. S., McGregor, A. P., Shaw, P. J., and Dover, G. A. (2006). Evolutionary and functional analysis of the *tailless* enhancer in *Musca domestica* and *Drosophila melanogaster. Evol. Dev.* **8,** 6–15.

Wray, G. A., Hahn, M. W., Abouheif, E., Balhoff, J. P., Pizer, M., Rockman, M. V., and Romano, L. A. (2003). The evolution of transcriptional regulation in eukaryotes. *Mol. Biol. Evol.* **20,** 1377–1419.

Wülbeck, C., and Simpson, P. (2000). Expression of *achaete-scute* homologues in discrete proneural clusters on the developing notum of the medfly *Ceratitis capitata*, suggests a common origin for the stereotyped bristle patterns of higher Diptera. *Development* **127,** 1411–1420.

Wülbeck, C., and Simpson, P. (2002). The expression of *pannier* and *achaete-scute* homologues in a mosquito suggests an ancient role of *pannier* as a selector gene in the regulation of the dorsal body pattern. *Development* **129,** 3861–3871.

Yuan, D., Ma, X., and Ma, J. (1996). Sequences outside the homeodomain of bicoid are required for protein-protein interaction. *J. Biol. Chem.* **271,** 21660–21665.

Yuan, D., Ma, X., and Ma, J. (1999). Recognition of multiple patterns of DNA sites by *Drosophila* homeodomain protein Bicoid. *J. Biochem. (Tokyo).* **125,** 809–817.

Zdobnov, E. M., von Mering, C., Letunic, I., Torrents, D., Suyama, M., Copley, R. R., Christophides, G. K., Thomasova, D., Holt, R. A., Subramanian, G. M., Mueller, H., *et al.* (2002). Comparative genome and proteome analysis of *Anopheles gambiae* and *Drosophila melanogaster. Science* **298,** 149–159.

Zhao, C., Dave, V., Yang, F., Scarborough, T., and Ma, J. (2000). Target selectivity of bicoid is dependent on nonconsensus site recognition and protein-protein interaction. *Mol. Cell. Biol.* **20,** 8112–8123.

Zhao, C., York, A., Yang, F., Forsthoefel, D. J., Dave, V., Fu, D., Zhang, D., Corado, M. S., Small, S., Seeger, M. A., and Ma, J. (2002). The activity of the *Drosophila* morphogenetic protein Bicoid is inhibited by a domain located outside its homeodomain. *Development* **129,** 1669–1680.

Zhao, C., Fu, D., Dave, V., and Ma, J. (2003). A composite motif of the *Drosophila* morphogenetic protein bicoid critical to transcription control. *J. Biol. Chem.* **278,** 43901–43909.

4

β-Globin Regulation and Long-Range Interactions

Robert-Jan Palstra, Wouter de Laat, and Frank Grosveld
Department of Cell Biology and Genetics, Erasmus MC, 3000 CA Rotterdam, The Netherlands

I. Introduction
II. The β-Globin Locus
    A. The β-globin LCR
    B. The chromatin structure of the β-globin locus
    C. Developmental regulation of β-globin expression
III. Models of Long-Range Control of Gene Expression by Enhancers
    A. The looping model
    B. The tracking model
    C. The linking model
    D. Relocation models
IV. Long-Range Activation by the β-Globin LCR
    A. The LCR is in close proximity to the promoter
    B. How are LCR-promoter contacts established?
    C. How does the β-globin LCR increase transcription efficiency?
V. Enhancement of Transcription by the β-Globin LCR: Rate-Limiting Steps
    A. Promoter remodeling
    B. Transcription initiation
    C. Promoter escape and elongation
VI. The Concept of an Active Chromatin Hub
VII. Future Directions
    Acknowledgments
    References

Advances in Genetics, Vol. 61
Copyright 2008, Elsevier Inc. All rights reserved.

0065-2660/08 $35.00
DOI: 10.1016/S0065-2660(07)00004-1

ABSTRACT

Transcriptional activation in higher eukaryotes frequently involves the long-range action of a number of regulatory DNA elements. One of the main questions in transcriptional regulation is how *cis*-regulatory elements communicate with the promoter of a gene over large distances. There has been a lively debate in recent years whether this communication takes place via a noncontact mechanism (linking, tracking) or via a contact mechanism (looping). The demonstration that the major regulatory element of the β-globin locus, the locus control region (LCR), is in close proximity to the active β-globin genes validates the contact model for long-range activation. Here, we will review the β-globin locus as a model system to study long-range activation, briefly describe the different models for long-range activation, and summarize the recent findings that the LCR of the β-globin locus is in close proximity to the active promoters.

   Although it is now firmly established that looping takes place within the β-globin locus (and other loci), it is not clear how these long-range contacts are established and what the precise role is of the LCR. We will argue that the main action of the LCR takes place at the promoter and open reading frame of the gene itself and we will discuss key rate-limiting steps in transcriptional activation and the possible mechanisms by which they are influenced by the LCR.   © 2008, Elsevier Inc.

I. INTRODUCTION

Stringent regulation of gene expression is crucial for the proper development and survival of eukaryotic organisms. The eukaryotic genome has evolved many DNA elements that ensure the proper temporal and spatial expression of genes. Vast distances, reaching up to megabases, can separate metazoan *cis*-regulatory elements. How these *cis*-regulatory elements communicate over these vast distances has intrigued researchers since their discovery and whether this communication takes place via a noncontact mechanism (linking, tracking) or via a contact mechanism (looping) has been the subject of intense debate.

   The β-globin locus has played a key role as a useful model system. For example, the β-globin genes were among the first to be studied by biochemical and molecular biological methods because of the easy availability of globin protein and globin mRNA from erythroid cells. More recently, the β-globin locus became a paradigm in the study of tissue-specific and developmentally regulated transcription. In the context of this chapter, the β-globin locus has been at the basis of elucidating the mechanisms of long-range transcriptional activation.

First, the large body of data will be summarized that relate to the many aspects of β-globin gene regulation. The β-globin locus will then be introduced as a model system for long-range activation. The most common models for long-range activation will be described briefly, and the recent studies that show that the locus control region (LCR) of the β-globin locus is in close proximity to the active promoters will be summarized. Recently, it has been shown that the β-globin locus also contacts other chromosomal regions; however, these findings are beyond the scope of this chapter. Finally, we will speculate on the mechanisms for establishing long-range interactions and the mode of action of the regulatory elements of the β-globin locus once they are in close proximity to the β-globin gene.

II. THE β-GLOBIN LOCUS

The mammalian β-globin loci, especially the mouse and human β-globin loci, are highly conserved between species. The mouse β-globin locus is located on chromosome 7 and is embedded in an array of olfactory receptor (OR) genes (Bulger et al., 1999). The locus contains four functional genes arranged in the order of their temporal expression. At the 5′ side of the mouse β-globin locus, the embryonic/fetal εy and βh1 genes are found, followed more downstream by the adult β-major and β-minor genes (Fig. 4.1). Upstream of the locus lies a cluster of erythroid-specific cis-regulatory elements specified by DNase I hypersensitive sites (HSs), the so-called LCR (Fig. 4.1). Additional erythroid-specific HS are found in the mouse β-globin locus. 3′ HS1 is found 68 kb downstream of the εy cap site (Tuan et al., 1985), and two sets of recently discovered HSs are located 85/84 kb (5′ HS-84/-85) and 62/60 kb (5′ HS-60/-62) upstream of the εy cap site (Bulger et al., 2003; Farrell et al., 2000). Therefore, the locus spans to our current knowledge over 150 kb.

The human β-globin locus is located on chromosome 11 and has a similar genomic organization as the mouse β-globin locus (Fig. 4.1). The human locus contains five functional genes arranged in the order 5′ ε-^{G}γ -^{A}γ-δ-β 3′. Like the mouse, it contains a 3′ HS1 and a set of HS constituting the LCR (Grosveld et al., 1987; Tuan et al., 1985). The cis-regulatory DNA elements and genes are highly conserved between both species and both loci have a GC-content of ∼40% and an almost identical percentage of repetitive DNA (Bulger et al., 1999; Hardison et al., 1997; Moon and Ley, 1990). In humans, a homologue of the mouse 5′ HS-62 was found at 110 kb and an additional site at 107 kb upstream of ε (5′ HS-107/-110), but the cis-regulatory element corresponding to the mouse 5′ HS-84/-85 is absent (Bulger et al., 2003).

The β-globin genes are relatively small genes spanning ∼1500 bp and encode proteins of roughly 146 amino acids. The β-like globin genes have three coding regions (exons) separated by two intervening sequences (introns).

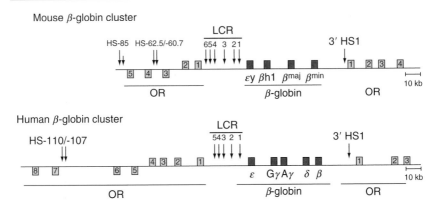

Figure 4.1. The mouse and human β-globin loci: The mouse (top) and human (bottom) β-globin locus are shown schematically. Erythroid-specific β-like globin genes are represented as black boxes while *cis*-regulatory DNA elements corresponding with erythroid cell-specific DNaseI hypersensitive sites are indicated by arrows. Both the mouse and human β-globin locus contain multiple β-like globin genes arranged along the linear chromatin template in the order of their temporal expression (see text for details). Upstream of the genes a cluster of erythroid-specific hypersensitive sites, collectively known as the locus control region, is found. More distal hypersensitive sites are located up and downstream of the mouse and human β-globin locus and both loci are embedded in a cluster of olfactory receptor genes (OR; grey boxes), which are silent in erythroid cells.

Regulatory elements can be detected as DNase I HSs in the chromatin of erythroid cells (Groudine *et al.*, 1983) and include promoters, enhancers, and silencers, which flank the genes in *cis* and are sufficient and necessary for tissue-specific expression and developmental timing of the individual genes (Behringer *et al.*, 1987; Bodine and Ley, 1987; Cao *et al.*, 1989a,b; Chada *et al.*, 1986; Dillon and Grosveld, 1991; Kollias *et al.*, 1987; Lacy *et al.*, 1983; Magram *et al.*, 1985; Ronchi *et al.*, 1996).

The globin promoters are located within 200 bp of the respective transcriptional start sites and contain in addition to a TATA box different recognition sites for erythroid-specific transcription factors like GATA-1, Nuclear Factor-erythroid 2(NF-E2), and erythroid Krüppel-like factor (EKLF), which provide part of the tissue-specificity to the promoter.

A. The β-globin LCR

While early transgenic experiments in mice showed that the proximal regulatory elements of the adult human β-globin gene were sufficient for tissue-specificity and developmental timing of expression, it appeared that the expression levels of the transgene were very low and severely influenced by the site of integration (Behringer *et al.*, 1987; Kollias *et al.*, 1986, 1987).

A Dutch γδβ thalassemia patient provided the first indication of the existence of a distally located cis-regulatory DNA element. This patient carried a large 100-kb deletion on one chromosome resulting in the absence of the ε- and γ-genes, while the β-globin gene together with its proximal cis-regulatory DNA elements was unaffected. The affected chromosome failed to express any β-like globin genes and importantly DNase I sensitivity and methylation experiments showed that the affected β-globin gene was present in an inactive configuration in vivo. Since the β-like globin genes from the unaffected chromosome are expressed normally, it was concluded that a deletion far from the β-globin gene caused the suppression of its activity rather than a defect in trans-acting environment (Kioussis et al., 1983).

Further analysis of the deleted region showed that several erythroid-specific DNase I HSs were located upstream of the globin genes (Forrester et al., 1986; Tuan et al., 1985). Linkage of these HSs to a β-globin gene resulted in tissue-specific, position-independent, and copy number-dependent expression in transgenic mice (Grosveld et al., 1987).

The human β-globin LCR was the first LCR to be identified and was functionally defined as a DNA element that provides high levels of tissue-specific expression to a cis-linked gene in a copy number-dependent manner, which is independent of the integration site in the host genome (Grosveld et al., 1987). The mouse β-globin LCR was subsequently identified on the basis of sequence homology between the human and mouse loci (Moon and Ley, 1990). Although most data regarding LCRs have come from the studies on the human and mouse β-globin LCR, several other LCR-containing loci have been identified (reviewed in Li et al., 2002).

The human β-globin LCR consists of five erythroid-specific, developmentally stable HSs each encompassing about 200–300 bp and containing binding sites for several transcription factors (Hardison et al., 1997). These transcription factors include GATA-1, NF-E2, EKLF, and Sp1 and binding to their recognition sites is required for hypersensitivity (Goodwin et al., 2001). Formation of the HSs appears to precede β-globin transcription (Blom van Assendelft et al., 1989), but the LCR needs to be linked to an active promoter to stay hypersensitive (Guy et al., 1996; Reitman et al., 1993; Tewari et al., 1996).

Deletion of any single hypersensitive site from the human β-globin LCR in a transgene abolishes position-independent expression, suggesting that all the HSs act together as a single entity or holocomplex (Ellis et al., 1993; Milot et al., 1996). Analysis of β-globin premRNA transcripts in single cells supported the notion of a holocomplex by demonstrating that the β-globin genes are alternately transcribed, that is, only one gene is transcribed at any given moment (Gribnau et al., 1998; Wijgerde et al., 1995). A similar alternating transcription was observed for the mouse β-globin genes (Trimborn et al., 1999). Experiments in transgenic mice and cell lines, which tested the individual contributions of single hypersensitive sites, showed that their functional properties are not equivalent though (reviewed in Harju et al., 2002).

In mice, the different properties of the individual HSs are less well defined and they seem to be redundant (Bender et al., 2001; Fiering et al., 1995; Hug et al., 1996). The discrepancies found between human and mouse may reflect differences in functional assays used (transgenic mice vs targeted deletions), the differences in number, structure, and affinities of the individual factor binding sites, or the existence of redundant elements in the endogenous mouse locus, capable of chromatin opening, which are missing from the human β-globin transgenic constructs or the complete endogenous human locus.

B. The chromatin structure of the β-globin locus

A domain of erythroid-specific DNase I sensitivity extends from ~10 kb upstream of 5′ HS-60/-62 to a few kilobase downstream of 3′ HS1 in the endogenous mouse locus. The borders of this DNase I sensitivity domain are not marked by any regulatory sequences (Bulger et al., 2003). Acetylation of histone H3 and histone H4 and methylation at lysine 4 of histone H3 (H3 meK4) are found in a domain overlapping with, but not identical to, the domain of DNase I sensitivity (Bulger et al., 2003; Forsberg et al., 2000). At *cis*-regulatory DNA elements within the active β-globin locus peaks of erythroid cell-specific histone H3 and H4 hyperacetylation and methylation of lysine 4 of histone H3 (H3meK4) are found (Bulger et al., 2003; Forsberg et al., 2000; Kiekhaefer et al., 2002). Additionally histone H3 at the active β-major promoter, but not at inactive promoters or HS2 of the LCR, is methylated at lysine 79 (H3meK79). In contrast in EryP cells, which express the embryonic genes, H3meK79 is found in a domain spanning at least from 3′ of the εy gene to the (active) $\beta h1$ promoter (Im et al., 2003). These chromatin modification patterns are governed by hematopoietic-specific activators like GATA-1 and NF-E2 (Im et al., 2003; Kiekhaefer et al., 2002).

Chromatin analyses of mouse erythroid precursor cells, which express the globin genes at basal levels (Delassus et al., 1999; Hu et al., 1997), showed that 5′ HS3 and the β-major promoter are already accessible to DNase I concurrent with low to moderate acetylation levels at the β-major promoter. Furthermore, in fully differentiated cells, which express the genes at high levels, acetylation and accessibility is further increased (Bottardi et al., 2003).

The human locus, like the mouse β-globin locus, is also more sensitive to DNase I than "bulk" DNA in erythroid cells (Weintraub and Groudine, 1976). In a Hispanic patient with thalassemia, a deletion was found that extended from 9.5 to 39 kb upstream of the ε-globin gene, removing 5′ HS2–5 of the LCR but leaving the distally located HSs (i.e., 5′ HS-107/-110 and 3′ HS1) and all the β-globin genes intact. This deletion results in the inactivation of the intact genes, making the locus DNase I-resistant and the remaining HSs are not formed (Driscoll *et al.*, 1989; Forrester *et al.*, 1990). Acetylation levels of the human locus were determined in mouse erythroid leukemia (MEL) cell hybrids carrying human chromosome 11 with a wild-type locus containing recombinase sites around 5′ HS2–5 of the LCR. In this experimental setup, the human β-globin locus showed similar DNase I sensitivity and acetylation levels throughout the locus before as well as after recombinase-mediated deletion of 5′ HS2–5. Conversely, the locus of the thalassemic Hispanic deletion patient was DNase I insensitive and hypoacetylated (Reik *et al.*, 1998; Schubeler *et al.*, 2000). These findings and experiments with the human β-globin locus in transgenic mice (Bulger and Groudine, 1999; Fraser and Grosveld, 1998) suggested that the LCR is required for chromatin opening but not for maintenance in the human locus. However, these conclusions are in apparent contradiction with the data on the endogenous mouse locus, where the LCR is dispensable for the establishment of an open chromatin domain (Grosveld, 1999; Higgs, 1998).

Recent data on erythroid precursor cells of humans and mice might give an explanation for the observed functional differences between the mouse and human LCR. The authors show that the β-globin locus is subject to epigenetic potentiation in hematopoietic progenitor cells. The epigenetic control mechanisms differ between the human and mouse β-globin locus, and this difference is maintained when the human locus is introduced in transgenic mice, suggesting that the primary DNA sequence rather than the organism determines the difference in this epigenetic code (Bottardi *et al.*, 2003). Furthermore, lineage-specific transcriptional activators and RNApolII bind to regulatory elements in the β-globin locus prior to activation and participate in this potentiation (Bottardi *et al.*, 2006; Levings *et al.*, 2006).

C. Developmental regulation of β-globin expression

In all species that contain β-globin genes, a switch in globin gene expression coincides with changes in morphology of the erythroid cell, the site of erythropoiesis, and hemoglobin composition (for review see, Stamatoyannopoulos and Grosveld, 2001). The order of the mouse and human globin genes along the linear chromatin fiber reflects their expression pattern during ontogeny (Stamatoyannopoulos and Grosveld, 2001). The expression of the embryonic genes occurs in primitive cells derived from the embryonic yolk sac. In contrast, the fetal/adult genes are expressed in definitive cells, which originate from stem cells derived from the Aorta-Gonad-Mesonephros (AGM) region of the developing embryo (Ling and Dzierzak, 2002; Muller *et al.*, 1994). A single switch from embryonic (εy/βh1) gene expression to adult (β-major/β-minor) gene expression is observed in mice while in humans, two switches are observed; one switch from embryonic (ε) to fetal (γ) globin gene expression and a second switch from fetal (γ) to adult (δ/β) globin expression.

A dual mechanism of hemoglobin gene switching has been proposed: autonomous gene control and gene competition for a direct interaction with the LCR (Hanscombe et al., 1991; Peterson and Stamatoyannopoulos, 1993). The concept of autonomous gene control is largely based on experiments with human globin genes with only proximal cis-regulatory DNA elements (i.e., promoters, enhancers, and silencers) in transgenic mice. These experiments showed that individual human transgenes express tissue-specifically and with the correct developmental timing, albeit in a position-dependent manner (Behringer et al., 1987; Chada et al., 1986; Kollias et al., 1986; Magram et al., 1985). Transgenes of the embryonic/fetal human ε- and γ-globin genes, including their proximal cis-regulatory DNA elements, linked to an LCR are properly silenced at later stages of development (Dillon and Grosveld, 1991; Raich et al., 1992), and it has been shown that these silencing elements reside in their promoter regions.

The first indications for gene competition emerged from hereditary persistence of fetal hemoglobin patients, which have mutations in the γ-globin genes resulting in an increased γ-gene expression in adult life. Importantly, this leads to a decrease of β-globin gene expression from the same allele, while expression of the nonmutated allele is unaffected (Giglioni et al., 1984). Several experiments confirmed this competition of the genes for LCR function and showed that it was based on gene order and relative distance. Introduction of a "marked" β-gene (βm) close to the LCR resulted in a competitive transcription advantage of this proximal βm-gene over the more distant β-gene. When the βm-gene was positioned close to the β-gene, thus positioning both genes at a similarly large distance from the LCR, the competitive advantage of the βm-gene over the β-gene was diminished (Dillon et al., 1997). In addition, inverting the gene order of the locus, thereby altering relative distances, activated the β-gene at early stages and abolished ε-gene expression (Tanimoto et al., 1999). Analysis of β-globin pre-mRNA transcripts in single cells supported the competition mechanism by demonstrating that the actual switching between γ- and β-gene transcription is a continuous dynamic process (a flip-flop mechanism) in which both genes can be alternately transcribed until the γ-genes are autonomously silenced (Wijgerde et al., 1995). The trans-acting environment plays a crucial role in autonomous gene silencing and hemoglobin gene switching. As such, trans-acting factors can favor activation or silencing of specific β-globin genes (Peterson and Stamatoyannopoulos, 1993). Several hematopoietic transcription factors that regulate β-globin expression have been identified (Harju et al., 2002). The most important factors for β-globin expression are GATA-1, NF-E2, and EKLF, which can be part of multiprotein complexes (reviewed in

Mahajan and Weissman, 2006). Additionally, ubiquitous transcription factors such as Sp1, TFII-I, and USF are involved in control of β-globin gene expression (Crusselle-Davis et al., 2006; Feng and Kan, 2005).

III. MODELS OF LONG-RANGE CONTROL OF GENE EXPRESSION BY ENHANCERS

The current view of transcription activation is primarily based on the concept of "recruitment" (Ptashne and Gann, 1997). According to this view, various factors participating in transcription are recruited to promoters via protein–protein interactions with an activator protein. In essence, activator proteins increase the local concentration of the transcriptional machinery near the promoter. However, binding of activator proteins to an enhancer that is separated in space from the promoter does not automatically lead to an increase in concentration of the transcriptional machinery at the promoter. Therefore, other strategies or mechanisms have to function for metazoan enhancers that are often separated by tens of kilobases up to megabases away from the promoter. Several models have been proposed for enhancer action over a distance.

A. The looping model

The looping model proposes a direct interaction of enhancer-bound activator proteins with proteins at the promoter by bringing them in close proximity in the nuclear space. The intervening chromatin loops out and does not participate in the activation process (Ptashne, 1986). DNA looping is a common way of communicating among distantly positioned DNA sequences in prokaryotes (Matthews and Nichols, 1998). Several proteins, for example, Bach1 and Sp1, are able to form looped structures between their DNA binding sites *in vitro* (Mastrangelo et al., 1991; Su et al., 1991; Yoshida et al., 1999). The SWI/SNF chromatin-remodeling complex is also able to form loops in an ATP-independent fashion in *in vitro* reconstituted nucleosomal arrays (Bazett-Jones et al., 1999).

 The looping model can readily explain a number of observations that have been made in eukaryotic systems. Transvection is a naturally occurring process in *Drosophila* whereby an enhancer on one chromosome activates a promoter on another chromosome (Morris et al., 1999; Wu and Morris, 1999). This phenomenon, together with experiments that show that an enhancer on one DNA molecule can activate a promoter in *trans* on another DNA molecule (Dunaway and Droge, 1989; Mahmoudi et al., 2002; Mueller-Storm et al., 1989), demonstrate that a *cis* configuration of a promoter and an enhancer is not an absolute prerequisite for transcriptional activation. Gene competition for a single regulator (de Villiers et al., 1983; Hanscombe et al., 1991;

Wasylyk and Chambon, 1983), leading to alternate transcription by a stochastic "flip-flop" mechanism (Gribnau _et al._, 1998; Trimborn _et al._, 1999; Wijgerde _et al._, 1995), is most easily explained by direct interactions of an enhancer with the promoter, especially because the competitive advantage of the enhancer proximal gene is lost when genes are closely spaced at a larger distance from the regulator (Dillon _et al._, 1997; Hanscombe _et al._, 1991; Heuchel _et al._, 1989).

B. The tracking model

In the tracking model, the enhancer acts as a loading platform for a DNA-tracking protein, which will consequently move along the chromatin fiber in the direction of the promoter. Since the tracking protein is leaving the enhancer, subsequent proteins can be loaded and this eventually leads to accumulation of the tracking protein in the vicinity of the promoter. Tracking has been shown to be the mechanism in the enhancer action of the late genes of bacteriophage T4 (Kolesky _et al._, 2002). However, there is no conclusive evidence for tracking in eukaryotes and no activators are known that have to leave the enhancer to activate transcription, although it should be noted that transcription and replication themselves are tracking processes.

　　　The tracking model is superficially attractive because it can straight-forwardly explain the position dependence of enhancer-blockers. A protein bound to the enhancer-blocker could easily obstruct the progression of the tracking protein from enhancer to promoter. Therefore, the tracking model predicts that an enhancer-blocker only functions as such when placed between a promoter and an enhancer as is indeed observed in many experimental setups. The detection of intergenic transcripts that originate from enhancers (Gribnau _et al._, 2000; Kong _et al._, 1997; Tuan _et al._, 1992) has renewed interest in this model and suggest that RNA polymerase II (RNApolII) might play a role in this process particularly with respect to the first activation of a locus.

C. The linking model

As in the tracking model, the linking model proposes that the enhancer acts as a loading platform for a DNA-binding protein. However, now the bound protein facilitates polymerization of proteins in the direction of the promoter, thereby coating the chromatin fiber (Bulger and Groudine, 1999; Dorsett, 1999). The linking model was proposed to explain the properties of the _Drosophila_ Chip protein. Chip cannot bind to DNA directly, but can interact with numerous transcription factors and facilitates their action over a distance _in vivo_ (Morcillo _et al._, 1997; Torigoi _et al._, 2000). It was proposed that Chip is recruited by an activator bound at an enhancer where it functions as a protein "bridge" between the activator bound at the enhancer and other proteins having multiple weak binding sites between an enhancer and a promoter (Dorsett, 1999; Gause et al., 2001). If distances between enhancer and promoter are large, a considerable amount of protein is needed to coat the chromatin fiber; therefore, it has been proposed that the spreading induces formation of small loops (~1.5 kb) between multiple small activator binding sites. However, recent experiments have shown that Ldb1, the mouse equivalent of Chip, forms large complexes of transcription factors that appear to bind specific sites in the genome and that Ldb1 complexes are more likely to play a role in looping than in linking-type processes (Meier et al., 2006; Wadman et al., 1997).

D. Relocation models

Another function that has been proposed for enhancers/LCRs is relocation in the nucleus, which is not directly based on recruitment. Early observations showed that active genes move away from nuclear landmarks that are considered to be incompatible with transcription like the nuclear periphery, heterochromatin, and tend to loop out of chromosome territories. Therefore, it was suggested that a gene has to relocate to compartments within the nucleus that are more favorable for transcription to be activated (reviewed in Lanctot et al., 2007). The observation that repositioning of a transgenic β-globin gene away from centromeric heterochromatin requires an intact enhancer (HS2) and correlates with stable transcription suggested that an enhancer activates a gene by relocating the locus to a different nuclear compartment (Francastel et al., 1999). This model has gained renewed attention now that it has been demonstrated that the β-globin locus relocalizes to discrete foci containing RNApolII, so-called "transcription factories" (Osborne et al., 2004). Efficient relocation of the β-globin locus and its efficient association with transcription factories were shown to require the LCR (Ragoczy et al., 2006).

IV. LONG-RANGE ACTIVATION BY THE β-GLOBIN LCR

Analysis of the β-globin locus has often been at the fore front of revealing new mechanisms of gene regulation and proposed models for enhancer/LCR action (linking, tracking, looping, relocation) were to a large extend based on observations made in the β-globin locus. Given the vast amount of knowledge obtained in the past on the subject of transcriptional regulation of the β-globin locus, the organization of its cis-regulatory elements, and its primary sequence, it is perhaps not surprising that the β-globin locus was the first locus of which detailed information regarding its spatial organization was obtained.

A. The LCR is in close proximity to the promoter

Two studies investigating the spatial organization of the β-globin locus provided unequivocal evidence in support of the contact model of enhancer function (Carter et al., 2002; Tolhuis et al., 2002). The first study cleverly modified RNA-FISH to tag and recover chromatin in the immediate vicinity of the actively transcribed β-globin gene. The authors found that not only the chromatin containing the active β-globin gene was tagged and recovered but also the classical enhancer element; HS2 of the LCR was tagged and recovered with high efficiency indicating close proximity (Carter et al., 2002). In the second study, the spatial organization of a 200-kb region spanning the mouse β-globin locus was analyzed, at an unprecedented level of resolution, in expressing erythroid and nonexpressing brain tissue using chromosome conformation capture (3C). 3C-technology involves quantitative PCR-analysis of cross-linking frequencies between two given DNA restriction fragments, and gives a measure of their interaction frequency in the nuclear space. The technique was originally developed to analyze the conformation of chromosomes in yeast (Dekker et al., 2002).

It was found that the inactive β-globin locus in brain tissue adopts a seemingly linear conformation, while in erythroid cells, the DNase I HSs of the β-globin LCR are in close spatial proximity to the active genes (Fig. 4.2). The 40–60 kb of intervening chromatin containing the inactive globin genes loops out. Surprisingly, two distant hypersensitive regions, being 130 kb apart from each other, were found to participate in these interactions. This spatial clustering of transcriptional regulatory elements is referred to as an active chromatin hub (ACH; Tolhuis et al., 2002).

Both studies clearly demonstrate, using independent unrelated techniques, that two distant *cis*-regulatory elements, the β-globin LCR and β-major promoter, are in close proximity in expressing cells, which strongly suggests they directly interact as was predicted by the original looping model. In recent years, contacts between enhancers/LCRs and gene promoters have been described in several other gene loci and even *trans*-interactions have been described (e.g., Liu and Garrard, 2005; Lomvardas et al., 2006; Spilianakis and Flavell, 2004). Subsequent analysis of the spatial organization of the β-globin locus in primitive erythroid cells, a developmental stage when the embryonic globin genes are active and the adult genes are silent, linked transcriptional status to gene-LCR proximity (Palstra et al., 2003). A core of the ACH containing the β-globin regulatory elements is developmentally conserved, while the globin genes switch their interaction with this cluster during development, and this correlates with the switch in their transcriptional activity (Fig. 4.2). A transgenic human β-globin locus in mice adopts a similar configuration as was observed for the mouse β-globin locus (Palstra et al., 2003).

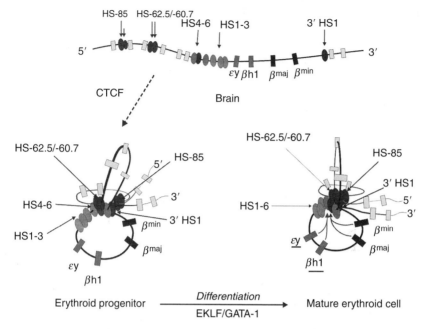

Figure 4.2. The *cis*-regulatory elements of the β-globin locus interact and create an active chromatin hub (ACH). A two-dimensional presentation of three-dimensional interactions that occur between regulatory DNA elements (ovals) and β-globin genes (dark rectangles) within the 160 kb β-globin locus. In fetal brain cells (top), where the β-globin locus is inactive and hypersensitive sites (HSs) are not detected, the locus adopts a seemingly linear conformation. In erythroid progenitors (bottom left), which are committed to but do not yet express β-globin, a substructure called the chromatin hub is present, which is formed through interactions between CTCF-bound *cis*-regulatory elements (black ovals); the upstream HS-85, HS-60.5/-62.7, the downstream 3′ HS1 and HS at the 5′ side of the locus control region (LCR; HS4–HS6). Interaction between these HSs is dependent on the presence of the transcription factor CTCF (indicated by the dashed arrow). During erythroid differentiation, the remaining HSs of the LCR (HS1–HS3) and the β-globin gene that will be activated stably interact with this substructure to form a functional ACH (bottom right). The core of the ACH is erythroid-specific and developmentally stable; a developmental switch occurs in β-like globin genes participating in ACH formation. In primitive erythroid cells, εy and βh1 (underlined) enter the core of the ACH while in definitive erythroid cells β-major and β-minor enter the core of the ACH (bottom right). The inactive β-like globin genes and olfactory receptor genes (light grey squares) loop out. In definitive erythroid cells, progression from the chromatin hub to a fully functional ACH, including the HS at the 3′ side of the LCR and the active β-globin gene, is dependent on the presence of erythroid Krüppel-like factor (EKLF) and GATA-1.

In erythroid progenitors that are committed to, but do not yet express, globin, only a subset of regulatory sites cluster and form a structure called a chromatin hub (CH). Upon erythroid differentiation, a fully functional ACH is formed, containing all regulatory elements and the gene that becomes activated (Fig. 4.2; Palstra *et al.*, 2003).

Several studies addressed the role of transcription factors in β-globin ACH formation. In fetal livers obtained from EKLF knockout mice, a conformation of the locus very similar to that observed in erythroid progenitors was found, indicating that EKLF is needed for the progression to and/or stabilization of a fully functional ACH (Drissen *et al.*, 2004). Similarly GATA-1 and its cofactor FOG-1 are required for the establishment of proximity between the β-globin LCR and the β-major promoter. Kinetic studies showed that GATA-1-induced loop formation correlates with the onset of β-globin transcription strongly suggesting the importance of LCR-gene interaction for high-level expression (Vakoc *et al.*, 2005). This shows that erythroid-specific transcription factors, known to be essential for proper β-globin expression, play an essential role in the three-dimensional organization of the β-globin locus (Fig. 4.2). However, another erythroid-specific transcription factor, p45 NF-E2, that was implicated in β-globin expression and transfer of RNApolII from HS2 to the promoter (Johnson *et al.*, 2001; Vieira *et al.*, 2004) is dispensable for establishing LCR-gene contacts within the β-globin locus (Kooren *et al.*, 2007).

The subset of regulatory sites that cluster and form the poised CH in erythroid progenitors all contain a binding site for CCCTC-binding factor (CTCF), a transcription factor implicated in gene insulation and boundary formation at transcription competent domains. Targeted deletions of some of these elements alone, and in combination, do not affect transcription of the β-globin genes (Bender *et al.*, 2006). Conditional deletion of CTCF and targeted disruption of the CTCF binding site of 3′ HS1 were shown to destabilize the long-range interactions that form the CH and cause a local loss of histone acetylation and gain of histone methylation, apparently without affecting transcription at the locus (Fig. 4.2; Splinter *et al.*, 2006). These data suggest that these distal loops do not play a role in gene expression and do not support the generality of boundaries demarcating expression domains. Given the fact that CTCF-dependent chromatin loops are tissue-specific and evolutionary conserved between mouse and human a function may exist that currently escapes our attention. Possibly their nature is only structural and evolution selects against sites forming chromatin loops within a gene locus, resulting in positioning of these structural elements outside the β-globin locus (Dillon and Sabbattini, 2000).

The HSs located in the LCR of the β-globin locus are more crucial for globin expression and investigations into their role in the establishment of long-range interactions between the LCR and the promoter are in progress.

Deletion of individual HSs from the β-globin LCR results in diminished β-globin expression, although the extent is somewhat different between the mouse and human β-globin locus, which may be due to differences between the sites in mouse and human. In addition the levels depend on the precise nature of the deletion made (reviewed in Harju et al., 2002). When the β-globin promoter or HS3 is deleted from a transgenic construct containing the complete human β-globin locus in the context of a full LCR, the ACH is maintained with the β-major gene remaining in proximity to the LCR. Deletion of hypersensitive site HS3 of the LCR in combination with deletion of the β-major promoter resulted in a strong destabilization of the ACH and a complete switch down of transcription, demonstrating that multiple interactions between the LCR and the β-globin gene are required to maintain the appropriate spatial configuration in vivo (Patrinos et al., 2004). Interestingly, when only the (small) core of HS3 is deleted, long-range interactions within the human β-globin locus are disrupted even without the need of the additional β-globin promoter (Fang et al., 2007). These results are in agreement with the idea that the LCR forms a holocomplex and suggest a mass action model for ACH formation whereby the likelihood of establishing a stable ACH structure depends on a critical number of interactions to take place (de Laat and Grosveld, 2003).

Recent studies of the spatial organization of the β-globin locus and several other loci have now firmly established that enhancers and LCRs are in close proximity to the genes when they are active. However, virtually nothing is known as to how these long-range interactions are established. Furthermore, concepts of how an enhancer/LCR functions when it is in close proximity to a promoter are sketchy at best.

B. How are LCR-promoter contacts established?

The work reviewed above clearly demonstrates that distant cis-regulatory elements within the β-globin locus physically interact to control gene activity, as predicted by the original looping models. How this looped structure is formed and maintained remains elusive though. Random collision (by Brownian motion) between distal elements in the active β-globin locus currently seems to be the most likely hypothesis. However, theoretical calculations (Rippe, 2001) and experimental measurements of site-specific recombination between sites separated by several kilobases in mammalian cells (Ringrose et al., 1999) suggested that random diffusion alone is not sufficient to establish contacts between sites separated by several kilobases on a chromatin fiber. However, many nuclear processes, for example, V(D)J recombination and transcription, have been shown to involve DNA looping occurring over large distances (e.g., Sayegh et al., 2005; Skok et al., 2007; Spilianakis and Flavell, 2004; Tolhuis et al., 2002). In addition, recombination based on the bacterial lox/Cre system works efficiently over distance in eukaryotic cells, as exemplified in an experiment that inverts the order of the five human β-globin genes over a distance similar to the distance of promoter-LCR contact (Tanimoto *et al.*, 1999). These data show that establishment of loops in chromatin might be less problematic than previously thought. It has been suggested that loop formation and the required flexibility may be facilitated by acetylation of chromosome domains (Li *et al.*, 2006). Recombination events are thought to require short-lived single interactions, while transcription of, for example, the β-globin genes requires longer contact times estimated to be in the range from 45 to 80 min for the β-globin–LCR interaction in definitive erythrocytes (Wijgerde *et al.*, 1995). Stabilization and maintenance rather than initiation of interaction resulting from random collision might therefore be the more important functions of the proteins involved in the looping process. Transcription factors that bind DNA and subsequently multimerize could function as stabilizing factors. Indeed, transcription factors have been shown to play a pivotal role in the establishment or maintenance of promoter-LCR contacts. For example, contacts between the LCR and genes are lost in the absence of the erythroid-specific transcription factors EKLF and GATA-1 (Drissen *et al.*, 2004). EKLF and GATA-1 bind to DNA, but they are not known to homomultimerize (like, e.g., CTCF, Bach1, or GAGA); therefore, it is unlikely that these factors are directly responsible for stabilizing interactions between distant DNA elements. GATA-1 and EKLF interact with large proteins or protein complexes like p300/CBP and Mediator. These complexes are able to interact simultaneously with factors bound at enhancers and promoters and could in this way be suitable candidates for stabilization of the promoter-LCR contact (Chan and La Thangue, 2001; Kuras *et al.*, 2003). Thus, promoter-LCR contact must be thought of as a multicomponent entity containing many different proteins and DNA elements possibly via a self-organizing principle. Self-organization involves the physical interaction of molecules in a steady state structure, in which components are freely exchanged, and is a principle used for building many cellular structures (Cook, 2002; Misteli, 2001a). Removal of one of the crucial components like, for example, EKLF will destabilize the multicomponent structure and as a result promoter-LCR contacts will not be firmly established.

A group of ubiquitously expressed proteins, facilitator factors, interact with other *trans*-acting factors and may link the interacting *cis*-regulatory DNA elements to further stabilize promoter-LCR contacts. Facilitator factors were originally identified in *Drosophila* (Dorsett, 1999) and have mammalian homologues like LIM-domain binding protein 1 (Ldb1) and Idn3 (Morcillo *et al.*, 1997; Rollins *et al.*, 1999). Indeed, in erythroid cells Ldb-1 has been found in a large complex with erythroid cell-specific transcription factors, including GATA-1 (Meier *et al.*, 2006; Wadman *et al.*, 1997), that would be able to bind several different DNA sequences simultaneously.

Other mechanisms might exist that establish enhancer-promoter contact and simultaneously create a chromatin loop. To find the promoter, an enhancer-bound activator can "scan" the surrounding DNA regions by a "hopping" and "scanning" mechanism as proposed for translocation of the *lac* repressor on DNA (Berg *et al.*, 1981; Winter *et al.*, 1981). Such a "looping-scanning" mechanism has also been proposed for eukaryotic enhancers in the facilitated tracking model (Blackwood and Kadonaga, 1998). Some evidence for the facilitated tracking model has come from a study of the HNF4α enhancer (Hatzis and Talianidis, 2002). In this study, a ChIP approach was used to show that enhancer-bound proteins (C/EBPα and HNF-3β) could be cross-linked to the 6.5 kb of spacer DNA separating the enhancer and promoter as well as to the enhancer itself. This cross-linking to spacer DNA can only be detected after activation of the enhancer but before actual transcription of the gene at which stage the enhancer-bound proteins are cross-linked to the promoter. However, the enhancer takes about 80 h to track along the 6.5 kb of spacer DNA and therefore it seems unlikely that this process takes place in larger loci like the β-globin locus or when fast transcriptional activation is required. It is important to keep in mind that the failure to detect by 3C intermediate states within the β-globin locus, whereby the LCR preferentially interacts with the intervening chromatin fiber, does not exclude the possibility of a facilitated tracking mechanism. The contact between the LCR and the promoter is probably quite stable and can therefore readily be identified in the 3C assay, while the many transitory interactions formed as the LCR tracks along the intervening chromatin fiber will probably be too transient to be identified with this assay.

The observation that intergenic transcripts emanate from the β-globin LCR and many other regulatory elements in several loci has led to the suggestion that transcription underlies the establishment of long-range interactions between regulatory elements (Ling *et al.*, 2004; Tuan *et al.*, 1992; Zhao and Dean, 2004). In this model, the LCR is translocated to the promoter by piggy-backing on an elongating RNApolII complex. As a variation on this model, it was recently proposed that active transcription regulatory elements, like the LCR, relocate to a preformed transcription factory. This would be followed by the reeling in of the chromatin fiber until a promoter is encountered and as a consequence chromatin loops are generated (Chakalova *et al.*, 2005; West and Fraser, 2005).

However, models that invoke such a role for the intervening chromatin fiber in establishing LCR-promoter contacts are difficult to reconcile with phenomena like gene competition for a single regulator (de Villiers *et al.*, 1983; Hanscombe *et al.*, 1991; Wasylyk and Chambon, 1983), which forms the basis of alternate transcription (Gribnau *et al.*, 1998; Trimborn *et al.*, 1999; Wijgerde *et al.*, 1995). Other models propose that long-range interactions are established through colocalization of transcribed regions in the same transcription factory

(Cook, 2003) or accumulation of transcribing polymerases through molecular crowding (Marenduzzo *et al.*, 2006a,b). These models predict that inhibition of transcription changes DNA folding and eliminates looping (Faro-Trindade and Cook, 2006). All of these models implying RNApolII and factories have the inherent weakness that they do not explain how the LCR and/or a gene end up at a factory in the first place by a mechanism that would be fundamentally different from one based on stochastic events and mass action. Moreover, treatment of erythroid cells with the transcription inhibitor α-amanitin inhibits genic and intergenic transcription and severely reduces RNApolII loading at the β-major promoter and HS2 of the LCR, while interactions between the β-globin LCR and the β-major promoter remain intact (Palstra *et al.*, manuscript in preparation). This suggests that (inter-)genic transcription and the presence of RNApolII at regulatory elements are not necessary for maintenance of long-range interactions.

C. How does the β-globin LCR increase transcription efficiency?

LCRs were originally termed dominant control regions to indicate that they would overcome the silencing of a (trans)gene. They were thought to be "dominant" because such a set of DNA elements provided high expression levels to a *cis*-linked gene in a tissue-specific, copy number-dependent, and integration site-independent manner (Grosveld *et al.*, 1987). The name was changed to accommodate the term locus activating region, which was used to indicate that this region may be the region that activates a locus (Forrester *et al.*, 1987). Subsequent data generated using transgenic constructs of the human β-globin LCR introduced at ectopic genomic sites demonstrated that deletion of HSs results in variegated expression and suggesting a domain opening activity of the LCR (reviewed in Harju *et al.*, 2002). These studies were interpreted such that the LCR counteracts silencing of the β-globin genes by creating open chromatin domains or translocating the gene to transcription permissive regions in the nucleus.

However, targeted deletion of the mouse LCR shows that even after germline passage of the deletion, the locus is still DNase I sensitive and that the remaining HSs are normally formed. This shows that in the mouse β-globin locus, the LCR is dispensable to initiate an open chromatin conformation and suggests that elements elsewhere in the locus are sufficient to establish and maintain an open chromatin domain (Bender *et al.*, 2000a), although it should be remembered that globin transgenes inserted at random positions in the genome are expressed at similar levels (Kollias *et al.*, 1986; Magram *et al.*, 1985). Acetylation of nucleosomes in the region between the LCR and the β-major and β-minor genes is only modestly reduced in the LCR deletion as compared to wild type (Epner *et al.*, 1998; Schubeler *et al.*, 2001). Similarly, hyperacetylation of histones at the β-major and β-minor promoter was similar as in the wild-type locus and β-globin promoter remodeling occurs in the absence of the LCR (Bender *et al.*, 2000b). Transcription of the β-major gene in the LCR deletion was reduced to about 4% of the wild-type level and subsequent analysis of transcription at the single cell level showed that the level of transcription was reduced in all cells (Schubeler *et al.*, 2001). A similar low-level transcription is found in erythroid progenitor cells where the LCR is not yet in close proximity to the β-globin promoter (Kooren *et al.*, 2007). These data suggest that in its native context, the mouse β-globin LCR mainly functions as an enhancer by increasing the transcriptional output of the globin genes. In contrast, LCR deletions in the human locus result in a chromatin inactive β-globin gene *in vivo* (Kioussis *et al.*, 1983).

Addition of a transcriptional enhancer to a gene results in an increase in RNA levels in a pool of cells. Interpretations of how this increase is achieved differ widely. The classical view is that the enhancer elevates the rate of transcription in every cell (rheostat model). Data obtained by observing pools of cells supports this view but cannot exclude the opposing view that enhancers increase the chance that a gene will be expressed but when it is expressed it will express at the maximal rate (a binary model). A recent study, visualizing ongoing transcription of a developmental regulated gene in live cells, shows that eukaryotic transcription occurs in burst (Chubb *et al.*, 2006) as was suggested before based on RNA-FISH experiments (Wijgerde *et al.*, 1995). In addition, it was found that once a cell produced a pulse of RNA it was more likely to produce a second pulse (Chubb *et al.*, 2006). It can therefore be imagined that enhancers increase the pulse length and/or reduce the time span between pulses, although it should be noted that loci containing all of the elements are not always activated to transcribe (de Krom *et al.*, 2002), suggesting such loci are either off or on full. Recent experiments suggest that the basal transcription machinery can respond both in an on/off or graded response. Furthermore, a graded response can be converted to an on/off switch depending on the relative abundance of transactivators and transrepressors competing for the same DNA regulatory element (Rossi *et al.*, 2000).

Relocation of activated loci from nuclear landmarks like pericentromeric heterochromatin, chromatin territories, and the nuclear periphery is often observed, and it is thought that this brings the locus in a compartment that is more favorable to transcription (reviewed in Lanctot *et al.*, 2007). Claims regarding the nuclear repositioning of the β-globin locus and the role of the LCR herein exist but are not consistent. Relocation of a transgenic β-globin gene away from pericentromeric heterochromatin was observed in a cell-line and this required an intact enhancer (HS2) (Francastel *et al.*, 1999). However, the same group also analyzed MEL cell hybrids carrying a human chromosome 11 containing either a wild-type β-globin locus or a locus containing LCR deletions and found that location away from centromeric heterochromatin was independent of the LCR although relocation is required to achieve general hyperacetylation and an open chromatin structure of the locus (Schubeler *et al.*, 2000). In yet another study, they showed that the murine β-globin locus and a wild-type β-globin locus on human chromosome 11 in MEL cell hybrids tend to loop out of their chromosome territory prior to activation which is, at least for the human locus in the MEL cell hybrids, LCR dependent (Ragoczy *et al.*, 2003). However in a particularly careful analysis focusing on the nuclear position of the endogenous mouse β-globin locus at various stages of erythroid differentiation, it was shown that the locus always remains inside its territory (Brown *et al.*, 2006).

Recently the relocation model of enhancer/LCR function gained renewed attention by the observation that the β-globin locus relocates to sites of nascent RNA production marked by discrete foci containing RNApolII, so-called "transcription factories" (Osborne *et al.*, 2004). This led to the proposal that the LCR enhances transcription by targeting the β-globin locus to pre-formed transcription factories (Chakalova *et al.*, 2005; Fraser, 2006). The concept of preassembled transcription factories is primarily based on immuno-cytochemistry, and it should therefore be noted that the number of foci and their size differ depending on the RNApolII antibody used and the fixation procedure. Importantly, the high mobility of RNApolII in the nucleus (Kimura *et al.*, 2002) does not particularly support the concept of preassembled transcription factories, and (preassembled) transcription factories still have not been detected in live cells. It may be more plausible that the activated β-globin locus functions as an assembly site for a focus of RNApolII or splicing factors in agreement with its high transcription and processing rate. The β-globin locus relocates from the nuclear periphery to the interior upon erythroid differentiation, which is LCR dependent, correlating with a change in nuclear RNApolII distribution. Transcriptional activity of the β-globin locus does not appear to depend on an internal nuclear position since one-third of the active loci are still located at the nuclear periphery. This suggests that relocation is a consequence of transcription and not a prerequisite.

Transcription can take place anywhere within the nucleus, and many studies suggest that relocation of loci does not enhance transcription per se. If transcription levels of genes would be controlled by targeting them to a compartment, which is more favorable to transcription, then all genes in the neighborhood that potentially can be active would "profit" and also increase their transcription. This is clearly not the case in the human β-globin locus, because the two adult genes β-major and β-minor are transcribed alternately (Gribnau *et al.*, 1998; Wijgerde *et al.*, 1995) and introduction of an additional β-major gene results in gene competition (Dillon *et al.*, 1997; Wai *et al.*, 2003).

Introduction of a human β-globin LCR in two different orientations in an ectopic, gene dense locus indeed results in increased looping out of the locus from its territory. Each LCR orientation enhances transcription of an overlapping but different subset of genes (Noordermeer et al., in press) independent of the position of these genes relative to the chromosomal territory, suggesting that enhancement of gene expression is a local affair and that nuclear position of the locus plays at best a facilitating role. Transcriptional activation of a promoter includes several rate-limiting steps that might be controlled by the β-globin LCR to increase the transcription efficiency of the β-globin genes and they will be discussed next.

V. ENHANCEMENT OF TRANSCRIPTION BY THE β-GLOBIN LCR: RATE-LIMITING STEPS

Transcription is a highly complex process requiring the coordinate action of many proteins (reviewed in Li et al., 2007; Svejstrup, 2004). Chromatin is thought to be highly restrictive to processes like transcription. Therefore, first the promoter region has to be made accessible via chromatin remodeling and histone eviction, which is mediated by chromatin-remodeling enzymes and histone modifiers that are recruited by sequence-specific transcription factors. Only then can several general transcription factors (GTFs) assemble at the promoter and form a preinitiation complex (PIC) and recruit the enzyme RNA-polII. RNApolII is able to incorporate the first nucleotides in the nascent RNA after the PIC forms an open complex resulting in promoter escape. After a brief pause in which the integrity of the complex is checked, elongation commences fully. When RNApolII reaches the end of the gene, transcription is terminated and RNApolII is reprocessed to be able to start another round of transcription. This whole process is highly regulated and many proteins have been identified that influence the individual steps. Recently, it is becoming clear that all individual steps in the transcription process are intimately linked and influence each other. The whole process is further complicated by the fact that in eukaryotes in vivo transcription takes place on a chromatin template that needs to be remodeled and modified before transcription can take place (Li et al., 2007). Clearly, the β-globin LCR could in principle be acting at each of the above steps (or even several steps simultaneously).

A. Promoter remodeling

To start PIC assembly, the chromatin at the promoter has to be made accessible to the general transcription machinery. It is believed that sequence-specific transcription factors recruit chromatin-remodeling complexes and histone-modifying enzymes to the promoter. Several transcription factors that bind the β-globin promoter, for example, EKLF and GATA-1, have been shown to recruit chromatin-remodeling complexes and histone-modifying enzymes (Blobel *et al.*, 1998; Zhang and Bieker, 1998; Zhang *et al.*, 2001). All these sequence-specific transcription factors also bind the cores of the HSs of the LCR. Close proximity of the β-globin LCR to the promoters of β-globin genes could greatly enhance the chromatin remodeling of the promoter. However, in the absence of the LCR, formation of a hypersensitive site at the mouse β-globin promoter does not seem to be affected and acetylation levels of histone H3 and H4 are similar to the levels found in the wild-type locus (Bender *et al.*, 2000b; Schubeler *et al.*, 2001). This is likely mediated by chromatin modifiers recruited by GATA-1 and p45 NF-E2 since these factors occupy the β-major promoter independently of the LCR (Bottardi *et al.*, 2006; Vakoc *et al.*, 2005). Of course, many chromatin marks exist and it cannot be excluded that the LCR recruits chromatin-modifying enzymes that modify histones at the β-globin promoter in a manner not yet detected. In addition, it is important to remember that the human β-globin gene is not sensitive to DNase I and methylated in patients with a deletion of the LCR. This discrepancy between the mouse and human situation may be mass action. Mass action would act fast to provide a stable situation with factors bound and chromatin modified through clustering of regulatory sites (as discussed below in "the concept of an ACH"). With just a promoter, this would still happen but presumably much slower. In the mouse, there may be just sufficient sites or a sufficiently high concentration of erythroid-specific factors to get a stable modification of the locus whereas this may not be the case in humans. One of these factors may be just at the other side of the equilibrium and not have sufficient "mass for action," that is, occupied sites to create a stable situation without the human LCR.

B. Transcription initiation

The next step in transcriptional activation is the assembly of a functional preinitiation complex. PIC formation generally occurs in a stepwise fashion, whereby the TATA box is first recognized by TBP and its associated factors (TFIID). Then factors such as TFIIA, TFIIB, TFIIF, RNApolII, TFIIE, and TFIIH enter the promoter. Different intermediates are stable *in vitro*, but *in vivo* the stable binding of TBP to the TATA box needs the function of other GTFs, like TFIIB and Mediator (reviewed in Svejstrup, 2004). TBP, TFIIB, and RNApolII are recruited to the β-globin LCR (Johnson *et al.*, 2003; Levings *et al.*, 2006; Vieira *et al.*, 2004). *In vitro*, RNApolII can be transferred from immobilized LCR constructs to a β-globin promoter in an NF-E2-dependent fashion (Vieira *et al.*, 2004). However, knockout mice lacking the erythroid-specific NF-E2 subunit p45 express the β-globin genes at near maximal levels (Shivdasani and Orkin, 1995). Recruitment of RNApolII and GTFs has been observed at the enhancers of other loci and it has been suggested that LCRs/enhancers function as nucleation centers for the assembly of the PIC (Szutorisz et al., 2005).

Mediator is required for activator-mediated recruitment of RNApolII and is therefore essential for activator-dependent (or regulated) transcription (Wu et al., 2003). In yeast, enhancers of active genes are associated with the Mediator complex, and binding to the enhancer is mediated via different subunits of Mediator as the ones interacting with the basic promoter complex. The physical interaction between enhancer and promoter that is established in this way would make Mediator a primary conduit of regulatory information from enhancers to promoters (Kuras et al., 2003). Interestingly, it was recently found that the Mediator complex functions as a coactivator for GATA-1 in erythropoiesis and Mediator components are detected at GATA-1-occupied enhancer sites (Stumpf et al., 2006), although it is currently unclear if Mediator is bound at the β-globin locus and if it plays a role in the regulation of the β-globin locus. Furthermore, PIC assembly is strongly coordinated with chromatin modification that is regulated by the cooperative binding and direct interaction between Mediator and p300 (Black et al., 2006). The highly related transcriptional activators p300/CBP act as protein bridges connecting different sequence-specific transcription factors to the transcription apparatus. p300/CBP also possess histone acetyl transferase activity that allows them to acetylate histones, transcription factors, other nonhistone chromatin-associated proteins, and themselves (Chan and La Thangue, 2001). p300/CBP interacts with EKLF, GATA-1, and NF-E2 and has been shown to be crucial for long-range activation by the β-globin LCR (Blobel et al., 1998; Cheng et al., 1997; Forsberg et al., 1999; Zhang and Bieker, 1998).

An important mechanism to increase the efficiency of transcription is reinitiation. Recent in vivo analysis of the RNApolII transcription cycle has shown that most transcription units are unoccupied by RNA polymerases for three-quarters of their time (Kimura et al., 2002). Initiation of transcription is very rapid (seconds) and elongation takes minutes. In contrast, the assembly of the PIC by the stochastic exchange of freely mobile components takes up 50–85% of the duration of a transcription cycle. Transcription reinitiation, the cyclic process of RNA synthesis from active genes, bypasses several protein-DNA association steps and the rate of new transcription cycles is increased with respect to the first transcription round (Dieci and Sentenac, 2003). A reinitiation intermediate that includes transcription factors TFIID, TFIIA, TFIIH, TFIIE, and Mediator has been identified, and this intermediate is stabilized in the presence of an activator (Asturias, 2004; Yudkovsky et al., 2000). Reinitiation would be further enhanced by the recently detected gene looping, whereby the 3' end of the gene loops back bringing it in close proximity to the promoter

(O'Sullivan *et al.*, 2004). Therefore proximity of the LCR, containing transcription factors that interact with Mediator and p300/CBP, to the β-globin promoter could result in a highly stabilized PIC at this promoter leading to increased frequencies of transcription (re)initiation.

C. Promoter escape and elongation

RNApolII enters the PIC in a hypo-phosphorylated state. Phosphorylation of the C-terminal repeat domain (CTD) of RNApolII at serine 5 by TFIIH and at serine 2 by pTEFb triggers promoter clearance and marks the transition of initiation to elongation. The DNA helicases of TFIIH open the promoter allowing RNApolII to incorporate the first nucleotides in the nascent RNA. Phosphorylation of the CTD leads to disruption of the strong bond between the RNApolII and the PIC. Simultaneously RNA maturation factors and other mRNA processing factors like Elongator are loaded onto RNApolII. Within the first 20–45 nucleotides, RNApolII pauses, which represents a general rate-limiting step (Krumm *et al.*, 1995). This brief pause appears to be a critical checkpoint to ensure the integrity of the produced pre-mRNA and warrants its proper capping. Promoter proximal pausing is subsequently overcome by recruitment of several elongation factors and pTEFb that phosphorylates serine 2 of the CTD after which elongation fully commences. To efficiently elongate, the CTD of RNApolII has to remain phosphorylated during transcription along the entire length of the gene. For example, at the c-fos gene this is at least in part accomplished by the active recruitment of pTEFb (Ryser *et al.*, 2007). Several studies have demonstrated that activators and enhancers are able to influence RNApolII elongation (Brown *et al.*, 1998; Kadener *et al.*, 2002; Yankulov *et al.*, 1994), and it has been suggested that the main mode of action of the β-globin LCR is enhancing the transition from transcription initiation to elongation (Sawado *et al.*, 2003).

Upon transcriptional induction of the β-globin gene, transcription factor recruitment, PIC assembly, and RNApolII recruitment are induced, correlating with enhanced promoter-LCR interactions, leading to enhanced transcription (Kooren *et al.*, 2007; Sawado *et al.*, 2003). Deletion of the LCR results in an approximately twofold decrease in PIC assembly and RNApolII recruitment while phosphorylation of the RNApolII CTD was decreased fourfold. The presence of (phosphorylated) RNApolII was even further reduced in the 3rd exon of the β-globin gene in agreement with the reduced levels of transcription (Sawado *et al.*, 2003). The authors concluded that PIC formation is largely LCR independent while the phosphorylation of the CTD of RNApolII at the transition of initiation to elongation, which is necessary for efficient transcription, is LCR dependent. In this model, the LCR would either directly recruit CTD kinases, for example, the pTEFb-associated kinase cdk9, or stimulate their activity. However, the results can equally well be explained by the LCR having a stimulatory effect on reinitiaton efficiency of the PIC at the β-globin promoter. Moreover, because of the relative long cross-linking times used in ChIP experiments, it is notoriously difficult to address issues of PIC stability and binding dynamics using this technique. Therefore, one cannot rule out that in the absence of the LCR the PIC is destabilized but that these unstable complexes become trapped during the fixation times used in ChIP.

VI. THE CONCEPT OF AN ACTIVE CHROMATIN HUB

The overall concentration of many proteins in the nucleus is estimated to be in the range of, or lower than, the dissociation constant for many protein–protein or protein–DNA interactions (Chambeyron and Bickmore, 2004). Chromatin associated proteins find their binding sites by diffusion and three-dimensional scanning of the genome space (Phair et al., 2004), while the formation of productive transcription complexes on DNA is expected to take place via a "stop and go" mechanism (Misteli, 2001b). In this mechanism, a factor binds to its binding site and will drop off if another factor does not bind within its residence time. This makes protein concentration a very sensitive limiting factor in the efficiency of transcription. Moreover, transcription factors move rapidly through the nucleus. This begs the question of how a high local concentration of these factors at a genomic site is achieved to ensure efficient transcription? This would be most easily achieved by a spatial clustering of their cognate binding sites (Droge and Muller-Hill, 2001).

    The clustering of the cis-regulatory elements and active genes of the β-globin locus in an ACH would result in such an increase in the concentration of transcription factors essential for the high transcription rate of the β-globin genes (Palstra et al., 2003). Several studies indicate that RNApolII and other basal transcription factors are recruited to LCR core HS elements and that these can be subsequently transferred to the β-globin promoter (Johnson et al., 2002, 2003; Vieira et al., 2004). This would be most easily achieved by a concentration of DNA templates, RNA polymerases, transcription factors, and transcription units in specific clusters or compartments resulting in a mass action mechanism. A high concentration of factors would ensure efficient binding to the DNA templates and dissociating factors could rapidly reassociate to the same or neighboring sequences, thus significantly facilitating RNApolII recycling and high-frequency reutilization of stable preinitiation complexes. This would enable the production of extremely high levels of transcript. The β-globin ACH may function as such a nuclear compartment specifically dedicated to efficient RNApolII-mediated transcription of the β-globin genes (Palstra et al., 2003). It is tempting to speculate that instead of the β-globin locus moving to a preassembled factory, the de novo assembly and subsequent accumulation of RNApolII molecules at the β-globin ACH explains why a transcription factory is frequently observed at the β-globin locus (Osborne *et al.*, 2004; Ragoczy *et al.*, 2006).

The ability of LCRs and other combinations of *cis*-regulatory elements to form an ACH may also underlie the position-independent, copy number-dependent expression levels that are obtained when transgenes are linked to these elements. ACH formation also provides a mechanistic framework to understand how overlapping gene loci can set up independent tissue-specific expression profiles (reviewed in de Laat and Grosveld, 2003).

VII. FUTURE DIRECTIONS

The analysis of the spatial organization of the mouse β-globin locus and other loci has led to the conclusion that *cis*-regulatory elements in a locus cluster in the nuclear space when the genes are active. However, many key questions concerning gene regulation and chromatin folding in the spatial context of the nucleus remain unanswered. It will be a challenge to determine the mechanism by which the *cis*-regulatory elements find each other in the nucleus and to identify the factors involved in this process. The functionality of these interactions is not clear yet; for example, is an LCR-promoter interaction crucial for high-level transcription or just a consequence of transcriptional activity? How does the LCR proximity affect the molecular events at the promoter and within the open reading frame of a gene? Genome wide knockout/down approaches in combination with 3C and ChIP experiments, although difficult to set up, can potentially reveal factors that are crucial for long-range interactions within the loci. Alternatively, knockout/down of candidate genes or the use of small molecule inhibitors to factors suspected to be involved in long-range interactions could be revealing.

The detailed knowledge gathered in the past regarding the β-globin locus in combination with the new molecular toolboxes and inducible erythroid progenitor cell lines will ensure that it remains an important model system to study the spatial organization of multigene loci. It is expected that analysis of the β-globin locus will keep a leading role in revealing the mechanisms of gene regulation and enhancing our understanding of the functional significance of chromatin folding *in vivo*.

Acknowledgments

We thank J. Kooren for critical reading of the manuscript and discussion. This work was financially supported by grants from the Dutch Scientific Organization (NWO) to F.G. (912–03–009) and to W.L. (912–04–082). We apologize to investigators whose work we did not discuss owing to space limitations.

References

Asturias, F. J. (2004). RNA polymerase II structure, and organization of the preinitiation complex. *Curr. Opin. Struct. Biol.* **14,** 121–129.

Bazett-Jones, D. P., Cote, J., Landel, C. C., Peterson, C. L., and Workman, J. L. (1999). The SWI/SNF complex creates loop domains in DNA and polynucleosome arrays and can disrupt DNA-histone contacts within these domains. *Mol. Cell. Biol.* **19,** 1470–1478.

Behringer, R. R., Hammer, R. E., Brinster, R. L., Palmiter, R. D., and Townes, T. M. (1987). Two 3′ sequences direct adult erythroid-specific expression of human beta-globin genes in transgenic mice. *Proc. Natl. Acad. Sci. USA* **84,** 7056–7060.

Bender, M. A., Bulger, M., Close, J., and Groudine, M. (2000a). Beta-globin gene switching and DNase I sensitivity of the endogenous beta-globin locus in mice do not require the locus control region. *Mol. Cell* **5,** 387–393.

Bender, M. A., Mehaffey, M. G., Telling, A., Hug, B., Ley, T. J., Groudine, M., and Fiering, S. (2000b). Independent formation of DNase I hypersensitive sites in the murine beta-globin locus control region. *Blood* **95,** 3600–3604.

Bender, M. A., Roach, J. N., Halow, J., Close, J., Alami, R., Bouhassira, E. E., Groudine, M., and Fiering, S. N. (2001). Targeted deletion of 5′HS1 and 5′HS4 of the beta-globin locus control region reveals additive activity of the DNase I hypersensitive sites. *Blood* **98,** 2022–2027.

Bender, M. A., Byron, R., Ragoczy, T., Telling, A., Bulger, M., and Groudine, M. (2006). Flanking HS-62. 5 and 3′ HS1, and regions upstream of the LCR, are not required for beta-globin transcription. *Blood* **108,** 1395–1401.

Berg, O. G., Winter, R. B., and von Hippel, P. H. (1981). Diffusion-driven mechanisms of protein translocation on nucleic acids. 1. Models and theory. *Biochemistry* **20,** 6929–6948.

Black, J. C., Choi, J. E., Lombardo, S. R., and Carey, M. (2006). A mechanism for coordinating chromatin modification and preinitiation complex assembly. *Mol. Cell* **23,** 809–818.

Blackwood, E. M., and Kadonaga, J. T. (1998). Going the distance: A current view of enhancer action. *Science* **281,** 61–63.

Blobel, G. A., Nakajima, T., Eckner, R., Montminy, M., and Orkin, S. H. (1998). CREB-binding protein cooperates with transcription factor GATA-1 and is required for erythroid differentiation. *Proc. Natl. Acad. Sci. USA* **95,** 2061–2066.

Blom van Assendelft, G., Hanscombe, O., Grosveld, F., and Greaves, D. R. (1989). The beta-globin dominant control region activates homologous and heterologous promoters in a tissue-specific manner. *Cell* **56,** 969–977.

Bodine, D. M., and Ley, T. J. (1987). An enhancer element lies 3′ to the human A gamma globin gene. *EMBO J.* **6,** 2997–3004.

Bottardi, S., Aumont, A., Grosveld, F., and Milot, E. (2003). Developmental stage-specific epigenetic control of human beta-globin gene expression is potentiated in hematopoietic progenitor cells prior to their transcriptional activation. *Blood* **102,** 3989–3997.

Bottardi, S., Ross, J., Pierre-Charles, N., Blank, V., and Milot, E. (2006). Lineage-specific activators affect beta-globin locus chromatin in multipotent hematopoietic progenitors. *EMBO J.* **25,** 3586–3595.

Brown, S. A., Weirich, C. S., Newton, E. M., and Kingston, R. E. (1998). Transcriptional activation domains stimulate initiation and elongation at different times and via different residues. *EMBO J.* **17,** 3146–3154.

Brown, J. M., Leach, J., Reittie, J. E., Atzberger, A., Lee-Prudhoe, J., Wood, W. G., Higgs, D. R., Iborra, F. J., and Buckle, V. J. (2006). Coregulated human globin genes are frequently in spatial proximity when active. *J. Cell Biol.* **172,** 177–187.

Bulger, M., and Groudine, M. (1999). Looping versus linking: Toward a model for long-distance gene activation. *Genes Dev.* **13,** 2465–2477.

Palstra et al.

Bulger, M., von Doorninck, J. H., Saitoh, N., Telling, A., Farrell, C., Bender, M. A., Felsenfeld, G., Axel, R., and Groudine, M. (1999). Conservation of sequence and structure flanking the mouse and human beta-globin loci: The β-globin genes are embedded within an array of odorant receptor genes. *Proc. Natl. Acad. Sci. USA* **96,** 5129–5134.

Bulger, M., Schubeler, D., Bender, M. A., Hamilton, J., Farrell, C. M., Hardison, R. C., and Groudine, M. (2003). A complex chromatin landscape revealed by patterns of nuclease sensitivity and histone modification within the mouse beta-globin locus. *Mol. Cell. Biol.* **23,** 5234–5244.

Cao, S. X., Gutman, P. D., Dave, H. P., and Schechter, A. N. (1989a). Identification of a transcriptional silencer in the 5′-flanking region of the human epsilon-globin gene. *Proc. Natl. Acad. Sci. USA* **86,** 5306–5309.

Cao, S. X., Gutman, P. D., Dave, H. P., and Schechter, A. N. (1989b). Negative control of the human epsilon-globin gene. *Prog. Clin. Biol. Res.* **316A,** 279–289.

Carter, D., Chakalova, L., Osborne, C. S., Dai, Y. F., and Fraser, P. (2002). Long-range chromatin regulatory interactions *in vivo. Nat. Genet.* **32,** 623–626.

Chada, K., Magram, J., and Costantini, F. (1986). An embryonic pattern of expression of a human fetal globin gene in transgenic mice. *Nature* **319,** 685–689.

Chakalova, L., Debrand, E., Mitchell, J. A., Osborne, C. S., and Fraser, P. (2005). Replication and transcription: Shaping the landscape of the genome. *Nat. Rev. Genet.* **6,** 669–677.

Chambeyron, S., and Bickmore, W. A. (2004). Does looping and clustering in the nucleus regulate gene expression? *Curr. Opin. Cell Biol.* **16,** 256–262.

Chan, H. M., and La Thangue, N. B. (2001). p300/CBP proteins: HATs for transcriptional bridges and scaffolds. *J. Cell Sci.* **114,** 2363–2373.

Cheng, X., Reginato, M. J., Andrews, N. C., and Lazar, M. A. (1997). The transcriptional integrator CREB-binding protein mediates positive cross talk between nuclear hormone receptors and the hematopoietic bZip protein p45/NF-E2. *Mol. Cell. Biol.* **17,** 1407–1416.

Chubb, J. R., Trcek, T., Shenoy, S. M., and Singer, R. H. (2006). Transcriptional pulsing of a developmental gene. *Curr. Biol.* **16,** 1018–1025.

Cook, P. R. (2002). Predicting three-dimensional genome structure from transcriptional activity. *Nat. Genet.* **32,** 347–352.

Cook, P. R. (2003). Nongenic transcription, gene regulation and action at a distance. *J. Cell Sci.* **116,** 4483–4491.

Crusselle-Davis, V. J., Vieira, K. F., Zhou, Z., Anantharaman, A., and Bungert, J. (2006). Antagonistic regulation of beta-globin gene expression by helix-loop-helix proteins USF and TFII-I. *Mol. Cell. Biol.* **26,** 6832–6843.

de Krom, M., van de Corput, M., von Lindern, M., Grosveld, F., and Strouboulis, J. (2002). Stochastic patterns in globin gene expression are established prior to transcriptional activation and are clonally inherited. *Mol. Cell* **9,** 1319–1326.

de Laat, W., and Grosveld, F. (2003). Spatial organization of gene expression: The active chromatin hub. *Chromosome Res.* **11,** 447–459.

de Villiers, J., Olson, L., Banerji, J., and Schaffner, W. (1983). Analysis of the transcriptional enhancer effect. *Cold Spring Harb. Symp. Quant. Biol.* **47**(Pt. 2), 911–919.

Dekker, J., Rippe, K., Dekker, M., and Kleckner, N. (2002). Capturing chromosome conformation. *Science* **295,** 1306–1311.

Delassus, S., Titley, I., and Enver, T. (1999). Functional and molecular analysis of hematopoietic progenitors derived from the aorta-gonad-mesonephros region of the mouse embryo. *Blood* **94,** 1495–1503.

Dieci, G., and Sentenac, A. (2003). Detours and shortcuts to transcription reinitiation. *Trends Biochem. Sci.* **28,** 202–209.

Dillon, N., and Grosveld, F. (1991). Human gamma-globin genes silenced independently of other genes in the beta-globin locus. *Nature* **350,** 252–254.

Dillon, N., and Sabbattini, P. (2000). Functional gene expression domains: Defining the functional unit of eukaryotic gene regulation. *Bioessays* **22**, 657–665.

Dillon, N., Trimborn, T., Strouboulis, J., Fraser, P., and Grosveld, F. (1997). The effect of distance on long-range chromatin interactions. *Mol. Cell* **1**, 131–139.

Dorsett, D. (1999). Distant liaisons: Long-range enhancer-promoter interactions in Drosophila. *Curr. Opin. Genet. Dev.* **9**, 505–514.

Driscoll, M. C., Dobkin, C. S., and Alter, B. P. (1989). Gamma delta beta-thalassemia due to a *de novo* mutation deleting the 5′ beta-globin gene activation-region hypersensitive sites. *Proc. Natl. Acad. Sci. USA* **86**, 7470–7474.

Drissen, R., Palstra, R.-J., Gillemans, N., Splinter, E., Grosveld, F., Philipsen, S., and de Laat, W. (2004). The active spatial organization of the beta-globin locus requires the transcription factor EKLF. *Genes Dev.* **18**, 2485–2490.

Droge, P., and Muller-Hill, B. (2001). High local protein concentrations at promoters: Strategies in prokaryotic and eukaryotic cells. *Bioessays* **23**, 179–183.

Dunaway, M., and Droge, P. (1989). Transactivation of the Xenopus rRNA gene promoter by its enhancer. *Nature* **341**, 657–659.

Ellis, J., Talbot, D., Dillon, N., and Grosveld, F. (1993). Synthetic human beta-globin 5′HS2 constructs function as locus control regions only in multicopy transgene concatamers. *EMBO J.* **12**, 127–134.

Epner, E., Reik, A., Cimbora, D., Telling, A., Bender, M. A., Fiering, S., Enver, T., Martin, D. I., Kennedy, M., Keller, G., and Groudine, M. (1998). The beta-globin LCR is not necessary for an open chromatin structure or developmentally regulated transcription of the native mouse beta-globin locus. *Mol. Cell* **2**, 447–455.

Fang, X., Xiang, P., Yin, W., Stamatoyannopoulos, G., and Li, Q. (2007). Cooperativeness of the higher chromatin structure of the beta-globin locus revealed by the deletion mutations of DNase I hypersensitive site 3 of the LCR. *J. Mol. Biol.* **365**, 31–37.

Faro-Trindade, I., and Cook, P. R. (2006). Transcription factories: Structures conserved during differentiation and evolution. *Biochem. Soc. Trans.* **34**, 1133–1137.

Farrell, C. M., Grinberg, A., Huang, S. P., Chen, D., Pichel, J. G., Westphal, H., and Felsenfeld, G. (2000). A large upstream region is not necessary for gene expression or hypersensitive site formation at the mouse beta-globin locus. *Proc. Natl. Acad. Sci. USA* **97**, 14554–14559.

Feng, D., and Kan, Y. W. (2005). The binding of the ubiquitous transcription factor Sp1 at the locus control region represses the expression of beta-like globin genes. *Proc. Natl. Acad. Sci. USA* **102**, 9896–9900.

Fiering, S., Epner, E., Robinson, K., Zhuang, Y., Telling, A., Hu, M., Martin, D. I., Enver, T., Ley, T. J., and Groudine, M. (1995). Targeted deletion of 5′HS2 of the murine beta-globin LCR reveals that it is not essential for proper regulation of the beta-globin locus. *Genes Dev.* **9**, 2203–2213.

Forrester, W. C., Thompson, C., Elder, J. T., and Groudine, M. (1986). A developmentally stable chromatin structure in the human beta-globin gene cluster. *Proc. Natl. Acad. Sci. USA* **83**, 1359–1363.

Forrester, W. C., Takegawa, S., Papayannopoulou, T., Stamatoyannopoulos, G., and Groudine, M. (1987). Evidence for a locus activation region: The formation of developmentally stable hypersensitive sites in globin-expressing hybrids. *Nucleic. Acids Res.* **15**, 10159–10177.

Forrester, W. C., Epner, E., Driscoll, M. C., Enver, T., Brice, M., Papayannopoulou, T., and Groudine, M. (1990). A deletion of the human beta-globin locus activation region causes a major alteration in chromatin structure and replication across the entire beta-globin locus. *Genes Dev.* **4**, 1637–1649.

Forsberg, E. C., Johnson, K., Zaboikina, T. N., Mosser, E. A., and Bresnick, E. H. (1999). Requirement of an E1A-sensitive coactivator for long-range transactivation by the beta-globin locus control region. *J. Biol. Chem.* **274**, 26850–26859.

Forsberg, E. C., Downs, K. M., Christensen, H. M., Im, H., Nuzzi, P. A., and Bresnick, E. H. (2000). Developmentally dynamic histone acetylation pattern of a tissue-specific chromatin domain. *Proc. Natl. Acad. Sci. USA* **97,** 14494–14499.

Francastel, C., Walters, M. C., Groudine, M., and Martin, D. I. (1999). A functional enhancer suppresses silencing of a transgene and prevents its localization close to centrometric heterochromatin. *Cell* **99,** 259–269.

Fraser, P. (2006). Transcriptional control thrown for a loop. *Curr. Opin. Genet. Dev.* **16,** 490–495.

Fraser, P., and Grosveld, F. (1998). Locus control regions, chromatin activation and transcription. *Curr. Opin. Cell. Biol.* **10,** 361–365.

Gause, M., Morcillo, P., and Dorsett, D. (2001). Insulation of enhancer-promoter communication by a gypsy transposon insert in the Drosophila cut gene: Cooperation between suppressor of hairy-wing and modifier of mdg4 proteins. *Mol. Cell. Biol.* **21,** 4807–4817.

Giglioni, B., Casini, C., Mantovani, R., Merli, S., Comi, P., Ottolenghi, S., Saglio, G., Camaschella, C., and Mazza, U. (1984). A molecular study of a family with Greek hereditary persistence of fetal hemoglobin and beta-thalassemia. *EMBO J.* **3,** 2641–2645.

Goodwin, A. J., McInerney, J. M., Glander, M. A., Pomerantz, O., and Lowrey, C. H. (2001). In vivo formation of a human beta-globin locus control region core element requires binding sites for multiple factors including GATA-1, NF-E2, erythroid Kruppel-like factor, and Sp1. *J. Biol. Chem.* **276,** 26883–26892.

Gribnau, J., de Boer, E., Trimborn, T., Wijgerde, M., Milot, E., Grosveld, F., and Fraser, P. (1998). Chromatin interaction mechanism of transcriptional control in vivo. *EMBO J.* **17,** 6020–6027.

Gribnau, J., Diderich, K., Pruzina, S., Calzolari, R., and Fraser, P. (2000). Intergenic transcription and developmental remodeling of chromatin subdomains in the human beta-globin locus. *Mol. Cell* **5,** 377–386.

Grosveld, F. (1999). Activation by locus control regions? *Curr. Opin. Genet. Dev.* **9,** 152–157.

Grosveld, F., van Assendelft, G. B., Greaves, D. R., and Kollias, G. (1987). Position-independent, high-level expression of the human beta-globin gene in transgenic mice. *Cell* **51,** 975–985.

Groudine, M., Kohwi-Shigematsu, T., Gelinas, R., Stamatoyannopoulos, G., and Papayannopoulou, T. (1983). Human fetal to adult hemoglobin switching: Changes in chromatin structure of the beta-globin gene locus. *Proc. Natl. Acad. Sci. USA* **80,** 7551–7555.

Guy, L. G., Kothary, R., DeRepentigny, Y., Delvoye, N., Ellis, J., and Wall, L. (1996). The beta-globin locus control region enhances transcription of but does not confer position-independent expression onto the lacZ gene in transgenic mice. *EMBO J.* **15,** 3713–3721.

Hanscombe, O., Whyatt, D., Fraser, P., Yannoutsos, N., Greaves, D., Dillon, N., and Grosveld, F. (1991). Importance of globin gene order for correct developmental expression. *Genes Dev.* **5,** 1387–1394.

Hardison, R., Slightom, J. L., Gumucio, D. L., Goodman, M., Stojanovic, N., and Miller, W. (1997). Locus control regions of mammalian beta-globin gene clusters: Combining phylogenetic analyses and experimental results to gain functional insights. *Gene* **205,** 73–94.

Harju, S., McQueen, K. J., and Peterson, K. R. (2002). Chromatin structure and control of beta-like globin gene switching. *Exp. Biol. Med. (Maywood)* **227,** 683–700.

Hatzis, P., and Talianidis, I. (2002). Dynamics of enhancer-promoter communication during differentiation-induced gene activation. *Mol. Cell* **10,** 1467–1477.

Heuchel, R., Matthias, P., and Schaffner, W. (1989). Two closely spaced promoters are equally activated by a remote enhancer: Evidence against a scanning model for enhancer action. *Nucleic Acids Res.* **17,** 8931–8947.

Higgs, D. R. (1998). Do LCRs open chromatin domains? *Cell* **95,** 299–302.

Hu, M., Krause, D., Greaves, M., Sharkis, S., Dexter, M., Heyworth, C., and Enver, T. (1997). Multilineage gene expression precedes commitment in the hemopoietic system. *Genes Dev.* **11,** 774–785.

Hug, B. A., Wesselschmidt, R. L., Fiering, S., Bender, M. A., Epner, E., Groudine, M., and Ley, T. J. (1996). Analysis of mice containing a targeted deletion of beta-globin locus control region 5′ hypersensitive site 3. *Mol. Cell. Biol.* **16**, 2906–2912.

Im, H., Park, C., Feng, Q., Johnson, K. D., Kiekhaefer, C. M., Choi, K., Zhang, Y., and Bresnick, E. H. (2003). Dynamic regulation of histone H3 methylated at lysine 79 within a tissue-specific chromatin domain. *J. Biol. Chem.* **278**, 18346–18352.

Johnson, K. D., Christensen, H. M., Zhao, B., and Bresnick, E. H. (2001). Distinct mechanisms control RNA polymerase II recruitment to a tissue-specific locus control region and a downstream promoter. *Mol. Cell* **8**, 465–471.

Johnson, K. D., Grass, J. A., Boyer, M. E., Kiekhaefer, C. M., Blobel, G. A., Weiss, M. J., and Bresnick, E. H. (2002). Cooperative activities of hematopoietic regulators recruit RNA polymerase II to a tissue-specific chromatin domain. *Proc. Natl. Acad. Sci. USA* **99**, 11760–11765.

Johnson, K. D., Grass, J. A., Park, C., Im, H., Choi, K., and Bresnick, E. H. (2003). Highly restricted localization of RNA polymerase II within a locus control region of a tissue-specific chromatin domain. *Mol. Cell. Biol.* **23**, 6484–6493.

Kadener, S., Fededa, J. P., Rosbash, M., and Kornblihtt, A. R. (2002). Regulation of alternative splicing by a transcriptional enhancer through RNA pol II elongation. *Proc. Natl. Acad. Sci. USA* **99**, 8185–8190.

Kiekhaefer, C. M., Grass, J. A., Johnson, K. D., Boyer, M. E., and Bresnick, E. H. (2002). Hemato-poietic-specific activators establish an overlapping pattern of histone acetylation and methylation within a mammalian chromatin domain. *Proc. Natl. Acad. Sci. USA* **99**, 14309–14314.

Kimura, H., Sugaya, K., and Cook, P. R. (2002). The transcription cycle of RNA polymerase II in living cells. *J. Cell Biol.* **159**, 777–782.

Kioussis, D., Vanin, E., deLange, T., Flavell, R. A., and Grosveld, F. G. (1983). Beta-globin gene inactivation by DNA translocation in gamma beta-thalassaemia. *Nature* **306**, 662–666.

Kolesky, S. E., Ouhammouch, M., and Geiduschek, E. P. (2002). The mechanism of transcriptional activation by the topologically DNA-linked sliding clamp of bacteriophage T4. *J. Mol. Biol.* **321**, 767–784.

Kollias, G., Wrighton, N., Hurst, J., and Grosveld, F. (1986). Regulated expression of human A gamma-, beta-, and hybrid gamma beta-globin genes in transgenic mice: Manipulation of the developmental expression patterns. *Cell* **46**, 89–94.

Kollias, G., Hurst, J., deBoer, E., and Grosveld, F. (1987). The human beta-globin gene contains a downstream developmental specific enhancer. *Nucleic Acids Res.* **15**, 5739–5747.

Kong, S., Bohl, D., Li, C., and Tuan, D. (1997). Transcription of the HS2 enhancer toward a cis-linked gene is independent of the orientation, position, and distance of the enhancer relative to the gene. *Mol. Cell. Biol.* **17**, 3955–3965.

Kooren, J., Palstra, R. J., Klous, P., Splinter, E., von Lindern, M., Grosveld, F., and de Laat, W. (2007). Beta-globin active chromatin Hub formation in differentiating erythroid Cells and in p45 NF-E2 knock-out mice. *J. Biol. Chem.* **282**, 16544–16552.

Krumm, A., Hickey, L. B., and Groudine, M. (1995). Promoter-proximal pausing of RNA polymerase II defines a general rate-limiting step after transcription initiation. *Genes Dev.* **9**, 559–572.

Kuras, L., Borggrefe, T., and Kornberg, R. D. (2003). Association of the Mediator complex with enhancers of active genes. *Proc. Natl. Acad. Sci. USA* **100**, 13887–13891.

Lacy, E., Roberts, S., Evans, E. P., Burtenshaw, M. D., and Costantini, F. D. (1983). A foreign beta-globin gene in transgenic mice: Integration at abnormal chromosomal positions and expression in inappropriate tissues. *Cell* **34**, 343–358.

Lanctot, C., Cheutin, T., Cremer, M., Cavalli, G., and Cremer, T. (2007). Dynamic genome architecture in the nuclear space: Regulation of gene expression in three dimensions. *Nat. Rev. Genet.* **8**, 104–115.

Levings, P. P., Zhou, Z., Vieira, K. F., Crusselle-Davis, V. J., and Bungert, J. (2006). Recruitment of transcription complexes to the beta-globin locus control region and transcription of hypersensitive site 3 prior to erythroid differentiation of murine embryonic stem cells. _FEBS J._ **273,** 746–755.

Li, B., Carey, M., and Workman, J. L. (2007). The role of chromatin during transcription. _Cell_ **128,** 707–719.

Li, Q., Peterson, K. R., Fang, X., and Stamatoyannopoulos, G. (2002). Locus control regions. _Blood_ **100,** 3077–3086.

Li, Q., Barkess, G., and Qian, H. (2006). Chromatin looping and the probability of transcription. _Trends Genet._ **22,** 197–202.

Ling, K. W., and Dzierzak, E. (2002). Ontogeny and genetics of the hemato/lymphopoietic system. _Curr. Opin. Immunol._ **14,** 186–191.

Ling, J., Ainol, L., Zhang, L., Yu, X., Pi, W., and Tuan, D. (2004). HS2 enhancer function is blocked by a transcriptional terminator inserted between the enhancer and the promoter. _J. Biol. Chem._ **279,** 51704–51713.

Liu, Z., and Garrard, W. T. (2005). Long-range interactions between three transcriptional enhancers, active Vkappa gene promoters, and a 3′ boundary sequence spanning 46 kilobases. _Mol. Cell. Biol._ **25,** 3220–3231.

Lomvardas, S., Barnea, G., Pisapia, D. J., Mendelsohn, M., Kirkland, J., and Axel, R. (2006). Interchromosomal interactions and olfactory receptor choice. _Cell_ **126,** 403–413.

Magram, J., Chada, K., and Costantini, F. (1985). Developmental regulation of a cloned adult beta-globin gene in transgenic mice. _Nature_ **315,** 338–340.

Mahajan, M. C., and Weissman, S. M. (2006). Multi-protein complexes at the beta-globin locus. _Brief. Funct. Genomic. Proteomic._ **5,** 62–65.

Mahmoudi, T., Katsani, K. R., and Verrijzer, C. P. (2002). GAGA can mediate enhancer function in trans by linking two separate DNA molecules. _EMBO J._ **21,** 1775–1781.

Marenduzzo, D., Finan, K., and Cook, P. R. (2006a). The depletion attraction: An underappreciated force driving cellular organization. _J. Cell Biol._ **175,** 681–686.

Marenduzzo, D., Micheletti, C., and Cook, P. R. (2006b). Entropy-driven genome organization. _Biophys. J._ **90,** 3712–3721.

Mastrangelo, I. A., Courey, A. J., Wall, J. S., Jackson, S. P., and Hough, P. V. (1991). DNA looping and Sp1 multimer links: A mechanism for transcriptional synergism and enhancement. _Proc. Natl. Acad. Sci. USA_ **88,** 5670–5674.

Matthews, K. S., and Nichols, J. C. (1998). Lactose repressor protein: Functional properties and structure. _Prog. Nucleic Acid Res. Mol. Biol._ **58,** 127–164.

Meier, N., Krpic, S., Rodriguez, P., Strouboulis, J., Monti, M., Krijgsveld, J., Gering, M., Patient, R., Hostert, A., and Grosveld, F. (2006). Novel binding partners of Ldb1 are required for haematopoietic development. _Development_ **133,** 4913–4923.

Milot, E., Strouboulis, J., Trimborn, T., Wijgerde, M., de Boer, E., Langeveld, A., Tan-Un, K., Vergeer, W., Yannoutsos, N., Grosveld, F., and Fraser, P. (1996). Heterochromatin effects on the frequency and duration of LCR-mediated gene transcription. _Cell_ **87,** 105–114.

Misteli, T. (2001a). The concept of self-organization in cellular architecture. _J. Cell Biol._ **155,** 181–185.

Misteli, T. (2001b). Protein dynamics: Implications for nuclear architecture and gene expression. _Science_ **291,** 843–847.

Moon, A. M., and Ley, T. J. (1990). Conservation of the primary structure, organization, and function of the human and mouse beta-globin locus-activating regions. _Proc. Natl. Acad. Sci. USA_ **87,** 7693–7697.

Morcillo, P., Rosen, C., Baylies, M. K., and Dorsett, D. (1997). Chip, a widely expressed chromosomal protein required for segmentation and activity of a remote wing margin enhancer in Drosophila. Genes Dev. 11, 2729–2740.

Morris, J. R., Geyer, P. K., and Wu, C. T. (1999). Core promoter elements can regulate transcription on a separate chromosome in trans. Genes Dev. 13, 253–258.

Mueller-Storm, H. P., Sogo, J. M., and Schaffner, W. (1989). An enhancer stimulates transcription in trans when attached to the promoter via a protein bridge. Cell 58, 767–777.

Muller, A. M., Medvinsky, A., Strouboulis, J., Grosveld, F., and Dzierzak, E. (1994). Development of hematopoietic stem cell activity in the mouse embryo. Immunity 1, 291–301.

Noordermeer, D., Branco, M. R., Splinter, E., Klous, P., van IJcken, W., Swagemakers, S., Koutsourakis, M., van der Spek, P., Pombo, A., and de Laat, W. (in Press). Transcription and chromatin organization of a housekeeping gene cluster containing an integrated beta-globin Locus Control Region.

Osborne, C. S., Chakalova, L., Brown, K. E., Carter, D., Horton, A., Debrand, E., Goyenechea, B., Mitchell, J. A., Lopes, S., Reik, W., and Fraser, P. (2004). Active genes dynamically colocalize to shared sites of ongoing transcription. Nat. Genet. 36, 1065–1071.

O'Sullivan, J. M., Tan-Wong, S. M., Morillon, A., Lee, B., Coles, J., Mellor, J., and Proudfoot, N. J. (2004). Gene loops juxtapose promoters and terminators in yeast. Nat. Genet. 36, 1014–1018.

Palstra, R. J., Tolhuis, B., Splinter, E., Nijmeijer, R., Grosveld, F., and de Laat, W. (2003). The beta-globin nuclear compartment in development and erythroid differentiation. Nat. Genet. 35, 190–194.

Patrinos, G. P., de Krom, M., de Boer, E., Langeveld, A., Imam, A. M., Strouboulis, J., de Laat, W., and Grosveld, F. G. (2004). Multiple interactions between regulatory regions are required to stabilize an active chromatin hub. Genes Dev. 18, 1495–1509.

Peterson, K. R., and Stamatoyannopoulos, G. (1993). Role of gene order in developmental control of human gamma- and beta-globin gene expression. Mol. Cell. Biol. 13, 4836–4843.

Phair, R. D., Scaffidi, P., Elbi, C., Vecerova, J., Dey, A., Ozato, K., Brown, D. T., Hager, G., Bustin, M., and Misteli, T. (2004). Global nature of dynamic protein-chromatin interactions in vivo: Three-dimensional genome scanning and dynamic interaction networks of chromatin proteins. Mol. Cell. Biol. 24, 6393–6402.

Ptashne, M. (1986). Gene regulation by proteins acting nearby and at a distance. Nature 322, 697–701.

Ptashne, M., and Gann, A. (1997). Transcriptional activation by recruitment. Nature 386, 569–577.

Ragoczy, T., Telling, A., Sawado, T., Groudine, M., and Kosak, S. T. (2003). A genetic analysis of chromosome territory looping: Diverse roles for distal regulatory elements. Chromosome Res. 11, 513–525.

Ragoczy, T., Bender, M. A., Telling, A., Byron, R., and Groudine, M. (2006). The locus control region is required for association of the murine beta-globin locus with engaged transcription factories during erythroid maturation. Genes Dev. 20, 1447–1457.

Raich, N., Papayannopoulou, T., Stamatoyannopoulos, G., and Enver, T. (1992). Demonstration of a human epsilon-globin gene silencer with studies in transgenic mice. Blood 79, 861–864.

Reik, A., Telling, A., Zitnik, G., Cimbora, D., Epner, E., and Groudine, M. (1998). The locus control region is necessary for gene expression in the human beta-globin locus but not the maintenance of an open chromatin structure in erythroid cells. Mol. Cell. Biol. 18, 5992–6000.

Reitman, M., Lee, E., Westphal, H., and Felsenfeld, G. (1993). An enhancer/locus control region is not sufficient to open chromatin. Mol. Cell. Biol. 13, 3990–3998.

Ringrose, L., Chabanis, S., Angrand, P. O., Woodroofe, C., and Stewart, A. F. (1999). Quantitative comparison of DNA looping in vitro and in vivo: Chromatin increases effective DNA flexibility at short distances. EMBO J. 18, 6630–6641.

Rippe, K. (2001). Making contacts on a nucleic acid polymer. Trends Biochem. Sci. 26, 733–740.

Rollins, R. A., Morcillo, P., and Dorsett, D. (1999). Nipped-B, a Drosophila homologue of chromosomal adherins, participates in activation by remote enhancers in the cut and Ultrabithorax genes. *Genetics* **152**, 577–593.

Ronchi, A., Berry, M., Raguz, S., Imam, A., Yannoutsos, N., Ottolenghi, S., Grosveld, F., and Dillon, N. (1996). Role of the duplicated CCAAT box region in gamma-globin gene regulation and hereditary persistence of fetal haemoglobin. *Embo J.* **15**, 143–149.

Rossi, F. M., Kringstein, A. M., Spicher, A., Guicherit, O. M., and Blau, H. M. (2000). Transcriptional control: Rheostat converted to on/off switch. *Mol. Cell* **6**, 723–728.

Ryser, S., Fujita, T., Tortola, S., Piuz, I., and Schlegel, W. (2007). The rate of c-fos transcription *in vivo* is continuously regulated at the level of elongation by dynamic stimulus-coupled recruitment of positive transcription elongation factor b. *J. Biol. Chem.* **282**, 5075–5084.

Sawado, T., Halow, J., Bender, M. A., and Groudine, M. (2003). The beta-globin locus control region (LCR) functions primarily by enhancing the transition from transcription initiation to elongation. *Genes Dev.* **17**, 1009–1018.

Sayegh, C., Jhunjhunwala, S., Riblet, R., and Murre, C. (2005). Visualization of looping involving the immunoglobulin heavy-chain locus in developing B cells. *Genes Dev.* **19**, 322–327.

Schubeler, D., Francastel, C., Cimbora, D. M., Reik, A., Martin, D. I., and Groudine, M. (2000). Nuclear localization and histone acetylation: A pathway for chromatin opening and transcriptional activation of the human beta-globin locus. *Genes Dev.* **14**, 940–950.

Schubeler, D., Groudine, M., and Bender, M. A. (2001). The murine beta-globin locus control region regulates the rate of transcription but not the hyperacetylation of histones at the active genes. *Proc. Natl. Acad. Sci. USA* **98**, 11432–11437.

Shivdasani, R. A., and Orkin, S. H. (1995). Erythropoiesis and globin gene expression in mice lacking the transcription factor NF-E2. *Proc. Acad. Sci. USA* **92**, 8690–8694.

Skok, J. A., Gisler, R., Novatchkova, M., Farmer, D., de Laat, W., and Busslinger, M. (2007). Reversible contraction by looping of the Tcra and Tcrb loci in rearranging thymocytes. *Nat. Immunol.* doi:10. 1038/ni1448.

Spilianakis, C. G., and Flavell, R. A. (2004). Long-range intrachromosomal interactions in the T helper type 2 cytokine locus. *Nat. Immunol.* **5**, 1017–1027.

Splinter, E., Heath, H., Kooren, J., Palstra, R. J., Klous, P., Grosveld, F., Galjart, N., and de Laat, W. (2006). CTCF mediates long-range chromatin looping and local histone modification in the beta-globin locus. *Genes Dev.* **20**, 2349–2354.

Stamatoyannopoulos, G., and Grosveld, F. (2001). Hemoglobin switching. *In* "The Molecular Basis of Blood Diseases" (G. Stamatoyannopoulos, P. Majerus, R. Perlmutter, and H. Varmus, eds.), pp. 135–182. W. B. Saunders, Philadelphia.

Stumpf, M., Waskow, C., Krotschel, M., van Essen, D., Rodriguez, P., Zhang, X., Guyot, B., Roeder, R. G., and Borggrefe, T. (2006). The mediator complex functions as a coactivator for GATA-1 in erythropoiesis via subunit Med1/TRAP220. *Proc. Natl. Acad. Sci. USA* **103**, 18504–18509.

Su, W., Jackson, S., Tjian, R., and Echols, H. (1991). DNA looping between sites for transcriptional activation: Self-association of DNA-bound Sp1. *Genes. Dev.* **5**, 820–826.

Svejstrup, J. Q. (2004). The RNA polymerase II transcription cycle: Cycling through chromatin. *Biochim. Biophys. Acta.* **1677**, 64–73.

Szutorisz, H., Canzonetta, C., Georgiou, A., Chow, C. M., Tora, L., and Dillon, N. (2005). Formation of an active tissue-specific chromatin domain initiated by epigenetic marking at the embryonic stem cell stage. *Mol. Cell. Biol.* **25**, 1804–1820.

Tanimoto, K., Liu, Q., Bungert, J., and Engel, J. D. (1999). Effects of altered gene order or orientation of the locus control region on human beta-globin gene expression in mice. *Nature* **398**, 344–348.

Tewari, R., Gillemans, N., Harper, A., Wijgerde, M., Zafarana, G., Drabek, D., Grosveld, F., and Philipsen, S. (1996). The human beta-globin locus control region confers an early embryonic erythroid-specific expression pattern to a basic promoter driving the bacterial lacZ gene. *Development* **122**, 3991–3999.

Tolhuis, B., Palstra, R. J., Splinter, E., Grosveld, F., and de Laat, W. (2002). Looping and interaction between hypersensitive sites in the active beta-globin locus. *Mol. Cell* **10**, 1453–1465.

Torigoi, E., Bennani-Baiti, I. M., Rosen, C., Gonzalez, K., Morcillo, P., Ptashne, M., and Dorsett, D. (2000). Chip interacts with diverse homeodomain proteins and potentiates bicoid activity *in vivo*. *Proc. Natl. Acad. Sci. USA* **97**, 2686–2691.

Trimborn, T., Gribnau, J., Grosveld, F., and Fraser, P. (1999). Mechanisms of developmental control of transcription in the murine alpha- and beta-globin loci. *Genes Dev.* **13**, 112–124.

Tuan, D., Solomon, W., Li, Q., and London, I. M. (1985). The "beta-like-globin" gene domain in human erythroid cells. *Proc. Natl. Acad. Sci. USA* **82**, 6384–6388.

Tuan, D., Kong, S., and Hu, K. (1992). Transcription of the hypersensitive site HS2 enhancer in erythroid cells. *Proc. Natl. Acad. Sci. USA* **89**, 11219–11223.

Vakoc, C. R., Letting, D. L., Gheldof, N., Sawado, T., Bender, M. A., Groudine, M., Weiss, M. J., Dekker, J., and Blobel, G. A. (2005). Proximity among distant regulatory elements at the beta-globin locus requires GATA-1 and FOG-1. *Mol. Cell* **17**, 453–462.

Vieira, K. F., Levings, P. P., Hill, M. A., Crusselle, V. J., Kang, S. H., Engel, J. D., and Bungert, J. (2004). Recruitment of transcription complexes to the beta-globin gene locus *in vivo* and *in vitro*. *J. Biol. Chem.* **279**, 50350–50357.

Wadman, I. A., Osada, H., Grutz, G. G., Agulnick, A. D., Westphal, H., Forster, A., and Rabbitts, T. H. (1997). The LIM-only protein Lmo2 is a bridging molecule assembling an erythroid, DNA-binding complex which includes the TAL1, E47, GATA-1 and Ldb1/NLI proteins. *EMBO J.* **16**, 3145–3157.

Wai, A. W., Gillemans, N., Raguz-Bolognesi, S., Pruzina, S., Zafarana, G., Meijer, D., Philipsen, S., and Grosveld, F. (2003). HS5 of the human beta-globin locus control region: A developmental stage-specific border in erythroid cells. *EMBO J.* **22**, 4489–4500.

Wasylyk, B., and Chambon, P. (1983). Potentiator effect of the SV40 72-bp repeat on initiation of transcription from heterologous promoter elements. *Cold Spring Harb. Symp. Quant. Biol.* **47**(Pt. 2), 921–934.

Weintraub, H., and Groudine, M. (1976). Chromosomal subunits in active genes have an altered conformation. *Science* **193**, 848–856.

West, A. G., and Fraser, P. (2005). Remote control of gene transcription. *Hum. Mol. Genet.* **14** (Spec. No. 1), R101–R111.

Wijgerde, M., Grosveld, F., and Fraser, P. (1995). Transcription complex stability and chromatin dynamics *in vivo*. *Nature* **377**, 209–213.

Winter, R. B., Berg, O. G., and von Hippel, P. H. (1981). Diffusion-driven mechanisms of protein translocation on nucleic acids. 3. The *Escherichia coli* lac repressor–operator interaction: Kinetic measurements and conclusions. *Biochemistry* **20**, 6961–6977.

Wu, C. T., and Morris, J. R. (1999). Transvection and other homology effects. *Curr. Opin. Genet. Dev.* **9**, 237–246.

Wu, S. Y., Zhou, T., and Chiang, C. M. (2003). Human mediator enhances activator-facilitated recruitment of RNA polymerase II and promoter recognition by TATA-binding protein (TBP) independently of TBP-associated factors. *Mol. Cell. Biol.* **23**, 6229–6242.

Yankulov, K., Blau, J., Purton, T., Roberts, S., and Bentley, D. L. (1994). Transcriptional elongation by RNA polymerase II is stimulated by transactivators. *Cell* **77**, 749–759.

Yoshida, C., Tokumasu, F., Hohmura, K. I., Bungert, J., Hayashi, N., Nagasawa, T., Engel, J. D., Yamamoto, M., Takeyasu, K., and Igarashi, K. (1999). Long range interaction of cis-DNA elements mediated by architectural transcription factor Bach1. *Genes Cells* **4**, 643–655.

Yudkovsky, N., Ranish, J. A., and Hahn, S. (2000). A transcription reinitiation intermediate that is stabilized by activator. *Nature* **408,** 225–229.

Zhang, W., and Bieker, J. J. (1998). Acetylation and modulation of erythroid Kruppel-like factor (EKLF) activity by interaction with histone acetyltransferases. *Proc. Natl. Acad. Sci. USA* **95,** 9855–9860.

Zhang, W., Kadam, S., Emerson, B. M., and Bieker, J. J. (2001). Site-specific acetylation by p300 or CREB binding protein regulates erythroid Kruppel-like factor transcriptional activity via its interaction with the SWI-SNF complex. *Mol. Cell. Biol.* **21,** 2413–2422.

Zhao, H., and Dean, A. (2004). An insulator blocks spreading of histone acetylation and interferes with RNA polymerase II transfer between an enhancer and gene. *Nucleic Acids Res.* **32,** 4903–4919.

5

Long-Range Regulation of α-Globin Gene Expression

Douglas R. Higgs, Douglas Vernimmen, and Bill Wood
MRC Molecular Haematology Unit, Weatherall Institute
of Molecular Medicine, John Radcliffe Hospital, Headington,
Oxford OX3 9DS, United Kingdom

I. Introduction
II. The Normal Structure and Evolution of the α-Globin Cluster
III. Functional Analysis of the α-Globin Regulatory Domain
IV. Structure of the Upstream Regulatory Elements and the Promoters
V. Transcription Factors Involved in Erythropoiesis
VI. Cellular Resources for Studying the Key Stages of Hematopoiesis
VII. Transcription Factor Binding to the Upstream Regulatory Elements
VIII. Transcription Factor Binding to the Promoter
IX. The Recruitment of RNA Polymerase and GTFs to the α-Globin Cluster
X. What Role Do the Remote Regulatory Elements Play?
XI. How Do the Upstream Elements Interact with the Promoter?
XII. Sequential Activation of the α-Globin Gene Cluster During Differentiation
XIII. Conclusions, Speculation, and Future Directions
Acknowledgments
References

Advances in Genetics, Vol. 61
0065-2660/08 $35.00
DOI: 10.1016/S0065-2660(07)00005-3

ABSTRACT

Over the past 20 years, there has been an increasing awareness that gene expression can be regulated by multiple *cis*-acting sequences located at considerable distances (10–1000 kb) from the genes they control. Detailed investigation of a few specialized mammalian genes, including the genes controlling the synthesis of hemoglobin, provide important models to understand how such long-range regulatory elements act. In general, these elements contain a high density of evolutionarily conserved, transcription factor-binding sites and in many ways resemble the upstream regulatory elements found adjacent to the promoters of genes in simpler organisms, differing only in the distance over which they act. We have investigated in detail how the remote regulatory elements of the α-globin cluster become activated as hematopoietic stem cells (HSCs) undergo commitment, lineage specification, and differentiation to form red blood cells. In turn, we have addressed how, during this process, the upstream elements control the correct spatial and temporal expression from the α-gene promoter which lies ∼60 kb downstream of these elements. At present too few loci have been studied to determine whether there are general principles underlying long-range regulation but some common themes are emerging. © 2008, Elsevier Inc.

I. INTRODUCTION

With completion of the sequence and initial annotation of the human genome, we now know that there are 20–25,000 genes (International Human Genome Sequencing Consortium, 2004). The next major question in biology is to understand how these genes are switched on and off during development and differentiation to control cell fate and to specify different lineages. Ultimately, it is these processes, encoded in the DNA sequence, which regulate production of the hundreds of different cell types that make up an organism. In outline, we now know that gene expression is regulated by complex networks involving *cis*-acting elements (e.g., promoters, enhancers, locus control regions, silencers, and insulators) and *trans*-acting factors [e.g., general transcription factors (GTFs), activators, and co-activators], which interact to recruit and/or activate RNA Polymerase II (Pol II) and, consequently, increase gene transcription (called the transcriptional network) (Maston *et al.*, 2006). The situation is further complicated by the fact that all of these processes take place in the context of DNA packaged into chromatin, and the various states of this packaging determine the accessibility of the *cis*-elements to the *trans*-acting factors (Felsenfeld and

Groudine, 2003). Since the state of chromatin packaging is also determined by a complex network of proteins (called the epigenetic network), integrating and understanding the transcriptional and epigenetic networks that control patterns of gene expression (so-called systems biology) will be an important but inordinately complex task.

In the medium term, a more realistic aim is to understand in detail how individual genes are switched on and off in the context of the transcriptional and epigenetic networks that emerge during differentiation and development: in other words, to elucidate the principles and mechanisms driving one unit of the "system." At first sight, this appears to be relatively straightforward task. However, we now know that expression of an individual gene may be controlled by several regulatory elements (each ~200–500 bp) linked to a corresponding promoter, forming a functional unit of chromatin which may include elements dispersed over tens or hundreds of kilobases of DNA (Dillon and Sabbattini, 2000; Kleinjan and van Heyningen, 2005). Furthermore, it has emerged that the cis-elements are not bound by single activators but by complexes that often contain a large number of proteins associated with DNA wrapped in nucleosomes, which themselves are multiprotein structures. Ultimately, these multiprotein complexes, bound at a variety of cis-acting elements, interact to recruit and activate the preinitiation complex (PIC) (including RNA Pol II, and the GTF) at the promoter of the gene in question (Maston et al., 2006).

In outline, we need to identify all of the cis-acting sequences controlling expression of that gene (the regulatory domain) and then identify all of the trans-acting factors and cofactors that bind these sequences and finally understand how these protein–DNA complexes interact within the context of chromatin to influence transcription of the gene. The timing of expression during differentiation and development is also important, and, therefore, an additional, important issue is to determine the hierarchy and order of events that lead to activation. This can be achieved either by analyzing the gene in question in purified primary cells or by studying cell lines that can be manipulated to mimic development and differentiation as it occurs in vivo.

Therefore, in summary, depending on the level of detail sought, understanding the mechanism by which a single gene is activated represents a complex but potentially tractable problem in the foreseeable future. To date, very few mammalian genes have been studied in appropriate depth, so we do not know whether there are general principles governing the regulation of gene expression from complex loci in vivo or whether all genes, to some extent, differ in the way that they achieve this. Among all mammalian genes, the α- and β-globin loci have been two of the most intensively studied, and both appear to be regulated at the level of transcriptional initiation. In this chapter, we review how the α-globin genes are regulated during the synthesis of red blood cells (erythropoiesis), and in Chapter 4 the mechanisms underlying β-globin expression are discussed.

II. THE NORMAL STRUCTURE AND EVOLUTION OF THE α-GLOBIN CLUSTER

Before we can analyze how the transcription factor program is played out on the α-globin cluster, it is important to identify all of the *cis*-acting sequences that contribute to α-globin expression. The α-globin cluster has now been sequenced in more than 22 species spanning 500 million years of evolution, and in all of these there is an embryonic gene (ζ), at least one α gene and most species also have two minor globin genes (αD and θ) (Hughes *et al.*, 2005). In human, the α-globin cluster lies near the tip of chromosome 16 (16p13.3) with the structure telomere-ζ $-\alpha$D$-\alpha$2$-\alpha$1$-\theta$-centromere (Fig. 5.1) (Hughes *et al.*, 2005). Comparative sequence analysis has identified a region of $\sim$135 kb containing the α-globin cluster as well as several widely expressed, flanking genes arranged in the same order (conserved synteny) in all 22 species studied (Fig. 5.1). Although in many species this region is located close to a telomere (as in human chromosome 16), in mouse, it lies at an interstitial region (Hughes *et al.*, 2005; Tan and Whitney, 1993). This observation suggests that the translocated, conserved, syntenic segment (135 kb) contains all of the *cis*-acting elements required to obtain fully regulated tissue- and developmental stage-expression of the α-globin cluster.

Having delimited the region containing all *cis*-acting sequences (the regulatory domain), it is now generally accepted that the most efficient way to identify critical *cis*-acting elements is to search for highly conserved, noncoding DNA sequences [Multispecies conserved sequences (MCS)] and to map the positions of DNAse 1 hypersensitive sites (DHSs) in chromatin, *in vivo* (ENCODE Project Consortium, 2007). These two features often (but not always) coincide, and the tissue-specific pattern of the DHS reflects when the underlying *cis*-elements are active. Using these approaches to analyze the human and mouse α-globin clusters has revealed a set of conserved, erythroid-specific elements (Fig. 5.1 and Table 5.1) intermingled with many constitutive DHSs associated with the widely expressed genes in this area (Fig. 5.1) (Flint *et al.*, 1997; Higgs *et al.*, 1990; Hughes *et al.*, 2005; Vyas *et al.*, 1992). These erythroid-specific elements include the promoters of the ζ- and α-globin genes and four MCS elements (MCS-R1 to 4) lying 30–70 kb upstream of the α-globin genes, suggesting that these long-range elements may be involved in the regulation of globin gene expression. In mouse, an additional, nonconserved region associated with an erythroid-specific DHS (HS-12) has been identified (Anguita *et al.*, 2004; Hughes *et al.*, 2005; Kielman *et al.*, 1996).

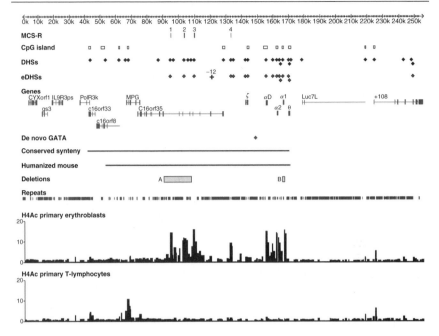

Figure 5.1. Overview of the human α-globin cluster. In the center, the line marked "genes" shows the annotated arrangement of the α-globin genes and the surrounding widely expressed genes. Above this, the position of all DNAse 1 hypersensitive sites (DHSs) and erythroid-specific DHSs (eDHS) are shown. A cross marks the position of a species-specific DHS (HS-12) seen only in mouse. Above this, the positions of MCS-elements (MCS-R1 to 4) are shown. Below the cluster, the region of conserved synteny and the region recombined into the "humanized mouse" model are shown. Two key mutants that define the essential upstream elements (mutant A, Viprakasit et al., 2006) and promoter elements (Mutant B, Poodt et al., 2006) are shown. Below examples of the profile (using ChIP on chip, De Gobbi et al., 2007) of histone H4 acetylation in primary erythroid and non-erythroid (T cells) are shown.

III. FUNCTIONAL ANALYSIS OF THE α-GLOBIN REGULATORY DOMAIN

The most valuable resource for relating structure to function has come from analysis of the natural mutants of the endogenous human locus, which cause downregulation of α-globin expression and give rise to the common human genetic disease α thalassemia. Alpha thalassemia most frequently results from more than 30 large deletions (~3 to >100 kb), which remove one or both adult α-globin genes (α2 and α1) (Higgs, 2001). But one rare, small deletion (970 bp, Poodt et al., 2006) which causes α thalassemia delimits the critical sequences in

Table 5.1. Transcription Factors Involved in Erythropoiesis

TF	Class	Binding sequence	Expression	Comment	Key References
GATA1	Zinc Finger (C2C2)	WGATAR	Prog, E, Meg, Eos, and Mast	Gene targeting studies have shown that GATA1 is essential for normal erythropoiesis. It binds the promoters and/or enhancers of virtually all erythroid-specific genes.	(Cantor and Orkin, 2002; Orkin, 2000)
GATA2	Zinc Finger (C2C2)	WGATAR	Prog, Meg, and Mast	GATA2 may initiate the erythroid program in early progenitors to be replaced later by GATA1. GATA2 also plays a role in expansion and maintenance of hematopoietic progenitors.	(Cantor and Orkin, 2002; Orkin, 2000)
SCL	bHLH	CAGGTG (E box)	E, Meg, and Mast	SCL heterodimerises with E proteins, such as E2A, and plays an essential role in early hematopoiesis. SCL also plays an important role in erythropoiesis often binding with GATA1 as part of an erythroid pentameric complex.	(Cantor and Orkin, 2002; Wadman et al., 1997)
E2A	bHLH	CAGGTG (E box)	Ubiquitous		
p18 (NF-E2)	b-Zip	TGCTGACTCAT	Ubiquitous	NF-E2 is a heterodimer of p45 and p18 (MafK) which binds the enhancers of many erythroid-specific genes. Surprisingly, a p45 knockout does not affect erythropoiesis, suggesting other proteins may play a similar redundant role.	(Andrews, 1998; Motohashi et al., 2002)
p45 (NF-E2)	b-Zip	TGCTGACTCAT	Prog, E, Meg, Gra, and Mast		

(Continues)

Table 5.1. (*Continued*)

TF	Class	Binding sequence	Expression	Comment	Key References
EKLF	Zinc Finger (C2H2)	CACC	Erythroid	Like GATA1, EKLF binds many erythroid-restricted genes. Mice in which EKLF is absent die from severe anemia in fetal life.	(Bieker, 2001)
Sp1	Zinc Finger (C2H2)	GGGGCGGGG	Ubiquitous	These are widely expressed transcription factors that are thought to play an important, but possibly redundant, role at the promoters and enhancers of erythroid genes.	(Suske, 1999)
Sp3	Zinc Finger (C2H2)	GGGGCGGGG	Ubiquitous		(Suske, 1999; van Loo et al., 2003)
ZBP-89	Zinc Finger (C2H2)	<<G-rich box>>	Ubiquitous	ZBP-89 is a ubiquitously expressed Kruppel-type zinc finger transcription factor that binds GC-rich elements and interacts with GATA1. Disruption of ZBP-89 markedly impairs definitive erythropoiesis.	(Woo et al., 2005)
NF-Y	Heterodimeric CCAAT factor	CCAAT	Ubiquitous	NF-Y is a ubiquitous factor that binds a subset of CCAAT boxes. It has been shown to have a role in recruitment of Pol II.	(Kabe et al., 2005) (Mantovani, 1999)
FOG-1	Multi-type Zinc finger	Cofactor	Prog, E, Meg, and Mast	Specifically binds to the amino terminal finger of GATA1. Mice lacking FOG-1 have a very similar phenotype to the GATA1 knockout mice and die at mid-gestation from anemia.	(Cantor and Orkin, 2002; Orkin, 2000)

(*Continues*)

Table 5.1. (*Continued*)

TF	Class	Binding sequence	Expression	Comment	Key References
LMO2	LIM domain	Cofactor	Prog and E	Loss of function produces an identical phenotype to loss of SCL, with which it physically interacts, as part of the pentameric erythroid complex.	(Nam and Rabbitts, 2006)
Ldb-1	LIM domain	Cofactor	Widely expressed	Another member of the pentameric erythroid complex and a binding partner for general LIM domain transcription factors.	(Visvader et al., 1997)

Abbreviations: b-Zip, Basic Leucine Zipper factors; bHLH, Basic Helix-Loop-Helix factors; Prog, Progenitors; E, Erythroid; Meg, Megacaryocyte; Eos, Eosinophil; Gra, Granulocyte; Mast, Mast Cells.

and around the promoter required for α-globin activation (Deletion B Fig. 5.1). There are also 13 rare deletions which remove two or more of the remote MCS elements but leave the α-globin genes intact (Viprakasit *et al.*, 2003, 2006). The smallest of these deletions (Deletion A in Fig. 5.1) (Viprakasit *et al.*, 2006) removes MCS-R1 and -R2 (HS-48 and HS-40 in human), demonstrating that one or both of these elements are essential for activating the globin genes.

Naturally occurring mutants rely on serendipitous discovery, so α-globin expression has also been analyzed in experimental systems using transient and stable transfection and by the analysis of transgenic mice containing human constructs (summarized in Higgs *et al.*, 1998). In all of these experiments, human α-globin expression occurs at very low levels ($<1\%$ of the level seen in normal erythroblasts) unless the α gene is linked to the upstream elements (MCS-R1 to -R4 in Fig. 5.1). In these transgenes, of all the elements tested, MCS-R2 (which corresponds to HS-40 in human and HS-26 in mouse) is the only one capable of enhancing α-globin expression on its own. Furthermore, no consistent, additional activity is seen when the other upstream elements are added to such constructs (Sharpe *et al.*, 1993). However, such experiments are often unsatisfactory and difficult to interpret because there are variations in transgene copy number, and often it is not possible to fully map the transgenes. Assessing copy number dependence and position independence [locus control region (LCR) activity] in such experiments is technically challenging.

To circumvent these problems, we recently used homologous recombination to replace what was considered to be the entire mouse α-globin domain (85 kb) with the entire human domain (117 kb), producing a "humanized" locus with just a single copy of the regulatory domain in a permissive chromosomal environment (Wallace *et al.*, 2007). Even under these ideal circumstances, although the human α genes were expressed in the correct tissue- and developmental stage-specific pattern, their level of expression was suboptimal ($\sim40\%$ of normal). As discussed elsewhere (De Gobbi *et al.*, 2007), it seems unlikely that this is due to any missing elements; rather, changes in the structure or recognition sequences of the key transcription factors have altered during evolution of the two species such that the binding or stability of mouse transcription factors on human sequences is suboptimal. These findings may be of general importance when developing mouse models to understand gene expression in other species or to understand human diseases. Despite these reservations, this is currently the best experimental model for analyzing human α-globin expression.

Removal of just MCS-R2 from the humanized mouse model severely downregulates α-globin expression (Wallace *et al.*, 2007). Together with analysis of natural mutants (Viprakasit *et al.*, 2006), these observations show that MCS-R1, -R3, and -R4 on their own are not sufficient to enhance α-globin expression above ~2–3% of normal. Do these elements provide any enhancement of α-globin expression over and above that of MCS-R2 or are they indeed necessary but not sufficient for α-globin expression? It is also possible that they serve some non-globin role. The full answer to this awaits removal of each element from the humanized locus. However, using hybrids containing human chromosome 16 (α globin) in a mouse erythroid (MEL) background, it is clear that a large deletion of MCS-R1, -R2, and -R3 causes a more complete silencing of α-globin expression ($<0.1\%$) than seen by just deleting MCS-R2 (2–3%) (Vernimmen *et al.*, 2007 and unpublished observations, also see below).

Like many tissue-specific genes, α-globin expression is controlled via highly conserved elements at the promoter and long-range, distal regulatory elements, which in the case of α globin includes at least one strong tissue-specific enhancer (MCS-R2).

IV. STRUCTURE OF THE UPSTREAM REGULATORY ELEMENTS AND THE PROMOTERS

As set out above, there is strong evidence supporting a role for MCS-R2 (associated with HS-40 in human and HS-26 in mouse) as a long-range enhancer of α-globin gene expression. Although the other MCS elements are highly conserved, have been shown to bind transcription factors, and undergo chromatin modification during erythropoiesis (Table 5.1), their precise role *in vivo* remains unknown. Therefore, of the upstream elements, this section will focus on MCS-R2.

The starting point for characterizing this element is the evolutionary sequence analysis. Of all the MCS elements lying upstream of the α-globin cluster that are associated with erythroid-specific DHSs, MCS-R2 is the most highly conserved. Alignment of all known orthologues of this element reveals a conserved core ($\sim$90 bp) containing a GATA-binding site (CAGATAAC), followed by a pair of AP1/NF-E2-binding sites (CATGACTCAG.......... TGCTGAGTCAT) and a second variant GATA site (GCTGATTA) (Fig. 5.2). Both the sequences and the spacing between elements are conserved in this core (Hughes *et al.*, 2005). Surrounding this, there are less well-conserved GATA- and CACC-binding sites (Fig. 5.2). All of these elements have been shown to bind the appropriate transcription factors *in vitro* and footprinting together with chromatin immunoprecipitation (ChIP) experiments (see below) confirms that the core elements are bound *in vivo* (Jarman *et al.*, 1991; Strauss *et al.*, 1992; Zhang *et al.*, 1995).

The core promoter and associated, local (proximal) regulatory elements (extending $\sim$150 bp from the transcriptional start site) of the α-globin genes have now been sequenced in 27 species spanning 500 million years of evolution (Hughes *et al.*, 2005 and unpublished observations) (Fig. 5.3). Nearly all of these sequences contain a variant of the TATA box (CATAAA), a conserved G-rich element (GGGCGTGCCC) and an extended CCAAT box (CCAGCCAATGAGC).

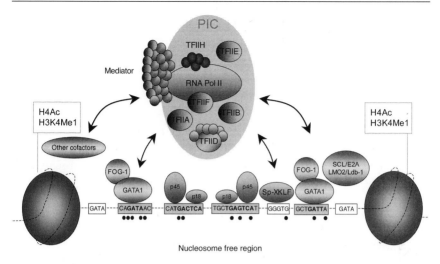

Figure 5.2. The regulatory elements associated with the human eDHS-40 (MCS-R2) summarizes the most highly conserved sequences within this region (grey boxes) and other binding sites (white boxes). Nucleotides associated with *in vivo* footprints are marked with a black spot. Associated, predominant chromatin modifications are summarized above the schematic nucleosomes. Factors shown are components of the PIC (including TFIIA,B, D,E,F,H, and Pol II), the Mediator complex and upstream activators [Sp-XKLF (EKLF), p18 and p45NF-E2, GATA1 and its co-activator FOG-1]. TATA-associated factors (TAFs) are shown with TFIID. The pentameric complex (SCL/E2A/LMO2/Ldb-1/ GATA1) has been shown to bind to this region and is likely to attach via its interaction with GATA1. Although these factors have been shown to bind to this region by ChIP, their binding at each corresponding site is inferred. Arrows allude to the idea that these factors may interact directly or indirectly with components of the PIC, which is recruited independently to MCS-R2.

A conserved element (AGACTCAGAAAGAA) lying between the transcription start site (usually AC) and the initiation codon (ATG) may be important in transcription, processing, and/or translation of the α gene and its mRNA. Beyond these elements, there is very little conservation around the α-globin promoter, and this is consistent with analysis in transient expression assays which suggest that most of the promoter activity is encoded in these conserved, core sequences (Mellon *et al.*, 1981; Pondel *et al.*, 1995; Rombel *et al.*, 1995).

V. TRANSCRIPTION FACTORS INVOLVED IN ERYTHROPOIESIS

Alpha-globin gene transcription (as judged by *in situ* RNA analysis) is undetectable in most haematopoietic stem cells (HSCs) and progenitors and first becomes detectable in proerythroblasts, reaching a maximum in intermediate

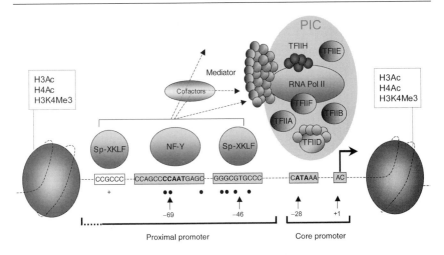

Figure 5.3. The human α-globin promoter summarizes the most highly conserved sequences within this region (grey boxes) and other binding sites (white boxes). Distances (nucleotide) from the transcription start site (+1) are shown. Nucleotides associated with *in vivo* footprint are marked with a black spot. Associated chromatin modifications are summarized above the schematic nucleosomes. Factors shown are the components of the PIC (including TFIIA,B,D,E,F,H, and Pol II), the Mediator complex and upstream activators (Sp-XKLF and NF-Y). TATA-associated factors (TAFs) are shown with TFIID. Arrows allude to the idea that they may interact directly or indirectly with components of the PIC.

erythroblasts (Fig. 5.4), when nascent globin mRNA can be detected from both alleles in >90% of cells (Brown *et al.*, 2006). During the later stages of erythroid cell maturation, the amount of RNA per cell and the rate of protein synthesis decline but the relative stability of globin mRNA ensures that globin becomes the predominant polypeptide made in late erythroblasts and reticulocytes. The amount of globin mRNA reaches ~20,000 molecules per cell in late erythroblasts.

Over the past ten years, global patterns of gene expression have been extensively analyzed in both human and mouse erythropoiesis (Bruno *et al.*, 2004; Gubin *et al.*, 1999; Welch *et al.*, 2004), including the changes that occur in erythroid-affiliated transcription factors. In particular, there are a handful of transcription factors that are known to play a role in regulating expression of typical erythroid genes such as globin (Table 5.2 and Fig. 5.4). In addition to these lineage-affiliated transcription factors, many widely expressed transcription factors (TFs) are involved, including the GTFs that mediate expression of all Pol II-driven genes. Other transcription factors probably remain to be identified.

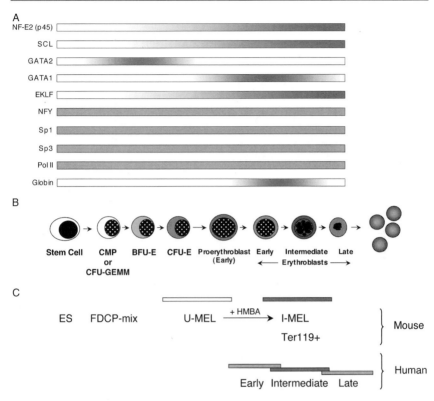

Figure 5.4. An outline of erythropoiesis, the cell lines used, and the expression pattern of the key and general transcription factors (GTFs) binding the α-globin locus. (A) summarizes the expression of key transcription factors during hematopoiesis. Below this is shown the stages of differentiation as haematopoietic stem cell (HSCs) differentiate to mature erythrocytes (B) and the corresponding primary cells and cell lines (C) used in our studies. As a source of pluripotent mouse cells, we used embryonic stem cells (ES). Mouse FDCP-mix cells most clearly resemble CMP (Spooncer *et al.*, 1986). Mouse erythroleukaemia (MEL) cells are well-characterized, transformed erythroid cells that are blocked at the CFU-E or early proerythroblast stage of differentiation (U-MEL). Induction of MEL cells by HMBA (hexa-methylene bis-acetamide) gives rise to terminally differentiated, but still nucleated, erythroid cells that express α- and β-globin mRNA at high levels (I-MEL). Primary mouse erythroblasts (Ter119+) were isolated from the spleens of phenylhydrazine-treated mice (Spivak *et al.*, 1973). Primary human erythroblasts at different stages of differentiation (early, intermediate, and late) were isolated as previously described (Brown *et al.*, 2006). Abbreviations used: FDCP, Factor Dependent Cell Patterson; ES, Embryonic Stem cells; CMP, Common Myeloid Progenitors; CFU-GEMM, colony forming unit-granulocyte, erythrocyte, monocyte, megakaryocyte; CFU-E, Colony Forming Units-erythroid; U-MEL, Uninduced Mouse Erythroleukemia cells; I-MEL, Induced MEL cells.

Table 5.2. Comparison of Histone Modifications and Transcription Factor Binding at the Human and Mouse α-Globin Loci in Mature Erythroblasts

	Human	Mouse	Human	Mouse	Human	Mouse	Human	Mouse	Human	Mouse	Human	Mouse
	MCS-R1		MCS-R2		MCS-R3				MCS-R4		α-globin gene promoters	
	HS-48	HS-31	HS-40	HS-26	HS-33	HS-21	hoHS-12	HS-12	HS-10	HS-8		
H4ac	++	++	++	++	++	+++	−	++	++	++	++	++
H3ac	+	+	+	+	+	+	−	+	++	+	++	++
H3K4me1	++	++	++	++	++	++	−	++	++	++	++	++
H3K4me2	+	+	+	+	+	+	−	+	+	+	++	++
H3K4me3	−	−	−	−	−	−	−	−	−	−	+++	+++
GATA1	+	+	+	+	+	+	−	+	+	+	+	+
SCL	+	+	+	+	+	+	−	+	−	+	−	−
NF-E2	+	+	+	+	−	−	−	+	+	+	−	−
Pol II	+	+	+	+	+	+	−	−	+	+	++++	++++
NFY	−	−	−	−	−	−	−	−	−	−	+	+
Sp1	−	−	−	−	−	−	−	−	−	−	+	+
Sp3	−	−	−	−	−	−	−	−	−	−	+	+
EKLF	+	−	+	+	+	−	−	−	−	−	+	+
ZBP-89	−	−	−	−	−	−	−	−	−	+	−	+

Notes: The table summarizes ChIP results from Anguita *et al.* (2004), de Gobbi *et al.* (2007), and Vernimmen *et al.* (2007) at erythroid-specific DHSs and α-globin gene promoters in expressing erythroblasts. Differences between mouse and human are shown in grey shadow boxes. hoHS-12 represents the human orthologous sequence corresponding the mouse HS-12 region.

VI. CELLULAR RESOURCES FOR STUDYING THE KEY STAGES OF HEMATOPOIESIS

Having described the key cis-active elements around the α-globin genes and the trans-acting factors involved in their activation, we next consider what cellular resources are available in which to study their interactions.

The differentiation process by which multipotent HSCs exit the self-renewal compartment to undergo commitment and lineage specification to mature blood cells (called hematopoiesis) provides a cellular model for analyzing how genes are switched on and off. The process of hematopoiesis appears to have been largely conserved throughout evolution and in mammals has been most extensively studied in mouse and in human (Cantor and Orkin, 2002; Orkin, 2000). In both species, primary cells are easily accessible (from bone marrow and blood) and cells at various stages of erythropoietic commitment can be separated on the basis of their well-characterized cell surface markers (Brown et al., 2006). Once isolated, the provenance of these sorted cells can be evaluated by analyzing their lineage potential in clonal assays, in vitro.

Since the α-globin genes are expressed only in the erythroid (red cell) lineage, we will concentrate on the pathway by which mature red cells are produced from HSCs (Fig. 5.4). Once HSCs are committed to differentiation, they give rise to multipotential progenitors (including the common myeloid progenitor, CMP, or colony forming unit-granulocyte, erythrocyte, monocyte, megakaryocyte, CFU-GEMM in Fig. 5.4) that can differentiate into a wide variety of mature blood cells, including erythroid. The earliest progenitors that are entirely restricted to the red cell lineage produce large erythroid colonies in vitro, consisting of several subunits known as burst forming units erythroid (so-called BFU-E, Fig. 5.4). Late erythroid progenitors (identified in clonal assays and called colony forming units, CFU-E, Fig. 5.4) correspond to the earliest recognizable erythroid precursor in the bone marrow (the proerythroblast). As these erythroid precursors progress through maturation (intermediate erythroblasts), the nucleus becomes progressively condensed (intermediate and late erythroblasts) and is eventually expelled producing the mature red cell (Fig. 5.4). Large numbers of primary erythroid cells (mouse or human) representing all stages of differentiation can either be isolated from erythroid tissues (bone marrow and, in the mouse, spleen) in vivo or be grown in vitro from progenitors isolated from the bone marrow or peripheral blood (Fibach et al., 1989; Spivak et al., 1973).

In addition to these primary cells, immortalized mouse erythroid cells [mouse erythroleukemia (MEL) cells], blocked at the proerythroblast stage of differentiation, can be induced by a variety of chemical agents to undergo the terminal stages of differentiation, thus providing a useful cellular model for this stage of erythropoiesis in mouse (Marks et al., 1987). Unfortunately, there are no simple, equivalent cellular models of erythropoiesis in human, although K562 cells are often used to represent an erythroid cell line. These are an unsatisfactory model since they are pseudo-triploid cells (originally derived from a patient with chronic myeloid leukemia) that produce very small amounts of embryonic globins.

HSCs and early progenitors are generally too rare for standard genetic and epigenetic analyses and, even when purified, are inevitably heterogeneous. On the other hand, hematopoietic cell lines, although relatively homogeneous, may not faithfully reflect all aspects of the corresponding cell *in vitro*. Therefore, where possible, it is best to analyze both primary cells and cell lines representing these early progenitors. As a source of pluripotent cells, we have analyzed human and mouse embryonic stem (ES) cells and we have also analyzed the factor-dependent cell Patterson (FDCP)—mix cells (Spooncer *et al.*, 1986), which are thought to be the equivalent of the CMP (Fig. 5.4). Using this mixture of primary cells and cell lines has enabled us to analyze globin gene expression at many stages of differentiation and, therefore, provides a way to investigate the order of events leading to gene activation.

A valuable additional resource is interspecific hybrids of MEL cells containing a human chromosome 16. On induction, these hybrids mimic the terminal stages of erythropoiesis, expressing not just the endogenous mouse globin genes but also the human α-globin genes (Deisseroth and Hendrick, 1978; Zeitlin and Weatherall, 1983). We have analyzed hybrids derived from normal individuals and from patients with previously characterized, natural mutations of the α cluster (Craddock *et al.*, 1995). These include mutant chromosomes in which one or more of the remote upstream elements have been deleted but the α genes remain intact (e.g., deletion A in Fig. 5.1), and a chromosome in which the α genes are deleted but the upstream elements remain intact (not shown).

With these various resources, we can examine the binding of transcription factors to the various *cis*-acting elements, not only at the time the globin genes are transcribed but also at previous steps in the differentiation pathway. In this way, we can obtain a sequential picture of the order of events required to set up the α-globin domain for transcription.

VII. TRANSCRIPTION FACTOR BINDING TO THE UPSTREAM REGULATORY ELEMENTS

ChIP has shown that MCS-R2 is bound in both human and mouse late erythroid cells by GATA1, which appears to nucleate a complex containing stem cell leukaemia factor (SCL), E2A, Lmo2, and Ldb-1 [the so-called erythroid pentameric complex (Wadman *et al.*, 1997), Fig. 5.2]. In addition, this region binds the NF-E2 heterodimer (including p45 and p18) and members of the Sp-XKLF family of proteins (Anguita *et al.*, 2004; Vernimmen *et al.*, 2007). MCS-R2, like other enhancers, is therefore bound by a closely grouped cluster of transcription factor complexes that work cooperatively to enhance transcription (so-called enhancesome). In many ways, long-range enhancers like MCS-R2 are structurally similar and are thought to work in a similar way, to proximal promoter elements, but are influencing a promoter which can be tens or hundreds of kb away. It is necessary for transcription factors to gain access to these elements *in vivo* and, consistent with this, MCS-R2 is associated with an erythroid-specific DHS and the closely associated nucleosomes acquire the histone modifications associated with activation in erythroid cells (Anguita *et al.*, 2004; de Gobbi *et al.*, 2007) (see below). It is becoming clear, as more long-range enhancers are analyzed, that these regions also recruit GTFs and Pol II (Szutorisz *et al.*, 2005b Fig. 5.2), and we have recently shown that this is the case for MCS-R2 (Vernimmen *et al.*, 2007).

What of the other conserved upstream elements (MCS-R1, -R3, and -R4)? Although we have not yet elucidated their functional role(s) *in vivo*, they appear very similar to MCS-R2 in that, when activated, they are associated with DHSs, they are bound by multiprotein transcription factor complexes, and ultimately recruit the GTFs and Pol II (Table 5.1) (Anguita *et al.*, 2004; de Gobbi *et al.*, 2007).

VIII. TRANSCRIPTION FACTOR BINDING TO THE PROMOTER

In vivo, the core promoter is regarded as the docking site for the GTFs (GTFs in Fig. 5.3) which together with Pol II form a PIC. Different GTFs are involved in positioning the PIC on the TATA box (TBP, TFIIA, and TFIIB), melting the core promoter DNA (TFIIE), promoter clearance (TFIIH), and elongation [TFIIF and FACT (Maston *et al.*, 2006; Vernimmen *et al.*, 2007)]. The extended TATA box of the core promoter (CATAAA), which is highly conserved in all α genes but not in other core promoters containing a TATA box, suggests that it may confer some specificity in the "general" factors that are recruited to this core promoter. It has been suggested that such specificity might lie in the TBP-associated factors (TAFs) and that this might determine which promoters do or do not respond to upstream elements (Chen and Manley, 2003).

The current model of transcription (established in yeast) views the process as a cycle in which complete PIC assembly occurs only once. After Pol II escapes from the promoter, a scaffold composed of TFIID, TFIIE, TFIIH, and another multiprotein complex (Mediator Fig. 5.3) remains on the core promoter. Subsequent reinitiation of transcription only requires recruitment of Pol II, TFIIF, and TFIIB. *In vitro*, the assembly of the PIC on the core promoter directs only low levels of transcription (basal transcription), but transcription can be stimulated by activators bound to adjacent sequences (proximal promoter region, Fig. 5.3). Such activators are thought to work synergistically to stimulate one or more steps of the transcription cycle via direct interactions with the GTFs (reviewed in Maston *et al.*, 2006). *In vitro* binding experiments, together with *in vivo* footprinting, point to conserved sequences in the promoter that may be bound by activators (Pondel *et al.*, 1995; Zhang *et al.*, 1995, 2006). These include the CCAAT box, whose extended consensus again suggests some specificity to the protein(s) binding here, and a conserved GC-rich motif. Other potential binding sites for erythroid (e.g., GATA1) and ubiquitously expressed factors (e.g., NF1, Rein *et al.*, 1995) have been identified adjacent to the core promoter in some species but not others.

For the PIC and transcriptional activators to gain access to the core promoter *in vivo* requires that this DNA element is accessible in chromatin and, indeed, it has been shown that the α-globin promoter coincides with a DHS in committed erythroid cells, even prior to globin expression (Higgs *et al.*, 1990; Yagi *et al.*, 1986). Careful analysis of this nuclease-sensitive region with micrococcal nuclease and specific endonucleases has defined what appears to be a nucleosome free region ($\sim +50$ to -275, with $+1$ being the transcriptional start site, unpublished), specifically in human erythroid cells (Fig. 5.3). How this nucleosome free region is created is currently unknown although it is clear that transcriptional activators may themselves recruit co-activator complexes that covalently modify chromatin [e.g., histone acetyl transferase (HAT) and histone de-acetylase (HDAC)] (Berger, 2007) and ATP-dependent chromatin remodeling complexes (Havas *et al.*, 2001). In most mammalian species, the α-globin promoter lies within a constitutively unmethylated CpG island (Bird *et al.*, 1987); such structures are found associated with about half of all known promoters. However, the CpG island associated with the mouse α-globin promoter has been eroded during evolution (Hughes *et al.*, 2005).

Recently, using ChIP assays, we have shown that in differentiating erythroid cells (but in no other cell type) the core α-globin promoter is bound by NF-Y (a CCAAT-binding protein) and Sp-XKLF proteins (which bind GC-rich sequences) (Vernimmen *et al.*, 2007). In mouse, the promoter also binds GATA1 and ZBP-89 (Vernimmen *et al.*, 2007). Finally, in erythroid cells, nucleosomes in close association with the core promoter acquire the histone modifications associated with activation (see below). These changes presumably help to create the chromatin environment required to recruit the GTFs, Mediator, and Pol II. But these can only be demonstrated by ChIP in cells transcribing the α-globin genes (Fig. 5.3 and Table 5.1).

IX. THE RECRUITMENT OF RNA POLYMERASE AND GTFs TO THE α-GLOBIN CLUSTER

As discussed above, gene expression requires recruitment, release, and elongation of RNA Pol II. Alpha-globin expression does not start until the very last few divisions of terminal differentiation and maturation. Critically, the PIC (including Pol II) is not recruited to the upstream elements or the promoter in progenitor cells (even though the chromatin is accessible), but only bind late in erythropoiesis, when globin expression is first seen by mRNA *in situ* studies (Brown *et al.*, 2006).

Since immediately preceding this the cluster is otherwise already poised for expression (with transcription factors bound and chromatin modifications made) an important question is to know what triggers recruitment of the PIC to the enhancer and promoter at this late stage of erythropoiesis? This could involve the expression or recruitment of another transcription factor, the post-translational modification of resident factors, or both. Therefore, it is of interest that Sp-XKLF factors (known to interact with GATA1) (Merika and Orkin, 1993) that are expressed relatively early or throughout hematopoiesis only become recruited to MCS-R2 and the promoter late in erythropoiesis, in parallel with recruitment of the PIC. Sp1 and Sp3 are recruited to the promoter, whereas EKLF is recruited to the promoter and to MCS-R2 (Vernimmen *et al.*, 2007). The Sp-XKLF transcription factors, therefore, are good candidates for being involved in the transition from a poised to an active transcriptional state.

X. WHAT ROLE DO THE REMOTE REGULATORY ELEMENTS PLAY?

In the absence of the upstream elements, there is very little transcription of the α-globin genes either in experimental systems or *in vivo*. In human, most of the enhancer activity is concentrated in MCS-R2, whereas in mouse the activity clearly involves other sequences as well (discussed in Anguita *et al.*, 2002). To investigate the mechanism by which the upstream elements enhance α-globin expression, we have analyzed interspecific hybrids (chromosome 16 × MEL).

Using these hybrids, we have shown that the formation of DHSs occurs independently at the α-globin promoters and the upstream elements (Craddock *et al.*, 1995). Furthermore, key tissue-restricted transcription factors (NF-E2 and SCL) still bind the upstream elements when the α-globin genes are deleted. Similarly, the α-globin promoter still binds NF-Y and Sp-XKLF factors in the absence of the upstream elements (Vernimmen *et al.*, 2007). These findings suggest that recruitment of transcription factors, and the associated chromatin modifications required to create the poised state, occur independently at the promoter and upstream elements and require no interaction between them. So the upstream elements appear to play no significant role in recruiting the specific transcription factors or in modifying the associated chromatin at the promoter.

Do the upstream elements influence recruitment of the PIC to the α-globin promoters? In the absence of the α-globin promoters, Pol II and other tested components of the PIC are still recruited to the upstream elements showing that these sequences specifically and independently recruit the PIC during erythropoiesis. By contrast, when all of the upstream elements are deleted, very little, if any, PIC is recruited to the α-globin promoters (Vernimmen et al., 2007). Therefore, it appears that a major role of the upstream elements is to facilitate recruitment of the GTFs and Pol II to the promoters of the α genes.

The role of MCS-R2 in this process is interesting. When this element alone is deleted from the α cluster, GTFs appear to be recruited to the promoter at relatively normal levels, whereas Pol II recruitment was severely reduced (Vernimmen et al., 2007), commensurate with the associated decrease in mRNA expression (~3% normal) (Bernet et al., 1995). These findings suggest that recruitment of the complete PIC (GTFs and Pol II) at the promoter depends on the presence of more than just MCS-R2 and points to complementary roles for the other MCS elements, even though they do not act as enhancers on their own.

XI. HOW DO THE UPSTREAM ELEMENTS INTERACT WITH THE PROMOTER?

The mechanism by which the long-range, upstream elements regulate recruitment of the PIC to the α-globin promoters provides a well-characterized example of an increasingly recognized general question in mammalian genetics, namely, how do long-range *cis*-elements interact? It seems reasonable to assume that protein complexes assembled at one element physically contact those at other elements.

To investigate this at the α-globin cluster, we used a modified version of the chromosome conformation capture (3C) technology (Dekker et al., 2002). In these experiments, formaldehyde is used to cross-link protein–DNA and protein–protein interactions in intact nuclei. The cross-linked chromatin is then digested by a restriction enzyme followed by ligation. If an interaction between a remote regulatory sequence and a promoter occurs, new, hybrid fragments containing these two elements are generated (at the ligation step), and polymerase chain reaction (PCR) reactions can be used to detect these newly combined elements. We and others (Splinter et al., 2006; Vernimmen et al., 2007) recently used a Taqman system to measure these interactions. Accurate quantitation is important because background interactions, associated with specific PCR products, are nearly always detected, even between fragments separated by over 400 kb. Using 3C, we have shown that the α-globin genes (in human and mouse) interact with all of the upstream MCS elements in late erythroid cells but not in early hematopoietic cells, pluripotent ES cells, or non-erythroid cells (Fig. 5.5). These late, erythroid-specific 3C interactions

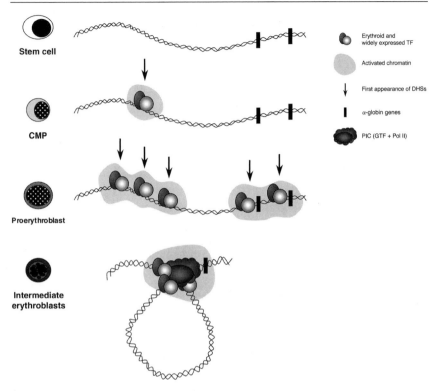

Figure 5.5. A model proposing how complexes form and interact at the α-globin locus during erythropoiesis. In multipotent cells (CMP), the cluster is primed in the upstream region (MCS-R2) by multiprotein complexes containing SCL and NF-E2 nucleated by GATA2. In committed erythroid progenitors (U-MEL, proerythroblast stage), additional remote regulatory sequences are bound by multiprotein complexes containing various combinations of SCL and NF-E2 and GATA1 which replaces previously bound GATA2. At this stage, the α-globin promoter is also occupied by a combination of factors including NF-Y and is poised for expression. In differentiating erythroid cells, the PIC, including Pol II, is recruited to the enhancers in a cooperative manner but independently of the promoter. Kruppel-like transcription factors are also recruited, independently of the upstream elements, to the promoter. At this final stage, the α-globin promoter is now occupied by a multiprotein complex that represents a docking site for the recruitment of PIC, which is entirely dependent on the presence of the upstream elements (for details, see Anguita et al., 2004 and Vernimmen et al., 2007). During terminal maturation, the upstream elements interact with the promoter forming a loop.

appear to coincide with the onset of transcription (Vernimmen et al., 2007), although some preexisting structure may exist as has been shown in the β-globin cluster (see Chapter 4 by Palstra et al., this volume).

Three types of model have been proposed to explain how such interactions may occur in the nucleus (reviewed in Li et al., 2002). The first proposes that activators/PIC are first bound at the enhancer and then some or all of these proteins disengage from the enhancer and track along the DNA to the promoter [tracking model (Hatzis and Talianidis, 2002)]. It should be noted that this model requires no physical interaction between the enhancer and the promoter. The detection of transcripts originating from enhancers and LCRs has led to the suggestion that Pol II is the tracking motor. Although there is evidence of transcription from many long-range elements, to date we have detected no transcripts originating in the α-globin MCS elements that track toward the α-globin promoters.

A second model proposes that chromatin loops in such a way that multiprotein complexes bound to the upstream elements and to the promoters are able to interact (looping model). Variants of such models suggest that the interactions may have different affinities, depending on the proteins present, and may be transient. These features are exemplified in the "flip-flop" model (Wijgerde et al., 1995) to explain interaction between the β-globin LCR and its target promoters (see Chapter 4 by Palstra et al., this volume) (Palstra et al., 2003).

A third type of model proposes that protein–DNA complexes remain attached at cis-elements and move toward each other. These models (referred to as facilitated tracking) are all similar to that proposed by Wang et al. (2005) and involve scanning the intervening chromatin or DNA, progressively forming a loop; presumably the upstream elements could track toward the promoter, or vice-versa, or both could track toward each other. The molecular details of how this might occur have not been clearly set out. However, several variants of this model link tracking with transcription of the intervening DNA, although it seems equally possible that other molecular motors could drive the movement of such DNA protein complexes to scan without associated transcription. Whatever the details, both looping and facilitated tracking models result in displacement of the chromatin lying between the cis-acting elements so that the resulting chromatin structure would look identical in both models.

Two important questions underlying this mechanism arise from these models. The first is which proteins mediate these interactions? Clearly, this is related to how an enhancer distinguishes its target promoter(s) from others and why the affinity for competing promoters may differ (e.g., in a developmental-stage-specific manner). Although some proteins with the potential for mediating such interactions have been proposed (e.g., GATA1 and FOG-1) (Vakoc et al., 2005), it has not been possible to separate looping and gene activation and they may not be separable.

The second issue is to understand what a newly formed enhancer/ promoter complex does that the promoter cannot do on its own. In the absence of MCS-R2 alone, the remaining interactions can recruit the GTFs to the α-globin promoter, but not Pol II, suggesting that one role of the upstream elements is to recruit the GTFs and a second role, in which MCS-R2 is important, is to recruit Pol II (Vernimmen et al., 2007). Is the Pol II recruited to the enhancer transferred to the promoter or do the upstream elements localize the cluster in a transcription factory (Osborne et al., 2004) and recruit the promoter into this nuclear subcompartment? It appears at the α locus that Pol II recruited at the enhancer may be in an inactive form, whereas after looping the Pol II is active. Something in this process appears to be activating Pol II and allowing transcription to proceed; however, as yet, the rate-limiting stage in the transcription cycle that is affected by this interaction is unknown and this raises important questions for the future.

XII. SEQUENTIAL ACTIVATION OF THE α-GLOBIN GENE CLUSTER DURING DIFFERENTIATION

By analyzing either primary cells or cell lines representing different stages of erythropoiesis (Fig. 5.4 and see above), and knowing the key transcription factors and cofactors that bind the conserved cis-acting elements, it has been possible to reconstruct the order of events that occur, as a chromosomal domain is "set up" and the genes contained within it become activated during differentiation and development.

In pluripotent ES cells (Fig. 5.4), in which the key hematopoietic transcription factors are not readily detectable, none of the MCS elements are associated with DHSs or bound by any of the transcription factors tested. By contrast in multipotent hematopoietic cells, in which some early stage hematopoietic factors are expressed, MCS-R2 is bound by GATA2, NF-E2, and the pentameric SCL complex, and thus starts to become "primed" for expression even though these cells retain the potential to form cell types in which globin gene expression is silenced. By the time the cell is fully committed as a unilineage erythroid progenitor (proerythroblast, Fig. 5.5), GATA1 has replaced GATA2 at MCS-R2 (possibly facilitated by FOG-1) and binding is also seen at MCS-R1 and -R4. At this stage, the α-globin promoters also become activated as judged by the appearance of DHSs (Anguita et al., 2004). During terminal differentiation and maturation, the levels of GATA1, NF-E2, and SCL mRNA and binding increase and MCS-R3 is bound and activated. Thus, at the earliest stages of terminal differentiation, the α-globin cluster appears to be poised for action, even though little or no globin is made in these cells. It is interesting to note that binding and activation appear to start at the most upstream elements and proceed toward the promoters and similar polarity has been observed at other loci regulated by long-range enhancers and LCRs (Hatzis and Talianidis, 2002; Levings et al., 2006; Szutorisz et al., 2005a; Wang et al., 2005).

Modifications of the N-terminal tails of histones associated with activation appear in parallel with the binding of transcription factors in erythroid cells. These include acetylation of histone H3 and H4 (H3Ac and H4Ac) together with mono-, di-, and tri-methylation of histone H3 at lysine 4 (H3K4me1, -2, and -3) (De Gobbi *et al.*, 2007) (Figs. 5.2 and 5.3 and Table 5.2). Toward the final stages of maturation, the entire α-globin domain becomes acetylated (Fig. 5.1) with peaks of acetylation associated with each MCS element and the promoters of the globin genes. It is of interest that this domain of acetylation is entirely contained within the previously described region of synteny (see above) consistent with this region containing all of the regulatory elements controlling α-globin expression (Anguita *et al.*, 2001).

When examined in detail, the pattern of histone modification is clearly different at the upstream elements compared to the promoters (De Gobbi *et al.*, 2007). Acetylation of histone H4 is equally strong at the upstream elements and the promoters in intermediate and late erythroid cells. This could be explained most simply by the recruitment of histone acetylases by transcription factors at all *cis*-acting elements. By contrast, acetylation of histone H3 is much less prominent at the upstream elements than at the promoters. Furthermore, while H3K4me1 is equally enriched at the upstream elements and the promoters, H3K4me2 is more prominent at the promoters, and H3K4me3 is still further enriched at the promoters compared with the upstream elements. It has previously been shown that patterns of histone H3K4 methylation may be influenced by the state of the associated Pol II (Berger, 2007). This model, which cannot be discussed in detail here, would suggest that Pol II recruited at the upstream elements is inactive (associated with H3K4me1), whereas the Pol II at the promoters is engaged (associated with H3K4me3) and poses the question of how Pol II recruited to the remote upstream elements and the promoter during differentiation.

XIII. CONCLUSIONS, SPECULATION, AND FUTURE DIRECTIONS

Some simple observations from the endogenous, intact globin loci of various animals point to some principles by which upstream elements may work *in vivo* and raise some unanswered questions. First, in all mammals studied, the α-like globin genes lie downstream of their regulatory elements [MCS-embryonic (ζ) and adult (α)] and it appears that activation of the locus (transcription factor binding and chromatin modification) during differentiation starts at the upstream elements and moves toward the promoters, although further studies are required to refine these observations. Furthermore, in all mammals studied, the major globin genes are arranged in the order in which they are expressed in development, and all the α-like globin genes are transcribed from the same strand of DNA (Hughes *et al.*, 2005). Together, these findings suggest that there may be some polarity in the mechanism underlying activation of the α-globin cluster starting at the upstream MCS elements and moving to the promoter. Paradoxically, it is generally accepted that enhancers function independently of their orientation with respect to the promoter. And yet, inversion of the β-globin LCR, within the context of the entire locus, severely down-regulates β-globin expression, suggesting that in their natural chromosomal environment, orientation of the upstream elements may be important (Tanimoto *et al.*, 1999). These observations question the importance of enhancer orientation and polarity with respect to the genes they regulate *in vivo*, and this is currently being tested in the α-globin cluster.

   Comparing the α-globin loci from different animals, remote elements (~40 kb upstream) are found in all species studied but there is considerable variation in the linear distances between MCS elements and their target promoters. Furthermore, even within a single species (human), normal variation, due to insertion or deletion (copy number variants), means that in some individuals MCS-R2 lies 50 kb from the promoter, whereas in others it is 70 kb from the promoter and yet this is not associated with any major alteration in gene expression (Winichagoon *et al.*, 1982). Given that some *cis*-acting elements may lie up to 1 Mb from the genes they regulate, it will be interesting to determine the effect of moving remote regulatory elements closer or further away from their promoter to establish the relationship, if any, between the distance separating such elements and regulation of expression. The use of homologous recombination in the humanized mouse model of α-globin gene expression will allow us to examine this issue in a systematic way.

   The α-globin regulatory elements were among the first to be found in the introns of an adjacent gene (c16orf35) unrelated to globins, and surrounded by promoters other than those associated with the globin genes. Subsequent annotation of the genome has revealed that this situation is by no means uncommon and therefore raises the question of what determines which promoters the elements activate. Presumably, this involves specific protein–protein interactions, but at present, the molecular basis for this specificity is not clear. It should be noted, however, that genes contained within an active domain may be activated in a nonspecific way as a "bystander" (Cajiao *et al.*, 2004), and preliminary evidence suggests that some widely expressed genes within the α-globin regulatory domain may be upregulated in erythroid cells (Karen Lower, unpublished). Another unsolved issue is how an enhancer bound by a multiprotein complex activates gene expression while being transcribed as part of the intron of another gene. One possibility is that transcription of these elements and their participation in promoter activation are mutually exclusive. However, preliminary data from mRNA *in situ* analysis suggests that both α genes and both copies of the gene containing the upstream regulatory elements (c16orf35) can be simultaneously transcribed in a single cell (Brown, unpublished).

Enhancers acting in *cis* are often required to choose between different promoter targets and may even ignore some promoters. For example, in the globin clusters current evidence suggests that the same upstream elements regulate expression of the embryonic, fetal, and adult globin genes in erythroid cells at different stages of development. One possibility is that changes in stage-specific proteins alter the affinities of the promoters for the upstream elements so that, at any particular stage, the appropriate promoter outcompetes the others for an interaction. There is good evidence for such promoter competition in the β-globin cluster (Wijgerde *et al.*, 1995). In the α cluster, three sets of observations are consistent with a model in which the α genes compete for the upstream element(s). First, when a new active promoter, for example, a selectable marker gene, is inserted (by homologous recombination) between the upstream elements and the α-gene promoters, in some (but not all) positions, it appears to outcompete the α promoters and this leads to α thalassemia (Anguita *et al.*, 2002; Esperet *et al.*, 2000). Removal of these selectable markers (e.g., by Cre-mediated recombination) restores normal α-globin expression. The second observation comes from a natural mutation (a regulatory SNP, referred to as *de novo* GATA in Fig. 5.1) that creates a new promoter sequence between the ζ and $\psi\zeta$ genes (Fig. 5.1). When mutated, this region becomes transcriptionally active and this appears to downregulate expression of the α genes in *cis* (causing α thalassemia) by outcompeting and "stealing" activity from the promoters (De Gobbi *et al.*, 2006). Finally, it has been noted that although the duplicated α genes in most species have similar or identical promoters, the gene closest to the upstream elements is usually expressed at a higher level (reviewed in Higgs *et al.*, 1989). When more than two genes are present, the additional genes lying further downstream appear to be expressed at ever decreasing levels (Vestri *et al.*, 1994). None of these observations unequivocally discriminate between looping and tracking models but given the large distance between the upstream elements and the relatively closely spaced competing α-globin promoters, they suggest that there is at least some element of tracking or scanning involved rather than free looping.

    This also raises the issue of what limits enhancer–promoter interactions to one chromosome. Some recent evidence suggests that the human (and to a lesser extent mouse) α-globin loci frequently co-localize in erythroid cells (Brown *et al.*, 2006) but there is no evidence for trans-activation of the α-globin promoter by an enhancer on the homologous chromosome, although long-range interactions between chromosomes have been described for other loci (Osborne *et al.*, 2004; Spilianakis and Flavell, 2004; Spilianakis *et al.*, 2005). Clearly, any form of tracking would limit interactions to the same chromosome as opposed to free looping of chromatin which would predict that, at a certain distance, an enhancer would be as likely to interact in *trans* as in *cis*.

A remaining puzzle is why the mammalian genome evolved long-range regulation. Why are all the elements not simply arranged around the gene itself as in lower organisms? In the case of globin gene expression, one explanation might center around the timing of activation during differentiation. Globin expression is needed only during the last few divisions of maturation; expression earlier in erythropoiesis would be extremely deleterious. However, when globin expression is needed, it is required at fully activated levels. It seems possible that sequential, long-range activation might play a role in preparing the cluster for high-level expression and yet not activate the genes until the last moment. Clearly, this could occur by simply building up the complex at the gene itself, so why long-range regulation? Is it possible that there is a time-dependent polarity of activation along a regulatory domain and the positions of these elements along this activation time line determines when the final complex is ready for action? How such a system would evolve is completely unclear, but these are fascinating lines for us to follow in the future.

Acknowledgments

We are grateful to all members of the laboratory who have contributed to this work. We also thank Dr. R. Gibbons for his critical reading of the manuscript. We thank Dr. J. Hughes for help preparing some of the data and Ms. L. Rose and Ms. N. Gray for help in preparing the manuscript. This work is supported by the Medical Research Council (UK).

References

Andrews, N. C. (1998). The NF-E2 transcription factor. Int. J. Biochem. Cell Biol. 30, 429–432.

Anguita, E., Johnson, C. A., Wood, W. G., Turner, B. M., and Higgs, D. R. (2001). Identification of a conserved erythroid specific domain of histone acetylation across the alpha-globin gene cluster. Proc. Natl. Acad. Sci. USA 98, 12114–12119.

Anguita, E., Sharpe, J. A., Sloane-Stanley, J. A., Tufarelli, C., Higgs, D. R., and Wood, W. G. (2002). Deletion of the mouse alpha-globin regulatory element (HS-26) has an unexpectedly mild phenotype. Blood 100, 3450–3456.

Anguita, E., Hughes, J., Heyworth, C., Blobel, G. A., Wood, W. G., and Higgs, D. R. (2004). Globin gene activation during haemopoiesis is driven by protein complexes nucleated by GATA-1 and GATA-2. EMBO J. 23, 2841–2852.

Berger, S. L. (2007). The complex language of chromatin regulation during transcription. Nature 447, 407–412.

Bernet, A., Sabatier, S., Picketts, D. J., Ouazana, R., Morle, F., Higgs, D. R., and Godet, J. (1995). Targeted inactivation of the major positive regulatory element (HS-40) of the human alpha-globin gene locus. Blood 86, 1202–1211.

Bieker, J. J. (2001). Kruppel-like factors: Three fingers in many pies. J. Biol. Chem. 276, 34355–34358.

Bird, A. P., Taggart, M. H., Nicholls, R. D., and Higgs, D. R. (1987). Non-methylated CpG-rich islands at the human alpha-globin locus: Implications for evolution of the alpha-globin pseudo-gene. EMBO J. 6, 999–1004.

Brown, J. M., Leach, J., Reittie, J. E., Atzberger, A., Lee-Prudhoe, J., Wood, W. G., Higgs, D. R., Iborra, F. J., and Buckle, V. J. (2006). Coregulated human globin genes are frequently in spatial proximity when active. *J. Cell Biol.* **172,** 177–187.

Bruno, L., Hoffmann, R., McBlane, F., Brown, J., Gupta, R., Joshi, C., Pearson, S., Seidl, T., Heyworth, C., and Enver, T. (2004). Molecular signatures of self-renewal, differentiation, and lineage choice in multipotential hemopoietic progenitor cells *in vitro. Mol. Cell. Biol.* **24,** 741–756.

Cajiao, I., Zhang, A., Yoo, E. J., Cooke, N. E., and Liebhaber, S. A. (2004). Bystander gene activation by a locus control region. *EMBO J.* **23,** 3854–3863.

Cantor, A. B., and Orkin, S. H. (2002). Transcriptional regulation of erythropoiesis: An affair involving multiple partners. *Oncogene* **21,** 3368–3376.

Chen, Z., and Manley, J. L. (2003). Core promoter elements and TAFs contribute to the diversity of transcriptional activation in vertebrates. *Mol. Cell. Biol.* **23,** 7350–7362.

Craddock, C. F., Vyas, P., Sharpe, J. A., Ayyub, H., Wood, W. G., and Higgs, D. R. (1995). Contrasting effects of alpha and beta globin regulatory elements on chromatin structure may be related to their different chromosomal environments. *EMBO J* **14,** 1718–1726.

De Gobbi, M., Viprakasit, V., Hughes, J. R., Fisher, C., Buckle, V. J., Ayyub, H., Gibbons, R. J., Vernimmen, D., Yoshinaga, Y., de Jong, P., Cheng, J.-F., Rubin, E. M., *et al.* (2006). A regulatory SNP causes a human genetic disease by creating a new transcriptional promoter. *Science* **312,** 1215–1217.

De Gobbi, M., Anguita, E., Hughes, J., Sloane-Stanley, J. A., Sharpe, J. A., Koch, C. M., Dunham, I., Gibbons, R. J., Wood, W. G., and Higgs, D. R. (2007). Tissue-specific histone modification and transcription factor binding in α globin gene expression. *Blood* **110,** 4503–4510.

Deisseroth, A., and Hendrick, D. (1978). Human alpha-globin gene expression following chromosomal dependent gene transfer into mouse erythroleukemia cells. *Cell* **15,** 55–63.

Dekker, J., Rippe, K., Dekker, M., and Kleckner, N. (2002). Capturing chromosome conformation. *Science* **295,** 1306–1311.

Dillon, N., and Sabbattini, P. (2000). Functional gene expression domains: Defining the functional unit of eukaryotic gene regulation. *Bioessays* **22,** 657–665.

ENCODE Project Consortium (2007). Identification and analysis of functional elements in 1% of the human genome by the ENCODE pilot project. *Nature* **447,** 799–816.

Esperet, C., Sabatier, S., Deville, M. A., Ouazana, R., Bouhassira, E. E., Godet, J., Morle, F., and Bernet, A. (2000). Non-erythroid genes inserted on either side of human HS-40 impair the activation of its natural alpha-globin gene targets without being themselves preferentially activated. *J. Biol. Chem.* **275,** 25831–25839.

Felsenfeld, G., and Groudine, M. (2003). Controlling the double helix. *Nature* **421,** 448–453.

Fibach, E., Manor, D., Oppenheim, A., and Rachmilewitz, E. A. (1989). Proliferation and maturation of human erythroid progenitors in liquid culture. *Blood* **73,** 100–103.

Flint, J., Thomas, K., Micklem, G., Raynham, H., Clark, K., Doggett, N. A., King, A., and Higgs, D. R. (1997). The relationship between chromosome structure and function at a human telomeric region. *Nat. Genet.* **15,** 252–257.

Gubin, A. N., Njoroge, J. M., Bouffard, G. G., and Miller, J. L. (1999). Gene expression in proliferating human erythroid cells. *Genomics* **59,** 168–177.

Hatzis, P., and Talianidis, I. (2002). Dynamics of enhancer-promoter communication during differentiation-induced gene activation. *Mol. Cell* **10,** 1467–1477.

Havas, K., Whitehouse, I., and Owen-Hughes, T. (2001). ATP-dependent chromatin remodeling activities. *Cell. Mol. Life. Sci.* **58,** 673–682.

Higgs, D. R. (2001). Molecular mechanisms of α thalassemia. *In* "Disorders of Hemoglobin" (M. H. Steinberg, B. G. Forget, D. R. Higgs, and R. L. Nagel, eds.), pp. 405–430. Cambridge University Press, Cambridge.

Higgs, D. R., Vickers, M. A., Wilkie, A. O., Pretorius, I. M., Jarman, A. P., and Weatherall, D. J. (1989). A review of the molecular genetics of the human alpha-globin gene cluster. *Blood* **73,** 1081–1104.

Higgs, D. R., Wood, W. G., Jarman, A. P., Sharpe, J., Lida, J., Pretorius, I. M., and Ayyub, H. (1990). A major positive regulatory region located far upstream of the human alpha-globin gene locus. *Genes Dev.* **4,** 1588–1601.

Higgs, D. R., Sharpe, J. A., and Wood, W. G. (1998). Understanding alpha globin gene expression: A step towards effective gene therapy. *Semin. Hematol.* **35,** 93–104.

Hughes, J. R., Cheng, J. F., Ventress, N., Prabhakar, S., Clark, K., Anguita, E., de Gobbi, M., de Jong, P., Rubin, E., and Higgs, D. R. (2005). Annotation of cis-regulatory elements by identification, subclassification, and functional assessment of multispecies conserved sequences. *Proc. Natl. Acad. Sci. USA* **102,** 9830–9835.

International Human Genome Sequencing Consortium (2004). Finishing the euchromatic sequence of the human genome. *Nature* **431,** 931–945.

Jarman, A. P., Wood, W. G., Sharpe, J. A., Gourdon, G., Ayyub, H., and Higgs, D. R. (1991). Characterization of the major regulatory element upstream of the human alpha-globin gene cluster. *Mol. Cell. Biol.* **11,** 4679–4689.

Kabe, Y., Yamada, J., Uga, H., Yamaguchi, Y., Wada, T., and Handa, H. (2005). NF-Y is essential for the recruitment of RNA polymerase II and inducible transcription of several CCAAT box-containing genes. *Mol. Cell. Biol.* **25,** 512–522.

Kielman, M. F., Smits, R., Hof, I., and Bernini, L. F. (1996). Characterization and comparison of the human and mouse Dist1/alpha-globin complex reveals a tightly packed multiple gene cluster containing differentially expressed transcription units. *Genomics* **32,** 341–351.

Kleinjan, D. A., and van Heyningen, V. (2005). Long-range control of gene expression: Emerging mechanisms and disruption in disease. *Am. J. Hum. Genet.* **76,** 8–32.

Levings, P. P., Zhou, Z., Vieira, K. F., Crusselle-Davis, V. J., and Bungert, J. (2006). Recruitment of transcription complexes to the beta-globin locus control region and transcription of hypersensitive site 3 prior to erythroid differentiation of murine embryonic stem cells. *Febs J* **273,** 746–755.

Li, Q., Peterson, K. R., Fang, X., and Stamatoyannopoulos, G. (2002). Locus control regions. *Blood* **100,** 3077–3086.

Mantovani, R. (1999). The molecular biology of the CCAAT-binding factor NF-Y. *Gene* **239,** 15–27.

Marks, P. A., Sheffery, M., and Rifkind, R. A. (1987). Induction of transformed cells to terminal differentiation and the modulation of gene expression. *Cancer Res.* **47,** 659–666.

Maston, G. A., Evans, S. K., and Green, M. R. (2006). Transcriptional regulatory elements in the human genome. *Annu. Rev. Genomics. Hum. Genet.* **7,** 29–59.

Mellon, P., Parker, V., Gluzman, Y., and Maniatis, T. (1981). Identification of DNA sequences required for transcription of the human alpha 1-globin gene in a new SV40 host-vector system. *Cell* **27,** 279–288.

Merika, M., and Orkin, S. H. (1993). DNA-binding specificity of GATA family transcription factors. *Mol. Cell. Biol.* **13,** 3999–4010.

Motohashi, H., O'Connor, T., Katsuoka, F., Engel, J. D., and Yamamoto, M. (2002). Integration and diversity of the regulatory network composed of Maf and CNC families of transcription factors. *Gene* **294,** 1–12.

Nam, C. H., and Rabbitts, T. H. (2006). The role of LMO2 in development and in T cell leukemia after chromosomal translocation or retroviral insertion. *Mol. Ther.* **13,** 15–25.

Orkin, S. H. (2000). Diversification of haematopoietic stem cells to specific lineages. *Nat. Rev. Genet.* **1,** 57–64.

Osborne, C. S., Chakalova, L., Brown, K. E., Carter, D., Horton, A., Debrand, E., Goyenechea, B., Mitchell, J. A., Lopes, S., and Reik, W. (2004). Active genes dynamically colocalize to shared sites of ongoing transcription. *Nat. Genet.* **36,** 1065–1071.

Palstra, R. J., Tolhuis, B., Splinter, E., Nijmeijer, R., Grosveld, F., and de Laat, W. (2003). The beta-globin nuclear compartment in development and erythroid differentiation. *Nat. Genet.* **35,** 190–194.

Pondel, M. D., Murphy, S., Pearson, L., Craddock, C., and Proudfoot, N. J. (1995). Sp1 functions in a chromatin-dependent manner to augment human alpha-globin promoter activity. *Proc. Natl. Acad. Sci. USA* **92,** 7237–7241.

Poodt, J., Martens, H. A., Walsh, I. B., Felix-Schollaart, B., and Hermans, M. H. (2006). A newly identified deletion of 970 bp at the alpha-globin locus that removes the promoter region of the alpha1 gene. *Hemoglobin* **30,** 471–477.

Rein, T., Forster, R., Krause, A., Winnacker, E. L., and Zorbas, H. (1995). Organization of the alpha-globin promoter and possible role of nuclear factor I in an alpha-globin-inducible and a non-inducible cell line. *J. Biol. Chem.* **270,** 19643–19650.

Rombel, I., Hu, K. Y., Zhang, Q., Papayannopoulou, T., Stamatoyannopoulos, G., and Shen, C. K. (1995). Transcriptional activation of human adult alpha-globin genes by hypersensitive site-40 enhancer: Function of nuclear factor-binding motifs occupied in erythroid cells. *Proc. Natl. Acad. Sci. USA* **92,** 6454–6458.

Sharpe, J. A., Summerhill, R. J., Vyas, P., Gourdon, G., Higgs, D. R., and Wood, W. G. (1993). Role of upstream DNase I hypersensitive sites in the regulation of human alpha globin gene expression. *Blood* **82,** 1666–1671.

Spilianakis, C. G., and Flavell, R. A. (2004). Long-range intrachromosomal interactions in the T helper type 2 cytokine locus. *Nat. Immunol.* **5,** 1017–1027.

Spilianakis, C. G., Lalioti, M. D., Town, T., Lee, G. R., and Flavell, R. A. (2005). Interchromosomal associations between alternatively expressed loci. *Nature* **435,** 637–645.

Spivak, J. L., Toretti, D., and Dickerman, H. W. (1973). Effect of phenylhydrazine-induced hemolytic anemia on nuclear RNA polymerase activity of the mouse spleen. *Blood* **42,** 257–266.

Splinter, E., Heath, H., Kooren, J., Palstra, R. J., Klous, P., Grosveld, F., Galjart, N., and de Laat, W. (2006). CTCF mediates long-range chromatin looping and local histone modification in the beta-globin locus. *Genes Dev.* **20,** 2349–2354.

Spooncer, E., Heyworth, C. M., Dunn, A., and Dexter, T. M. (1986). Self-renewal and differentiation of interleukin-3-dependent multipotent stem cells are modulated by stromal cells and serum factors. *Differentiation* **31,** 111–118.

Strauss, E. C., Andrews, N. C., Higgs, D. R., and Orkin, S. H. (1992). *In vivo* footprinting of the human alpha-globin locus upstream regulatory element by guanine and adenine ligation-mediated polymerase chain reaction. *Mol. Cell. Biol.* **12,** 2135–2142.

Suske, G. (1999). The Sp-family of transcription factors. *Gene* **238,** 291–300.

Szutorisz, H., Canzonetta, C., Georgiou, A., Chow, C. M., Tora, L., and Dillon, N. (2005a). Formation of an active tissue-specific chromatin domain initiated by epigenetic marking at the embryonic stem cell stage. *Mol. Cell. Biol.* **25,** 1804–1820.

Szutorisz, H., Dillon, N., and Tora, L. (2005b). The role of enhancers as centres for general transcription factor recruitment. *Trends. Biochem. Sci.* **30,** 593–599.

Tan, H., and Whitney, J. B., 3rd (1993). Genomic rearrangement of the alpha-globin gene complex during mammalian evolution. *Biochem. Genet.* **31,** 473–484.

Tanimoto, K., Liu, Q., Bungert, J., and Engel, J. D. (1999). Effects of altered gene order or orientation of the locus control region on human beta-globin gene expression in mice. *Nature* **398,** 344–348.

Vakoc, C. R., Letting, D. L., Gheldof, N., Sawado, T., Bender, M. A., Groudine, M., Weiss, M. J., Dekker, J., and Blobel, G. A. (2005). Proximity among distant regulatory elements at the beta-globin locus requires GATA-1 and FOG-1. *Mol. Cell.* **17,** 453–462.

Van Loo, P. F., Bouwman, P., Ling, K. W., Middendorp, S., Suske, G., Grosveld, F., Dzierzak, E., Philipsen, S., and Hendriks, R. W. (2003). Impaired hematopoiesis in mice lacking the transcription factor Sp3. *Blood* **102,** 858–866.

Vernimmen, D., de Gobbi, M., Sloane-Stanley, J. A., Wood, W. G., and Higgs, D. R. (2007). Long-range chromosomal interactions regulate the timing of the transition between poised and active gene expression. *EMBO J.* **26,** 2041–2051.

Vestri, R., Pieragostini, E., and Ristaldi, M. S. (1994). Expression gradient in sheep alpha alpha and alpha alpha alpha globin gene haplotypes: mRNA levels. *Blood* **83,** 2317–2322.

Viprakasit, V., Kidd, A. M., Ayyub, H., Horsley, S., Hughes, J., and Higgs, D. R. (2003). *De novo* deletion within the telomeric region flanking the human alpha globin locus as a cause of alpha thalassaemia. *Br. J. Haematol.* **120,** 867–875.

Viprakasit, V., Harteveld, C. L., Ayyub, H., Stanley, J. S., Giordano, P. C., Wood, W. G., and Higgs, D. R. (2006). A novel deletion causing alpha thalassemia clarifies the importance of the major human alpha globin regulatory element. *Blood* **107,** 3811–3812.

Visvader, J. E., Mao, X., Fujiwara, Y., Hahm, K., and Orkin, S. H. (1997). The LIM-domain binding protein Ldb1 and its partner LMO2 act as negative regulators of erythroid differentiation. *Proc. Natl. Acad. Sci. USA* **94,** 13707–13712.

Vyas, P., Vickers, M. A., Simmons, D. L., Ayyub, H., Craddock, C. F., and Higgs, D. R. (1992). Cis-acting sequences regulating expression of the human alpha-globin cluster lie within constitutively open chromatin. *Cell* **69,** 781–793.

Wadman, I. A., Osada, H., Grutz, G. G., Agulnick, A. D., Westphal, H., Forster, A., and Rabbitts, T. H. (1997). The LIM-only protein Lmo2 is a bridging molecule assembling an erythroid, DNA-binding complex which includes the TAL1, E47, GATA-1 and Ldb1/NLI proteins. *EMBO J.* **16,** 3145–3157.

Wallace, H. A., Marques-Kranc, F., Richardson, M., Luna-Crespo, F., Sharpe, J. A., Hughes, J., Wood, W. G., Higgs, D. R., and Smith, A. J. (2007). Manipulating the mouse genome to engineer precise functional syntenic replacements with human sequence. *Cell* **128,** 197–209.

Wang, Q., Carroll, J. S., and Brown, M. (2005). Spatial and temporal recruitment of androgen receptor and its coactivators involves chromosomal looping and polymerase tracking. *Mol. Cell* **19,** 631–642.

Welch, J. J., Watts, J. A., Vakoc, C. R., Yao, Y., Wang, H., Hardison, R. C., Blobel, G. A., Chodosh, L. A., and Weiss, M. J. (2004). Global regulation of erythroid gene expression by transcription factor GATA-1. *Blood* **104,** 3136–3147.

Wijgerde, M., Grosveld, F., and Fraser, P. (1995). Transcription complex stability and chromatin dynamics *in vivo. Nature* **377,** 209–213.

Winichagoon, P., Higgs, D. R., Goodbourn, S. E., Lamb, J., Clegg, J. B., and Weatherall, D. J. (1982). Multiple arrangements of the human embryonic zeta globin genes. *Nucleic Acids Res.* **10,** 5853–5868.

Woo, A. J., Moran, T. B., Choe, S. K., Schindler, Y. L., Sullivan, M. R., Fujiwara, Y., Paw, B. H., and Cantor, A. B. (2005). Identification of zfp148 (ZBP-89) as a novel GATA-1 associated transcription factor involved in megakaryopoiesis and definitive erythropoiesis. *Blood* **106,** 828a.

Yagi, M., Gelinas, R., Elder, J. T., Peretz, M., Papayannopoulou, T., Stamatoyannopoulos, G., and Groudine, M. (1986). Chromatin structure and developmental expression of the human alpha-globin cluster. *Mol. Cell. Biol.* **6,** 1108–1116.

Zeitlin, H. C., and Weatherall, D. J. (1983). Selective expression within the human alpha globin gene complex following chromosome-dependent transfer into diploid mouse erythroleukaemia cells. *Mol. Biol. Med.* **1,** 489–500.

Zhang, Q., Rombel, I., Reddy, G. N., Gang, J. B., and Shen, C. K. (1995). Functional roles of *in vivo* footprinted DNA motifs within an alpha-globin enhancer. Erythroid lineage and developmental stage specificities. *J. Biol. Chem.* **270,** 8501–8505.

Zhang, X., Yazaki, J., Sundaresan, A., Cokus, S., Chan, S. W., Chen, H., Henderson, I. R., Shinn, P., Pellegrini, M., and Jacobsen, S. E. (2006). Genome-wide high-resolution mapping and functional analysis of DNA methylation in arabidopsis. *Cell* **126,** 1189–1201.

6 Global Control Regions and Regulatory Landscapes in Vertebrate Development and Evolution

Francois Spitz* and Denis Duboule[†,‡]

*Developmental Biology Unit, EMBL, 69117 Heidelberg, Germany
[†]NCCR "Frontiers in Genetics" and Department of Zoology and Animal Biology, University of Geneva, 1211 Geneva, Switzerland
[‡]School of Life Sciences, Federal Polytechnic School, CH-1015 Lausanne, Switzerland

 I. Introduction
 II. Global Controls
   A. Clusters of co-expressed developmental genes
   B. Shared regulatory elements between neighboring genes
   C. Global regulation in *Hox* cluster
 III. Co-Expression Chromosomal Territories, Regulatory Landscapes, and Global Control Regions
   A. Extending away from the *Hox* cluster: *Evx2* and *Lnp*
   B. *Gremlin/Formin*
   C. Co-expression territories
 IV. Mechanisms of Underlying Global Regulation
 V. Co-Expression Chromosomal Territories, Regulatory Landscapes: Bystander Effects or Functional Operons?
 VI. Evolutionary Implications of Global Gene Control
 VII. Global Regulation, Chromosomal Architecture, and Genetic Disorders
 VIII. Concluding Remarks
   References

Advances in Genetics, Vol. 61
0065-2660/08 $35.00
DOI: 10.1016/S0065-2660(07)00006-5

ABSTRACT

During the course of evolution, many genes that control the development of metazoan body plans were co-opted to exert novel functions, along with the emergence or modification of structures. Gene amplification and/or changes in the *cis*-regulatory modules responsible for the transcriptional activity of these genes have certainly contributed in a major way to evolution of gene functions. In some cases, these processes led to the formation of groups of adjacent genes that appear to be controlled *by both* global and shared mechanisms. © 2008, Elsevier Inc.

I. INTRODUCTION

The proper implementation of the genetic program controlling cell differentiation and, ultimately, metazoan development requires the highly coordinated actions of multiple genes. Consequently, these genes need to be tightly regulated in both time and space, and with respect to their quantitative outputs. Small changes in a single gene expression pattern can lead to severe morphological alterations, as exemplified by haplo-insufficiencies in many human syndromes (e.g., Guris *et al.*, 2006; Slager *et al.*, 2003), ectopic gene expression in cancers (e.g., Hayday *et al.*, 1984), or even slight heterochronic variations (e.g., Juan and Ruddle, 2003; Zakany *et al.*, 1997). Yet gene products rarely act alone, but usually interact with partners either to form large multiproteins complexes (e.g., Polycomb-group complexes, muscle contractile apparatus) or to be part of a sequential metabolic pathway (e.g., retinoic acid). Therefore, it is important that genes whose products are part of the same functional pathway are expressed at the same time and in the same cells.

In 1960, François Jacob and Jacques Monod showed that *LacZ*, *LacY*, and *LacA*, the genes required to degrade lactose in *Escherichia coli*, are aligned sequentially along the bacterial chromosome. Their expression is controlled by a repressor molecule produced by the *lacI* gene which blocks their transcriptional activity by binding to an operator element localized upstream of this small gene cluster (Jacob *et al.*, 1960). Furthermore, these genes are transcribed as a single RNA molecule, reenforcing their tight coordination. Such an organization of genes within "operons" provided a paradigm for the integration of gene expression, function, and structural organization, and indeed, many prokaryotic genes are organized similarly.

Operons are also present in eukaryotes, even though they usually differ slightly from the canonical prokaryotic version. For instance, up to 15% of *Caenorhabditis elegans* genes are organized as operons (Blumenthal *et al.*, 2002). They are transcribed as polycistronic pre-mRNAs that are subsequently processed by polyadenylation and *trans*-splicing to form mature, single gene coding mRNAs. As for prokaryotes, some of these "operons" may contain genes whose products are involved in the same biological process (Blumenthal et al., 2002). However, it is unclear how ancestral these operons are, since they do not seem to be conserved in another nematode (Lee and Sommer, 2003). In mammals, only a few bicistronic systems have been described and it is as yet unclear whether the functions of the genes concerned are related in any way (Gray et al., 1999; Lee, 1991). Therefore, our current view of eukaryotic gene regulation tends to consider genes as individual units, generally controlled by private sets of regulators. However, if we consider the original concept of the operon, initially regarded as "a grouping of adjacent structural genes controlled by a common operator" (Francois Jacob, in his Nobel Lecture), several recently described eukaryotic genetic loci may fulfill such a definition. Indeed, recent approaches involving either specific loci or genome-wide gene analysis tools tell us that a significant proportion of adjacent eukaryotic genes display related expression patterns, suggesting shared regulatory mechanisms.

In eukaryotes, many functionally related genes are found in clusters. In most cases, these situations derive from tandem duplications of an ancestral gene, leading to series of contiguous genes structurally related to each other, rather than distinct genes displaying related functions. In many instances, genes belonging to such clusters are expressed similarly, regardless of their general functional classification as "structural" or "regulatory" genes. For example, clustering may concern not only genes such as keratins, olfactory receptors, or proto-cadherins (Glusman et al., 2001; Hesse et al., 2004; Wu and Maniatis, 1999), but also transcription factors such as homeobox genes (Brooke et al., 1998; Duboule et al., 1986; Jagla et al., 2001) or signaling molecules such as Wnts and Fgfs (Katoh, 2002; Nusse, 2001). While some of these clusters emerged recently during evolution, others can be traced down to the common metazoan ancestor and have since been maintained. This conservation of clustering illustrates the existence of constraints, one of them likely being the shared regulation of several gene members in a given cluster, suggesting that potent regulatory mechanisms act globally to impose a common regulation at the level of the cluster, rather than acting upon individual genes. Recent work on different model systems has identified such global mechanisms of gene expression, and these are discussed below.

Using genome-wide approaches, several studies have reported that adjacent, yet otherwise unrelated, genes share common expression specificities, leading to large co-expression territories (Boutanaev et al., 2002; Spellman and Rubin, 2002). In vertebrates, while global surveys have mainly revealed clusters of housekeeping genes (Lercher et al., 2002), it is quite clear that similar situations occur around developmental genes, as illustrated by a growing list of examples (Crackower et al., 1996; Holmes et al., 2003; Maas and Fallon, 2004; Spitz et al., 2003; Zuniga et al., 2004). While it is as yet unclear whether "regulatory landscapes" fulfill particular biological function and/or correspond to a functional partition of the genome, these findings have important practical and conceptual implications for our understanding of the relationships between gene regulation and genome structural organization and evolution.

Comparisons between different metazoan genomes have revealed that most of the genes involved in generating animal body forms were already present in their common ancestors (e.g., Carroll, 2006), indicating that evolution proceeded by redeploying a rather limited repertoire of genes. The co-option of genes [or of whole regulatory circuitries (Davidson, 2006)] in parallel with the evolution of novel tasks, and the resulting pleiotropy, was naturally accompanied by what can be seen as side effects, such as compensatory mechanisms and redundancy (Duboule and Wilkins, 1998; Kirschner and Gerhart, 1998). While most gene families were arguably present in basal animals, the number of available genes subsequently varied. In vertebrates, for instance, this repertoire expanded following both whole genome and local duplications. These two successive rounds of genome duplications account for the observation that many genes are found in three to four copies in mammals, whereas they are unique in lophotrochozoans and ecdysozoans (Levine and Tjian, 2003; McLysaght et al., 2002). Alternatively, or in parallel, gene families have also been expanded through tandem duplications, leading to the formation of multigenic clusters. Some of these clusters are rather young and have evolved rapidly, such as pheromone receptors (Rodriguez, 2005), whereas others are found conserved in insects and mammals, suggesting that these peculiar organizations are functionally constrained.

II. GLOBAL CONTROLS

A. Clusters of co-expressed developmental genes

Developmental genes are often grouped into multigenic complexes containing two or more closely related genes. Interestingly, this seems to mostly involve transcription factors [e.g., the *Fox* (Wotton and Shimeld, 2006), *Hox* (Krumlauf, 1994), *Irx* (Gomez-Skarmeta and Modolell, 2002), *Six* (Gallardo et al., 1999), *NK-Lbx* (Luke et al., 2003) gene families], though some signaling molecules, such as *Fgf3/4/19* and *Wnt1/6/10* (Katoh, 2002; Nusse, 2001), are also organized in clusters. Quite often, genes within a cluster tend to share some expression features during embryogenesis or in adult tissues.

The *Irx/Iroquois* gene clusters are rather typical in this respect. *Irx/Iroquois* genes encode homeobox-containing transcription factors and are organized in clusters, containing three genes separated by large intergenic regions (from 300 kb to more than 1 Mb in human). In mammals, two *Irx* clusters have been reported (*Irx1/2/4* and *Irx3/5/6*) (Peters *et al.*, 2000), which may have evolved independently from the single one described in *Drosophila* (*Iro-C*) (Gomez-Skarmeta and Modolell, 2002). The expression of vertebrate *Irx* genes has been studied in several vertebrate species (Christoffels *et al.*, 2000; Houweling *et al.*, 2001; Lecaudey *et al.*, 2005; Mummenhoff *et al.*, 2001; Zulch *et al.*, 2001) and revealed that, within a cluster, adjacent genes are expressed in similar structures during embryogenesis. For example, both *Irx1* and *Irx2* show strong and comparable expression domains during mouse brain development (ventral mesencephalon and telencephalon, otic vesicle), in the neural tube and in the condensing cartilage of developing limb buds, as well as in other sites (Houweling *et al.*, 2001). Likewise, *Irx3* and −5 are co-expressed in domains that may overlap with those of *Irx1/2*, yet with important differences in the neural tube, in motor neurons as well as in developing limbs. In both clusters, the gene located at the 3' extremity (*Irx4* or *Irx6*) displays usually a more divergent pattern, which does not correlate with a greater intergenic spacing, for the distance between *Irx1* and *Irx2* is larger than that between *Irx2* and *Irx4*. Some expression domains are nonetheless shared by all three members of each cluster, such as, for example, in kidneys or in mammary glands (Houweling *et al.*, 2001). Both the structural organization and some of these expression features are conserved between mammalian, amphibian, and teleost *Irx* genes.

Co-expression of tandemly duplicated genes can derive from different processes. For example, an ancestral gene and its associated regulatory elements (promoter, enhancers, and silencers) are duplicated all together, leading to a situation where both copies have their own regulatory elements. As these elements have a common origin, the duplicated genes may control expression in a similar way, before each copy evolves separate regulatory capacities. Such a situation does not require genes to maintain a topographic link. Alternatively, coding regions may be duplicated without some of their associated regulatory regions, or one set of the regulatory regions duplicated together with the coding regions might be subsequently lost, before evolving unique properties that would favor its preservation (Force *et al.*, 1999). In such cases where co-expression of the duplicated genes in a subset of domains is of selective advantage, regulations must be shared, and therefore the tandem organization of genes and *cis*-regulatory elements will be maintained over long evolutionary periods. In this context, several blocs of highly conserved DNA sequences have been found within the large intergenic regions of the *Irx* gene clusters (de la Calle-Mustienes *et al.*, 2005; McEwen *et al.*, 2006). Some of these potential *cis*-regulatory elements have been shown to drive reporter gene expression in domains that are shared between adjacent genes (de la Calle-Mustienes *et al.*, 2005; Pennacchio *et al.*, 2006). Interestingly, sequence comparisons have revealed that among these enhancers, some of them are related to each other, indicating that they were probably part of the duplicated material (Fig. 6.1).

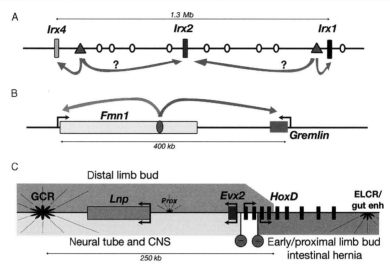

Figure 6.1. Global and shared regulations. (A) The *Irx124* cluster. Two homologous evolutionarily conserved sequences (triangles) were described within the complex and shown to activate gene expression in domains related to *Irx1*, *−2*, *and −4* expression patterns (de la Calle-Mustienes *et al.*, 2005). These elements probably derived from an ancestral enhancer that was duplicated along with the rearrangement(s) that led to the formation of the cluster as related elements are also found at orthologous positions in the sister *Irx356* cluster. However, because of the large number of other associated enhancers present within the cluster and the possibility of redundancy, it is unclear whether these duplicated elements are shared between the different *Irx* genes and account for their highly similar expression profiles. (B) The *Formin1–Gremlin* locus. The limb bud expression of *Gremlin* is controlled by an enhancer localized within the adjacent *Formin1* gene (Zuniga *et al.*, 2004). This enhancer also activates *Fmn1* in the limb bud, even though no function has been associated with this evolutionarily conserved expression. Both *Gremlin* and *Formin1* are also expressed autonomously in distinct domains. (C) The *Lnp–Evx2–Hoxd* regulatory landscapes. Several overlapping regulatory landscapes (shaded regions) coexist around the *Lunapark* (*Lnp*), *Evx2*, and *Hoxd* gene cluster. *Lnp*, *Evx2*, and the posterior *Hoxd* genes are included within a large region where all genes are expressed in the distal part of the limb and genital buds. This landscape is defined by a remote Global Control Region (GCR) which reinforces the intrinsically more restricted action of another enhancer element (*Prox*) (Gonzalez *et al.*, 2007; Spitz *et al.*, 2003). The GCR also contributes to the neural tube and CNS expression domains shared by *Lnp* and *Evx2*. At the other end of the *Hoxd* complex, a gut enhancer and an Early Limb Control Region (ELCR) localized 3′ of the cluster activate the *Hoxd1-Hoxd11* genes in the intestinal hernia, and in early stages and proximal parts of the limb, respectively (Spitz *et al.*, 2005; Zakany *et al.*, 2004). The activities of the GCR in the neural/CNS and of the gut enhancer are somehow restricted by boundary elements, whereas in the distal limb bud, it progressively declines with the distance to the 5′ end of the *Hoxd* cluster (Kmita *et al.*, 2000b, 2002a,b).

Therefore, the *Irx* clusters may illustrate both processes by which co-expression of tandemly duplicated genes can be maintained, with the sharing of regulatory elements, and duplicate ones, all of them contributing to co-expression of the *Irx* genes.

While several gene clusters show co-expression among neighboring genes, it is noteworthy that others do not, as exemplified by the Krüppel associated box (KRAB) gene family. These genes encode zinc-finger transcription factors and are organized in clusters (Huntley *et al.* 2006). They form a rather young and dynamic gene family, since mouse and human clusters arose probably from independent duplication events in rodents and primates (Huntley *et al.*, 2006; Shannon *et al.*, 2003). Despite this recent evolution, mammalian KRAB genes within a given cluster do not clearly share expression domains, indicating that tandem duplication does not necessarily imply the subsequent use of the same regulations. These young KRAB clusters might still be evolving, with their fate (either breaking down to individual genes or developing shared regulation) yet to be established.

B. Shared regulatory elements between neighboring genes

The observation that a given enhancer may be shared by neighboring genes calls for more than the mere contemplation of expression domains, requiring that experimental investigations be carried out *in vivo*. Using homologous recombination in ES cells and production of genetically modified mice, the endogenous locus can be deleted, so as to observe the effects upon both, or all potential target genes. Alternatively, the locus of interest can be transferred onto a large "transgene" such as Bacterial Artificial Chromosomes (BACs). The increasing availability of BAC resources (a list can be accessed at http://www.genome.gov/10001852), their accessibility to manipulation, and the development of rapid protocols to modify them at will have greatly facilitated the studies of large-scale gene regulation over the past years. BACs can contain large genomic fragments (up to 200 kb) and are thus able to incorporate entire loci or multigenic clusters. In addition, they are less prone to uncontrolled recombination than YACs and easier to expand and purify. The introduction of "recombineering strategies" in *E. coli* (Lee *et al.*, 2001; Muyrers *et al.*, 1999; Warming *et al.*, 2005) has opened the door to all kinds of modifications, such as the insertion of reporter genes, the deletion of *cis*-regulatory elements, or the introduction of point mutations. In this way, shared regulatory elements were successfully identified, for example, in the β-globin (see chapter 4), the *Il4–13–5* (Loots *et al.*, 2000), the *Dlx* (Sumiyama *et al.*, 2002), the *Myf5/Mrf4* (Carvajal *et al.*, 2001), or the *Hox* gene clusters (Chiu *et al.*, 2000; Spitz *et al.*, 2001, 2003). In this context, *Hox* clusters are of special interest as they illustrate all possible mechanisms that may be involved in the global regulation of gene clusters during development.

Mammals have 39 *Hox* genes, present in four different complementation groups, each containing from 9 to 11 genes. These four *Hox* clusters (*HoxA* to *HoxD*) are derived from an elusive cluster present in an early chordate, as may possibly be illustrated by the single *Hox* cluster described in the cephalochordate *Amphioxus* [see, e.g., Garcia-Fernandez (2005)]. A cluster of *Hox* genes must have existed in ancestral bilaterian animals, at the base of the radiation between vertebrates, lophotrochozoans, and ecdysozoans, and remnants of such a cluster can be found in various species within these large groups (Lemons and McGinnis, 2006). Generally, the expression domains of *Hox* genes along the anterior to posterior (AP) axis provide positional cues to the developing embryo which contribute to defining the fate of axial structures (Krumlauf, 1994). Interestingly, the anterior most limits of these expression domains in somitic mesoderm and in the neural tube correlate with the relative position of the genes within their clusters, such that genes localized at the 3′ end of the cluster are activated first and in anterior structures, whereas more 5′ located genes are expressed later and in more posterior domains. Different models have been proposed to account for this phenomenon of "colinearity," that is, to explain how the topography of these genes may relate to their expressions in time and space (Kmita and Duboule, 2003).

During mammalian development, *Hox* gene expression is controlled by multiple mechanisms. For example, Gould, Sharpe, Krumlauf, and colleagues reported the presence of enhancers within the *HoxB* cluster and shared between two adjacent genes (Gould *et al.*, 1997). They showed that the adjacent *Hoxb3* and *Hoxb4* genes are co-expressed in the hindbrain, with a sharp boundary between rhombomeres 6 and 7, and identified a conserved DNA sequence between these two genes that conferred hindbrain expression up to the r6 to r7 level (Gould *et al.*, 1997). When this enhancer was deleted from the transgene, expression no longer reached this rostral level, suggesting that both *Hoxb3* and *Hoxb4* require this regulation to extend to their proper expression boundary. A similar analysis on the adjacent *Hoxb4*, *Hoxb5*, *and Hoxb6* region used a two-reporters transgenic strategy, allowing the transcriptional activities of neighboring genes to be monitored in parallel (Sharpe *et al.*, 1998). In this way, several tissue-specific enhancer elements were localized within this short genomic interval, including a shared enhancer that activates simultaneously *Hoxb4* and *Hoxb5* in somites 7 to 8. In contrast, other enhancers appeared to act on a single gene only, either due to promoter selectivity or due to promoter competition. Enhancer sharing among adjacent mammalian *Hox* genes was further demonstrated by targeted alterations of the *HoxD* cluster. Deletion of several evolutionarily conserved intergenic elements led to selective alterations in the expression of several genes, either in time or in space (Gerard *et al.*, 1996; Zakany *et al.*, 1997). This was particularly well illustrated by the deletion of the RXI element, leading to a concomitant posteriorization of both *Hoxd10* and *Hoxd11* expressions in the trunk of embryos (Gerard *et al.*, 1996).

C. Global regulation in *Hox* cluster

Extensive sharing of local *cis*-regulatory elements between adjacent groups of *Hox* genes provides an attractive model to account for both the related expression patterns and the maintenance of clustered organization (Duboule, 1998). However, studies on the HoxD cluster showed that this model is incomplete, with other elements and mechanisms being involved. In particular, the early phase of *Hox* genes expression could not be properly reproduced with transgenes, which often showed delayed activation. Accordingly, a large transgene containing most of the human HOXD cluster was shown to contain those sequences required for collinear expression in late phases, but could not clearly reproduce early expression patterns (Spitz et al., 2001). Since several local *Hoxb* enhancers are controlled by HOX products themselves (Gould et al., 1997; Maconochie et al., 1997; Popperl and Featherstone, 1992; Tumpel et al., 2007), it is possible that shared local enhancers are used to maintain, reenforce, or secure collinear gene expression, through auto- and cross-regulatory loops. The lack of maintenance of appropriate expression patterns associated with a large deletion within the HoxD cluster (Spitz et al., 2001) indeed suggests that sequences within the cluster are important to properly maintain a collinear pattern possibly setup by controls localized outside of the cluster itself.

1. Shared remote enhancers acting on contiguous set of *Hox* genes

In addition to their expression in somitic mesoderm and in the neural tube, *Hox* genes have acquired a variety of functional tasks along with the emergence of vertebrates and the concomitant cluster amplification. As a consequence, while all four clusters are functionally important in the ancestral expression domains, more recent functionalities are associated with either one or a few specific clusters. Good examples of this are given by the HoxD cluster, where the seven most anterior genes (from *Hoxd1* to *Hoxd11*) are co-expressed in the intestinal hernia, a region of the gut that corresponds to the future transition between the ileocaecal region and the colon and which develops outside the fetal abdomen (Roberts et al., 1995; Zakany and Duboule, 1999). Likewise, the four contiguous genes localized 5'-most in the cluster (from *Hoxd13* to *Hoxd10*) are co-expressed in future digits of the developing limb buds (Dolle et al., 1989). Two genes, *Hoxd11* and *Hoxd10*, thus show expression in both the developing limbs and the intestinal hernia. However, neither transgenes containing a single *Hoxd* gene and its associated promoter region nor an entire cluster were able to recapitulate expression in any of these domains (Gerard et al., 1993; Herault et al., 1998; Renucci et al., 1992; Spitz et al., 2001; van der Hoeven et al., 1996). Conversely, when the full HoxD cluster was deleted but replaced by a single reporter gene, the latter displayed expression in these structures, indicating that elements localized inside the cluster are dispensable for these particular expression domains (Spitz et al., 2001).

These results suggest that *Hox* genes respond to regulatory cues that are in part imposed by elements localized outside the clusters (remote enhancers), and furthermore, such regulations appear to bear little promoter specificity, since expression of a given gene largely depends upon its relative position within a cluster (Herault et al., 1999; Tarchini and Duboule, 2006; van der Hoeven et al., 1996; Zakany et al., 2004). Accordingly, whenever foreign transcription units were inserted within the *HoxD* cluster, they adopted an expression pattern corresponding to the site of insertion (Herault et al., 1999; Kmita et al., 2000a; van der Hoeven et al., 1996).

2. Controlling enhancer activity: relay, silencer, boundary, and tethering elements

Which mechanisms can control or distribute such global activities? Experimental approaches *in vivo* have shown that the mere relative position of a gene within the cluster, that is, its relative distances from either the 3′ or the 5′ ends are important cues (Tarchini and Duboule, 2006), possibly reflecting a progressive reduction of enhancer efficiency as the distance (number of interspersed promoters) increases. Yet, regardless of how such basic mechanism is implemented, it is helped or refined by elements of well-defined functions in fine-tuning the responses of particular (groups of) genes to this global effect. For example, any *Hoxd* gene localized naturally or artificially at the 5′ extremity of the cluster is always expressed with maximal efficiency when compared with its 3′ located neighbors in developing digits. This regulatory preference for a gene localized at the end of the cluster is a consequence of both the proximity and the presence of a helper element located nearby (Kmita et al., 2002a).

Likewise, the intestinal hernia enhancer does not regulate the two most-posterior *Hoxd* genes because of the presence of an element downstream of *Hoxd12* that isolates *Hoxd12* and *Hoxd13* from this global regulation. This block has a polarity since, upon inversion of this element *in vivo*, *Hoxd12* and *Hoxd13* are activated in the hernia, whereas the isolation is observed on regulations coming from the other side of the cluster (Kmita et al., 2000b). Therefore, the relative extent of global enhancer activities clearly depends on the intricate interplay between the position of the gene within the cluster (distance from 5′ and 3′ ends and the number of promoters in between), and a combination of local tethering or boundary elements that will either favor or restrict the action of these enhancers to a specific set of target genes.

III. CO-EXPRESSION CHROMOSOMAL TERRITORIES, REGULATORY LANDSCAPES, AND GLOBAL CONTROL REGIONS

A. Extending away from the *Hox* cluster: *Evx2* and *Lnp*

One way to precisely assign global enhancer sequences to one or other side of a gene cluster relies upon splitting it into two subclusters. This was done via an engineered chromosomal inversion that separated the *HoxD* cluster into two independent subclusters (Spitz *et al.*, 2005). Expression analyses of embryos carrying such inversions demonstrated that the digit and the genital bud enhancers lie centromeric to (5′ from-) the *HoxD* cluster, whereas the intestinal hernia enhancer lies telomeric to *HoxD* (Spitz *et al.*, 2005). Interestingly, *Evx2* and *Lunapark* (*Lnp*), two genes located centromeric to *HoxD*, are also expressed in developing digits and genital bud with the same specificity as the neighboring posterior *Hoxd* genes (Bastian *et al.*, 1992; Dolle *et al.*, 1994; Spitz *et al.*, 2003), suggesting an extensive sharing of regulatory elements among them.

    Evx2 codes for a homeobox-containing transcription factor somewhat related to HOX proteins (Gauchat *et al.*, 2000), and its physical association with *Hox* genes must be an ancestral feature since it is observed in both *HoxA* and *HoxD* vertebrate clusters (Dolle *et al.*, 1994), in the lancelet (Minguillon and Garcia-Fernandez, 2003), and in coral (Miller and Miles, 1993). However, the murine *Evx* genes do not seem to endorse any *Hox*-related function, but rather seem to be required for the specification of specific pools of interneurons within the developing neural tube (Herault *et al.*, 1996; Moran-Rivard *et al.*, 2001).

    The case of the *Lnp* gene is even more striking as co-expression with *Hox* genes is observed despite a larger genomic distance. *Lnp* is indeed further away from the *HoxD* cluster than *Evx2* (90 kb against 8 kb), and encodes a protein of unknown function that is conserved in plants, yeasts, and all animals. It is structurally unrelated to homeobox-containing transcription factors, hence *Lnp* bears no relationship with *Hox* genes whatsoever, other than being adjacent to the *HoxD* cluster. Yet *Lnp* is expressed in developing limbs of mice and chicken in a distal posterior domain virtually identical to those of either *Hoxd13* or *Evx2* (Spitz *et al.*, 2003). Besides limbs, *Hoxd*, *Evx2*, and *Lnp* are also co-expressed in the developing external genital organs, and *Evx2* and *Lnp* have overlapping expressions in several domains of the developing central nervous system (CNS). The expression patterns of these genes are nevertheless not fully equivalent in all tissues and, unlike *Hox* genes, neither *Evx2* nor *Lnp* are transcribed in the trunk. Likewise, *Hoxd* genes are not expressed in those neural derivatives where both *Evx2* and *Lnp* are transcribed. Furthermore, *Lnp* is expressed both in the heart and in the eyes, where neither *Evx2* nor any *Hoxd* gene was shown expressed (Spitz *et al.*, 2003).

The existence of such global expression patterns on the top of more "gene-specific" features suggests that the *cis*-acting elements underlying the former may confer the associated regulation in a complete gene-independent fashion, that is, to any transcription unit lying within the realm of action of a hitherto qualified "global enhancer." The extent of the DNA interval within which various promoters will respond to a given global regulation was defined as a "regulatory landscape" (Spitz *et al.*, 2003). In the example described above, two overlapping regulatory landscapes are considered: a "limb landscape," which encompasses *Lnp*, *Evx2* and the posterior part of the *HoxD* cluster, and a "CNS" landscape restricted to *Lnp* and *Evx2* (Fig. 6.1C).

The search for, and identification of, *cis*-acting sequences involved in such global regulations is complicated by their intrinsic property to work at a distance. One method for identifying these elements was designed so as to isolate the enhancer responsible for the above-mentioned limb regulatory landscape. A contig of BAC clones covering the *HoxD* locus and flanking sequences was used as a starting point for the random transposition of a Tn7-based transposon containing a minimal promoter-reporter system (Spitz *et al.*, 2003). In this way, a region was identified that conferred expression to the reporter gene in both distal limb mesenchyme, the genital bud, and the neural tubes, that is, domains where either *Lnp*, *Evx2*, or *Hoxd* genes (or a combination thereof) are normally expressed. This region lies at ca. 200 kb from the *HoxD* cluster, beyond *Lnp*, at the border of a gene desert.

The molecular understanding of the mouse *Ulnaless* (*Ul*) mutation (Davisson and Cattanach, 1990) provided further evidence for the role of this element in controlling gene expression within these regulatory landscapes. *Ulnaless* is an X-ray induced inversion of a ca. 770-kb fragment that includes both *Evx2* and the *HoxD* cluster, with a breakpoint within *Lnp* (Spitz *et al.*, 2003). As a consequence of this inversion, the enhancer element is separated from both *Evx2* and *Hoxd* genes. As expected, these genes show a loss of expression in distal limb mesenchyme as well as in some dorsal neurons, all structures where the enhancer was shown active in a transgenic context (Herault *et al.*, 1997; Peichel *et al.*, 1997; Spitz *et al.*, 2003). Altogether, these various forms of evidence showed that this region contained various regulatory elements, defining at least two overlapping (limb and neural) regulatory landscapes. The term "Global Control Region (GCR)" was created to describe such DNA regions containing several long-range regulations acting on multiple genes (Spitz *et al.*, 2003).

B. *Gremlin/Formin*

Subsequently, regulatory elements of similar nature have been reported, as exemplified by studies of the *limb deformity* (*ld*) locus (Zuniga *et al.*, 2004). The *ld* mutation (Woychik *et al.*, 1985), as well as several spontaneous or engineered mutant stocks allelic to *ld*, shows a reduction in the number and size of digits, a concurrent loss of digit identities, and a fusion between the ulna and the radius. Positional cloning narrowed down the *ld* gene to a small region on mouse chromosome 2, and in several *ld* alleles, alterations were identified at the 3′ end of the *formin* gene (the gene formerly identified to cause the *ld* phenotype), leading to presumptive proteins with a truncated C-terminus (Mass *et al.*, 1990). *Formin* is indeed expressed in distal limb mesenchyme and its expression is diupted in several *ld* alleles, supporting a role in limb morphogenesis (Zeller *et al.*, 1989). It was subsequently established that the BMP-antagonist *Gremlin*, a critical factor for limb development involved in both *Shh* and *Fgf* signaling feedbacks, was absent in *ld* mutant limbs. Analysis of *Gremlin* expression in control and *ld* animals, as well as genetic analysis, supported a model whereby *Formin/ld* controls the expression of *Gremlin* (Zuniga *et al.*, 1999).

However, additional engineered alleles of the *formin* gene with deletion of the 5′ region failed to reproduce the *ld* phenotypes (Michos *et al.*, 2004) and, conversely, some *ld* alleles were shown to be associated with mutations in the *Gremlin* gene, rather than in *formin*, the two genes lying 80 kb from one another (Zuniga *et al.*, 2004). This contrasting picture was solved by using BAC transgenes, showing that the *ld* alleles affecting the 3′ part of the *formin* gene were not only deleting the protein encoded by this gene but also removing a critical *cis*-acting element required for the expression of *Gremlin* in distal limb mesenchyme (Zuniga *et al.*, 2004). Therefore, much like in the case of the *HoxD* limb regulatory landscape, a *cis*-acting element can work concomitantly on different promoters over a distance of ca. 400 kb, associated either with a gene involved in actin skeleton remodeling (*formin*) or with a BMP antagonist (*gremlin*) (Fig. 6.1B).

C. Co-expression territories

In addition to the *Lnp/Evx2/Hoxd* and *Fmn1/Gremlin* regulatory landscapes, cases have been described of co-expression of adjacent genes that are otherwise not related in any way. For example, *Lmx1b* and its immediate neighbor ALC (Holmes *et al.*, 2003), or the clustered *Dlx5-Dlx6* and adjacent *Dss1* (Crackower *et al.*, 1996) genes, are co-expressed. In the case of the *hGH* and *CD79b* genes, this co-expression was also shown to be imposed by a shared enhancer sequence (Cajiao *et al.*, 2004). However, the question as to whether regulatory landscapes and co-expression territories are widespread, or are restricted to a few anecdotal situations, remains to be firmly established. The development of various technologies to study genome-wide gene expression patterns, such as microarrays, SAGE, or MPSS libraries, has fostered various studies looking for correlation between gene expression profiling on the one hand and chromosomal position on the other. In *Drosophila* and *C. elegans*, several studies (Boutanaev *et al.*, 2002;

Roy *et al.*, 2002; Spellman and Rubin, 2002) have shown that adjacent genes often share expression specificities, forming what Spellman and Rubin described as "co-expression territories."

It is as yet unclear how such co-expression territories relate to regulatory landscapes and how many of them are defined by sequences analogous to GCRs. A key difference between the two notions is that co-expression territories are defined by similarities in expression across multiple tissues or experimental conditions, using semiquantitative approaches, whereas regulatory landscapes are defined by co-expression, as assayed by qualitative *in situ* hybridization, within one tissue or embryonic structure. Therefore, these terms may correspond to rather different situations, underlined by distinct molecular mechanisms.

In vertebrates, only a few co-expression territories have been described (Lercher *et al.*, 2002; Su *et al.*, 2004), and most of them correspond either to clusters of housekeeping genes or to tandem arrays of duplicated genes expressed in a very few tissues, such as clusters of olfactory receptors. This failure to detect tissue-specific co-expression territories and/or regulatory landscapes by genome-wide approaches, for example, in human or mice, may not reflect their absence but, instead, the inadequacy of the currently used "global profiling" methodology. As noted above, control of gene expression in mammals is mostly the result of modular, *cis*-acting elements. Consequently, GCRs are expected to work on top of other gene-specific expression specificities, and the use of "global co-expression" as a criterium would certainly disqualify the *HoxD/Lnp* regulatory landscape because differences between expression patterns would overcome the expression similarities in the limb. We conclude that global profiling may be quite effective in detecting co-expression of genes displaying either a broad (ubiquitous) or restricted (e.g., a single tissue) transcriptional activity. In contrast, it might fall short in cases involving genes sharing a global enhancer sequence for only one particular aspect of otherwise quite divergent expression patterns.

IV. MECHANISMS OF UNDERLYING GLOBAL REGULATION

Gene expression is generally seen as a rather tightly controlled process, yet the observation that adjacent genes often share expression specificities raises interesting questions regarding the mechanism(s) that could potentially be involved (Fig. 6.2).

The description of large co-expression territories in flies, *C. elegans*, and human has led to some mechanistic proposals: for example, in nematodes, most clusters are associated with multicistronic transcripts (operons *sensu stricto*) and are thus defined by the cotranscription, on a single precursor RNA molecule, of several genes (Lercher *et al.*, 2003). In *Drosophila*, co-expression of adjacent genes often involves one strongly expressed gene, whereas others are expressed at much

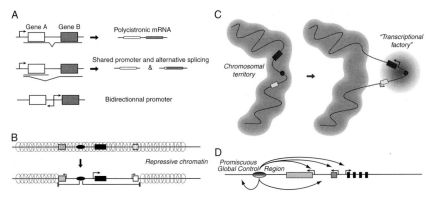

Figure 6.2. Mechanisms associated with co-expression of adjacent genes. (A) Local action. Through polycistronic transcription, alternative splicing, or even bidirectional promoters, two adjacent genes can share regulatory control by the same transcription factors. (B) Large-scale remodeling of chromatin structure. Enhancers can bind transcription factors which then recruit chromatin remodeling complexes that could act over relatively large distances. Genes in their vicinity might become more accessible to the basal transcription machinery, leading to their loose expression. (C) Global relocation of locus into the nuclear space. Enhancers, together with their target gene, might move away from compacted and repressive chromosomal territories to reach "transcription factories," and thus drag along the adjacent genes. (D) Promiscuous global—and usually long-range—enhancers contact all available promoters within their range of action. These elements might act in combination with other enhancers to drive expression in a variety of tissues. Their action depends on the architecture of the locus (distances, number of genes), and might be limited by boundary elements. These different schematized models are not mutually exclusive, and probably correspond to different facets of the mechanisms involved in global and long-range regulation of gene expression.

lower levels (Spellman and Rubin, 2002). Accordingly, it was proposed that these latter genes are transcribed simply because of their close proximity to a strongly transcribed unit, thus appearing as a bystander effect. Such bystander regulation may either involve a shared enhancer, like the GCR, which may contact promoters surrounding the strongly transcribed unit eagerly, with no restricted specificity, or from the spreading of transcriptionally permissive chromatin structures around active genes.

Genomic DNA is usually packaged around histones as well as in larger chromatin structures. It is generally believed that this compaction of DNA into dense chromatin structures is associated with repression of gene expression by prohibiting transcription factors and RNA polymerases to access promoter and enhancer sequences. A release in chromatin structure is necessary for gene activation, and consequently, chromatin found around active genes may be somewhat relaxed, thus increasing the chance for basal transcription factors to access promoter regions, leading to basal transcription of those genes located in the proximity of strongly expressed genes (for a more complete discussion about this topic, see Sproul *et al.*, 2005). Differential posttranslational modifications of histones, in particular through acetylation or methylation of different residues, have been associated with either an increase or a decrease of transcriptional activity (for a recent review, see Li *et al.*, 2007). Propagation of repressive chromatin from a nucleation center was shown to spread from specific DNA sequences [e.g., Polycomb Response Elements (Papp and Muller, 2006; Schwartz and Pirrotta, 2007), see also Rest/NRSF elements (Lunyak *et al.*, 2002)], and the counteraction of these mechanisms, for example via trithorax group proteins, might appear to be sufficient to keep the locus in a poised/permissive configuration that could allow basal transcription to proceed, not only for the "main" gene but also for the adjacent transcription units. However, genome-wide studies have revealed that these histone marks associated with gene activity are usually confined to discrete *cis*-acting regions (promoter, enhancer), rather than being spread over broad domains, with the striking exception of *Hox* clusters (Bernstein *et al.*, 2005). There is, in general, therefore no straightforward link between large-scale co-expression loci and locus-wide distribution of a specific chromatin/histone pattern, and the possible underlying mechanisms/factors required remain, in most cases, elusive (Sproul *et al.*, 2005).

Several reports have also underlined that the position of a given locus within the nucleus might impact on its expression. As initially shown in the yeast and, subsequently, in vertebrates (reviewed in Kosak and Groudine, 2004), chromosomal regions containing transcriptionally active genes tend to loop out of the condensed chromosomal territory, coming close to the periphery of the nucleus in so-called "transcription factories," that is, regions of the nucleus enriched in components of the transcription machinery. Genes that are co-expressed, but separated by several megabases, have indeed been shown to colocalize in these discrete nuclear compartments (Osborne *et al.*, 2004). It is conceivable that genes localized around actively transcribed genes would be carried along and "passively" benefit from the changes induced, leading to their spurious transcriptional activation. Several loci have evolved boundary elements that limit such bystander effects. For example, the formation of the Active Chromatin Hub at the β-globin locus serves a dual function. It favors an efficient interaction between remote enhancer elements and the associated globin genes (see chapter 4), but it also excludes the neighboring olfactory receptor genes from the influence of this interactive hub (Bulger *et al.*, 2000, 2003). While boundary elements, or other regions with chromatin organizing activity, provide some molecular mechanisms to compartment the genome in distinct functional domains, the extensive embedding of enhancers and genes over large regions might somehow increase the complexity in setting up such ordered organization, and would therefore allow leaky expression of more or less large sets of adjacent genes.

Bidirectional promoters may also lead to the co-expression of neighboring genes, and several eukaryotic promoters have this property (Engstrom et al., 2006). In the majority of cases, bona fide mRNAs are produced together with noncoding, likely nonfunctional, transcripts (Engstrom et al., 2006). However, such bidirectional activity sometimes correlates with genuine co-expression of functional transcripts (Bellizzi et al., 2007; Chen et al., 2007). As recent surveys of gene transcription start sites suggest the existence of a much more complex and extended organization than anticipated, bidirectional promoter sharing might have been overlooked and may be one cause of co-expression (Carninci et al., 2006; Engstrom et al., 2006).

However, promoter sharing can hardly account for those coregulated genes that are organized in a tail-to-tail manner, for example, the Formin–Gremlin pair, or various pairs of Dlx genes. Likewise, it cannot explain the coregulation of the Lnp–Evx2–Hoxd genes in developing limbs. While in the latter example bidirectional promoters likely do exist in the DNA interval between the genes (Sessa et al., 2007), experimental data argue against such a simple mechanism playing a really prominent role. First, the deletion of the region where such a promoter should lie (between Evx2 and Hoxd13) did not impair the activity of the remaining genes (Kmita et al., 2002a). Hox gene co-expression has also been proposed to arise through multicistronic or alternatively spliced transcripts. Indeed, some evidence for such phenomena have been described in crustaceans (Shiga et al., 2006), and in mammals, several hybrid transcripts can be found for all four vertebrate Hox clusters in ESTs databases. Nevertheless, transgene insertions targeted into this DNA region are still expressed in digits regardless of their orientation (Herault et al., 1999; van der Hoeven et al., 1996), and an engineered inversion of the Hoxd12 to Hoxd11 segment did not impact greatly upon the expression of these two genes in developing limbs and genitals (Kmita et al., 2000b). Therefore, within the Lnp–Hoxd regulatory landscapes, shared expression patterns are observed irrespective of transcriptional orientation.

In this case, the GCR and associated elements behave rather like a promiscuous enhancer, which is able to contact various promoters within the same cell at the same time. It is as yet unclear whether all the genes in this landscape are transcribed simultaneously or sequentially, as shown for the β-globin genes (Dillon et al., 1997; Wijgerde et al., 1995). Also, the associated chromatin structure has not been studied in detail, even though it appears that chromatin around the Lnp/Hoxd interval is decondensed in E9.5 limb buds (Morey et al., 2007). While the GCR is not yet fully active in limbs at this stage (Lnp, Hoxd13, or GCR-linked reporter genes start to be expressed at E10.5), these results suggest that some global chromatin modifications may be necessary beforehand. Whether such modifications are determined by the GCR is unknown; however, it is clear that the GCR does not only work via a global modification of the surrounding chromatin, and that its activity is required for proper gene transcription, as observed from the effect of either the deletion or the addition of genes in the landscape (Kmita et al., 2002a; Monge et al., 2003). In such cases, the resulting up- and downregulation of resident genes of the landscape indicate transcriptional reallocations that are at odds with a mere modification of the chromatin structure. Instead, this competition in transcription argues for an active role of the GCR in contacting the different promoters. The identification of the transcription factors bound to the GCR and of the protein complexes involved in the long-range action of this element would help to decipher the precise molecular mechanisms at play in these complex gene regulations.

V. CO-EXPRESSION CHROMOSOMAL TERRITORIES, REGULATORY LANDSCAPES: BYSTANDER EFFECTS OR FUNCTIONAL OPERONS?

Regardless of the nature of the underlying mechanisms, the existence of co-expressed gene clusters raises the issue of both their biological significance and their functional relevance. In this context, clusters containing similar genes, that is, derived from ancestral gene duplication event(s), and clusters containing genes unrelated in their phylogenies may have to be considered separately.

In the former case, for example, the Hox or Dlx gene clusters, genes may often encode proteins containing similar functional domains, which are thus more susceptible to endorse related—if not similar—functions, such as activating or repressing the same set of target genes. For example, the Dlx5 and Dlx6 genes are functionally redundant and mice with a targeted inactivation of either the one or the other have normal limbs, whereas the combined deletion led to a severe ectrodactyly (Robledo et al., 2002). Likewise, the inactivation of Hoxd13, Hoxd12, or Hoxd11 in mice led to malformed hands and feet, implying that all these genes are somewhat involved in the patterning of appendages (Davis and Capecchi, 1994; Dolle et al., 1993; Favier et al., 1995; Kondo et al., 1996). In this case however, the defects were much more pronounced whenever Hoxd13 was mutated, indicating that within the cluster, the functionalities of each gene are not equal, either for extrinsic reasons, such as their expression levels, or for intrinsic reasons, such as their various capacities to compensate for one another's function (Kmita et al., 2002a; Tarchini et al., 2006). This example shows that genes within a co-expression cluster can be involved in similar pathways, yet with very different functional impacts.

In the case of clustering of unrelated genes, there is little evidence that co-expression is associated with related functions and, beyond some anecdotal situations [e.g., Ellis-van Creveld syndrome (Ruiz-Perez et al., 2003)], the function associated with the global expression is generally achieved by only one gene, the others being dispensable. Neither the *Lnp*, *Evx2* nor *Formin* genes seem to have important function in limb morphogenesis despite their strong expression in the developing limb bud, which results from their position within either the *Hoxd* or the *Gremlin* regulatory landscapes, respectively (Herault *et al.*, 1996; Spitz *et al.*, 2003; Trumpp *et al.*, 1992; Zuniga *et al.*, 2004). In the *hGF/CD79b* locus, both genes are transcriptionally active in the pituitary, yet only hGF was detected at the protein level, while *CD79b* is translated in B cells, where it is important for signal transduction from the B-cell receptor, but not in the pituitary (Cajiao *et al.*, 2004). In this case, expression was proposed to be a nonfunctional consequence of the genes' localization within the highly acety-lated chromatin domain defined by the enhancer of the growth-hormone genes. These examples suggest that in most reported cases, co-expression may be a collateral effect of the mechanism implemented to ensure proper transcription in a given place and time of the one gene functionally important within the locus. The observation that the immediate neighboring genes of *Lmx1b* in both mice and chick are expressed in dorsal limb mesenchyme much like *Lmx1b* despite the fact that the genes are not orthologous between these two species (Holmes *et al.*, 2003) also argues that bystander gene activation might occur in most co-expression regions.

If co-expressed clustered genes do not share any functional outcome, which constraint kept them together throughout evolution? If co-expression is judged based on either EST or SAGE or microarrays data, and considering the possible bias of the data set, as mentioned previously, it seems that these gene clusters were maintained more than it should be expected either by chance or by the inertia of the system (Semon and Duret, 2006; Singer *et al.*, 2005).

As discussed previously, clustering may reflect the highly intertwined organization of genes and their associated regulatory elements, which make rearrangements of the cluster difficult. In the cases of the *Formin/Gremlin* and *Lnp/HoxD* gene clusters, breaking the genomic linkage between these pairs of genes would also lead to the separation of a remote enhancer and the gene it must control. Such cases where synteny is conserved because of the complex architecture of the loci are frequent and do not necessarily associate with co-expression, as exemplified with the *Pax6*, *Myf5*, or *Shh* loci (Hadchouel *et al.*, 2000; Kleinjan *et al.*, 2002; Lettice *et al.*, 2002). In these cases, co-expression may occur, in particular if it is a parcimonious way of building the architecture and if expression of the neighboring gene is not deleterious to the embryo. In addition, this would save the organism the difficult task of evolving complex arrays of silencers and boundaries elements to restrict the action of global regulatory elements.

On the other hand, repressive or tethering elements may be used to fine-tune gene expression in complex loci, although these delicate mechanisms might not be as frequent as anticipated, since the potential benefit of their implementation has to be balanced by their own side effects. The expression of the *Shh* gene in the posterior limb bud, which defines the anterior to posterior asymmetry of the limbs, is controlled by an enhancer localized ca. 800 kb away, lying within the intron of the *Lmbr1* gene (Lettice *et al.*, 2003; Sagai *et al.*, 2005). In early stages of limb development, neither *Lmbr1* nor *Rnf32*, a gene located in between *Shh* and *Lmbr1*, is expressed in the Zone of Polarizing Activity (ZPA), implying that the enhancer is specifically acting on *Shh*. However, at later stages, *Lmbr1* expression is detected, at least in chicken, in the posterior limb together with *Shh* (Maas and Fallon, 2004), suggesting a relaxation of the tethering mechanisms that ensure a preferential interaction between this enhancer and the *Shh* gene during early limb development.

It is also possible that the maintenance of co-expression be functionally constrained, despite the absence of a functional protein produced from the adjacent genes. This is conceivable in the case where the function of a remote enhancer might be facilitated whenever immediate neighboring genes are transcriptionally active. This may help to provide a permissive chromatin configuration around the enhancer and make this region more accessible to transcription factors. In this view, promoter regions of genes localized in between a remote enhancer and a functional target might act as relays to help in setting up the proper interaction.

Besides such purely mechanistic help provided by co-expressed genes, the presence of neighboring promoters may be important to fine-tune the expression level of the functionally active target gene, not by acting in *trans* but rather by interactions in *cis*, for instance through the titration of the regulatory input. As mentioned earlier, the deletion or addition of genes within the *Hoxd* cluster lead to reallocation of the GCR activity and preventing the action of the GCR on *Lnp* and *Evx2* might also redistribute the regulation on the remaining *Hoxd* genes in a way deleterious for limb development. A similar argument might account for the expression of *Lmbr1* under the control of the *Shh* enhancer in chicken limbs, despite any detectable function. In such a view, the apparent absence of *Lmbr1* expression in the mouse ZPA raises the possibility that this differential behavior contributes to the different morphologies of rodent and bird autopods.

Various examples of loss or gain of expression have been reported following transgene insertion at the proximity of regulatory elements (Olson *et al.*, 1996; Qin *et al.*, 2004; Sharpe *et al.*, 1999), probably caused by complex titration-like mechanisms of either positive or negative regulations. In addition, few examples exist of gain of expression caused either by a deletion near a gene promoter (Kmita *et al.*, 2002a) or by a microdeletion removing several hundred kilobases (Kokubu *et al.*, 2003). Therefore, the importance of these titration mechanisms in the regulation of our transcriptome is still elusive and additional examples are required to draw up general rules on this issue.

VI. EVOLUTIONARY IMPLICATIONS OF GLOBAL GENE CONTROL

Such mechanisms may also have contributed to the evolution of gene function. The number and content of gene families are not grossly different among various metazoans, which indicates that novel function were generally not associated with the evolution of novel genes, but rather emerged by the functional redeployment of preexisting transcription units. Novel functionalities may have resulted from changes in the coding sequences of a gene by altering the properties of the protein or by modifying those regulatory modules that control the expression of a given gene. The latter possibility, suggested 30 years ago by King and Wilson (1975), has now received ample experimental validation in different systems (reviewed by Carroll, 2005).

These changes generally occurred either by slight alterations of existing regulations or by the loss of ancestral *cis*-acting elements, and the dissection of the regulatory mechanisms that control highly multifunctional genes has revealed that their expression usually results from the addition of small independent modules. These modules individually define specific aspects of the overall expression pattern and can be localized quite far away from the promoter they will interact with. Altogether, this suggests that newly evolved enhancer sequences generally correspond to novel functions; however, the mechanisms underlying the emergence of these sequences and their recruitment by target promoters remain unknown.

In this respect, regulatory landscapes (as defined by the presence of global enhancers) might provide a context wherein regulatory innovations may occur at an interesting frequency. First, GCR-like sequences (DNA segments containing several enhancer sequences) already have the capacity to interact with many genes, over large distances. Should a GCR evolve a modified activity, for example, through random drift of its DNA sequence, this novel activity could be tested on a variety of potential target genes. Also, some GCRs seem to synergize with other enhancers' sequences, which themselves may have more limited ranges of action either in terms of distance, number of target genes, or in their sensitivity to boundary elements (Gonzalez *et al.*, 2007), to further extend the GCR's realm of action. It is therefore conceivable that regulatory landscapes, while not favoring the emergence of enhancers, can facilitate their further association with an appropriate target. Some steps in the evolution of *Hoxd* gene regulation in tetrapods may have followed such a scenario, taking advantage of the intrinsic long-range property of a remote enhancer, originally acting in the CNS to control *Lnp* and *Evx2* expression, to gradually evolve limb enhancer activity, leading to the progressive establishment of a limb-specific regulatory landscape (Gonzalez *et al.*, 2007; Spitz *et al.*, 2003).

The promiscuous action of GCRs, or of other long-range acting enhancers, can thus contribute significantly to the evolution of gene function by increasing the opportunities to select for novel expression specificities.

Consequently, co-expression territories or regulatory landscapes may represent rather unstable situations (in an evolutionary time frame), prone to evolve into more specific gene-enhancer interactions, as suggested by both the persistence of the phenomenon and the relative lack of maintenance of these territories in different species. In this respect, the highly intermingled distribution of genes and enhancers along chromosomes perhaps testifies to the successive ancestral regulatory landscapes and promiscuous interactions that subsequently evolved into more exclusive interactions.

The situation may have been quite different when clusters of evolutionary related genes are considered. Indeed, in this case the enhancer likely existed prior to the duplication that led to the formation of the cluster and horizontal gene duplication could generate slightly different proteins that may be controlled in a similar way. Tandem duplications, by releasing some constraints on the duplicated genes, offer the possibility to have either one gene or both, evolving a new function, whereas the ancestral function is either retained by one of the duplicates or distributed between all copies. In such a situation, keeping duplicated genes under the same control mechanism provides the possibility to use their differentially evolved properties in a coherent manner.

The control of expression from a single shared regulatory mechanism, by distributing its activity either homogeneously or sequentially in time and space, can both provide this coherence and prevent potential dosage effects associated with overexpression, as might occur in the case where all gene duplicates were to have their own enhancer. An example is given by the *Dlx5–Dlx6* gene-pair: while the genes appear functionally redundant, the proteins evolved such that the domains required for chondrogenesis are significantly different. They might assemble on different target genes or, alternatively, form different complexes on the same targets (Hsu *et al.*, 2006), thus leading to near equivalent outputs. Such a situation provides robustness to the skeletal differentiation program by transforming an ancestral circuit into two parallel ones that use different plugs to connect on the same targets.

At the β-globin locus, the locus control region (LCR) is used to switch from an embryonic to adult isoform of a protein. Here the same mechanism is used both to ensure a suitable level of expression for the most appropriate isoform at a given time and to turn down the other genes. Similar temporal switches, implemented by a global mechanism, have been reported to involve a shared mechanism, even though the temporal overlap between the different genes of the cluster may be reduced to the minimal [e.g., α-fetoprotein/albumin, *Rhox* gene cluster (Maclean *et al.*, 2005)]. The *Hox* clusters are another example of this regulatory logic, whereby the distribution of the genes along the chromosomes is translated into a coherent spatiotemporal pattern of activity following the action of global mechanisms. Similarly, though at the single cell level, the exclusive expression of a single olfactory receptor gene in a given neuron from the olfactory bulb may also depend upon the use of a shared enhancer (Lomvardas et al., 2006) that can act not only on the cluster of receptors localized nearby on the same chromosome but also on those localized on different chromosomes.

VII. GLOBAL REGULATION, CHROMOSOMAL ARCHITECTURE, AND GENETIC DISORDERS

Because GCRs can act over long distances and on several genes, they likely contributed to the preservation of syntenies along the chromosomes of various different animals. Likewise, chromosomal rearrangements occurring within a regulatory landscape might affect genes at a distance from the breakpoint, potentially leading to both loss of function, whenever genes are moved away from the GCR, and gain of function, whenever genes are brought within the zone of influence of a particular GCR. In humans, several genetic disorders are associated with recurrent rearrangements, for example, the deletion between low copy repeats (Ji et al., 2000). In such cases, attention is generally focused on the genes directly affected by the rearrangements, which may lead to an incorrect understanding of the underlying mechanistic cause, as illustrated by the case of the ld mutation. The formin gene was considered as causative of the ld phenotype because chromosomal rearrangement disrupted this gene in multiple ld alleles, a hypothesis strengthened by the observed expression of formin in the embryonic structures affected by the mutation. Subsequently, it was shown that the gene responsible for the ld phenotype (Gremlin) is located 80 kb away from the various mutagenic events (Zuniga et al., 2004). Similarly, the dominant Ul mutation is primarily a disruption of the Lnp gene through the presence of an inversion breakpoint. Because this inversion disrupts the GCR–Lnp–Evx2–HoxD regulatory landscape, it also involves tissue-specific loss of function of Hoxd and Evx2, as well as gain of function of Hoxd genes in other domains of the developing limbs (Spitz et al., 2003). This latter case is a paradigm of the multiple, complex defects that can be obtained whenever regulatory landscapes are affected by large chromosomal rearrangements. It also nicely exemplifies the great difficulty in understanding the phenotype–genotype relationships of such complex mutations outside the context of global regulations.

VIII. CONCLUDING REMARKS

Over the past 30 years, our thinking of regulation of gene expression has been very much influenced by the pioneering work done on viral and bacterial systems, as well as in cultured eukaryotic cells. The general understanding was that genes and their regulations could be considered as separate units, and that the interactions between these units were mostly occurring in *trans*. In this ultimate genetic reductionism view, every gene is associated with (or responsible for) one particular function. This superficial view of gene function and regulation has persisted, due to both its mechanistic simplicity and paradigmatic value to account for orthodox (gradualist) evolutionary modifications; any character may indeed be "improved" in isolation since the underlying genetic circuitry can be taken in isolation.

However, recent progress in mouse molecular genetics and human genetics, following the sequencing of several genomes, is now revealing a slightly different picture, where various components are more interdependent on each other than previously anticipated, and where genes and their regulations can no longer be considered in isolation, out of their genomic context. While a large proportion of gene regulatory events are likely to occur in *trans*, several phenomena testify to the existence of global *cis*-mechanisms, which we will need to integrate in our future thinking of biological systems. Beside the mere mechanistic interest, both human genetics and evolutionary biology may rapidly benefit from these developments. While the concept of regulatory landscapes may help understand genetic syndromes for which no simple molecular etiology is available, it may also give us new avenues to think about the evolution of regulations, and hence the emergence of evolutionary novelties.

References

Bastian, H., Gruss, P., Duboule, D., and Izpisua-Belmonte, J. C. (1992). The murine even-skipped-like gene Evx-2 is closely linked to the Hox-4 complex, but is transcribed in the opposite direction. *Mamm. Genome* **3**, 241–243.

Bellizzi, D., Dato, S., Cavalcante, P., Covello, G., Di Cianni, F., Passarino, G., Rose, G., and De Benedictis, G. (2007). Characterization of a bidirectional promoter shared between two human genes related to aging: SIRT3 and PSMD13. *Genomics* **89**, 143–150.

Bernstein, B. E., Kamal, M., Lindblad-Toh, K., Bekiranov, S., Bailey, D. K., Huebert, D. J., McMahon, S., Karlsson, E. K., Kulbokas, E. J., III, Gingeras, T. R., Schreiber, S. L., and Lander, E. S. (2005). Genomic maps and comparative analysis of histone modifications in human and mouse. *Cell* **120**, 169–181.

Blumenthal, T., Evans, D., Link, C. D., Guffanti, A., Lawson, D., Thierry-Mieg, J., Thierry-Mieg, D., Chiu, W. L., Duke, K., Kiraly, M., and Kim, S. K. (2002). A global analysis of Caenorhabditis elegans operons. *Nature* **417**, 851–854.

Boutanaev, A. M., Kalmykova, A. I., Shevelyov, Y. Y., and Nurminsky, D. I. (2002). Large clusters of co-expressed genes in the *Drosophila* genome. *Nature* **420**, 666–669.

Brooke, N. M., Garcia-Fernandez, J., and Holland, P. W. (1998). The ParaHox gene cluster is an evolutionary sister of the Hox gene cluster. *Nature* **392**, 920–922.

Bulger, M., Bender, M. A., van Doorninck, J. H., Wertman, B., Farrell, C. M., Felsenfeld, G., Groudine, M., and Hardison, R. (2000). Comparative structural and functional analysis of the olfactory receptor genes flanking the human and mouse beta-globin gene clusters. *Proc. Natl. Acad. Sci. USA* **97**, 14560–14565.

Bulger, M., Schubeler, D., Bender, M. A., Hamilton, J., Farrell, C. M., Hardison, R. C., and Groudine, M. (2003). A complex chromatin landscape revealed by patterns of nuclease sensitivity and histone modification within the mouse beta-globin locus. *Mol. Cell. Biol.* **23,** 5234–5244.

Cajiao, I., Zhang, A., Yoo, E. J., Cooke, N. E., and Liebhaber, S. A. (2004). Bystander gene activation by a locus control region. *EMBO J.* **23,** 3854–3863.

Carninci, P., Sandelin, A., Lenhard, B., Katayama, S., Shimokawa, K., Ponjavic, J., Semple, C. A., Taylor, M. S., Engström, M. S., Frith, M. C., Forrest, A. R., Alkema, W. B., *et al.* (2006). Genome-wide analysis of mammalian promoter architecture and evolution. *Nat. Genet.* **38,** 626–635.

Carroll, S. B. (2005). Evolution at two levels: On genes and form. *PLoS Biol.* **3,** e245.

Carroll, S. B. (2006). "Endless Forms Most Beautiful: The New Science of Evo Devo." W. W. Norton, New York.

Carvajal, J. J., Cox, D., Summerbell, D., and Rigby, P. W. (2001). A BAC transgenic analysis of the Mrf4/Myf5 locus reveals interdigitated elements that control activation and maintenance of gene expression during muscle development. *Development* **128,** 1857–1868.

Chen, P. Y., Chang, W. S., Chou, R. H., Lai, Y. K., Lin, S. C., Chi, C. Y., and Wu, C. W. (2007). Two non-homologous brain diseases-related genes, SERPINI1 and PDCD10, are tightly linked by an asymmetric bidirectional promoter in an evolutionarily conserved manner. *BMC Mol. Biol.* **8,** 2.

Chiu, C. H., Amemiya, C. T., Carr, J. L., Bhargava, J., Hwang, J. K., Shashikant, C. S., Ruddle, F. H., and Wagner, G. P. (2000). A recombinogenic targeting method to modify large-inserts for cis-regulatory analysis in transgenic mice: Construction and expression of a 100-kb, zebrafish Hoxa-11b-lacZ reporter gene. *Dev. Genes Evol.* **210,** 105–109.

Christoffels, V. M., Keijser, A. G., Houweling, A. C., Clout, D. E., and Moorman, A. F. (2000). Patterning the embryonic heart: Identification of five mouse Iroquois homeobox genes in the developing heart. *Dev. Biol.* **224,** 263–274.

Crackower, M. A., Scherer, S. W., Rommens, J. M., Hui, C. C., Poorkaj, P., Soder, S., Cobben, J. M., Hudgins, L., Evans, L., and Tsui, L. C. (1996). Characterization of the split hand/split foot malformation locus SHFM1 at 7q21.3-q22.1 and analysis of a candidate gene for its expression during limb development. *Hum. Mol. Genet.* **5,** 571–579.

Davidson, E. H. (2006). "The Regulatory Genome: Gene Regulatory Networks in Development and Evolution". Academic Press/Elsevier, San Diego.

Davis, A. P., and Capecchi, M. R. (1994). Axial homeosis and appendicular skeleton defects in mice with a targeted disruption of hoxd-11. *Development* **120,** 2187–2198.

Davisson, M. T., and Cattanach, B. M. (1990). The mouse mutation ulnaless on chromosome 2. *J. Hered.* **81,** 151–153.

de la Calle-Mustienes, E., Feijoo, C. G., Manzanares, M., Tena, J. J., Rodriguez-Seguel, E., Letizia, A., Allende, M. L., and Gomez-Skarmeta, J. L. (2005). A functional survey of the enhancer activity of conserved non-coding sequences from vertebrate Iroquois cluster gene deserts. *Genome Res.* **15,** 1061–1072.

Dillon, N., Trimborn, T., Strouboulis, J., Fraser, P., and Grosveld, F. (1997). The effect of distance on long-range chromatin interactions. *Mol. Cell* **1,** 131–139.

Dolle, P., Izpisua-Belmonte, J. C., Falkenstein, H., Renucci, A., and Duboule, D. (1989). Coordinate expression of the murine Hox-5 complex homoeobox-containing genes during limb pattern formation. *Nature* **342,** 767–772.

Dolle, P., Dierich, A., LeMeur, M., Schimmang, T., Schuhbaur, B., Chambon, P., and Duboule, D. (1993). Disruption of the Hoxd-13 gene induces localized heterochrony leading to mice with neotenic limbs. *Cell* **75,** 431–441.

Dolle, P., Fraulob, V., and Duboule, D. (1994). Developmental expression of the mouse Evx-2 gene: Relationship with the evolution of the HOM/Hox complex. *Dev* (Suppl.), 143–153.

Duboule, D. (1998). Vertebrate hox gene regulation: Clustering and/or colinearity? *Curr. Opin. Genet. Dev.* **8,** 514–518.

Duboule, D., and Wilkins, A. S. (1998). The evolution of 'bricolage'. *Trends. Genet.* **14,** 54–59.

Duboule, D., Baron, A., Mahl, P., and Galliot, B. (1986). A new homeo-box is present in overlapping cosmid clones which define the mouse Hox-1 locus. *EMBO J.* **5,** 1973–1980.

Engstrom, P. G., Suzuki, H., Ninomiya, N., Akalin, A., Sessa, L., Lavorgna, G., Brozzi, A., Luzi, L., Lam Tan, S., Yang, L., Kunarso, G., Ng, E. L.-C., *et al.* (2006). Complex loci in human and mouse genomes. *PLoS Genet.* **2,** e47.

Favier, B., Le Meur, M., Chambon, P., and Dolle, P. (1995). Axial skeleton homeosis and forelimb malformations in Hoxd-11 mutant mice. *Proc. Natl. Acad. Sci. USA* **92,** 310–314.

Force, A., Lynch, M., Pickett, F. B., Amores, A., Yan, Y. L., and Postlethwait, J. (1999). Preservation of duplicate genes by complementary, degenerative mutations. *Genetics* **151,** 1531–1545.

Gallardo, M. E., Lopez-Rios, J., Fernaud-Espinosa, I., Granadino, B., Sanz, R., Ramos, C., Ayuso, C., Seller, M. J., Brunner, H. G., Bovolenta, P., and Rodríguez de Córdoba, S. (1999). Genomic cloning and characterization of the human homeobox gene SIX6 reveals a cluster of SIX genes in chromosome 14 and associates SIX6 hemizygosity with bilateral anophthalmia and pituitary anomalies. *Genomics* **61,** 82–91.

Garcia-Fernandez, J. (2005). The genesis and evolution of homeobox gene clusters. *Nat. Rev. Genet.* **6,** 881–892.

Gauchat, D., Mazet, F., Berney, C., Schummer, M., Kreger, S., Pawlowski, J., and Galliot, B. (2000). Evolution of Antp-class genes and differential expression of Hydra Hox/paraHox genes in anterior patterning. *Proc. Natl. Acad. Sci. USA* **97,** 4493–4498.

Gerard, M., Duboule, D., and Zakany, J. (1993). Structure and activity of regulatory elements involved in the activation of the Hoxd-11 gene during late gastrulation. *EMBO J.* **12,** 3539–3550.

Gerard, M., Chen, J. Y., Gronemeyer, H., Chambon, P., Duboule, D., and Zakany, J. (1996). *In vivo* targeted mutagenesis of a regulatory element required for positioning the Hoxd-11 and Hoxd-10 expression boundaries. *Genes Dev.* **10,** 2326–2334.

Glusman, G., Yanai, I., Rubin, I., and Lancet, D. (2001). The complete human olfactory subgenome. *Genome Res.* **11,** 685–702.

Gomez-Skarmeta, J. L., and Modolell, J. (2002). Iroquois genes: Genomic organization and function in vertebrate neural development. *Curr. Opin. Genet. Dev.* **12,** 403–408.

Gonzalez, F., Duboule, D., and Spitz, F. (2007). Transgenic analysis of Hoxd gene regulation during digit development. *Dev. Biol.* **306**(2), 847–859.

Gould, A., Morrison, A., Sproat, G., White, R. A., and Krumlauf, R. (1997). Positive cross-regulation and enhancer sharing: Two mechanisms for specifying overlapping Hox expression patterns. *Genes Dev.* **11,** 900–913.

Gray, T. A., Saitoh, S., and Nicholls, R. D. (1999). An imprinted, mammalian bicistronic transcript encodes two independent proteins. *Proc. Natl. Acad. Sci. USA* **96,** 5616–5621.

Guris, D. L., Duester, G., Papaioannou, V. E., and Imamoto, A. (2006). Dose-dependent interaction of Tbx1 and Crkl and locally aberrant RA signaling in a model of del22q11 syndrome. *Dev. Cell* **10,** 81–92.

Hadchouel, J., Tajbakhsh, S., Primig, M., Chang, T. H., Daubas, P., Rocancourt, D., and Buckingham, M. (2000). Modular long-range regulation of Myf5 reveals unexpected heterogeneity between skeletal muscles in the mouse embryo. *Development* **127,** 4455–4467.

Hayday, A. C., Gillies, S. D., Saito, H., Wood, C., Wiman, K., Hayward, W. S., and Tonegawa, S. (1984). Activation of a translocated human c-myc gene by an enhancer in the immunoglobulin heavy-chain locus. *Nature* **307,** 334–340.

Herault, Y., Hraba-Renevey, S., van der Hoeven, F., and Duboule, D. (1996). Function of the Evx-2 gene in the morphogenesis of vertebrate limbs. *EMBO J.* **15,** 6727–6738.

Herault, Y., Fraudeau, N., Zakany, J., and Duboule, D. (1997). Ulnaless (Ul), a regulatory mutation inducing both loss-of-function and gain-of-function of posterior Hoxd genes. *Development* **124,** 3493–3500.

Herault, Y., Beckers, J., Kondo, T., Fraudeau, N., and Duboule, D. (1998). Genetic analysis of a Hoxd-12 regulatory element reveals global versus local modes of controls in the HoxD complex. *Development* 125, 1669–1677.

Herault, Y., Beckers, J., Gerard, M., and Duboule, D. (1999). Hox gene expression in limbs: Colinearity by opposite regulatory controls. *Dev. Biol.* 208, 157–165.

Hesse, M., Zimek, A., Weber, K., and Magin, T. M. (2004). Comprehensive analysis of keratin gene clusters in humans and rodents. *Eur. J. Cell Biol.* 83, 19–26.

Holmes, G., Crooijmans, R., Groenen, M., and Niswander, L. (2003). ALC (adjacent to LMX1 in chick) is a novel dorsal limb mesenchyme marker. *Gene Expr. Patterns* 3, 735–741.

Houweling, A. C., Dildrop, R., Peters, T., Mummenhoff, J., Moorman, A. F., Ruther, U., and Christoffels, V. M. (2001). Gene and cluster-specific expression of the Iroquois family members during mouse development. *Mech. Dev.* 107, 169–174.

Hsu, S. H., Noamani, B., Abernethy, D. E., Zhu, H., Levi, G., and Bendall, A. J. (2006). Dlx5- and Dlx6-mediated chondrogenesis: Differential domain requirements for a conserved function. *Mech. Dev.* 123, 819–830.

Huntley, S., Baggott, D. M., Hamilton, A. T., Tran-Gyamfi, M., Yang, S., Kim, J., Gordon, L., Branscomb, E., and Stubbs, L. (2006). A comprehensive catalog of human KRAB-associated zinc finger genes: Insights into the evolutionary history of a large family of transcriptional repressors. *Genome Res.* 16, 669–677.

Jacob, F., Perrin, D., Sanchez, C., and Monod, J. (1960). Operon: A group of genes with the expression coordinated by an operator. *C. R. Hebd. Seances Acad. Sci.* 250, 1727–1729.

Jagla, K., Bellard, M., and Frasch, M. (2001). A cluster of *Drosophila* homeobox genes involved in mesoderm differentiation programs. *Bioessays* 23, 125–133.

Ji, Y., Eichler, E. E., Schwartz, S., and Nicholls, R. D. (2000). Structure of chromosomal duplicons and their role in mediating human genomic disorders. *Genome Res.* 10, 597–610.

Juan, A. H., and Ruddle, F. H. (2003). Enhancer timing of Hox gene expression: Deletion of the endogenous Hoxc8 early enhancer. *Development* 130, 4823–4834.

Katoh, M. (2002). WNT and FGF gene clusters (review). *Int. J. Oncol.* 21, 1269–1273.

King, M. C., and Wilson, A. C. (1975). Evolution at two levels in humans and chimpanzees. *Science* 188, 107–116.

Kirschner, M., and Gerhart, J. (1998). Evolvability. *Proc. Natl. Acad. Sci. USA* 95, 8420–8427.

Kleinjan, D. A., Seawright, A., Elgar, G., and van Heyningen, V. (2002). Characterization of a novel gene adjacent to PAX6, revealing synteny conservation with functional significance. *Mamm. Genome* 13, 102–107.

Kmita, M., and Duboule, D. (2003). Organizing axes in time and space; 25 years of colinear tinkering. *Science* 301, 331–333.

Kmita, M., van Der Hoeven, F., Zakany, J., Krumlauf, R., and Duboule, D. (2000a). Mechanisms of Hox gene colinearity: Transposition of the anterior Hoxb1 gene into the posterior HoxD complex. *Genes Dev.* 14, 198–211.

Kmita, M., Kondo, T., and Duboule, D. (2000b). Targeted inversion of a polar silencer within the HoxD complex re-allocates domains of enhancer sharing. *Nat. Genet.* 26, 451–454.

Kmita, M., Fraudeau, N., Herault, Y., and Duboule, D. (2002a). Serial deletions and duplications suggest a mechanism for the collinearity of Hoxd genes in limbs. *Nature* 420, 145–150.

Kmita, M., Tarchini, B., Duboule, D., and Herault, Y. (2002b). Evolutionary conserved sequences are required for the insulation of the vertebrate Hoxd complex in neural cells. *Development* 129, 5521–5528.

Kokubu, C., Wilm, B., Kokubu, T., Wahl, M., Rodrigo, I., Sakai, N., Santagati, F., Hayashizaki, Y., Suzuki, M., Yamamura, M., Abe, K., and Imai, K. (2003). Undulated short-tail deletion mutation in the mouse ablates Pax1 and leads to ectopic activation of neighboring Nkx2-2 in domains that normally express Pax1. *Genetics* 165, 299–307.

Kondo, T., Dolle, P., Zakany, J., and Duboule, D. (1996). Function of posterior HoxD genes in the morphogenesis of the anal sphincter. *Development* **122**, 2651–2659.

Kosak, S. T., and Groudine, M. (2004). Gene order and dynamic domains. *Science* **306**, 644–647.

Krumlauf, R. (1994). Hox genes in vertebrate development. *Cell* **78**, 191–201.

Lecaudey, V., Anselme, I., Dildrop, R., Ruther, U., and Schneider-Maunoury, S. (2005). Expression of the zebrafish Iroquois genes during early nervous system formation and patterning. *J. Comp. Neurol.* **492**, 289–302.

Lee, E. C., Yu, D., Martinez de Velasco, J., Tessarollo, L., Swing, D. A., Court, D. L., Jenkins, N. A., and Copeland, N. G. (2001). A highly efficient Escherichia coli-based chromosome engineering system adapted for recombinogenic targeting and subcloning of BAC DNA. *Genomics* **73**, 56–65.

Lee, S. J. (1991). Expression of growth/differentiation factor 1 in the nervous system: Conservation of a bicistronic structure. *Proc. Natl. Acad. Sci. USA* **88**, 4250–4254.

Lee, K. Z., and Sommer, R. J. (2003). Operon structure and trans-splicing in the nematode Pristionchus pacificus. *Mol. Biol. Evol.* **20**, 2097–2103.

Lemons, D., and McGinnis, W. (2006). Genomic evolution of Hox gene clusters. *Science* **313**, 1918–1922.

Lercher, M. J., Urrutia, A. O., and Hurst, L. D. (2002). Clustering of housekeeping genes provides a unified model of gene order in the human genome. *Nat. Genet.* **31**, 180–183.

Lercher, M. J., Blumenthal, T., and Hurst, L. D. (2003). Coexpression of neighboring genes in Caenorhabditis elegans is mostly due to operons and duplicate genes. *Genome Res.* **13**, 238–243.

Lettice, L. A., Horikoshi, T., Heaney, S. J. H., van Baren, M. J., van der Linde, H. C., Breedveld, G. J., Joosse, M., Akarsu, N., Oostra, B. A., Endo, N., Shibata, M., Suzuki, M., *et al.* (2002). Disruption of a long-range cis-acting regulator for Shh causes preaxial polydactyly. *Proc. Natl. Acad. Sci. USA* **99**, 7548–7553.

Lettice, L. A., Heaney, S. J. H., Purdie, L. A., Li, L., de Beer, P., Oostra, B. A., Goode, D., Elgar, G., Hill, R. E., and de Graaff, E. (2003). A long-range Shh enhancer regulates expression in the developing limb and fin and is associated with preaxial polydactyly. *Hum. Mol. Genet.* **12**, 1725–1735.

Levine, M., and Tjian, R. (2003). Transcription regulation and animal diversity. *Nature* **424**, 147–151.

Li, B., Carey, M., and Workman, J. L. (2007). The role of chromatin during transcription. *Cell* **128**, 707–719.

Lomvardas, S., Barnea, G., Pisapia, D. J., Mendelsohn, M., Kirkland, J., and Axel, R. (2006). Interchromosomal interactions and olfactory receptor choice. *Cell* **126**, 403–413.

Loots, G. G., Locksley, R. M., Blankespoor, C. M., Wang, Z. E., Miller, W., Rubin, E. M., and Frazer, K. A. (2000). Identification of a coordinate regulator of interleukins 4, 13, and 5 by cross-species sequence comparisons. *Science* **288**, 136–140.

Luke, G. N., Castro, L. F., McLay, K., Bird, C., Coulson, A., and Holland, P. W. (2003). Dispersal of NK homeobox gene clusters in amphioxus and humans. *Proc. Natl. Acad. Sci. USA* **100**, 5292–5295.

Lunyak, V. V., Burgess, R., Prefontaine, G. G., Nelson, C., Sze, S.-H., Chenoweth, J., Schwartz, P., Pevzner, P. A., Glass, C., Mandel, G., and Rosenfeld, M. G. (2002). Corepressor-dependent silencing of chromosomal regions encoding neuronal genes. *Science* **298**, 1747–1752.

Maas, S. A., and Fallon, J. F. (2004). Isolation of the chicken Lmbr1 coding sequence and characterization of its role during chick limb development. *Dev. Dyn.* **229**, 520–528.

Maclean, J. A., Chen, M. A., II, Wayne, C. M., Bruce, S. R., Rao, M., Meistrich, M. L., Macleod, C., and Wilkinson, M. F. (2005). Rhox: A new homeobox gene cluster. *Cell* **120**, 369–382.

Maconochie, M. K., Nonchev, S., Studer, M., Chan, S. K., Popperl, H., Sham, M. H., Mann, R. S., and Krumlauf, R. (1997). Cross-regulation in the mouse HoxB complex: The expression of Hoxb2 in rhombomere 4 is regulated by Hoxb1. *Genes Dev.* **11**, 1885–1895.

Mass, R. L., Zeller, R., Woychik, R. P., Vogt, T. F., and Leder, P. (1990). Disruption of formin-encoding transcripts in two mutant limb deformity alleles. *Nature* **346**, 853–855.

McEwen, G. K., Woolfe, A., Goode, D., Vavouri, T., Callaway, H., and Elgar, G. (2006). Ancient duplicated conserved noncoding elements in vertebrates: A genomic and functional analysis. *Genome Res.* **16**, 451–465.

McLysaght, A., Hokamp, K., and Wolfe, K. H. (2002). Extensive genomic duplication during early chordate evolution. *Nat. Genet.* **31**, 200–204.

Michos, O., Panman, L., Vintersten, K., Beier, K., Zeller, R., and Zuniga, A. (2004). Gremlin-mediated BMP antagonism induces the epithelial-mesenchymal feedback signaling controlling metanephric kidney and limb organogenesis. *Development* **131**, 3401–3410.

Miller, D. J., and Miles, A. (1993). Homeobox genes and the zootype. *Nature* **365**, 215–216.

Minguillon, C., and Garcia-Fernandez, J. (2003). Genesis and evolution of the Evx and Mox genes and the extended Hox and ParaHox gene clusters. *Genome Biol.* **4**, R12.

Monge, I., Kondo, T., and Duboule, D. (2003). An enhancer-titration effect induces digit-specific regulatory alleles of the HoxD cluster. *Dev. Biol.* **256**, 212–220.

Moran-Rivard, L., Kagawa, T., Saueressig, H., Gross, M. K., Burrill, J., and Goulding, M. (2001). Evx1 is a postmitotic determinant of v0 interneuron identity in the spinal cord. *Neuron* **29**, 385–399.

Morey, C., Da Silva, N. R., Perry, P., and Bickmore, W. A. (2007). Nuclear reorganisation and chromatin decondensation are conserved, but distinct, mechanisms linked to Hox gene activation. *Development* **134**, 909–919.

Mummenhoff, J., Houweling, A. C., Peters, T., Christoffels, V. M., and Ruther, U. (2001). Expression of Irx6 during mouse morphogenesis. *Mech. Dev.* **103**, 193–195.

Muyrers, J. P., Zhang, Y., Testa, G., and Stewart, A. F. (1999). Rapid modification of bacterial artificial chromosomes by ET-recombination. *Nucleic Acids Res.* **27**, 1555–1557.

Nusse, R. (2001). An ancient cluster of Wnt paralogues. *Trends Genet.* **17**, 443.

Olson, E. N., Arnold, H. H., Rigby, P. W., and Wold, B. J. (1996). Know your neighbors: Three phenotypes in null mutants of the myogenic bHLH gene MRF4. *Cell* **85**, 1–4.

Osborne, C. S., Chakalova, L., Brown, K. E., Carter, D., Horton, A., Debrand, E., Goyenechea, B., Mitchell, J. A., Lopes, S., Reik, W., and Fraser, P. (2004). Active genes dynamically colocalize to shared sites of ongoing transcription. *Nat. Genet.* **36**, 1065–1071.

Papp, B., and Muller, J. (2006). Histone trimethylation and the maintenance of transcriptional ON and OFF states by trxG and PcG proteins. *Genes Dev.* **20**, 2041–2054.

Peichel, C. L., Prabhakaran, B., and Vogt, T. F. (1997). The mouse Ulnaless mutation deregulates posterior HoxD gene expression and alters appendicular patterning. *Development* **124**, 3481–3492.

Pennacchio, L. A., Ahituv, N., Moses, A. M., Prabhakar, S., Nobrega, M. A., Shoukry, M., Minovitsky, S., Dubchak, I., Holt, A., Lewis, K. D., Plajzer-Frick, I., Akiyama, J., *et al.* (2006). *In vivo* enhancer analysis of human conserved non-coding sequences. *Nature* **444**, 499–502.

Peters, T., Dildrop, R., Ausmeier, K., and Ruther, U. (2000). Organization of mouse Iroquois homeobox genes in two clusters suggests a conserved regulation and function in vertebrate development. *Genome Res.* **10**, 1453–1462.

Popperl, H., and Featherstone, M. S. (1992). An autoregulatory element of the murine Hox-4.2 gene. *EMBO J.* **11**, 3673–3680.

Qin, Y., Kong, L. K., Poirier, C., Truong, C., Overbeek, P. A., and Bishop, C. E. (2004). Long-range activation of Sox9 in Odd Sex (Ods) mice. *Hum. Mol. Genet.* **13**, 1213–1218.

Renucci, A., Zappavigna, V., Zakany, J., Izpisua-Belmonte, J. C., Burki, K., and Duboule, D. (1992). Comparison of mouse and human HOX-4 complexes defines conserved sequences involved in the regulation of Hox-4.4. *EMBO J.* **11**, 1459–1468.

Roberts, D. J., Johnson, R. L., Burke, A. C., Nelson, C. E., Morgan, B. A., and Tabin, C. (1995). Sonic hedgehog is an endodermal signal inducing Bmp-4 and Hox genes during induction and regionalization of the chick hindgut. *Development* **121**, 3163–3174.

Robledo, R. F., Rajan, L., Li, X., and Lufkin, T. (2002). The Dlx5 and Dlx6 homeobox genes are essential for craniofacial, axial, and appendicular skeletal development. *Genes Dev.* **16,** 1089–1101.

Rodriguez, I. (2005). Remarkable diversity of mammalian pheromone receptor repertoires. *Proc. Natl. Acad. Sci. USA* **102,** 6639–6640.

Roy, P. J., Stuart, J. M., Lund, J., and Kim, S. K. (2002). Chromosomal clustering of muscle-expressed genes in Caenorhabditis elegans. *Nature* **418,** 975–979.

Ruiz-Perez, V. L., Tompson, S. W. J., Blair, H. J., Espinoza-Valdez, C., Lapunzina, P., Silva, E. O., Hamel, B., Gibbs, J. L., Young, I. D., Wright, M. J., and Goodship, J. A. (2003). Mutations in two nonhomologous genes in a head-to-head configuration cause Ellis-van Creveld syndrome. *Am. J. Hum. Genet.* **72,** 728–732.

Sagai, T., Hosoya, M., Mizushina, Y., Tamura, M., and Shiroishi, T. (2005). Elimination of a long-range cis-regulatory module causes complete loss of limb-specific Shh expression and truncation of the mouse limb. *Development* **132,** 797–803.

Schwartz, Y. B., and Pirrotta, V. (2007). Polycomb silencing mechanisms and the management of genomic programmes. *Nat. Rev. Genet.* **8,** 9–22.

Semon, M., and Duret, L. (2006). Evolutionary origin and maintenance of coexpressed gene clusters in mammals. *Mol. Biol. Evol.* **23,** 1715–1723.

Sessa, L., Breiling, A., Lavorgna, G., Silvestri, L., Casari, G., and Orlando, V. (2007). Noncoding RNA synthesis and loss of Polycomb group repression accompanies the colinear activation of the human HOXA cluster. *RNA* **13,** 223–239.

Shannon, M., Hamilton, A. T., Gordon, L., Branscomb, E., and Stubbs, L. (2003). Differential expansion of zinc-finger transcription factor loci in homologous human and mouse gene clusters. *Genome Res.* **13,** 1097–1110.

Sharpe, J., Nonchev, S., Gould, A., Whiting, J., and Krumlauf, R. (1998). Selectivity, sharing and competitive interactions in the regulation of Hoxb genes. *EMBO J.* **17,** 1788–1798.

Sharpe, J., Lettice, L., Hecksher-Sorensen, J., Fox, M., Hill, R., and Krumlauf, R. (1999). Identification of sonic hedgehog as a candidate gene responsible for the polydactylous mouse mutant Sasquatch. *Curr. Biol.* **9,** 97–100.

Shiga, Y., Sagawa, K., Takai, R., Sakaguchi, H., Yamagata, H., and Hayashi, S. (2006). Transcriptional read-through of Hox genes Ubx and Antp and their divergent post-transcriptional control during crustacean evolution. *Evol. Dev.* **8,** 407–414.

Singer, G. A., Lloyd, A. T., Huminiecki, L. B., and Wolfe, K. H. (2005). Clusters of co-expressed genes in mammalian genomes are conserved by natural selection. *Mol. Biol. Evol.* **22,** 767–775.

Slager, R. E., Newton, T. L., Vlangos, C. N., Finucane, B., and Elsea, S. H. (2003). Mutations in RAI1 associated with Smith-Magenis syndrome. *Nat. Genet.* **33,** 466–468.

Spellman, P. T., and Rubin, G. M. (2002). Evidence for large domains of similarly expressed genes in the Drosophila genome. *J. Biol.* **1,** 5.

Spitz, F., Gonzalez, F., Peichel, C., Vogt, T. F., Duboule, D., and Zakany, J. (2001). Large scale transgenic and cluster deletion analysis of the HoxD complex separate an ancestral regulatory module from evolutionary innovations. *Genes Dev.* **15,** 2209–2214.

Spitz, F., Gonzalez, F., and Duboule, D. (2003). A global control region defines a chromosomal regulatory landscape containing the HoxD cluster. *Cell* **113,** 405–417.

Spitz, F., Herkenne, C., Morris, M. A., and Duboule, D. (2005). Inversion-induced disruption of the Hoxd cluster leads to the partition of regulatory landscapes. *Nat. Genet.* **37,** 889–893.

Sproul, D., Gilbert, N., and Bickmore, W. A. (2005). The role of chromatin structure in regulating the expression of clustered genes. *Nat. Rev. Genet.* **6,** 775–781.

Su, A. I., Wiltshire, T., Batalov, S., Lapp, H., Ching, K. A., Block, D., Zhang, J., Soden, R., Hayakawa, M., Kreiman, G., Cooke, M. P., Walker, J. R., *et al.* (2004). A gene atlas of the mouse and human protein-encoding transcriptomes. *Proc. Natl. Acad. Sci. USA* **101,** 6062–6067.

Sumiyama, K., Irvine, S. Q., Stock, D. W., Weiss, K. M., Kawasaki, K., Shimizu, N., Shashikant, C. S., Miller, W., and Ruddle, F. H. (2002). Genomic structure and functional control of the Dlx3–7 bigene cluster. *Proc. Natl. Acad. Sci. USA* **99**, 780–785.

Tarchini, B., and Duboule, D. (2006). Control of Hoxd genes' collinearity during early limb development. *Dev. Cell* **10**, 93–103.

Tarchini, B., Duboule, D., and Kmita, M. (2006). Regulatory constraints in the evolution of the tetrapod limb anterior-posterior polarity. *Nature* **443**, 985–988.

Trumpp, A., Blundell, P. A., de la Pompa, J. L., and Zeller, R. (1992). The chicken limb deformity gene encodes nuclear proteins expressed in specific cell types during morphogenesis. *Genes Dev.* **6**, 14–28.

Tumpel, S., Cambronero, F., Ferretti, E., Blasi, F., Wiedemann, L. M., and Krumlauf, R. (2007). Expression of Hoxa2 in rhombomere 4 is regulated by a conserved cross-regulatory mechanism dependent upon Hoxb1. *Dev. Biol.* **302**, 646–660.

van der Hoeven, F., Zakany, J., and Duboule, D. (1996). Gene transpositions in the HoxD complex reveal a hierarchy of regulatory controls. *Cell* **85**, 1025–1035.

Warming, S., Costantino, N., Court, D. L., Jenkins, N. A., and Copeland, N. G. (2005). Simple and highly efficient BAC recombineering using galK selection. *Nucleic Acids Res.* **33**, e36.

Wijgerde, M., Grosveld, F., and Fraser, P. (1995). Transcription complex stability and chromatin dynamics *in vivo. Nature* **377**, 209–213.

Wotton, K. R., and Shimeld, S. M. (2006). Comparative genomics of vertebrate Fox cluster loci. *BMC Genomics* **7**, 271.

Woychik, R. P., Stewart, T. A., Davis, L. G., D'Eustachio, P., and Leder, P. (1985). An inherited limb deformity created by insertional mutagenesis in a transgenic mouse. *Nature* **318**, 36–40.

Wu, Q., and Maniatis, T. (1999). A striking organization of a large family of human neural cadherin-like cell adhesion genes. *Cell* **97**, 779–790.

Zakany, J., and Duboule, D. (1999). Hox genes and the making of sphincters. *Nature* **401**, 761–762.

Zakany, J., Gerard, M., Favier, B., and Duboule, D. (1997). Deletion of a HoxD enhancer induces transcriptional heterochrony leading to transposition of the sacrum. *EMBO J.* **16**, 4393–4402.

Zakany, J., Kmita, M., and Duboule, D. (2004). A dual role for Hox genes in limb anterior-posterior asymmetry. *Science* **304**, 1669–1672.

Zeller, R., Jackson-Grusby, L., and Leder, P. (1989). The limb deformity gene is required for apical ectodermal ridge differentiation and anteroposterior limb pattern formation. *Genes Dev.* **3**, 1481–1492.

Zulch, A., Becker, M. B., and Gruss, P. (2001). Expression pattern of Irx1 and Irx2 during mouse digit development. *Mech. Dev.* **106**, 159–162.

Zuniga, A., Haramis, A. P., McMahon, A. P., and Zeller, R. (1999). Signal relay by BMP antagonism controls the SHH/FGF4 feedback loop in vertebrate limb buds. *Nature* **401**, 598–602.

Zuniga, A., Michos, O., Spitz, F., Haramis, A.-P. G., Panman, L., Galli, A., Vintersten, K., Klasen, C., Mansfield, W., Kuc, S., Duboule, D., Dono, R., *et al.* (2004). Mouse limb deformity mutations disrupt a global control region within the large regulatory landscape required for Gremlin expression. *Genes Dev.* **18**, 1553–1564.

7

Regulation of Imprinting in Clusters: Noncoding RNAs Versus Insulators

Le-Ben Wan and Marisa S. Bartolomei

Department of Cell and Developmental Biology, University of Pennsylvania
School of Medicine, Philadelphia, Pennsylvania 19104

I. Introduction
II. Insulator Model of Regulation
   A. The *H19/Igf2* imprinted cluster
   B. Evidence for insulator-mediating silencing at the *Rasgrf1* locus
III. The ncRNA Model of Regulation
   A. The *Air* and *Igf2r* cluster
   B. ncRNA and the *Kcnq1* locus
IV. *Dlk1/Gtl2* Imprinted Cluster: A Bit of Everything
V. Conclusions
   Acknowledgments
   References

ABSTRACT

Genomic imprinting is an epigenetic mechanism of transcriptional regulation through which expression of a subset of mammalian genes is restricted to one parental allele. An intriguing characteristic of imprinted genes is that they often cluster in megabase-sized chromosomal domains, indicating that domain-specific mechanisms regulate imprinting. Detailed study of the known imprinted domains has revealed a number of common characteristics. First, all clusters have an imprinting control region (ICR) that is typically 1–5 kb in size and differentially methylated, and that regulates imprinting across the entire domain. Second, the clusters have at least one noncoding RNA (ncRNA) that is usually

Copyright 2008, Elsevier Inc. All rights reserved.
0065-2660/08 $35.00
DOI: 10.1016/S0065-2660(07)00007-7

expressed from the maternal allele and multiple paternally expressed protein-coding genes. Finally, the clusters are likely regulated by one of two mechanisms, transcription of a long ncRNA that silences expression of protein-coding genes bidirectionally in *cis* and blocking of shared enhancer elements by CCCTC binding factor (CTCF) binding insulators. More recent experiments may even suggest that both mechanisms operate at some clusters. In this chapter, we will describe what is known about imprinting at five well-studied imprinted loci and highlight some of the critical experiments that are required before a full understanding of imprinting mechanisms is achieved. © 2008, Elsevier Inc.

I. INTRODUCTION

A small number of autosomal genes in mammals are expressed exclusively or predominantly from a single parental allele. These genes are subject to genomic imprinting. The importance of proper imprinted gene expression for normal growth and development is underscored by the failure of embryos containing only maternal or paternal genomes to develop normally (Solter, 1988) and by development of human diseases and cancers, including Prader-Willi, Angelman, and Beckwith-Wiedemann syndromes, due to loss of imprinting (Nicholls and Knepper, 2001; Weksberg *et al.*, 2005).

In 1991, the first imprinted genes were identified in mouse. The first of these, *Igf2r* (Insulin-like growth factor type 2 receptor that is a "scavenger" receptor for the growth hormone *Igf2*), was identified as a maternally expressed imprinted gene (Barlow *et al.*, 1991). A few months later, the *Igf2* gene (Insulin-like growth factor type 2), which was known to function as a growth hormone, was identified as a paternally expressed imprinted gene (DeChiara *et al.*, 1991). Finally, the *H19* gene, an unusual noncoding RNA (ncRNA), was subsequently shown to be a maternally expressed imprinted gene (Bartolomei *et al.*, 1991). With the discovery of these first imprinted genes and the subsequent discovery of ~ 70 additional genes (for a complete listing, see http://www.mgu.har.mrc.ac.uk/research/imprinting), it was realized that imprinted genes are clustered over megabase regions in the genome and this clustering is essential to their imprinted gene regulation (Verona *et al.*, 2003). These clusters share a number of features including an ncRNA that is expressed from the opposite parental allele as the protein-coding genes and an imprinting control region (ICR). In cases that have been examined in detail, ICRs exhibit allele-specific epigenetic modifications (i.e., DNA methylation, chromatin modifications) that are likely conferred in the germline and subsequently maintained (Reik *et al.*, 2001). Furthermore, as their name suggests, ICRs control the imprinting of multiple genes in the cluster. The mechanism by which these clusters are regulated over long distances has been intensively studied in the last decade, with two prevailing models of regulation emerging. The first of these is the insulator model, in which imprinted genes share regulatory elements and the insulator controls access to these elements. In the second model, ncRNA mediates silencing of linked genes. In this chapter, we will discuss the mechanism by which various imprinted clusters are regulated.

II. INSULATOR MODEL OF REGULATION

A. The *H19/Igf2* imprinted cluster

As previously mentioned, the *H19* and *Igf2* genes were two of the first three identified imprinted genes. Both genes are also imprinted in humans, with many aspects of the imprinting regulation highly conserved. Although the imprinting status of *H19* and *Igf2* was tested for entirely different reasons, it soon became apparent that these genes reside next to each other (these experiments were performed in the pregenome sequence era) (Zemel *et al.*, 1992), initially raising the possibility that imprinted genes are clustered. Furthermore, the expression patterns of these genes are similar (Poirier *et al.*, 1991), and gene-targeting experiments ultimately demonstrated that the genes share regulatory elements and their imprinting is intimately connected. In the mid-90s, the only enhancer elements that had been mapped in the 150 kb region surrounding *H19* and *Igf2* were a pair of endodermal enhancers that were located 3′ of *H19* between +7.0 and +10 kb relative to the *H19* transcription start site (Yoo-Warren *et al.*, 1988). When the enhancer deletion is transmitted maternally, *H19* expression is eliminated in endodermal tissues and when transmitted paternally, *Igf2* expression is dramatically reduced (Leighton *et al.*, 1995), demonstrating that the two genes share enhancers. The two genes also share mesodermal enhancers located between +22 and +28 kb and the cardiac muscle enhancers between +35 and +130 kb (Ainscough *et al.*, 2000; Davies *et al.*, 2002; Kaffer *et al.*, 2000).

    The second element that was shown to be important for the imprinted expression of *H19* and *Igf2* is the ICR or differentially methylated domain (DMD). This element resides from −2 to −4 kb relative to the start of *H19* transcription and was initially identified as the sole region that was differentially methylated in the gametes and throughout development (Tremblay *et al.*, 1995, 1997). It was especially notable that the paternal-specific methylation is preserved in the early embryo, when there is massive and generalized demethylation (Morgan *et al.*, 2005), and that this same region is sensitive to nucleases in somatic tissues on the actively expressed maternal *H19* allele (Hark and Tilghman, 1998). Deletion of the region at the endogenous locus results in the loss of imprinting for both *H19* and *Igf2* on the maternal and paternal alleles (Thorvaldsen *et al.*, 1998, 2002). It was subsequently shown that the DMD binds

CTCF, a protein shown to mediate insulator activity at the *β-globin* locus (Bell *et al.*, 1999), and that the DMD itself functions as an insulator (Bell and Felsenfeld, 2000; Hark *et al.*, 2000; Kaffer *et al.*, 2000; Kanduri *et al.*, 2000; Szabo *et al.*, 2000). In this context, an insulator is defined as an element that blocks enhancer and promoter interactions when placed between them (Engel and Bartolomei, 2003). Thus, the model for imprinted gene expression at this locus is as follows (Fig. 7.1): on the maternal allele, CTCF binds to the DMD and blocks the access of *Igf2* to enhancers shared with *H19*. This allows *H19* exclusive access to the enhancers. On the paternal allele, the DMD acquires DNA methylation in the male germline, preventing CTCF from binding to it. Thus, on the paternal chromosome, *Igf2* interacts with the enhancers and is expressed from this chromosome. The presence of DNA methylation on the paternal DMD leads to secondary methylation of the *H19* promoter by an unknown mechanism and it becomes silenced on the paternal chromosome.

The question remains how insulators operate. Recently, chromosome conformation capture (3C) analysis has been used to detect long-range interactions within a given region (Kurukuti *et al.*, 2006; Murrell *et al.*, 2004; Yoon *et al.*, 2007). In this assay, predicted associations are tested systematically and attempts are made to quantify the interactions relative to each other, although the results appear to vary according to the methodology used by a given laboratory. These experiments have shown that active promoters at the *H19/Igf2* locus physically contact the enhancers (i.e., *Igf2* promoters on the paternal allele contact the

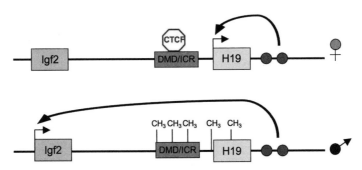

Figure 7.1. Regulation of imprinting at the mouse *H19/Igf2* locus. Shown are the maternally expressed *H19* gene, which encodes an ncRNA, and the paternally expressed *Igf2* gene, which encodes a fetal mitogen, with their shared enhancers (shaded circles). On the maternal chromosome, the unmethylated DMD/ICR binds the CTCF protein and forms an insulator that prevents the common enhancers from activating *Igf2*. Instead, the enhancers activate the nearby *H19* promoter. On the paternal chromosome, the methylated DMD/ICR cannot bind CTCF and an insulator does not form and the *Igf2* gene is expressed. Evidence suggests that the orthologous human locus is subject to the same regulation. Not drawn to scale.

paternal enhancers and the *H19* promoter on the maternal allele contacts the enhancers on the same allele), with stronger interactions in tissues with higher levels of transcription. Furthermore, the contacts between *Igf2* promoters and enhancers are prevented on the maternal allele when an active insulator is present. While one group suggests that the insulator (specifically CTCF-binding sites) interacts with a matrix attachment region and a differentially methylated region (DMR) within the *Igf2* gene to generate a tight loop around the maternal *Igf2* locus, which results in silencing (Kurukuti *et al.*, 2006), a second group has shown that the active insulator (DMD) appears to generate a novel interaction with the promoter and enhancers that it is blocking (Yoon *et al.*, 2007). Yoon *et al.* propose a model in which the insulator functions by interacting with the *Igf2* promoters and enhancers on the paternal allele in ways that do not favor the productive association of promoters and enhancers with each other. These alternative structures would prevent gene activation (Yoon *et al.*, 2007). Additional experiments will most definitely be required to understand how the insulator operates at this locus.

B. Evidence for insulator-mediating silencing at the *Rasgrf1* locus

The involvement of CTCF in the insulator model has led to the identification of CTCF-binding sites at other imprinted genes such as *Rasgrf1* and *Kcnq1ot1*, indicating that the insulator model may operate at other imprinted clusters (Fitzpatrick *et al.*, 2007; Yoon *et al.*, 2005). While the mechanism governing *Rasgrf1* imprinting is currently unknown, regulation at this locus presents a novel case that will be discussed here. The *Rasgrf1* imprinted locus on mouse chromosome 9 consists of a protein-coding gene conserved in humans and rats (*Rasgrf1*), and an ncRNA (*A19*) that is absent from the human locus (de la Puente *et al.*, 2002; Pearsall *et al.*, 1999; Plass *et al.*, 1996). The two genes are paternally expressed and transcribed in the same orientation, with *A19* upstream of *Rasgrf1*. Located between these two genes is a DMD followed closely by a conserved repetitive element.

The DMD is a CpG island that is methylated exclusively on the paternal allele in adult somatic tissue. In the germlines, however, methylation is not uniformly differential, and at least one region is methylated in both sperm and oocytes. This region contains the initially characterized differentially methylated *Not1* site and loses maternal CpG methylation gradually during preimplantation development (Shibata *et al.*, 1998). Paternal methylation of the DMD and paternal expression of *Rasgrf1* depend on the repetitive element in *cis*. When a deletion of the repetitive element is inherited paternally, methylation of the DMD and expression of *Rasgrf1* are reduced on the paternal allele (Yoon *et al.*, 2002). Sperm from mice without the repeat fail to acquire methylation on the DMD of the mutated chromosome, suggesting that the repetitive element is required for methylation establishment at the DMD. Moreover, a conditional deletion allele, when inherited paternally and activated at the one-cell stage, loses methylation at the DMD during preimplantation development, supporting a role of the repetitive element in methylation maintenance (Holmes et al., 2006). Together, these results are consistent with a hierarchical model of epigenetic regulation, with *Rasgrf1* expression being entirely dependent on DMD methylation, and DMD methylation being strongly influenced by the presence of the repetitive element.

The DMD functions as an insulator *in vitro* (Yoon et al., 2005). *In vivo*, it might prevent the *Rasgrf1* promoter from accessing putative upstream enhancers on the maternal chromosome. Studies of binding sites within the DMD suggest that CTCF mediates the function of the maternal DMD, whereas it cannot bind to the paternal DMD because of the presence of CpG methylation. This would be analogous to the *H19/Igf2* insulator model, but it does not account for the paternal expression of *A19*, or for the tissue specificity of *A19* and *Rasgrf1* imprinting. Given that enhancers of *A19* and *Rasgrf1* have yet to be found, it would be difficult to invoke a model analogous to the *H19/Igf2* model. Thus, more information, including the results of DMD deletion, is required to understand imprinting at this locus.

III. THE ncRNA MODEL OF REGULATION

A. The *Air* and *Igf2r* cluster

The maternally expressed mouse *Igf2r* gene resides in an imprinted cluster on proximal mouse chromosome 17 (Fig. 7.2). Two neighboring genes, *Slc22a2* and *Slc22a3* (solute carrier 22a2 and 22a3), are also expressed maternally. This region harbors one paternally expressed transcript, Air (antisense Igf2r RNA), that overlaps *Igf2r*. Several nonimprinted genes also reside within this cluster, including *Slc22a1*, *Mas1*, and *Plg* (*Plasminogen*). Only some aspects of the genomic organization and imprinting of this region of mouse 17A are conserved on human chromosome 6q26–27, in contrast to other imprinted clusters.

The goal to elucidate the mechanism by which this cluster is imprinted started with the investigation of differential methylation, as was discussed above for *H19* and *Igf2*. Two differentially methylated regions are found at the *Igf2r* locus. Region 1, which encompasses the *Igf2r* promoter, is methylated on the repressed paternal chromosome (Stöger et al., 1993). This methylation is not inherited from male germ cells but is acquired progressively during embryogenesis, suggesting that it is a consequence, rather than a cause, of silencing. Region 2, located in the second intron of *Igf2r*, is methylated on the expressed maternal chromosome. As this methylation is present in female germ cells and persists

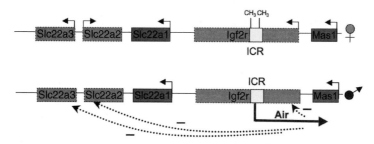

Figure 7.2. Imprinting on proximal mouse chromosome 17. *Igf2r*, *Slc22a2*, and *Slc22a3* are expressed from the maternal chromosome and *Air* is expressed from the paternal chromosome. Nonimprinted genes at this domain include *Mas1* and *Slc22a1*. The imprinting control region (ICR designated region 2 in text), which serves as the promoter to *Air*, is shown with a square box. The ICR is hypermethylated on the maternal strand, preventing transcription of *Air*, and allowing *Igf2r*, *Slc22a2*, and *Slc22a3* to be transcribed. On the paternal chromosome, the imprinting control region is unmethylated, *Air* is expressed, and surrounding genes (*Igf2r*, *Slc22a2*, and *Slc22a3*) are repressed (indicated with a dotted arrow). Transcriptional activity of a given gene is indicated by arrows. Not drawn to scale.

throughout embryonic development (Brandeis *et al.*, 1993; Lucifero *et al.*, 2002; Stöger *et al.*, 1993), it is a candidate for an imprinting mark (ICR) at this locus. Consistent with this possibility, deletion of region 2 leads to reactivation of paternal *Igf2r* expression (Wutz *et al.*, 1997), supporting a role for this region in the regulation of *Igf2r* imprinting.

Region 2 contains the promoter for the Air ncRNA, an unspliced and polyadenylated transcript that extends 108 kb in an antisense orientation to *Igf2r* (Lyle *et al.*, 2000; Wutz *et al.*, 1997). *Air* is reciprocally imprinted to *Igf2r* and expressed exclusively from the unmethylated paternal allele. As deletion of region 2 eliminates *Air* expression and causes loss of *Igf2r* imprinting, Air has been proposed to play a role in silencing *Igf2r* expression on the paternal chromosome. In fact, *Igf2r* is imprinted in almost all tissues where *Air* is expressed (Hu *et al.*, 1999). Surprisingly, deletion of region 2 on the paternal allele causes loss of imprinting of the neighboring *Slc22a2* and *Slc22a3* genes, in addition to *Igf2r* (Zwart *et al.*, 2001). Maternal expression of *Slc22a2* and *Slc22a3* is restricted to placenta at specific times during development, and no differential methylation has been detected around either *Slc22a2* or *Slc22a3* genes, indicating that imprinting of these genes may be regulated by elements located at a distance (Zwart *et al.*, 2001). Data from the region 2 deletion suggest that it contains an ICR for the entire cluster, or alternatively that the Air RNA itself or its transcription is required for silencing of *Igf2r*, *Slc22a2*, and *Slc22a3*. To distinguish between these two possible mechanisms, a mutant *Air* allele was generated that truncates the transcript to 4% of its length (*Air*-T) (Sleutels *et al.*, 2002).

This mutation does not perturb *Air* imprinting, as the expression of *Air-T* and the hypomethylation of the *Air* promoter in region 2 remain exclusively paternal. In contrast, *Igf2r* is biallelically expressed, with loss of paternal methylation in the *Igf2r* promoter region, directly implicating the Air RNA or its transcription in the repression of *Igf2r*. Unexpectedly, imprinting of *Slc22a2* and *Slc22a3* is also lost in midgestation placenta as a result of the *Air-T* mutation. This result was surprising, given that *Air* does not overlap *Slc22a2* and *Slc22a3* but is transcribed in the opposite direction. Overall, these results demonstrate that the full length *Air* transcript, or transcription through the region, is required for the silencing of imprinted genes on the paternal allele.

But how does *Air* transcription silence overlapping and nonoverlapping genes that are located in *cis*? While some of these silenced genes are several hundred kilobase away, genes more proximal escape the silencing. Multiple models have been posited to address this complex regulation (Pauler *et al.*, 2007; Verona *et al.*, 2003). First, an RNAi-based model is suggested in which the long antisense transcript forms a double-stranded RNA intermediate complementary to the silenced gene, which then triggers silencing by one of the strategies documented in other systems (Mattick and Makunin, 2006), namely, RNA degradation, translational repression or heterochromatin formation. This model, however, is not entirely plausible because it cannot explain silencing of nonoverlapping genes with no sequence similarity, nor can it account for silencing in *cis*. A second mechanism invokes a *Xist*-type model of X inactivation, in which the Xist RNA coats the inactive X chromosome (Thorvaldsen *et al.*, 2006). Here the silencing would be much more limited, and although there has not been a demonstration of these imprinted ncRNAs coating the silenced domain and the mechanism of Xist-mediated silencing remains incompletely understood, such a mechanism cannot be discounted. Finally, models invoking silencing by transcription through the locus have been proposed. In this case, it is the transcription rather than the product of transcription (i.e., ncRNA) that is important for silencing. Here, transcription could interfere with activators or activate repressors. This particular class of models requires the identity of *cis*-acting regulatory elements, which are currently unknown for the *Igf2r/Air* locus.

B. ncRNA and the *Kcnq1* locus

The *Kcnq1* locus, though less well studied from a mechanistic standpoint, appears to operate similarly to the *Air* locus. This domain contains one paternally expressed long (>60 kb) ncRNA and eight maternally expressed protein-coding genes. The promoter for the ncRNA is located in an intron of the *Kcnq1* gene and exhibits maternal-specific methylation (Lee *et al.*, 1999; Mitsuya *et al.*, 1999; Smilinich *et al.*, 1999). Deletion of the differentially methylated CpG island and promoter, designated KvDMR1, on the paternal allele results in activation of the eight genes that are normally only expressed on the maternal allele (Fitzpatrick et al., 2002; Mancini-Dinardo et al., 2006), suggesting that transcription of the ncRNA Kcnq1ot1 is essential to silence the eight protein-coding genes in cis. Furthermore, insertion of a transcriptional stop signal downstream of the promoter on the paternal allele results in the activation of the normally silenced protein-coding genes on that allele (Mancini-Dinardo et al., 2006). Thus, similar to what has been reported for the Igf2r/Air locus, transcription of Kcnq1ot1 or the transcript itself is required for bidirectional repression of genes in cis.

While the mechanism of imprinting at the Kcnq1 locus may be largely governed by the Kcnq1ot1 ncRNA, the locus has one further attribute that must be considered. Enhancer-blocking studies indicate that KvDMR1 also exhibits insulator or silencer activity (Kanduri et al., 2002). Moreover, Higgins and colleagues have defined two CTCF-binding sites within KvDMR1 that are occupied in vivo only on the unmethylated paternal allele (Fitzpatrick et al., 2007). Thus, it is possible that more than one mechanism is employed to regulate imprinting in this domain, perhaps regulating different subsets of genes or functioning at different times in or in different tissues during development. Consistent with this idea, DNA methylation appears to regulate allele-specific expression of the KvDMR1 domain in the embryo but less so in the placenta, where differential histone modifications may be more important (Lewis et al., 2004; Umlauf et al., 2004). One could envision a model where transcription of the Kcnq1ot1 ncRNA is sufficient to imprint linked genes in certain tissues, whereas the silencer or insulator activity located upstream from the promoter could be required to imprint other genes in the complex at distinct times. Here again, the characterization of these and other presently unknown regulatory elements is required to resolve these issues.

IV. *Dlk1/Gtl2* IMPRINTED CLUSTER: A BIT OF EVERYTHING

The imprinted Dlk1/Gtl2 locus on mouse chromosome 12 is conserved in sheep and humans (Paulsen et al., 2001). In all three species, protein-coding genes are paternally expressed, whereas ncRNAs are maternally expressed (Fig. 7.3). As initially described, the locus was considered to be analogous to the H19/Igf2 locus because a protein-coding gene that functions as a growth factor (Dlk1) is located upstream from an ncRNA (Gtl2, occasionally referred to as Meg3) (Kobayashi et al., 2000; Schmidt et al., 2000; Takada et al., 2000; Wylie et al., 2000). The two genes are separated by a germline DMR, designated IG-DMR, that is hypomethylated on the maternal allele and hypermethylated on the paternal allele, again reflecting the general organization of the H19/Igf2 locus (Takada et al., 2002). At the Dlk1/Gtl2 locus, however, several additional ncRNAs are expressed downstream of Gtl2, in the same orientation, and only from the maternal allele.

Maternal

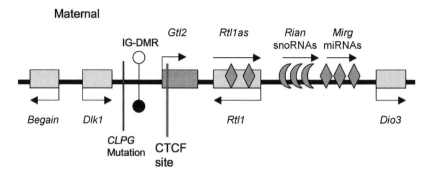

Figure 7.3. Structural organization of the imprinted *Dlk1/Gtl2* locus. Shown are maternally expressed ncRNAs (*Gtl2, Rtl1as, Rian,* and *Mirg*) and paternally expressed protein-coding genes (*Begain, Dlk1, Rtl1,* and *Dio3*). Arrows above and below the symbols indicate the relative orientations of maternally and paternally expressed genes, respectively. *Rtl1as* overlaps *Rtl1* in the antisense direction, and encodes multiple microRNAs (diamonds). Additional microRNAs and snoRNAs (diamonds and arcs) are shown downstream from *Rtl1as*. The germline IG-DMR is maternally hypomethylated (empty circle) and paternally hypermethylated (filled circle). Shown also are the relative locations of the *Callipyge* mutation and the conserved CTCF site within a 5′ intron of Gtl2. Not drawn to scale.

Among them are *Rtl1as/Peg11as*, which overlaps a paternally expressed protein-coding gene (*Rtl1/Peg11*) in the antisense direction; *Rian/Meg8*, which contains multiple C/D-type snoRNAs; and *Mirg/Meg9*, which harbors microRNAs (miR-NAs) (Cavaille *et al.*, 2002; Charlier *et al.*, 2001b; Seitz *et al.*, 2003, 2004). Further downstream from these genes is *Dio3*, and upstream from *Dlk1* is *Begain*, both of which are paternally expressed and protein-coding (Hernandez *et al.*, 2004; Smit *et al.*, 2005).

The *H19/Igf2* and *Dlk1/Gtl2* loci also differ in how their imprinting is influenced by their respective germline DMRs. When an IG-DMR deletion is inherited maternally, ncRNAs are silenced in *cis*, and protein-coding genes are activated in *cis* (Lin *et al.*, 2003). The effect of this deletion may be tissue-specific (Lin *et al.*, 2007). For example, placental tissue expresses ncRNAs when the IG-DMR is deleted maternally. However, in the embryo proper, the maternal allele has essentially switched to a paternal allele, exhibiting an absence of the ncRNAs and biallelic expression of the protein-coding genes. In contrast, when the IG-DMR deletion allele is inherited paternally, no changes are observed in *cis*. These results suggest that the ancestral state of gene expression for the locus is paternal; however, the mechanism by which an unmethylated IG-DMR changes this state is not known.

In addition to the germline IG-DMR, two additional DMRs, the *Dlk1*-DMR at the 3' end of *Dlk1* and the *Gtl2*-DMR at the 5' end of *Gtl2*, acquire methylation on the paternal allele postfertilization (Paulsen *et al.*, 2001). The *Gtl2*-DMR contains at least one conserved CTCF-binding site that is located within an intron of *Gtl2*. No CTCF-binding site has been found within the IG-DMR. Given the complex organization of the locus, a simple insulator model such as the *H19/Igf2* model is unlikely. However, it remains possible that an insulator acts locally to control a subset of genes within the cluster or in a cell-type-specific manner.

As stated above, several ncRNAs are located downstream of *Gtl2*, including *Rtl1as*, *Rian*, and *Mirg* (Fig. 7.3). *Rian* contains multiple copies of divergent C/D-type snoRNAs that lack complementarity to rRNA, and whose functions are unknown (Cavaille *et al.*, 2002). Multiple miRNAs are predicted to reside within *Mirg*, within *Rtl1as*, and in surrounding intergenic regions (Seitz *et al.*, 2004). While the targets for many of these remain elusive, five conserved miRNAs within *Rtl1as* have been shown to target the *Rtl1* sense strand for the RNAi cleavage pathway. This is the first demonstration of a *trans*-effect at an imprinted locus and also the first example of mammalian miRNAs having full complementarity to their targets, and thus directing mRNA cleavage rather than translational repression (Davis *et al.*, 2005). It is important to note that it is currently not known if *Gtl2*, *Rtl1as*, *Rian*, and *Mirg* represent stable processing intermediates of a longer transcript, and it is not known if their transcription and expression are silencing protein-coding genes in *cis*. If so, the mechanism of imprinting at the *Gtl2/Dlk1* locus may be analogous to what has been described at the *Igf2r* and *Kcnq1* loci, as discussed above. Experiments that truncate the ncRNA genes at the *Gtl2/Dlk1* locus would address the questions of whether there is one long ncRNA and whether transcription of this long transcript operates to silence transcription in *cis*.

Finally, it is worth mentioning an unusual mutation within the ovine *Dlk1/Gtl2* locus, as this mutation defines critical regulatory elements, important but poorly defined at other imprinted loci. This mutation is called *Callipyge* ("beautiful buttocks") because heterozygous animals have increased hindquarter muscle mass. The mutation shows an unusual pattern of inheritance termed polar overdominance, which means that only heterozygotes inheriting the mutation paternally show the phenotype, whereas homozygotes and heterozygotes inheriting the mutation maternally appear phenotypically normal (Cockett *et al.*, 1996). The mutation is a single A-to-G transition within a conserved dodecamer motif located between *Dlk1* and IG-DMR (Fig. 7.3; Smit *et al.*, 2003). When inherited paternally, all genes assayed are normally imprinted, but paternally expressed genes *Dlk1* and *Rtl1* are upregulated in *cis*. When the mutation is inherited maternally, imprinting is also normal, but maternally expressed genes *Gtl2*, *Rtl1as*, and *Rian* are upregulated in *cis* (Charlier *et al.*, 2001a). In both cases, changes in gene expression are accompanied by increased DNAse I hypersensitivity, decreased DNA methylation, and increased intergenic transcription at the *Callipyge* locus, consistent with the mutation affecting a long-range control element that is subordinate to the imprinting mechanism (Takeda *et al.*, 2006). The *Callipyge* phenotype may result from the overexpression of protein-coding genes when the mutation is inherited paternally. When inherited maternally, the mutation results in the overexpression of ncRNAs, and in the homozygous state, overexpressed ncRNAs might inhibit overexpressed protein-coding transcripts in trans. There is current interest in what paternally expressed genes might be contributing to the overgrowth phenotype, and what maternally derived products might be inhibiting these genes in the homozygous state (reviewed in Georges *et al.*, 2004). Nevertheless, this mutation resulted in the identification of a crucial regulatory element, which will ultimately contribute to our understanding of the mechanism of imprinting at this locus.

V. CONCLUSIONS

Two additional notable imprinting clusters have not been described here because of space limitations. These clusters (*Snprn* and *Gnas*) have all of the features described above, including paternally expressed protein-coding genes, maternally expressed ncRNAs, and DMRs that serve as ICRs. These clusters are reviewed in detail elsewhere (O'Neill, 2005; Verona *et al.*, 2003).

   Although much has been learned in the past 15+ years of studying imprinted genes and their regulation in clusters, an enormous amount of work remains to be done. One of the major surprises was the finding that the genes are clustered, with regulation conferred through an ICR. The ICR is typically differentially methylated, with hypermethylation repressing its activity. When active, the ICR serves as an insulator, as in the case of the *H19/Igf2* locus, or as a promoter for a long ncRNA, as observed in the *Air/Igf2r* and *Kcnq1* loci. In some cases, such as IG-DMR, the function of the ICR is unclear, whereas in others, such as KvDMR1, the ICR appears to have multiple elements with independent activities. Additionally, the determination of whether other ICRs function as insulators will require the identification of enhancers and other regulatory elements.

   Although we know that transcription of a long ncRNA is critical for regulation at some loci, it is unknown how this transcription represses multiple genes in *cis*, and in a bidirectional and nonoverlapping manner. Furthermore, while some ICRs function as enhancer-blockers *in vitro* and *in vivo*, it is still controversial whether enhancer-blockers serve as transcriptional regulatory elements, transcription factor tracking elements, or mediators of looping. Thus, much remains to be done.

Acknowledgments

Work in the authors' laboratory was supported by U.S. Public Health Service grants GM51279 and HD42026. L.-B. W. was supported by U.S. Public Health Service training grant GM08216.

References

Ainscough, J. F., John, R. M., Barton, S. C., and Surani, M. A. (2000). A skeletal muscle-specific mouse *Igf2* repressor lies 40 kb downstream of the gene. *Development* **127**, 3923–3930.

Barlow, D. P., Stoger, R., Herrmann, B. G., Saito, K., and Schweifer, N. (1991). The mouse insulin-like growth factor type-2 receptor is imprinted and closely linked to the *Tme* locus. *Nature* **349**, 84–87.

Bartolomei, M. S., Zemel, S., and Tilghman, S. M. (1991). Parental imprinting of the mouse *H19* gene. *Nature* **351**, 153–155.

Bell, A. C., and Felsenfeld, G. (2000). Methylation of a CTCF-dependent boundary controls imprinted expression of the *Igf2* gene. *Nature* **405**, 482–485.

Bell, A. C., West, A. G., and Felsenfeld, G. (1999). The protein CTCF is required for the enhancer blocking activity of vertebrate insulators. *Cell* **98**, 387–396.

Brandeis, M., Kafri, T., Ariel, M., Chaillet, J. R., McCarrey, J., Razin, A., and Cedar, H. (1993). The ontogeny of allele-specific methylation associated with imprinted genes in the mouse. *EMBO J* **12**, 3669–3677.

Cavaille, J., Seitz, H., Paulsen, M., Ferguson-Smith, A. C., and Bachellerie, J. P. (2002). Identification of tandemly-repeated C/D snoRNA genes at the imprinted human 14q32 domain reminiscent of those at the Prader-Willi/Angelman syndrome region. *Hum. Mol. Genet.* **11**, 1527–1538.

Charlier, C., Segers, K., Karim, L., Shay, T., Gyapay, G., Cockett, N., and Georges, M. (2001a). The callipyge mutation enhances the expression of coregulated imprinted genes in cis without affecting their imprinting status. *Nat. Genet.* **27**, 367–369.

Charlier, C., Segers, K., Wagenaar, D., Karim, L., Berghmans, S., Jaillon, O., Shay, T., Weissenbach, J., Cockett, N., Gyapay, G., and Georges, M. (2001b). Human-ovine comparative sequencing of a 250-kb imprinted domain encompassing the callipyge (clpg) locus and identification of six imprinted transcripts: DLK1, DAT, GTL2, PEG11, antiPEG11, and MEG8. *Genome Res.* **11**, 850–862.

Cockett, N. E., Jackson, S. P., Shay, T. L., Farnir, F., Berghmans, S., Snowder, G. D., Nielsen, D. M., and Georges, M. (1996). Polar overdominance at the ovine callipyge locus. *Science* **273**, 236–238.

Davies, K., Bowden, L., Smith, P., Dean, W., Hill, D., Furuumi, H., Sasaki, H., Cattanach, B., and Reik, W. (2002). Disruption of mesodermal enhancers for Igf2 in the minute mutant. *Development* **129**, 1657–1668.

Davis, E., Caiment, F., Tordoir, X., Cavaille, J., Ferguson-Smith, A., Cockett, N., Georges, M., and Charlier, C. (2005). RNAi-mediated allelic trans-interaction at the imprinted Rtl1/Peg11 locus. *Curr. Biol.* **15**, 743–749.

de la Puente, A., Hall, J., Wu, Y. Z., Leone, G., Peters, J., Yoon, B. J., Soloway, P., and Plass, C. (2002). Structural characterization of Rasgrf1 and a novel linked imprinted locus. *Gene* **291**, 287–297.

DeChiara, T. M., Robertson, E. J., and Efstratiadis, A. (1991). Parental imprinting of the mouse insulin-like growth factor II gene. *Cell* **64**, 849–859.

Engel, N., and Bartolomei, M. S. (2003). Mechanisms of insulator function in gene regulation and genomic imprinting. *Int. Rev. Cytol.* **232**, 89–127.

Fitzpatrick, G. V., Soloway, P. D., and Higgins, M. J. (2002). Regional loss of imprinting and growth deficiency in mice with a targeted deletion of *KvDMR1*. *Nat. Genet.* **32**, 426–431.

Fitzpatrick, G. V., Pugacheva, E. M., Shin, J. Y., Abdullaev, Z., Yang, Y., Khatod, K., Lobanenkov, V. V., and Higgins, M. J. (2007). Allele-Specific Binding of CTCF to the Multipartite Imprinting Control Region KvDMR1. *Mol. Cell. Biol.* **27,** 2636–2647.

Georges, M., Charlier, C., Smit, M., Davis, E., Shay, T., Tordoir, X., Takeda, H., Caiment, F., and Cockett, N. (2004). Toward molecular understanding of polar overdominance at the ovine callipyge locus. *Cold Spring Harb. Symp. Quant. Biol.* **69,** 477–483.

Hark, A. T., and Tilghman, S. M. (1998). Chromatin conformation of the *H19* epigenetic mark. *Hum. Mol. Genet.* **7,** 1979–1985.

Hark, A. T., Schoenherr, C. J., Katz, D. J., Ingram, R. S., Levorse, J. M., and Tilghman, S. M. (2000). CTCF mediates methylation-sensitive enhancer-blocking activity at the *H19/Igf2* locus. *Nature* **405,** 486–489.

Hernandez, A., Martinez, M. E., Croteau, W., and St Germain, D. L. (2004). Complex organization and structure of sense and antisense transcripts expressed from the DIO3 gene imprinted locus. *Genomics* **83,** 413–424.

Holmes, R., Chang, Y., and Soloway, P. D. (2006). Timing and Sequence Requirements Defined for Embryonic Maintenance of Imprinted DNA Methylation at Rasgrf1. *Mol. Cell. Biol.* **26,** 9564–9570.

Hu, J. F., Balaguru, K. A., Ivaturi, R. D., Oruganti, H., Li, T., Nguyen, B. T., Vu, T. H., and Hoffman, A. R. (1999). Lack of reciprocal genomic imprinting of sense and antisense RNA of mouse insulin-like growth factor II receptor in the central nervous system. *Biochem. Biophys. Res. Commun.* **257,** 604–608.

Kaffer, C. R., Srivastava, M., Park, K. Y., Ives, E., Hsieh, S., Batle, J., Grinberg, A., Huang, S. P., and Pfeifer, K. (2000). A transcriptional insulator at the imprinted *H19/Igf2* locus. *Genes Dev.* **14,** 1908–1919.

Kanduri, C., Pant, V., Loukinov, D., Pugacheva, E., Qi, C. F., Wolffe, A., Ohlsson, R., and Lobanenkov, V. V. (2000). Functional association of CTCF with the insulator upstream of the H19 gene is parent of origin-specific and methylation-sensitive. *Curr. Biol.* **10,** 853–856.

Kanduri, C., Fitzpatrick, G., Mukhopadhyay, R., Kanduri, M., Lobanenkov, V., Higgins, M., and Ohlsson, R. (2002). A differentially methylated imprinting control region within the Kcnq1 locus harbors a methylation-sensitive chromatin insulator. *J. Biol. Chem.* **277,** 18106–18110.

Kobayashi, S., Wagatsuma, H., Ono, R., Ichikawa, H., Yamazaki, M., Tashiro, H., Aisaka, K., Miyoshi, N., Kohda, T., Ogura, A., Ohki, M., Kaneko-Ishino, T., et al. (2000). Mouse Peg9/Dlk1 and human PEG9/DLK1 are paternally expressed imprinted genes closely located to the maternally expressed imprinted genes: Mouse Meg3/Gtl2 and human MEG3. *Genes Cells* **5,** 1029–1037.

Kurukuti, S., Tiwari, V. K., Tavoosidana, G., Pugacheva, E., Murrell, A., Zhao, Z., Lobanenkov, V., Reik, W., and Ohlsson, R. (2006). CTCF binding at the H19 imprinting control region mediates maternally inherited higher-order chromatin conformation to restrict enhancer access to Igf2. *Proc. Natl. Acad. Sci. USA* **103,** 10684–10689.

Lee, M. P., DeBaun, M. R., Mitsuya, K., Galonek, H. L., Brandenburg, S., Oshimura, M., and Feinberg, A. P. (1999). Loss of imprinting of a paternally expressed transcript, with antisense orientation to KvLQT1, occurs frequently in Beckwith-Wiedemann syndrome and is independent of insulin-like growth factor II imprinting. *Proc. Natl. Acad. Sci. USA* **96,** 5203–5208.

Leighton, P. A., Saam, J. R., Ingram, R. S., Stewart, C. L., and Tilghman, S. M. (1995). An enhancer deletion affects both *H19* and *Igf2* expression. *Genes Dev.* **9,** 2079–2089.

Lewis, A., Mitsuya, K., Umlauf, D., Smith, P., Dean, W., Walter, J., Higgins, M., Feil, R., and Reik, W. (2004). Imprinting on distal chromosome 7 in the placenta involves repressive histone methylation independent of DNA methylation. *Nat. Genet.* **36,** 1291–1295.

Lin, S. P., Youngson, N., Takada, S., Seitz, H., Reik, W., Paulsen, M., Cavaille, J., and Ferguson-Smith, A. C. (2003). Asymmetric regulation of imprinting on the maternal and paternal chromosomes at the Dlk1-Gtl2 imprinted cluster on mouse chromosome 12. *Nat. Genet.* **35,** 97–102.

Lin, S. P., Coan, P., da Rocha, S. T., Seitz, H., Cavaille, J., Teng, P. W., Takada, S., and Ferguson-Smith, A. C. (2007). Differential regulation of imprinting in the murine embryo and placenta by the Dlk1-Dio3 imprinting control region. *Development* **134**, 417–426.

Lucifero, D., Mertineit, C., Clarke, H. J., Bestor, T. H., and Trasler, J. M. (2002). Methylation dynamics of imprinted genes in mouse germ cells. *Genomics* **79**, 530–538.

Lyle, R., Watanabe, D., te Vruchte, D., Lerchner, W., Smrzka, O. W., Wutz, A., Schageman, J., Hahner, L., Davies, C., and Barlow, D. P. (2000). The imprinted antisense RNA at the Igf2r locus overlaps but does not imprint Mas1. *Nat. Genet.* **25**, 19–21.

Mancini-Dinardo, D., Steele, S. J., Levorse, J. M., Ingram, R. S., and Tilghman, S. M. (2006). Elongation of the Kcnq1ot1 transcript is required for genomic imprinting of neighboring genes. *Genes Dev.* **20**, 1268–1282.

Mattick, J. S., and Makunin, I. V. (2006). Non-coding RNA. *Hum. Mol. Genet.* **15**(Spec. No. 1), R17–R29.

Mitsuya, K., Meguro, M., Lee, M. P., Katoh, M., Schulz, T. C., Kugoh, H., Yoshida, M. A., Niikawa, N., Feinberg, A. P., and Oshimura, M. (1999). LIT1, an imprinted antisense RNA in the human KvLQT1 locus identified by screening for differentially expressed transcripts using monochromosomal hybrids. *Hum. Mol. Genet.* **8**, 1209–1217.

Morgan, H. D., Santos, F., Green, K., Dean, W., and Reik, W. (2005). Epigenetic reprogramming in mammals. *Hum. Mol. Genet.* **14**(Spec. No. 1), R47–R58.

Murrell, A., Heeson, S., and Reik, W. (2004). Interaction between differentially methylated regions partitions the imprinted genes Igf2 and H19 into parent-specific chromatin loops. *Nat. Genet.* **36**, 889–893.

Nicholls, R. D., and Knepper, J. L. (2001). Genome organization, function, and imprinting in Prader-Willi and Angelman syndromes. *Annu. Rev. Genomics. Hum. Genet.* **2**, 153–175.

O'Neill, M. J. (2005). The influence of non-coding RNAs on allele-specific gene expression in mammals. *Hum. Mol. Genet.* **14**(Spec. No. 1), R113–R120.

Pauler, F. M., Koerner, M. V., and Barlow, D. P. (2007). Silencing by imprinted noncoding RNAs: Is transcription the answer? *Trends Genet.* **23**, 284–292.

Paulsen, M., Takada, S., Youngson, N. A., Benchaib, M., Charlier, C., Segers, K., Georges, M., and Ferguson-Smith, A. C. (2001). Comparative sequence analysis of the imprinted Dlk1-Gtl2 locus in three mammalian species reveals highly conserved genomic elements and refines comparison with the Igf2-H19 region. *Genome Res.* **11**, 2085–2094.

Pearsall, R. S., Plass, C., Romano, M. A., Garrick, M. D., Shibata, H., Hyashizaki, Y., and Held, W. A. (1999). A direct repeat sequence at the *Rasgrf1* locus and imprinted expression. *Genomics* **55**, 194–201.

Plass, C., Shibata, H., Kalcheva, I., Mullins, L., Kotelevtseva, N., Mullins, J., Kato, R., Sasaki, H., Hirotsune, S., Okazaki, Y., Held, W. A., Hayashizaki, Y., *et al.* (1996). Identification of *Grf1* on mouse chromosome 9 as an imprinted gene by RLGS-M. *Nat. Genet.* **14**, 106–112.

Poirier, F., Chan, C.-T. J., Timmons, P. M., Robertson, E. J., Evans, M. J., and Rigby, P. W. J. (1991). The murine H19 gene is activated during embryonic stem cell differentiation in vitro and at the time of implantation in the developing embryo. *Development* **113**, 1105–1114.

Reik, W., Dean, W., and Walter, J. (2001). Epigenetic reprogramming in mammalian development. *Science* **293**, 1089–1093.

Schmidt, J. V., Matteson, P. G., Jones, B. K., Guan, X. J., and Tilghman, S. M. (2000). The Dlk1 and Gtl2 genes are linked and reciprocally imprinted. *Genes Dev.* **14**, 1997–2002.

Seitz, H., Youngson, N., Lin, S. P., Dalbert, S., Paulsen, M., Bachellerie, J. P., Ferguson-Smith, A. C., and Cavaille, J. (2003). Imprinted microRNA genes transcribed antisense to a reciprocally imprinted retrotransposon-like gene. *Nat. Genet.* **34**, 261–262.

Seitz, H., Royo, H., Bortolin, M. L., Lin, S. P., Ferguson-Smith, A. C., and Cavaille, J. (2004). A large imprinted microRNA gene cluster at the mouse Dlk1-Gtl2 domain. *Genome Res.* **14**, 1741–1748.

Shibata, H., Yoda, Y., Kato, R., Ueda, T., Kamiya, M., Hiraiwa, N., Yoshiki, A., Plass, C., Pearsall, R. S., Held, W. A., Muramatsu, M., Sasaki, H., et al. (1998). A methylation imprint mark in the mouse imprinted gene Grf1/Cdc25Mm locus shares a common feature with the U2afbp-rs gene: An association with a short tandem repeat and a hypermethylated region. Genomics 49, 30–37.

Sleutels, F., Zwart, R., and Barlow, D. P. (2002). The non-coding Air RNA is required for silencing autosomal imprinted genes. Nature 415, 810–813.

Smilinich, N. J., Day, C. D., Fitzpatrick, G. V., Caldwell, G. M., Lossie, A. C., Cooper, P. R., Smallwood, A. C., Joyce, J. A., Schofield, P. N., Reik, W., Nicholls, R. D., Weksberg, R., et al. (1999). A maternally methylated CpG island in KvLQT1 is associated with an antisense paternal transcript and loss of imprinting in Beckwith-Wiedemann syndrome. Proc. Natl. Acad. Sci. USA 96, 8064–8069.

Smit, M., Segers, K., Carrascosa, L. G., Shay, T., Baraldi, F., Gyapay, G., Snowder, G., Georges, M., Cockett, N., and Charlier, C. (2003). Mosaicism of Solid Gold supports the causality of a noncoding A-to-G transition in the determinism of the callipyge phenotype. Genetics 163, 453–456.

Smit, M. A., Tordoir, X., Gyapay, G., Cockett, N. E., Georges, M., and Charlier, C. (2005). BEGAIN: A novel imprinted gene that generates paternally expressed transcripts in a tissue- and promoter-specific manner in sheep. Mamm. Genome 16, 801–814.

Solter, D. (1988). Differential imprinting and expression of maternal and paternal genomes. Annual Review of Genetics 22, 127–146.

Stöger, R., Kubicka, P., Liu, C.-G., Kafri, T., Razin, A., Cedar, H., and Barlow, D. P. (1993). Maternal-specific methylation of the imprinted mouse Igf2r locus identifies the expressed locus as carrying the imprinting signal. Cell 73, 61–71.

Szabo, P. E., Tang, S.-H., Rentsendorj, A., Pfeifer, G. P., and Mann, J. R. (2000). Maternal-specific footprints at putative CTCF sites in the H19 imprinting control region give evidence for insulator function. Curr. Biol. 10, 607–610.

Takada, S., Tevendale, M., Baker, J., Georgiades, P., Campbell, E., Freeman, T., Johnson, M. H., Paulsen, M., and Ferguson-Smith, A. C. (2000). Delta-like and gtl2 are reciprocally expressed, differentially methylated linked imprinted genes on mouse chromosome 12. Curr. Biol. 10, 1135–1138.

Takada, S., Paulsen, M., Tevendale, M., Tsai, C. E., Kelsey, G., Cattanach, B. M., and Ferguson-Smith, A. C. (2002). Epigenetic analysis of the Dlk1-Gtl2 imprinted domain on mouse chromosome 12: Implications for imprinting control from comparison with Igf2-H19. Hum. Mol. Genet. 11, 77–86.

Takeda, H., Caiment, F., Smit, M., Hiard, S., Tordoir, X., Cockett, N., Georges, M., and Charlier, C. (2006). The callipyge mutation enhances bidirectional long-range DLK1-GTL2 intergenic transcription in cis. Proc. Natl. Acad. Sci. USA 103, 8119–8124.

Thorvaldsen, J. L., Duran, K. L., and Bartolomei, M. S. (1998). Deletion of the H19 differentially methylated domain results in loss of imprinted expression of H19 and Igf2. Genes Dev. 12, 3693–3702.

Thorvaldsen, J. L., Mann, M. R., Nwoko, O., Duran, K. L., and Bartolomei, M. S. (2002). Analysis of sequence upstream of the endogenous H19 gene reveals elements both essential and dispensable for imprinting. Mol. Cell. Biol. 22, 2450–2462.

Thorvaldsen, J. L., Verona, R. I., and Bartolomei, M. S. (2006). X-tra! X-tra! News from the mouse X chromosome. Dev. Biol. 298, 344–353.

Tremblay, K. D., Saam, J. R., Ingram, R. S., Tilghman, S. M., and Bartolomei, M. S. (1995). A paternal-specific methylation imprint marks the alleles of the mouse H19 gene. Nat. Genet. 9, 407–413.

Tremblay, K. D., Duran, K. L., and Bartolomei, M. S. (1997). A 5′ 2-kilobase-pair region of the imprinted mouse *H19* gene exhibits exclusive paternal methylation throughout development. *Mol. Cell. Biol.* **17**, 4322–4329.

Umlauf, D., Goto, Y., Cao, R., Cerqueira, F., Wagschal, A., Zhang, Y., and Feil, R. (2004). Imprinting along the Kcnq1 domain on mouse chromosome 7 involves repressive histone methylation and recruitment of Polycomb group complexes. *Nat. Genet.* **36**, 1296–1300.

Verona, R. I., Mann, M. R., and Bartolomei, M. S. (2003). Genomic imprinting: Intricacies of epigenetic regulation in clusters. *Annu. Rev. Cell Dev. Biol.* **19**, 237–259.

Weksberg, R., Shuman, C., and Smith, A. C. (2005). Beckwith-Wiedemann syndrome. *Am. J. Med. Genet. C Semin. Med. Genet.* **137**, 12–23.

Wutz, A., Smrzka, O. W., Schweifer, N., Schellander, K., Wagner, E. F., and Barlow, D. P. (1997). Imprinted expression of the *Igf2r* gene depends on an intronic CpG island. *Nature* **389**, 745–749.

Wylie, A. A., Murphy, S. K., Orton, T. C., and Jirtle, R. L. (2000). Novel imprinted *DLK1/GTL2* domain on human chromosome 14 contains motifs that mimic those implicated in *IGF2/H19* regulation. *Genome Res.* **10**, 1711–1718.

Yoon, B. J., Herman, H., Sikora, A., Smith, L. T., Plass, C., and Soloway, P. D. (2002). Regulation of DNA methylation of Rasgrf1. *Nat. Genet.* **30**, 92–96.

Yoon, B., Herman, H., Hu, B., Park, Y. J., Lindroth, A., Bell, A., West, A. G., Chang, Y., Stablewski, A., Piel, J. C., Loukinov, D. I., Lobanenkov, V. V., and Soloway, P. D. (2005). Rasgrf1 imprinting is regulated by a CTCF-dependent methylation-sensitive enhancer blocker. *Mol. Cell. Biol.* **25**, 11184–11190.

Yoon, Y. S., Jeong, S., Rong, Q., Park, K. Y., Chung, J. H., and Pfeifer, K. (2007). Analysis of the H19ICR Insulator. *Mol. Cell. Biol.* **27**, 3499–3510.

Yoo-Warren, H., Pachnis, V., Ingram, R. S., and Tilghman, S. M. (1988). Two regulatory domains flank the mouse H19 gene. *Mol. Cell. Biol.* **8**, 4707–4715.

Zemel, S., Bartolomei, M. S., and Tilghman, S. M. (1992). Physical linkage of two mammalian imprinted genes. *Nat. Genet.* **2**, 61–65.

Zwart, R., Sleutels, F., Wutz, A., Schinkel, A. H., and Barlow, D. P. (2001). Bidirectional action of the Igf2r imprint control element on upstream and downstream imprinted genes. *Genes Dev.* **15**, 2361–2366.

8

Genomic Imprinting and Imprinting Defects in Humans

Bernhard Horsthemke and Karin Buiting
Institut für Humangenetik, Universitätsklinikum Essen, Hufelandstrasse 55,
45122 Essen, Germany

I. Introduction
II. The Mechanisms of Genomic Imprinting
  A. The nature of genomic imprints
  B. Erasure, establishment, and maintenance of genomic imprints
III. Imprinting Defects
  A. Imprinting defects in 15q11-q13
  B. Imprinting defects in 11p15
IV. Conclusions
  References

ABSTRACT

In placental mammals some 100–200 genes are expressed only from the paternal or the maternal allele. This peculiar expression pattern is the result of genomic imprinting, an epigenetic process by which the male and the female germ line confer a parent-of-origin specific mark (imprint) on certain chromosomal regions. The size of imprinted regions ranges from several kilobases to several megabases. The process of genomic imprinting is controlled by *cis*-acting imprinting centers (IC) and *trans*-acting factors. IC mutations affect the establishment or maintenance of genomic imprints and hence the expression of all imprinted genes controlled by this IC. Imprinting defects play a causal role in several recognizable syndromes. © 2008, Elsevier Inc.

Advances in Genetics, Vol. 61
Copyright 2008, Elsevier Inc. All rights reserved.

0065-2660/08 $35.00
DOI: 10.1016/S0065-2660(07)00008-9

I. INTRODUCTION

Genomic imprinting is an epigenetic process by which the male and the female germ line of placental mammals (eutheria) confer a parent-of-origin specific mark (imprint) on certain chromosomal regions. In humans, imprinted regions have been found on chromosomes 6, 7, 11, 14, and 15. Owing to genomic imprinting, some 100–200 genes are expressed in a parent-of-origin specific manner, so that the paternal and the maternal genome are functionally nonequivalent. This was first shown by nuclear transplantation experiments in mice (McGrath and Solter, 1984; Surani et al., 1984), but it is also observed in humans. Occasionally, the haploid genome of an unfertilized egg undergoes duplication. Although such an egg has 46 chromosomes, it does not develop into a normal embryo, but into a benign ovarian teratoma, which contains tissues from all three germ layers, but no trophoblast. Similarly, an egg that was fertilized by two sperms but has lost its maternal genome does not undergo normal development; it develops into a hydatiform mole, which is degenerated trophoblast tissue.

The nonequivalence of the maternal and paternal genomes is also obvious from uniparental disomies. Uniparental disomy refers to the presence of two copies of a chromosome (or part of a chromosome) from one parent and none from the other (Engel, 1980). Uniparental disomy per se does not interfere with genomic imprinting or cause clinical problems. However, if the affected chromosome pair carries an imprinted gene, both alleles of this gene will be inactive or active, depending on the parental origin of the chromosomes. The presence of zero or two active copies of an imprinted gene, however, often interferes with normal development and growth.

II. THE MECHANISMS OF GENOMIC IMPRINTING

Genomic imprinting is a reversible process that does not affect the DNA sequence. Although there is some dispute on the nature of primary imprint, it has long been established that the parental copies of imprinted regions differ with respect to DNA methylation, histone modification, and the presence or absence of chromatin-binding proteins (Fig. 8.1).

A. The nature of genomic imprints

DNA methylation in the mammalian genome refers to the methylation of cytosine (5-methyl-cytosine, m^5C) within a CpG dinucleotide. CpG is a palindromic sequence, and typically the cytosines in both strands are methylated. 5-Methyl-cytosine is not incorporated into the DNA by the DNA polymerase

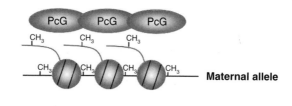

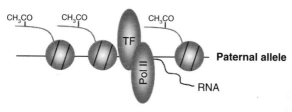

Figure 8.1. Epigenetic differences of parental alleles. The DNA (black solid line) is wrapped around nucleosomes (grey balls). For clarity, only some histone modifications and only one class of nonhistone proteins (PcGs) are shown. CH$_3$, methylation; CH$_3$CO, acetylation; PcG, polycomb group protein; TF, transcription factor; Pol II, DNA-dependent RNA polymerase II.

but is the result of posttranscriptional modification of cytosine by DNA methyltransferases. Three active DNA methyltransferases have been identified in human cells: DNMT1, DNMT3A, and DNMT3B. In addition, there is one putative DNA methyltransferase (DNMT2) and a protein that is highly similar to DNMT3A and B, but devoid of catalytic activity (DNMT3L). All methyltransferases use S-adenosyl-methionine as a methyl-donor. DNMT1 is ubiquitously expressed and is a maintenance methyltransferase, which methylates hemimethylated CpG dinucleotides in the nascent DNA strand after replication. It is essential for maintaining DNA methylation patterns in proliferating cells. DNMT3A and DNMT3B are regulated during development. They carry out *de novo* methylation and thus establish new DNA methylation patterns. DNMT3L cooperates with DNMT3A and DNMT3B to establish methylation imprints. The activity and function of DNMT2 remains undefined. For a recent review on eukaryotic DNA methyltransferases, see Goll and Bestor (2005).

DNA methylation can be removed by active or passive processes. Passive demethylation occurs by DNA replication in the absence of DNMT1. Recently, Gadd45a has been identified as a key regulator of active DNA methylation (Barreto *et al.*, 2007). Gadd45a is targeted to specific sites of demethylation and recruits the DNA repair machinery through the endonuclease xeroderma pigmentosum, group G (XPG). Methylated cytosines are then excised and replaced by unmethylated nucleotides. It is likely that other demethylation activities exist.

Histone modifications include the acetylation, ubiquitylation, and mono-, di-, and trimethylation of lysine, the mono- and dimethylation of arginine, and the phosophorylation of serine. Furthermore, histone ubiquitylation, histone sumoylation, and poly-ADP-ribosylation have been described. The pattern of modification determines the accessibility of promoter and enhancers to transcription factors and thus gene activity. For a recent review on histone modification, see Kubicek *et al.* (2006).

Many different combinations of histone modifications are possible, providing the cell with a wealth of different possibilities to regulate gene activity. Methylation of Lys4 of H3 (H3K4) and acetylation of Lys9 of H3 (H3K9), for example, have been associated with active gene expression, whereas the methylation of H3K9 has been associated with transcriptional silencing.

The degree of histone acetylation, that is, the acetylation of a lysine side chain, is determined by the activity of histone acetylases (HATs) and histone deacetylases (HDACs). Several transcription factors and coactivators have intrinsic HAT activity. The acetylation of a lysine residue changes the charge of the histone molecule and leads to an "open," transcriptionally permissive chromatin configuration. There are at least 18 HDACs in humans. Histone deacetylation leads to a "closed," repressive chromatin configuration.

In contrast to the acetylation of lysine residues, the methylation of lysine residues does not change the charge of the histone molecule. It does, however, change its basicity and hydrophobicity. Furthermore, histone methylation is considered to be more stable than histone acetylation and thus more relevant for epigenetic inheritance. The degree of histone methylation is determined by the activity of histone methyltransferases (HMTs) and histone demethylases (HDMs).

Like DNA methyltransferases, HMTs use *S*-adenosyl-methionine as a methyl-donor. The catalytic domain of lysine HMTs consists of ~130 amino acids and is called the SET domain. The acronym is derived from three drosophila proteins [Su(var)3–9, Enhancer-of zest, Trithorax]. At present, more than 300 proteins with a SET domain have been identified. The proteins also have bromo- and chromodomains. It is unclear, how mono-, di-, and trimethylation of lysine residues are regulated.

In addition to lysine residues, arginine residues can be methylated. This reaction is catalyzed by protein arginine methyltransferases. At present, seven such enzymes are known in mammals. Similar to the lysine HMTs, the arginine HMTs differ in their specificity.

Only recently, several histone demethylases have been identified, for example, the amine oxidase AOF2 (also called lysine-specific demethylase 1, LSD1) and the peptidylarginin-deiminase 4 (PADI4).

B. Erasure, establishment, and maintenance of genomic imprints

Genomic imprints are erased in primordial germ cells, newly established during later stages of germ cell development, and stably inherited through somatic cell divisions during postzygotic development (Fig. 8.2A). In somatic cells, the imprint is read by the transcription machinery and used to regulate parent-of-origin specific gene expression so that only the paternal or the maternal allele of a susceptible gene is active.

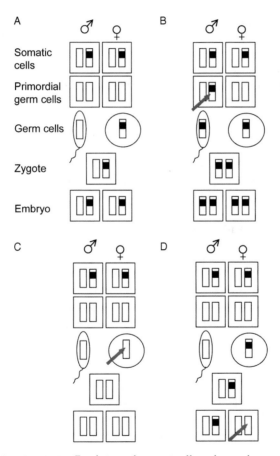

Figure 8.2. Genomic imprinting. For clarity, only one pair of homologous chromosomes (open bars) and a maternal imprint (black box) are shown. (A) Imprints are erased in the primordial germ cells, newly established during gametogenesis and maintained in the zygote and embryo. Imprinting defects can result from an error in imprint erasure (B), imprint establishment (C), or imprint maintenance (D).

Many imprinted genes are involved in regulating resource acquisition of the embryo and fetus. In fact, it has been proposed that imprinting coevolved with the placenta. In eutherian mammals, the fetus grows at the expense of the mother. As proposed by the genetic conflict theory (Wilkins and Haig, 2003), the paternal genome is "interested" in extracting as many resources from the mother as possible. This is because a male can spread his genes through many different females. By contrast, maternally inherited genes protect the mother from being exhausted by the fetus, because a female can spread her genes only through multiple pregnancies.

III. IMPRINTING DEFECTS

Errors in imprint erasure, establishment, or maintenance (imprinting defects, Fig. 8.2B–D) lead to aberrant gene expression and developmental disorders. The study of patients with an imprinting defect offers a unique opportunity to identify some of the factors and mechanisms involved in imprint erasure, establishment, and maintenance.

A. Imprinting defects in 15q11-q13

One cluster of imprinted genes with relevance to human disease is located in 15q11-q13 (Fig. 8.3). This chromosomal region is affected in the Prader-Willi syndrome (PWS) and the Angelman syndrome (AS). PWS and AS are distinct neurogenetic disorders and the first known examples of human diseases involving imprinted genes. PWS is characterized by neonatal muscular hypotonia and failure to thrive, hyperphagia and obesity starting in early childhood, hypogonadism, short stature, small hands and feet, sleep apnea, behavioral problems, and mild-to-moderate mental retardation. AS is characterized by microcephalus, ataxia, absence of speech, abnormal EEG pattern, severe mental retardation, and frequent laughing. Whereas PWS is caused by the loss of function of paternally expressed genes, AS is caused by the loss of function of the maternally expressed gene *UBE3A*.

Paternal-only expression of *MKRN3*, *NDN*, and *SNRPN* has been shown to be associated with differential DNA methylation. Whereas the promoter/exon 1 regions of these genes are unmethylated on the expressing paternal chromosome, the silent maternal alleles are methylated. In addition to DNA methylation, the parental copies of these genes also differ by histone modification.

The *SNRPN* gene encodes two proteins, SNURF and SNRPN, serves as a host for 76 C/D small nucleolar (sno) RNA genes, and overlaps, in an antisense orientation, the *UBE3A* gene. The snoRNAs are encoded within introns of the *SNRPN* gene. They are expressed from the paternal allele only, because they are

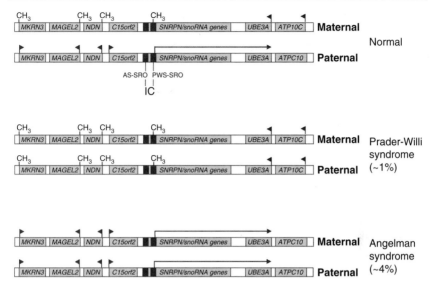

Figure 8.3. Normal and aberrant imprints in 15q11-q13. For clarity, the maps (from centromere, left, to telomere, right) are not drawn to scale, and not all genes are shown. Some genes show imprinted expression in some tissues only. Gene transcription is indicated by triangular flags. IC, imprinting centre. CH₃, DNA methylation. In PWS patients with an imprinting defect, both homologues carry a maternal imprint. In AS patients with an imprinting defect, both homologues carry a paternal imprint.

processed from the paternally expressed *SNRPN* sense/*UBE3A* antisense transcript during the splice process. Thus, imprinted expression of the snoRNAs is indirectly regulated through *SNRPN* methylation. Unlike other C/D box snoRNAs, the snoRNAs encoded within the *SNRPN* locus do not serve as guide RNAs for 2'-O-ribose methylation of nucleotides in rRNA. Their function remains to be determined.

C15orf2, which maps between *NDN* and *SNURF-SNRPN*, was initially described as a testis-specific intronless gene that is biallelically expressed (Färber *et al.*, 2000). In contrast to other genes in the PWS/AS region, *C15orf2* is found only in primates. On the basis of these findings, we had suggested that *C15orf2* may play a role in primate spermatogenesis. Reinvestigation of *C15orf2* revealed that this gene is also expressed in fetal brain and that expression in this tissue is monoallelic (Buiting *et al.*, 2007), suggesting that this gene is also subject to genomic imprinting.

In the majority of patients (~70%) with PWS, the disease is due to a 3–4 Mb *de novo* deletion of 15q11-q13. The second most common genetic abnormality in PWS (~30%) is a maternal uniparental disomy 15 [upd(15) mat], which most often arises from maternal meiotic nondisjunction followed by mitotic loss of the paternal chromosome 15 after fertilization. A small subset of patients with PWS (~1%) has apparently normal chromosomes 15 of biparental inheritance, but the paternal chromosome carries a maternal imprint (imprinting defect; Fig. 8.3). All three lesions lead to the lack of expression of imprinted genes that are active on the paternal chromosome only. This can easily be detected by DNA methylation analysis. Patients with PWS lack unmethylated alleles of several loci within 15q11-q13.

Based on the finding of point mutations in patients with AS, *UBE3A* has been identified as the gene affected. In contrast to the paternally active genes, the maternally active *UBE3A* gene lacks differential DNA methylation. Another striking difference is that imprinted *UBE3A* expression is tissue-specific. At present it is unclear how tissue-specific imprinting of *UBE3A* is regulated, but the paternally expressed *SNRPN* sense/*UBE3A* antisense transcript may be involved in silencing the paternal *UBE3A* allele (Rougeulle *et al.*, 1998; see below).

Similar to PWS, the major lesion in AS is a large deletion of 15q11-q13 (~70%), but in AS the deletion is on the maternal chromosome. AS can also result from upd(15)pat (~1% of cases), which most often arises from the postzygotic duplication of a zygote carrying only a paternal chromosome 15, or the lack of a maternal imprint on the maternal chromosome (imprinting defect; ~4% of cases; Fig. 8.3). All three lesions can be detected by DNA methylation analysis. The patients lack a methylated allele of maternally methylated loci within 15q11-q13. Five to 10 percent of patients have a mutation in the maternal *UBE3A* gene, and ~10–15% of patients with the clinical diagnosis of AS have a genetic defect of unknown nature. The latter two classes of patients cannot be detected by methylation analysis.

1. Imprinting defects caused by an imprinting center deletion

The identification of small deletions in a subgroup (8–15%) of patients with an imprinting defect has led to the definition of an imprinting center (IC) that regulates in *cis* imprint resetting and imprint maintenance in the whole imprinted domain (Buiting *et al.*, 1995; Sutcliffe *et al.*, 1994). So far, 19 deletions in patients with PWS and an imprinting defect, and 13 deletions in patients with AS and an imprinting defect have been reported (Fig. 8.4). These deletions define two different critical elements inside the IC, the AS-SRO and the PWS-SRO (Buiting *et al.*, 1995). All IC deletions found in patients with PWS affect the *SNURF-SNRPN* promoter/exon 1 region. In ~50% of IC deletions identified so far, the deletions are a familial mutation, associated with a 50% recurrence risk. These deletions can be transmitted silently through the female germ line, but lead to an incorrect, maternal imprint on the paternal chromosome when inherited from a male (Fig. 8.5). This explains why in some families only few and

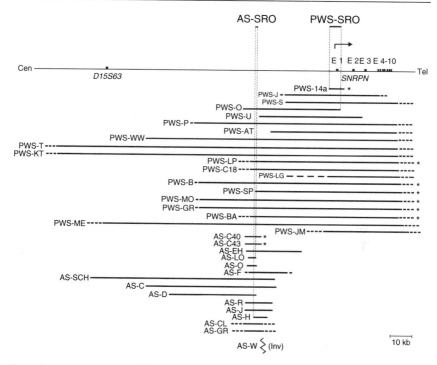

Figure 8.4. Imprinting center (IC) deletions in 15q11-q13. IC deletions in patients with PWS or AS and an imprinting defect. The extent of the deletions is shown as horizontal bars. *De novo* deletions are marked by *. The other deletions were inherited from the father (PWS) or the mother (AS). The smallest regions of deletion overlap in PWS and AS (PWS-SRO and AS-SRO) are indicated by broken lines. The inversion breakpoint in family AS-W (Buiting *et al.*, 2001) is shown by a zigzag line. Families PWS-S, PWS-O, PWS-U, AS-C, and AS-D are described in Buiting *et al.* (1995), families AS-SCH, AS-R, AS-J, and AS-H in Saitoh *et al.* (1996), family PWS-KT in Schuffenhauer *et al.* (1996), families PWS-14a, PWS-J, PWS-P, and PWS-T in Ohta *et al.* (1999a), families AS-O and AS-F in Ohta *et al.* (1999b), family AS-LO in Buiting *et al.* (1999), family PWS-LG in Buiting *et al.* (2000), family PWS-AT in El-Maarri *et al.* (2001), family PWS-WW in Bielinska *et al.* (2000), family PWS-ME in McEntagart *et al.* (2000), family PWS-JM in Ming *et al.* (2000), family AS-CL in Ronan *et al.* (submitted), and family AS-GR in Raca *et al.* (2004). IC deletions in families PWS-LP, PWS-C18, PWS-B, PWS-SP, PWS-MO, PWS-GR, PWS-BA, AS-C40, AS-C43, and AS-EH have been identified by us and are unpublished.

distantly related individuals are affected. In the case of a *de novo* deletion, the recurrence risk is not increased when it occurred after fertilization, but it can be up to 50% when the father has a germ line mosaic.

The shortest region of deletion overlap (PWS-SRO) is 4.1 kb (Ohta *et al.*, 1999a,b; Fig. 8.4). There is only one PWS family in which the deletion occurred postzygotically (Bielinska *et al.*, 2000). This family, where the father is mosaic for an

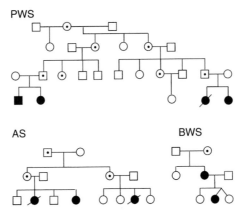

Figure 8.5. Segregation of imprinting center (IC) deletions. Unaffected carriers are indicated by a dot. Note that a deletion of the PWS-SRO of the 15q IC in the Prader-Willi syndrome family (family PWS-AT, El-Maarri *et al.*, 2001) is benign when transmitted through a female. After transmission through a male, the deletion chromosome acquires a maternal methylation pattern during early embryogenesis. A deletion of the AS-SRO element of the 15q IC in the AS family (unpublished) or a deletion of the 11p15 IC1 in the BWS family (family BWS-A, Sparago *et al.*, 2007) is benign when transmitted through a male. In the female germ line the deletion prevents the establishment of the maternal imprint.

IC deletion on his paternal chromosome, provided evidence that the deletion of this element leads to an epigenetic state that resembles the maternal imprint (Bielinska *et al.*, 2000). A similar deletion induced in chimeric mice generated from two independent embryonic stem cells (ES) leads to the same effect. Sperm DNA from males with a maternally inherited PWS IC deletion were found to have a normal paternal methylation imprint throughout human 15q11-q13, indicating that the imprint was correctly reset on the mutant maternal chromosome (El-Maarri *et al.*, 2001). Thus, the incorrect maternal methylation pattern on the mutant paternal chromosome of patients with PWS and an IC deletion must have occurred after fertilization. These findings indicate that the PWS-SRO is not (only) required for the establishment of the paternal imprint, but necessary for the maintenance of the paternal methylation imprint during early embryonic development.

Gene targeting in the mouse revealed that the murine gene cluster on chromosome 7 is regulated in a similar way. A paternal transmission of a 42-kb deletion (exons 1–6 of *Snurf-Snrpn* plus 23 kb upstream sequence) resulted in maternal imprint on the paternal chromosome and led to 100% postnatal lethality (Yang *et al.*, 1998). The mutant mice were smaller than their wild-type littermates, exhibited muscular hypotonia, and died within the first few days of life. In contrast, Bressler *et al.* (2001) found that a 0.9-kb deletion of the exon 1 region of *Snurf-Snrpn* had no effect, whereas a paternally transmitted 4.8-kb deletion led to mosaic imprinting defects and to postnatal lethality in about 50% of mutant mice. Surviving mutant mice were viable and healthy.

The shortest region of overlap in patients with AS and an IC deletion (AS-SRO) is 880 bp and maps 35 kb proximal to *SNURF-SNRPN* exon 1 (Buiting *et al.*, 1999; Fig. 8.4). In contrast to the PWS IC deletions, which sometimes include the AS-SRO, none of the IC deletions found in patients with AS affects the *SNURF-SNRPN* promoter/exon 1 region. Two out of 13 IC deletions in patients with AS have occurred *de novo* on the maternal chromosome, but in the majority of cases they have been inherited from the mother. A deletion of the AS-SRO prevents maternal imprinting of the mutated chromosome. A child inheriting this chromosome will develop AS, because the maternal *SNRPN* allele is unmethylated and expressed, and the maternal *UBE3A* allele is silenced. Since deletions of the AS-SRO element affect maternal imprinting only, they are silently transmitted through the paternal germ line (Fig. 8.5).

There is one familial case where a secondary epimutation is not due to an IC deletion but the result of an inversion spanning ~1.5 Mb with one breakpoint inside the IC (Buiting *et al.*, 2001). As a consequence, the IC is disrupted and the AS-SRO has been removed from the PWS-SRO in the center of the imprinted domain to its proximal border and is in an inverted orientation. When this inversion is transmitted through the female germ line, it prevents maternal imprinting in the whole domain, suggesting that the close proximity and/or the correct orientation of the AS-SRO and the PWS-SRO are necessary to establish a maternal imprint.

Up to now no murine homologue of the human AS-SRO has been identified. However, in a recent study two mouse mutations have been reported to result in defects similar to that seen in AS patients with a deletion of the AS-SRO (Wu *et al.*, 2006). An insertion/duplication mutation 13-kb upstream of *Snurf-Snrpn* exon 1 resulted in lack of methylation at the maternal *Snurf-Snrpn* promoter, activation of maternally repressed genes, and decreased expression of paternally repressed genes. A second mutation, an 80-kb deletion extending upstream of the first mutation, caused a similar imprinting defect with variable penetrance. These results suggest that there is a mouse functional equivalent to the human AS-IC.

From transgenic mouse models (Kantor *et al.*, 2004b; Perk *et al.*, 2002) there is some evidence that the maternal AS-SRO is essential for setting up the DNA methylation state and closed chromatin structure of the neighboring PWS-SRO. The general idea of this model was first proposed by Dittrich *et al.* (1996). On the basis of the finding that the AS-SRO contains upstream exons of the *SNURF-SNRPN* transcript, Dittrich *et al.* had suggested that transcripts containing these exons are a major factor for setting up the maternal imprint and that the AS-SRO acts as an imprintor on the imprint switch initiation site (the PWS-SRO) to establish the maternal imprint. Transgenic mouse studies by Shemer and colleagues supported a stepwise, unidirectional program in which imprinting at the AS-SRO brings about allele-specific repression of the maternal PWS-SRO, but challenged a role of alternative *SNURF–SNRPN* transcript in this process (Shemer et al., 2000). The authors showed that a mini-transgene with 1.0 kb of the human AS-SRO sequence fused to 0.2 kb of the mouse Snurf-Snrpn minimal promoter (homologous to the PWS-SRO) was appropriately imprinted after maternal and paternal transmission. The same group obtained tentative evidence that the AS-SRO contains one or more binding sites for trans-acting factors (Kantor et al., 2004a,b). Although the AS-SRO is not differentially methylated in adult human tissues (Buiting et al., 2003; Kantor et al., 2004a; Schumacher et al., 1998), the demonstration of increased nuclease hypersensitivity on the maternal chromosome in the AS-SRO region is consistent with the binding of trans-factors (Perk et al., 2002; Schweizer et al., 1999).

UBE3A displays predominant maternal expression in human fetal brain and adult frontal cortex. The mechanism of tissue-specific imprinting of UBE3A has not yet been conclusively resolved. In 2001, we reported that the SNURF-SNRPN locus, which is transcribed from centromere to telomere, overlaps the UBE3A gene, which is transcribed from telomere to centromere (Runte et al., 2001). The SNURF-SNRPN sense-UBE3A antisense transcription unit spans more than 460 kb and has more than 148 exons, including the previously identified IPW exons. In a follow-up study, we could demonstrate that splice forms of the SNURF-SNRPN transcript overlapping UBE3A in an antisense orientation are present in brain but barely detectable in blood (Runte et al., 2004). Our findings were compatible with the assumption that imprinted UBE3A expression is regulated through the SNURF-SNRPN sense/UBE3A antisense transcript by silencing the paternal UBE3A transcript in cis. This finding extends the results of Chamberlain and Brannan (2001), who have shown that a deletion of the murine IC results in the loss of the paternally expressed Ube3a antisense transcript (ATS). Recently, Marc Lalande's group reported similar findings in mice (Landers et al., 2004). They presented evidence that the murine Ube3a ATS is also a large (~1000 kb) transcript that is alternatively processed. Although Landers et al. could not connect the more upstream exons of the transcript with the most distal part, they also observed that expression of the more distal part of Ube3a-ATS, which is the host of the MBII-52 snoRNA gene cluster and which overlaps Ube3a, is brain-specific. In contrast to the human transcript, which was found to be initiated at the SNURF-SNRPN promoter region, Landers et al. showed that the murine transcripts are initiated from several alternative exons dispersed within a 500-kb region upstream of Snurf-Snrpn. The first evidence for the existence of such upstream exons in mice had been reported by Bressler et al. (2001). Similar to the distal part of Ube3a-ATS, these upstream (U) exon-containing transcripts are expressed at neuronal stages of differentiation (Landers et al., 2004). These findings suggest that brain-specific transcription of the murine Ube3a-ATS is regulated by the U exons rather than Snurf-Snrpn exon 1. In support of this hypothesis, Landers et al. detected U-Ube3a-ATS transcripts in which U exons are spliced to *Ube3a-ATS* with the exclusion of *Snurf-Snrpn*. In addition, the authors showed that there are at least nine U exons in the mouse. As is the case in human, the U exons appear to have arisen from genomic duplication of segments of the IC. These findings made by Landers *et al.* suggest that the U exons serve as initiation sites for IC-*Ube3a-ATS* transcripts in brain.

2. Imprinting defects not caused by an imprinting center deletion

IC deletions occur in only a small fraction of patients with PWS or AS and an imprinting defect. In 73 of 85 (86%) patients with PWS and 106 of 116 (92%) patients with AS, we found that the imprinting defect represents a primary epimutation that occurred without any underlying DNA sequence change (Buiting *et al.*, 2003, and unpublished).

a. Grandparental origin of the aberrantly imprinted Chromosome

To investigate the grandparental origin of the incorrectly imprinted chromosome, Buiting *et al.* (1998 and 2003) performed microsatellite analysis or used a combined methylation/RFLP test for the *SNURF-SNRPN* locus. In 19 informative PWS patients with a primary epimutation, the paternal chromosome carrying an incorrect maternal imprint was always inherited from the paternal grandmother (Buiting *et al.*, 2003). This bias is highly significant ($p = 0.000002$) and suggests that the (grand) maternal imprint was not erased in the father's germ line (Fig. 8.2B). This was the first demonstration of epigenetic inheritance in man. There is increasing evidence to suggest that epigenetic marks at some mammalian alleles are not completely erased from one generation to the next. We propose that epigenetic inheritance occurs at a low frequency in 15q11-q13 and possibly at other human loci as well.

In 18 informative AS patients with a primary epimutation, the maternal chromosome carrying an incorrect paternal imprint was inherited from the maternal grandfather in 11 patients and from the maternal grandmother in 7 patients (Buiting *et al.*, 2003). This finding suggests that the imprinting defect occurred after erasure of the parental imprints. Thus, the imprinting defect must either result from an error in imprint establishment (Fig. 8.2C) or imprint maintenance (Fig. 8.2D).

b. Somatic mosaicism

Approximately one third of patients with AS and a primary epimutation have somatic mosaicism, that is, cells with an imprinting defect as well as normal cells. Somatic mosaicism typically results from a postzygotic event (Fig. 8.2D). To narrow down the time at which the imprinting defect occurred, Nazlican et al. (2004) determined the pattern of X inactivation in the fibroblast clones. By studying the FMR1 locus, the authors found cells with an imprinting defect in which the paternal X chromosome was inactive and cells with an imprinting defect in which the maternal X chromosome was inactive. Assuming that the imprinting defect was a single event, this result means that the imprinting defect occurred before the onset of random X inactivation, that is, before the blastocyst stage.

In principle, imprint maintenance defects can occur at every cell division. After DNA replication, the methylation pattern on the template strand is recognized by the maintenance DNA methyltransferase and copied onto the daughter strand. A failure to recognize or copy this pattern results in a methylated and an unmethylated strand of DNA, which will segregate in subsequent cell divisions. However, there is one specific time during development where imprint maintenance may be especially prone to errors. This is the period of global DNA demethylation during the very first few days of development. The paternal genome is actively demethylated within the first few hours after fertilization, and the maternal genome is passively demethylated during subsequent cell divisions. The wave of global demethylation is followed by a wave of global remethylation, which is completed after the blastocyst stage (Reik et al., 2001). Gametic imprints survive the waves of global de- and remethylation, although it is unclear how they are protected against the global methylation changes. It is possible that the protection against demethylation occasionally fails so that a maternal imprint is lost in one cell. As imprints cannot be repaired in somatic cells, the daughter cells will inherit the imprinting defect.

To quantify the degree of mosaicism, Nazlican et al. developed a quantitative methylation-specific PCR assay. In 24 patients studied, the percentage of normally methylated cells ranged from <1 to 40%. Patients with a higher percentage tended to have milder symptoms, but the correlation was not statistically significant. Some of the mosaic patients had clinical signs reminiscent of PWS: they showed muscular hypotonia at birth, were obese but were able to speak (Gillessen-Kaesbach et al., 1999).

c. Cis- and *trans*-acting factors modifying the risk of imprinting defects

Genetic studies by us and others have suggested that the AS-SRO element interacts with the PWS-SRO element to establish the maternal imprint in the female germ line (Buiting et al., 2003; Dittrich et al., 1996; Perk et al., 2002). The molecular mechanisms are unknown but likely to involve *trans*-acting factors that bind to these two elements. Therefore it is possible that DNA sequence variations in the protein-binding sites of the IC might affect the binding of *trans*-acting factors and consequently the epigenetic state of the chromosomal domain. Zogel et al. (2006) have investigated whether common sequence variants in the bipartite IC are associated with an increased susceptibility to imprinting defects. The determination of the haplotype structure of the IC revealed that the PWS-SRO and AS-SRO lie on separate haplotype blocks. To identify susceptible IC sequence variants, a transmission disequilibrium test was used. While no preferential transmission of a paternal allele or haplotype in 41 PWS trios was observed, a trend for preferential maternal transmission of one AS-SRO haplotype (H-AS3) in 48 AS trios ($p = 0.058$) was found and two sequence variants in H-AS3 could be identified that are responsible for this effect.

It is possible that the spontaneous epimutation rate is also modified by *trans*-acting genetic factors. A good candidate for a *trans*-acting factor modifying the spontaneous epimutation rate is the 5,10-methylenetetrahydrofolate reductase (MTHFR) gene. MTHFR is a key regulatory enzyme in the one-carbon metabolism and plays an important role in folate metabolism, DNA methylation, and DNA synthesis. Zogel *et al.* (2006) obtained tentative evidence that homozygosity for the 677C > T variant of the *MTHFR* gene increases the risk of a maternal imprinting defect. The frequency of the 677TT genotype was significantly higher in the mothers of the patients with AS than in the patients' fathers or the general population ($p = 0.028$). These findings suggest that women with the IC haplotype H-AS3 or homozygosity for the *MTHFR* 677C > T variant have an increased risk of conceiving a child with an imprinting defect, although the absolute risk is low.

B. Imprinting defects in 11p15

Another cluster of imprinted genes with relevance to human disease is located in 11p15.5 (Fig. 8.6). It is affected in patients with Beckwith-Wiedemann syndrome (BWS) and ~30% of patients with Silver-Russell syndrome (SRS). BWS is an overgrowth syndrome characterized by high birth weight, macroglossia, hypoglycemia, exomphalos, and increased risk for Wilms tumors. It is highly variable, and in some patients only one side of the body is affected (hemihyperplasia). SRS is characterized by pre- and postnatal growth retardation and facial dysmorphism. Like BWS, SRS may affect only some parts of the body.

The majority of BWS cases are sporadic, although some familial cases have been described. BWS is caused by overexpression of *IGF2* and/or loss of expression of *CDKN1C*. Loss of *IGF2* expression causes SRS. *IGF2* is expressed from the paternal allele only, whereas *CDKN1C* are expressed from the maternal allele only. These genes map to the short arm of chromosome 11 (11p15.5), but are controlled by two different ICs, the *H19* DMR (IC1), which controls *IGF2* expression, and *LIT1/KCNQ1OT1* IC (IC2), which controls imprinting of *CDKN1C*.

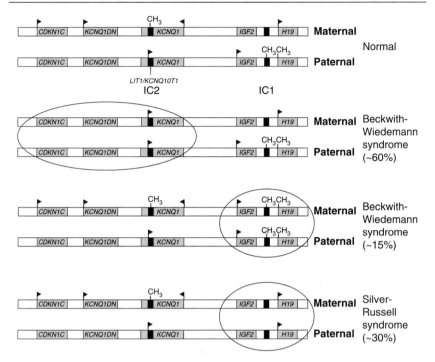

Figure 8.6. Normal and aberrant imprints in 11p15. For symbols and abbreviations, see legend to Fig. 8.3. The circles highlight the affected regions.

Approximately 20% of patients with BWS have paternal uniparental disomy of 11p15, which leads to overexpression of *IGF2* and loss of expression of *CDKN1C*. Rare patients have a paternal duplication or a balanced maternal translocation affecting this region. Approximately 5% of cases (40% of the familial cases) have a loss-of-function mutation of *CDKN1C*. In 50% of patients, *CDKN1C* is silenced by an IC2 imprinting defect. Approximately 15% have an IC1 imprinting defect leading to overexpression of *IGF2* gene and silencing of the adjacent *H19*. The latter patients have an increased risk for Wilms tumors. Most often the imprinting defect occurs without any underlying DNA mutation. However, in a few patients the imprinting defect is the result of an IC1 or IC2 mutation.

IC2 is located within the *KCNQ1* gene and includes the promoter of the non-protein coding *LIT1/KCNQ1OT1* transcript, which is transcribed from the paternal allele only. The promoter is unmethylated on the paternal allele and methylated on the maternal allele. It is unclear how IC2 controls imprinted expression of *CDKN1C*. So far, only one familial IC2 mutation has been described (Niemitz *et al.*, 2004; Fig. 8.7). In this family, maternal transmission

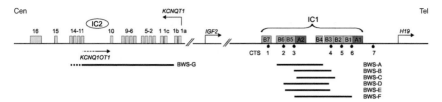

Figure 8.7. Imprinting center (IC) deletions in 11p15. The extent of the deletions is shown as horizontal bars. Families BWS-A, BWS-B, BWS-C, BWS-D, and BWS-E are described in Sparago et al. (2007), family BWS-F in Prawitt et al. (2005), and family BWS-G in Niemitz et al. (2004). Orientation of transcription is shown by horizontal bars. CTS, CTCF target site. Exons 1–16 of the *KCNQT1* gene are indicated by light grey boxes. The A- and B-repeat of IC1 are indicated by dark grey boxes.

of an IC2 deletion led to inactivation of the maternal *CDKN1C* gene and BWS in three children. Paternal transmission of the deletion was without any phenotypic effect. The transmission and gene expression pattern suggests that the deletion includes an enhancer for *CDKN1C*.

IC1 is located between *IGF2* and *H19* and has a bipartite structure. Each part consists of one A-repeat and several B-repeats (Fig. 8.7). In total, there are seven target sites for the CCCTC-binding factor (CTCF), which is a multifunctional protein. As a consequence of genomic imprinting, the CTCF targets sites (CTS) on the paternal chromosome (as well as the *H19* promoter) are methylated, while they are unmethylated on the maternal chromosome. Binding of CTCF to the unmethylated IC1 on the maternal chromosome isolates the *IGF2* gene from two enhancers, which it shares with *H19*. As a consequence, *IGF2* is silent, whereas *H19* is active. CTCF can-not bind to the methylated IC1 on the paternal chromosome. Therefore, the paternal *IGF2* allele is active, whereas the paternal *H19* allele is inactive. For more details, see the chapter by Bartolomei, this volume.

Aberrant methylation of the maternal CTCF-binding sites and *H19* promoter leads to activation of the maternal *IGF2* allele, two doses of IGF2, and BWS. Loss of methylation on the paternal chromosome allows binding of CTCF to the IC. As a consequence, the paternal *IGF2* allele is isolated from the enhancers and silenced. Lack of *IGF2* expression leads SRS.

To date, six IC1 mutations in BWS have been described (Prawitt et al., 2005; Sparago et al., 2004, 2007). The structure of IC1 and the extent of the deletions are shown in Fig. 8.7. Five deletions of 1.4–1.8 kb disrupt the bipartite structure of the IC, whereas one deletion of 2.2 kb retains the three CTS cluster organization. Maternal transmission of the first five deletions (BWS A–F) leads to hypermethylation of the remaining CTSs and biallelic expression of *IGF2* (Fig. 8.6). The finding of mosaic hypermethylation in the patients suggests that abnormal methylation occurs after fertilization. Most likely, the deletions impair protection of the IC against *de novo* methylation during early embryogenesis or in somatic cell lineages. Paternal transmission of the deletions is without any phenotypic effect. Maternal transmission of the 2.2-kb deletion, which retains the CTS cluster organization, does not lead to hypermethylation of the remaining CTSs and BWS. The patient in this family has an additional genetic event (duplication).

IV. CONCLUSIONS

To date, several imprinted gene clusters and IC controlling these clusters have been identified. They appear to use different mechanisms in establishing or maintaining imprints, but in most cases, the mechanistic details are unknown. In 15q, for example, the AS-SRO appears to establish maternal methylation at the PWS-SRO. Methylation then spreads to the genes upstream of *SNRPN*, although it is unclear how this happens (Fig. 8.8). With the help of transgenic animals, certain sequence elements within 15q IC have been detected (Kantor *et al.*, 2004b); the *trans*-acting factors binding to these elements remain to be identified.

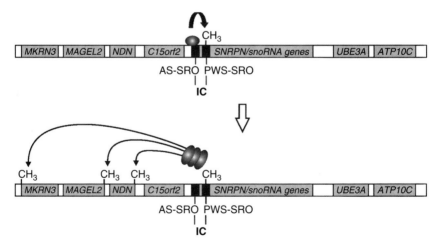

Figure 8.8. Function of the 15q IC. In the maternal germ line, the AS-SRO establishes methylation at the PWS-SRO. Methylation then spreads to the genes upstream of *SNRPN*. PWS-SRO methylation silences the *IC-SNURF-SNRPN* sense/*UBE3A* antisense transcript and indirectly activates *UBE3A*. *Trans*-acting factors are indicated by grey spheres.

IC mutations impair the establishment or maintenance of genomic imprints and lead to recognizable syndromes. It is likely that different imprinting clusters share the same *trans*-acting factors, so that a mutation affecting such a factor will probably be early lethal. Recently, mutations in *NALP7* have been detected in families with recurrent hydatiform moles (Murdoch *et al.*, 2006), but the role of NALP7, which is a member of the CATERPILLER protein family involved in inflammation and apoptosis, in the establishment or maintenance of genomic imprints is unclear.

Most imprinting defects occur without any underlying DNA sequence mutation. They are the result of sporadic errors of the enzymatic machinery involved in imprinting erasure, establishment, or maintenance. The error rate is probably affected by exogenous factors. Low levels of folate, for example, may play a role in this regard. It is also possible that assisted reproduction, which involves ovarian hyperstimulation and embryo culture, can have a deleterious effect on the establishment or stability of genomic imprints (for review, see Horsthemke and Ludwig, 2005). The identification of genetic and environmental risk factors for imprinting defects will be a major challenge in future.

References

Barreto, G., Schafer, A., Marhold, J., Stach, D., Swaminathan, S. K., Handa, V., Doderlein, G., Maltry, N., Wu, W., Lyko, F., and Niehrs, C. (2007). Gadd45a promotes epigenetic gene activation by repair-mediated DNA demethylation. *Nature* **445,** 671–675.

Bielinska, B., Blaydes, S. M., Buiting, K., Yang, T., Krajewska-Walasek, M., Horsthemke, B., and Brannan, C. I. (2000). *De novo* deletions of *SNRPN* exon 1 in early human and mouse embryos result in a paternal to maternal imprint switch. *Nat. Genet.* **25,** 74–78.

Bressler, J., Tsai, T. F., Wu, M. Y., Tsai, S. F., Ramirez, M. A., Armstrong, D., and Beaudet, A. L. (2001). The *SNRPN* promoter is not required for genomic imprinting of the Prader-Willi/Angelman domain in mice. *Nat. Genet.* **28,** 232–240.

Buiting, K., Saitoh, S., Gross, S., Dittrich, B., Schwartz, S., Nicholls, R. D., and Horsthemke, B. (1995). Inherited microdeletions in the Angelman and Prader-Willi syndromes define an imprinting centre on human chromosome 15. *Nat. Genet.* **9,** 395–400.

Buiting, K., Dittrich, B., Groß, S., Lich, C., Färber, C., Buchholz, T., Smith, E., Reis, A., Bürger, J., Nöthen, M. M., Barth-Witte, U., Janssen, B., *et al.* (1998). Sporadic imprinting defects in Prader-Willi syndrome and Angelman syndrome: Implications for imprint-switch models, genetic counseling, and prenatal diagnosis. *Am. J. Hum. Genet.* **63,** 170–180.

Buiting, K., Lich, C., Cottrell, S., Barnicoat, A., and Horsthemke, B. (1999). A 5-kb imprinting center deletion in a family with Angelman syndrome reduces the shortest region of deletion overlap to 880 bp. *Hum. Genet.* **105,** 665–666.

Buiting, K., Färber, C., Kroisel, P., Wagner, K., Brueton, L., Robertson, M. E., Lich, C., and Horsthemke, B. (2000). Imprinting centre (IC) deletions in two PWS families: Implications and strategies for diagnostic testing and genetic counselling. *Clin. Genet.* **58,** 284–290.

Buiting, K., Barnicoat, A., Lich, C., Pembrey, M., Malcolmm, S., and Horsthemke, B. (2001). Disruption of the bipartite imprinting center in a family with Angelman syndrome. *Am. J. Hum. Genet.* **68,** 1290–1294.

Buiting, K., Gross, S., Lich, C., Gillessen-Kaesbach, G., El-Maarri, O., and Horsthemke, B. (2003). Epimutations in Prader-Willi and Angelman syndromes: A molecular study of 136 patients with an imprinting defect. *Am. J. Hum. Genet.* **72,** 571–577.

Buiting, K., Nazlican, H., Galetzka, D., Wawrzik, M., Gross, S., and Horsthemke, B. (2007). *C15orf2* and a novel noncoding transcript from the Prader-Willi/Angelman syndrome region show monoallelic expression in fetal brain. *Genomics* **89,** 588–595.

Chamberlain, S. J., and Brannan, C. (2001). The Prader-Willi syndrome imprinting center activates the paternally expressed murine Ube3a antisense transcript but represses paternal *Ube3a*. *Genomics.* **73,** 316–322.

Dittrich, B., Buiting, K., Korn, B., Rickard, S., Buxton, J., Saitoh, S., Nicholls, R. D., Poustka, A., Winterpacht, A., Zabel, B., and Horsthemke, B. (1996). Imprint switching on human chromosome 15 may involve alternative transcripts of the *SNRPN* gene. *Nat. Genet.* **14,** 163–170.

El-Maarri, O., Buiting, K., Peery, E. G., Kroisel, P. M., Balaban, B., Wagner, K., Urman, B., Heyd, J., Lich, C., Brannan, C. I., Walter, J., and Horsthemke, B. (2001). Maternal methylation imprints on human chromosome 15 are established during or after fertilization. *Nat. Genet.* **27,** 341–344.

Engel, E. (1980). A new genetic concept: Uniparental disomy and its potential effect, isodisomy. *Am. J. Med. Genet.* **6,** 137–143.

Färber, C., Gross, S., Neesen, J., Buiting, K., and Horsthemke, B. (2000). Identification of a testis-specific gene (*C15orf2*) in the Prader-Willi syndrome region on chromosome 15. *Genomics.* **65,** 174–183.

Goll, M. G., and Bestor, T. H. (2005). Eukaryotic cytosine methyltransferases. *Annu. Rev. Biochem.* **74,** 481–514.

Gillessen-Kaesbach, G., Demuth, S., Thiele, H., Theile, U., Lich, C., and Horsthemke, B. (1999). A previously unrecognised phenotype characterised by obesity, muscular hypotonia, and ability to speak in patients with Angelman syndrome caused by an imprinting defect. *Eur. J. Hum. Genet.* **7,** 638–644.

Horsthemke, B., and Ludwig, M. (2005). Assisted reproduction—the epigenetic perspective. *Hum. Reprod. Update.* **11,** 473–482.

Kantor, B., Kaufman, Y., Makedonski, K., Razin, A., and Shemer, R. (2004a). Establishing the epigenetic status of the Prader-Willi/Angelman imprinting center in the gametes and embryo. *Hum. Mol. Genet.* **13,** 2767–2779.

Kantor, B., Makedonski, K., Green-Finberg, Y., Shemer, R., and Razin, A. (2004b). Control elements within the PWS/AS imprinting box and their function in the imprinting process. *Hum. Mol. Genet.* **13,** 751–762.

Kubicek, S., Schotta, G., Lachner, M., Sengupta, R., Kohlmaier, A., Perez-Burgos, L., Linderson, Y., Martens, J. H., O'Sullivan, R. J., Fodor, B. D., Yonezawa, M., Peters, A. H., *et al.* (2006). The role of histone modifications in epigenetic transitions during normal and perturbed development. *Ernst Schering Res. Found. Workshop.* **57,** 1–27.

Landers, M., Bancescu, D. L., Le Meur, E., Rougeulle, C., Glatt-Deeley, H., Brannan, C., Muscatelli, F., and Lalande, M. (2004). Regulation of the large (approximately 1000 kb) imprinted murine *Ube3a* antisense transcript by alternative exons upstream of *Snurf/Snrpn*. *Nucl. Acids Res.* **32,** 3480–3492.

McEntagart, M. E., Webb, T., Hardy, C., and King, M. D. (2000). Familial Prader-Willi syndrome: Case report and a literature review. *Clin. Genet.* **58,** 216–223.

McGrath, J., and Solter, D. (1984). Completion of mouse embryogenesis requires both the maternal and paternal genomes. *Cell.* **37,** 179–183.

Ming, J. E., Blagowidow, N., Knoll, J. H., Rollings, L., Fortina, P., McDonald-Mc-Ginn, D. M., Spinner, N. B., and Zackai, E. H. (2000). Submicroscopic deletion in cousins with Prader-Willi syndrome causes a grandmatrilineal inheritance pattern: Effects of imprinting. *Am. J. Med. Genet.* **92,** 19–24.

Murdoch, S., Djuric, U., Mazhar, B., Seoud, M., Khan, R., Kuick, R., Bagga, R., Kircheisen, R., Ao, A., Ratti, B., Hanash, S., Rouleau, G. A., et al. (2006). Mutations in NALP7 cause recurrent hydatidiform moles and reproductive wastage in humans. Nat. Genet. 38, 300–302.

Nazlican, H., Zeschnigk, M., Claussen, U., Michel, S., Boehringer, S., Gillessen-Kaesbach, G., Buiting, K., and Horsthemke, B. (2004). Somatic mosaicism in patients with Angelman syndrome and an imprinting defect. Hum. Mol. Genet. 13, 2547–2555.

Niemitz, E. L., DeBaun, M. R., Fallon, J., Murakami, K., Kugoh, H., Oshimura, M., and Feinberg, A. P. (2004). Microdeletion of LIT1 in familial Beckwith-Wiedemann syndrome. Am. J. Hum. Genet. 75, 844–849.

Ohta, T., Gray, T. A., Rogan, P. K., Buiting, K., Gabriel, J. M., Saitoh, S., Muralidhar, B., Bilienska, B., Krajewska-Walasek, M., Driscoll, D. J., Horsthemke, B., Butler, M. G., et al. (1999a). Imprinting-mutation mechanisms in Prader-Willi syndrome. Am. J. Hum. Genet. 64, 397–413.

Ohta, T., Buiting, K., Kokkonen, H., McCandless, S., Heeger, S., Leisti, H., Driscoll, D. J., Cassidy, S. B., Horsthemke, B., and Nicholls, R. D. (1999b). Molecular Mechanism of Angelman Syndrome in Two Large Families Involves an Imprinting Mutation. Am J Hum. Genet. 64, 385–396.

Perk, J., Makedonski, K., Lande, L., Cedar, H., Razin, A., and Shemer, R. (2002). The imprinting mechanism of the Prader-Willi/Angelman regional control center. Embo J. 21, 5807–5814.

Prawitt, D., Enklaar, T., Gartner-Rupprecht, B., Spangenberg, C., Oswald, M., Lausch, E., Schmidtke, P., Reutzel, D., Fees, S., Lucito, R., Korzon, M., Brozek, I., et al. (2005). Microdeletion of target sites for insulator protein CTCF in a chromosome 11p15 imprinting center in Beckwith-Wiedemann syndrome and Wilms' tumor. Proc. Natl. Acad. Sci. USA. 102, 4085–4090.

Raca, G., Buiting, K., and Das, S. (2004). Deletion analysis of the imprinting center region in patients with Angelman syndrome and Prader-Willi syndrome. Genetic Testing 8, 387–394.

Reik, W., Dean, W., and Walter, J. (2001). Epigenetic reprogramming in mammalian development. Science 293, 1089–1093.

Rougeulle, C., Cardoso, C., Fontes, M., Colleaux, L., and Lalande, M. (1998). An imprinted antisense RNA overlaps UBE3A and a second maternally expressed transcript. Nat. Genet. 19, 15–16.

Runte, M., Huttenhofer, A., Gross, S., Kiefmann, M., Horsthemke, B., and Buiting, K. (2001). The IC-SNURF-SNRPN transcript serves as a host for multiple small nucleolar RNA species and as an antisense RNA for UBE3A. Hum. Mol. Genet. 10, 2687–2700.

Runte, M., Kroisel, P. M., Gillessen-Kaesbach, G., Varon, R., Horn, D., Cohen, M. Y., Wagstaff, J., Horsthemke, B., and Buiting, K. (2004). SNURF-SNRPN and UBE3A transcript levels in patients with Angelman syndrome. Hum. Genet. 114, 553–561.

Saitoh, S., Buiting, K., Rogan, P. K., Buxton, J. L., Driscoll, D. J., Arnemann, J., König, R., Malcolm, S., Horsthemke, B., and Nicholls, R. D. (1996). Minimal definition of the imprinting center and fixation of a chromosome 15q11-q13 epigenotype by imprinting mutations. Proc. Natl. Acad. Sci. USA. 93, 7811–7815.

Schuffenhauer, S., Buchholz, T., Stengel-Rutkowski, S., Buiting, K., Schmidt, H., and Meitinger, T. (1996). A familial deletion in the Prader-Willi syndrome region including the imprinting control region. Hum. Mutation. 8, 288–292.

Schumacher, A., Buiting, K., Zeschnigk, M., Doerfler, W., and Horsthemke, B. (1998). Methylation analysis of the PWS/AS region does not support an enhancer-competition model. Nat. Genet. 19, 324–325.

Schweizer, J., Zynger, D., and Francke, U. (1999). In vivo nuclease hypersensitivity studies reveal multiple sites of parental origin-dependent differential chromatin conformation in the 150 kb SNRPN transcription unit. Hum. Mol. Genet. 8, 555–566.

Shemer, R., Hershko, A. Y., Perk, J., Mostoslavsky, R., Tsuberi, B., Cedar, H., Buiting, K., and Razin, A. (2000). The imprinting box of the Prader-Willi/Angelman syndrome domain. *Nat. Genet.* **26,** 440–443.

Sparago, A., Cerrato, F., Vernucci, M., Ferrero, G. B., Cirillo Silengo, M., and Riccio, A. (2004). Microdeletions in the human *H19* DMR result in loss of *IGF2* imprinting and Beckwith–Wiedemann syndrome. *Nat. Genet.* **36,** 958–960.

Sparago, A., Russo, S., Cerrato, F., Ferraiuolo, S., Castorina, P., Selicorni, A., Schwienbacher, C., Negrini, M., Ferrero, G. B., Silengo, M. C., Anichini, C., Larizza, L., *et al.* (2007). Mechanisms causing imprinting defects in familial Beckwith-Wiedemann syndrome with Wilms' tumour. *Hum. Mol. Genet.* **16,** 254–264.

Surani, M. A., Barton, S. C., and Norris, M. L. (1984). Development of reconstituted mouse eggs suggests imprinting of the genome during gametogenesis. *Nature* **308,** 548–550.

Sutcliffe, J. S., Nakao, M., Christian, S., Orstavik, K. H., Tommerup, N., Ledbetter, D. H., and Beaudet, A. L. (1994). Deletions of a differentially methylated CpG island at the *SNRPN* gene define a putative imprinting control region. *Nat. Genet.* **8,** 52–58.

Wilkins, J. F., and Haig, D. (2003). What good is genomic imprinting: The function of parent-specific gene expression. *Nat. Rev. Genet.* **45,** 359–368.

Wu, M. Y., Tsai, T. F., and Beaudet, A. L. (2006). Deficiency of *Rbbp1/Arid4a* and *Rbbp1l1/Arid4b* alters epigenetic modifications and suppresses an imprinting defect in the PWS/AS domain. *Genes Dev.* **20,** 2859–2870.

Yang, T., Adamson, T. E., Resnick, J. L., Leff, S., Wevrick, R., Francke, U., Jenkins, N. A., Copeland, N. G., and Brannan, C. I. (1998). A mouse model for Prader-Willi syndrome imprinting-centre mutations. *Nat. Genet.* **19,** 25–31.

Zogel, C., Bohringer, S., Gross, S., Varon, R., Buiting, K., and Horsthemke, B. (2006). Identification of cis- and trans-acting factors possibly modifying the risk of epimutations on chromosome 15. *Eur. J. Hum. Genet.* **14,** 752–758.

9

Epigenetic Gene Regulation in Cancer

Esteban Ballestar and Manel Esteller

Cancer Epigenetics Group, Molecular Pathology Programme,
Spanish National Cancer Centre (CNIO), 28029 Madrid, Spain

I. Introduction
II. Cancer Cells Show A Disruption of DNA Methylation Patterns
III. Disruption of the Histone Modification Profile in Cancer
IV. Cascades of Epigenetic Deregulation in Cancer
V. What Are the Mechanisms That Lead to Aberrant Methylation Patterns in Cancer?
VI. Epigenetic Therapy for Cancer Treatment
Acknowledgments
References

ABSTRACT

The observation that cancer cells suffer profound alterations in the DNA methylation profile, with functional consequences in the activity of key genes, together with the recognition that epigenetic alterations might be as important as genetic defects in the origin of cancers has started a new era in cancer research. In a few years, key discoveries have abruptly changed our vision of the determinants of cancer. Breakthroughs in the cancer epigenetics field include the finding of a tumor-type specificity of genes that suffer epigenetic deregulation at both DNA methylation and histone modifications, the interconnection between different epigenetic marks, the identification of mechanisms of targeting of epigenetic alterations, including the participation of Polycomb group (PcG) proteins, or the involvement of small RNAs, which regulate hundreds of target genes.

Advances in Genetics, Vol. 61
Copyright 2008, Elsevier Inc. All rights reserved.

0065-2660/08 $35.00
DOI: 10.1016/S0065-2660(07)00009-0

All these findings have multiple implications: first, they shed light on the mechanistic insights by which epigenetic defects complement genetic alterations in the development and progression of cancer; second, epigenetic alterations appear to play a prominent role in the initiation of cancer. In addition, because epigenetic changes are reversible, enzymes involved in their maintenance stand as targets for a variety of compounds for therapy. © 2008, Elsevier Inc.

I. INTRODUCTION

In the last decade, we have witnessed a fundamental change in the interpretation of the determinants of cancer. This is mainly due to the accumulation of evidence indicating that epigenetic deregulation of cells contributes and cooperates with genetic alterations in all stages of cancer development and progression. Most of the previous research efforts in the cancer field had been invested under the premise that genetic alterations occur at genes with a key role in different cellular processes including cell proliferation and differentiation, apoptosis, adhesion, etc. However, an increasingly strong line of evidence indicates that epigenetic changes have a significant contribution to the abnormal loss or gain of functions in cancer cells. Both genetic and epigenetic alterations affect the activity of genes. Genetic alterations leave a permanent print in the genome that can affect either the function of the protein, when located in the coding region of a gene, or its expression, when occurring in regulatory regions. In contrast, epigenetic alterations are reversible. Relevant epigenetic changes for cancer occur in regulatory regions of protein-coding genes, resulting in misexpression, or in repetitive sequences, where they have been proposed to be associated with chromosomal alterations. The enzymes responsible for the maintenance of epigenetic patterns are the potential target for a number of compounds opening the possibility of the design of new families of therapeutic agents.

But, what is the exact nature of epigenetic modifications? Epigenetic modifications are generally stable marks, resulting from the covalent modification of chromatin, that determine gene expression in a way that is heritable throughout mitotic divisions without involving a change in the DNA sequence. In contrast to genetic information, which is homogeneous in an organism regardless of the cell type, epigenetic modifications are characteristic of different cell types and, in fact, play a key role in defining the transcriptome, which ultimately determines the identity of each cell type (Fisher, 2002). Basically, cells store their epigenetic information by covalently modifying two different groups of molecules: DNA and histones.

In DNA, methylation of the 5' position of the cytosine in the context of CpG dinucleotides is the most common epigenetic modification (Miller *et al.*, 1974). Careful inspection of the proportion and distribution of CpGs shows that this dinucleotide has a lower frequency than the expected statistical value and a tendency to cluster in regions known as CpG islands (Gardiner-Garden and Frommer, 1987), many of which are coincident with the promoter of protein-coding genes. Most dispersed CpGs in the genome are methylated, unlike in CpG islands, where methylation occurs rarely in normal cells (Aissani and Bernardi, 1991; Gardiner-Garden and Frommer, 1987). Methylation of promoter CpG islands results in transcriptional repression (Keshet *et al.*, 1986), although this mechanism is restricted to a small number of genes, including imprinted genes (Reik *et al.*, 1987), whose expression is determined by the parent who contributed them, X chromosome genes in women (Wolf and Migeon, 1982), and a few tissue-specific genes (Shen and Maniatis, 1980), whose expression is required only for a short period.

On the contrary, histones store epigenetic information through a complex set of posttranslational modifications. Most of these modifications (lysine acetylation, arginine and lysine methylation, or serine phosphorylation, among others) occur at extremely well-conserved amino acid residues located in protruding N-terminal tails. These modifications have both direct effects and indirect architectural effects in a variety of nuclear processes, including gene transcription, DNA repair, or DNA replication. It has been proposed that distinct histone modifications, on one or more tails, act sequentially or in combination to form a "histone code" that is read by other proteins to bring about distinct downstream events (Strahl and Allis, 2000). The "histone code" has only just begun to be understood (Wang *et al.*, 2004). For instance, reversible acetylation of histone lysines is generally associated with transcriptional activation (Chahal *et al.*, 1980), although acetylation of specific lysines can have distinctive effects at different loci (Rundlett *et al.*, 1998). Methylation of histones can occur in lysine and arginine residues and the functional consequences depend on the type of residue and specific site that it modifies (Lachner *et al.*, 2001; Santos-Rosa *et al.*, 2002; Schotta *et al.*, 2004). For example, methylation of H3 at K4 (Santos-Rosa *et al.*, 2002) and R17 (Bauer *et al.*, 2002) is closely linked to transcriptional competence, whereas methylation of H3 at K9 or K27 and H4 at K20 is associated with transcriptional repression (Lachner *et al.*, 2001; Schotta *et al.*, 2004). The chemical groups introduced by different histone modification enzymes are recognized by different nuclear factors containing specialized structural domains, such as chromodomains or bromodomains, that specifically interact with the modified residues and act as true readers of the epigenetic code (de la Cruz *et al.*, 2005).

As mentioned above, besides a direct effect on nuclear processes, such as transcriptional activity, DNA methylation and histone modifications also play a key role in organizing nuclear architecture (see, for instance, Espada *et al.*, 2004; Esteller and Almouzni, 2005; Nakayama *et al.*, 2001), which, in turn, is also involved in regulating transcription and other nuclear processes.

Therefore, epigenetic modifications are essential for defining the cellular transcriptome at several levels. Aberrant changes in the pattern of epigenetic modifications would result in altered nuclear activity, and thereby an altered transcriptome, which would transform the identity of the cell.

II. CANCER CELLS SHOW A DISRUPTION OF DNA METHYLATION PATTERNS

The recognition of a general DNA hypomethylation was the first epigenetic alteration identified in cancer cells (Feinberg and Vogelstein, 1983; Gama-Sosa et al., 1983). At that time, analysis of the role of methylation using tissue-specific genes introduced into mammalian cells had led to a general consensus that DNA methylation results in the formation of nuclease-resistant chromatin and subsequent repression of gene activity (Keshet et al., 1986). Initially, global hypomethylation was thought to be the only significant methylation change in cancer and it was believed that it might lead to the massive overexpression of oncogenes. However, although there is a global decrease of 5-methylcytosine content, hypomethylation mainly occurs in isolated CpGs scattered throughout the genome.

Currently, it is accepted that hypomethylation of repetitive and parasitic DNA sequences correlates with a number of adverse outcomes in cancer. Since the vast majority of methylated CpGs in normal cells reside in fact within repetitive elements, global demethylation contributes to the reactivation of these parasitic sequences by transcription and movement. Also, decreased methylation of repetitive sequences in the satellite DNA of the pericentric region of chromosomes is associated with increased chromosomal rearrangements, a hallmark of cancer (Fig. 9.1). For instance, the finding that DNMT3b mutations, which occur in immunodeficiency, centromeric region instability, and facial anomalies (ICF) syndrome, cause centromeric instability is indicative of how global demethylation destabilizes overall chromatin organization. Furthermore, decreased methylation of proviral sequences can lead to reactivation and increased infectivity. In fact, one primary function of DNA methylation is the suppression of transcription and expansion of parasitic elements such as transposons (e.g., SINEs and LINEs) (Yoder et al., 1997).

Hypomethylation is not the only way in which methylation can contribute to cancer. The idea of the hypermethylation of CpG islands of tumor-suppressor genes as a mechanism of gene inactivation in cancer was proposed in 1994 (Herman et al., 1994), when methylation-dependent silencing of the Von Hippel-Lindau (VHL) gene was demonstrated to be a mechanism of gene inactivation in renal carcinoma. In the following years, parallel studies in the laboratories of Dr. Stephen Baylin and Dr. Peter Jones established that CpG

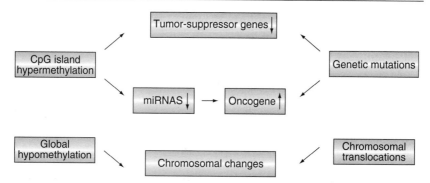

Figure 9.1. Cancer is the result of both genetic and epigenetic deregulation of key genes involved in different cellular processes. Hypermethylation of the promoter CpG island of tumor-suppressor genes results in stable inactivation. This mechanism also inactivates micro-RNAs (miRNAs), which, in turn, regulate oncogenes. Global hypomethylation has been associated with chromosomal instability. Genetic mutations account for both inactivation of tumor-suppressor genes and activation of oncogenes. Chromosomal translocations have a great impact in many different processes.

island hypermethylation is a common mechanism of gene inactivation in cancer. More recently, the systematic analysis of promoter hypermethylation of different tumor types led to the identification of a tumor-type-specific profile of CpG island hypermethylation (Esteller *et al.*, 2001a). The analysis of a few selected hypermethylated CpG islands can be so powerful that these classify tumors of unknown origin (Paz *et al.*, 2003a). Currently, CpG island hypermethylation of tumor-suppressor genes, which leads to their inactivation, is considered to be the major epigenetic alteration in cancer (Esteller, 2002) (Fig. 9.1). It has been proposed that epigenetic inactivation of tumor-suppressor genes by hypermethylation plays a key role by complementing genetic changes in the transformation from normal to malignant cells (Chen *et al.*, 1998; Esteller *et al.*, 2000b). The tumor-type-specific profile of CpG island hypermethylation is reflected, for instance, by the exclusivity of BRCA1 hypermethylation for breast and ovarian neoplasias (Esteller *et al.*, 2001b) or MLH1 hypermethylation for colon, endometrial, and gastric cancer (Esteller *et al.*, 1998, 1999; Fleisher *et al.*, 1999; Herman *et al.*, 1998).

The identification of CpG islands that become methylated in cancer has relied primarily on a candidate-gene approach. For this purpose, tumor-suppressor genes whose mutations have been associated with cancer have provided a useful source of candidates. This type of approach has served to identify many methylated genes in cancer for which key roles in tumorigenesis had previously been demonstrated.

The availability of genomic information since the completion of the human genome-sequencing projects has facilitated the development of new strategies intended to identify novel genes that become methylated in cancer. These genome-wide approaches include the use of microarrays (Suzuki et al., 2002), the use of methylation-sensitive restriction enzymes, and the use of two-dimensional gels (Costello et al., 2002), amplification of intermethylated sites (Paz et al., 2003b), and the combination of chromatin immunoprecipitation (ChIP) with genomic microarrays (ChIP-on-chip) (Ballestar et al., 2003). Recently, a novel technique based on the immunoprecipitation of 5-methylcytosine combined with hybridization in genomic microarrays has become available (Weber et al., 2005). This technique has become a powerful tool to obtain precise maps of methylation in normal and cancer cells.

There are several lines of evidence that imply an active role for hypermethylation of tumor-suppressor genes in the development of cancer. In the first place, hypermethylation is an early event in cancer. This is the case of p16INK4a, p14ARF, and MGMT (Esteller et al., 2000a,b) in colorectal adenomas and hMLH1 in endometrial hyperplasias (Esteller et al., 1999) and gastric adenomas (Fleisher et al., 2001). On the other hand, the comprehensive analysis of methylation in many different tumor types and gene promoters is evidence of the existence of the above-indicated tumor-type-specific profile. In theory, CpG islands should be the most "attractive" substrate for DNA methylation, since, by definition, they contain a high concentration of CpG-rich sequences. It has been speculated that there must be some factors that prevent unscheduled methylation at CpG islands. Many other questions arise: Why do CpG islands become methylated in cancer? Why do certain CpG islands become methylated while others do not? Is aberrant hypermethylation a targeted or a random process?

However, of the most important steps for conferring CpG island hypermethylation as a critical role in the origin and progression of a tumor is the demonstration of biological consequences of the inactivation of that particular gene. Another clue giving CpG island methylation significance is the fact that it should occur in the absence of gene mutations. Both events (genetic and epigenetic) abolish normal gene function and their coincidence in the same allele would be redundant from an evolutionary standpoint. The selective advantage of promoter hypermethylation in this context is provided by multiple examples but three are worth mentioning. First, the cell-cycle inhibitor p16INK4a in one allele of the HCT-116 colorectal cancer cell line has a genetic mutation, while the other is wild type: p16INK4a hypermethylation occurs only on the wild-type allele, while the mutated allele is kept unmethylated (Myohanen et al., 1998). The same selectivity for p16INK4a hypermethylation has also been observed in a bladder transitional cell carcinoma cell line (Yeager et al., 1998). A second example is that of tumor-suppressor gene adenomatous polyposis coli (APC), the gatekeeper of colorectal cancer, which is mutated in the vast majority of colon tumors. When APC methylation occurs in that type, it is clustered in the wild-type APC in 95% of the cases (Esteller *et al.*, 2000c). And finally we have recently demonstrated that in colorectal and breast tumors from families that harbor a germ line mutation in the tumor-suppressor genes hMLH1, BRCA1, or LKB1/STK11, only those tumors that still retain one wild-type allele undergo CpG island hypermethylation (Esteller *et al.*, 2001b). These results put CpG island hypermethylation on a par with gene mutation for accomplishing selective gene inactivation.

III. DISRUPTION OF THE HISTONE MODIFICATION PROFILE IN CANCER

A fundamental breakthrough in our understanding of the mechanisms of eukaryotic gene regulation has been the recognition of the role of chromatin, specifically histone modifications, in modulating gene activity. During the 1980s and early 1990s, researchers focused on the study and identification of transcription factors, which were considered to be major players in gene regulation. In eukaryotic cells, DNA is packaged in chromatin, whose repeating subunit is the nucleosome, a particle characterized by a protein core containing two copies of each of the histones H2A, H2B, H3, and H4. Whereas the central portion of these proteins is directly involved in the assembly of the histone octamer, their protruding N-terminal ends are characterized by the presence of many conserved residues that are target for a variety of posttranslational modifications. As introduced above, histone posttranslational modifications constitute the second group of epigenetic modifications. Nowadays, it seems quite obvious that the physical proximity between the methyl group introduced in the 5′ position of cytosines and the modifiable amino acid residues of histone tails has evolved in association with the existence of machineries that couple both types of modifications. Transcription factors associate with their target sites in the context of chromatin, where the establishment of interactions is modulated by DNA methylation and histone modifications, both of which are interconnected. During the past 10 years, increasing evidence has demonstrated the multiple connections between DNA methylation and histone modifications. These occur through different nuclear machineries, including DNA methyltransferases (DNMTs), that have been reported to recruit both histone deacetylases (Fuks *et al.*, 2001; Robertson *et al.*, 2000) and histone methyltransferases that modify K9 of histone H3 (Fuks *et al.*, 2003) or K27 of histone H3 (Vire *et al.*, 2006).

In addition, there are at least two groups of proteins that both bind methylated DNA and recruit histone-modifying enzymes. These include methyl-CpG-binding domain (MBD) proteins (Ballestar and Wolffe, 2001; Hendrich and Bird, 1998) and the Kaiso family of proteins (Filion *et al.*, 2006; Prokhortchouk *et al.*, 2001). The MBD family of proteins is the best characterized of these methylated DNA-binding proteins. The founding member of these proteins, MeCP2, was discovered in the early 1990s as a polypeptide with the ability to selectively interact with a symmetrically methylated CpG dinucleotide (Lewis *et al.*, 1992). During the next few years, two functional domains were identified: the MBD and the transcriptional repression domain (TRD). The first one was also found in other proteins by database inspection (Hendrich and Bird, 1998). Demonstration that the TRD was responsible for interaction between MeCP2 and histone deacetylase complexes constituted the first mechanistic connection between DNA methylation and transcriptional repression through the modification of chromatin (Jones *et al.*, 1998b; Nan *et al.*, 1998). This connection was later demonstrated for the other members of the MBD family of proteins (Ng *et al.*, 1999; Wade *et al.*, 1999).

The identification of cancer-associated changes in DNA methylation patterns together with the finding of multiple connections between DNA methylation and histone modifications led to the search of alterations in histone modifications in cancer (Fig. 9.2). The histone modification field had become part of the cancer epigenetics field!

Initially, aberrations in posttranslational modifications of histones were only shown to occur in cancer cells at individual promoters. These changes were reported to be associated with the presence of MBD proteins (Ballestar *et al.*, 2003; Magdinier and Wolffe, 2001). In this context, hypermethylation of

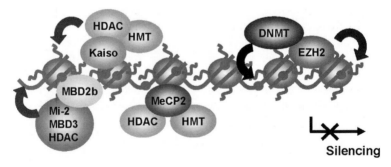

Figure 9.2. Many different nuclear factors participate in the establishment of an epigenetically silenced state in the promoter CpG island of tumor-suppressor genes. Histone octamers are represented by gray circles. DNA is represented on top of the histone octamers and methylated CpG dinucleotides are indicated (as small dark gray circles). Histone tails are lines protruding from octamers. Polycomb group (PcG) proteins, such as EZH2, have been recently proposed to premark genes that are going to become methylated. EZH2 recruits DNMTs to *de novo* methylate DNA. MBDs (MeCP2, MBD2) and Kaiso are recruited to methylated DNA and associate with histone deacetylases (HDAC) and histone methyltransferases (HMT).

the promoter CpG islands of tumor-suppressor genes was thought to be mechanistically linked to gene silencing through the recruitment of these methylated DNA-binding activities that would be followed by a change in the pattern of histone modifications that would lead to a change in the chromatin structure compatible with gene inactivation.

More recently, a global characterization of posttranslational modifications of histone H4 in a comprehensive panel of normal tissues, cancer cell lines, and primary tumors was performed (Fraga et al., 2005). In this study, a loss of monoacetylated and trimethylated forms of histone H4 was found to occur in cancer cells. Interestingly, these changes appear early and accumulate during the tumorigenic process, as shown in a mouse model of multistage skin carcinogenesis (Fraga et al., 2005). By using mass spectrometry, these losses were shown to occur predominantly at the acetylated K16 and trimethylated K20 residues of histone H4 and were associated with the well-characterized hypomethylation of DNA repetitive sequences. These data suggested that the global loss of monoacetylation and trimethylation of histone H4 is a common hallmark of human tumor cells, which parallels with global DNA hypomethylation.

Global analyses of histone modifications have provided markers with potential clinical implications. For instance, it has been shown that changes in global levels of individual histone modifications are predictive of clinical outcome of prostate cancer (Seligson et al., 2005). Through immunohistochemical staining of primary prostatectomy tissue samples, Seligson and colleagues showed two disease subtypes with distinct risks of tumor recurrence in patients with low-grade prostate cancer based on the differential staining for the histone acetylation and dimethylation of five residues in histones H3 and H4. These histone modification patterns have been considered to be predictors of outcome independently of tumor stage, preoperative prostate-specific antigen levels, and capsule invasion.

The identification of changes in the histone modification profile of cancer cells relies on the availability of techniques to study those changes. Mass spectrometry and specific antibodies raised against different histone modifications are nowadays the most powerful tools. Mass spectrometry studies would allow the identification of novel modifications. On the contrary, antibodies allow not only the determination of global changes but also, in the context of ChIP assays, the identification of specific changes at defined sequences. The recent availability of different types of genomic microarrays (tiling arrays, promoter sequences) allows a genome-wide screening of histone modifications as well as the identification of the association of dedicated histone-modifying activities. ChIP-on-chip performed with antibodies against specific histone modifications provides useful information of genes that are epigenetically deregulated (Kondo et al., 2004; Schlesinger et al., 2007) and maps of epigenetic deregulation at a chromosome level.

The map of the profile of histone modifications, or more in general, the epigenome, in cancer cells is a fundamental question to be solved. Different international consortia are now focused on the generation and integration of genome-wide data on epigenetic modifications. The mosaic of epigenetic and genetic alterations compose a sort of jigsaw puzzle that needs to be solved in two steps: first, it is necessary to obtain a complete description of the type and genomic location of both types of alterations; second, hierarchical relationships between different types of alterations have to be identified in order to distinguish the different extent in the impact of different alterations.

IV. CASCADES OF EPIGENETIC DEREGULATION IN CANCER

About 200 genes are reportedly mutated in human breast and colon cancers, with an average of 11 mutations for each tumor (Sjoblom et al., 2006). The vast majority of the genes identified in this study were not known to be genetically altered in tumors and were predicted to affect a wide range of cellular functions, including transcription, adhesion, and invasion, with a wide range of potential impact in contributing to cancer. Similarly, the average number of hypermethylated CpG islands in a particular cancer is a question of great interest. The answer could shed light on the relative contribution of genetic and epigenetic events in cancer development, and the synergy between them. Results obtained using various approaches indicate a range between 100 and 400 hypermethylated promoter CpG islands in a given tumor, although these numbers are likely to change as epigenomic studies are carried out across a wider range of tumor types. Similarly, because of the variable consequences of the genetic alteration of different genes, the functional consequences of the epigenetic silencing of certain genes will be more disadvantageous than that of genes with more restricted or specific functions. The epigenetic inactivation of certain genes has a specific effect in a single activity. For instance, epigenetic inactivation of the cyclin-dependent kinase inhibitor p15INK4b (Herman et al., 1996) or the heparan sulfate synthesis EXT1 (Ropero et al., 2004) genes are quite specific for leukemias and directly affect the expression levels and function of a single gene. However, when the gene that suffers epigenetic inactivation is a transcriptional regulator, such as the secreted frizzled-related protein 1 (SFRP1) (Fukui et al., 2005) or GATA4 (Akiyama et al., 2003), its silencing affects the levels of its target genes and therefore the epigenetic defect results in a cascade of epigenetic amplification.

Novel insights into the cascades of epigenetic deregulation have recently been uncovered. Some of these findings come from the study of microRNAs (miRNAs), an important group of transcriptional regulators that has recently attracted the attention of researchers in the epigenetics field. MiRNAs are short 22-nucleotide-long noncoding RNAs. Currently, there are over 300 miRNAs in the human genome, and each miRNA is predicted to control hundreds of gene targets. In mammals, miRNAs are transcribed by RNA polymerase II to form primary long transcripts, which after processing are transported to the cytoplasm where they are cleaved by the RNase III enzyme Dicer into mature miRNAs. Mature miRNAs then associate with RNA interference effector complex RISC (RNA-induced silencing complex), where miRNA downregulates specific gene products by translational repression via binding to partially complementary sequences in the 3′ untranslated region of the target mRNAs. Recent reports have shown that some miRNAs can be controlled by epigenetic inactivation similar to those occurring in protein-coding genes (Lujambio et al., 2007; Saito et al., 2006). In both reports, pharmacological (Saito et al., 2006) or genetic-dependent (Lujambio et al., 2007) genome-wide demethylation results in epigenetic activation of a number of miRNAs. The identification of specific oncogenic targets that are misregulated as a consequence of epigenetic deregulation of miRNAs shows the complexity and levels of aberrant expression patterns in cancer.

These reports show that one mechanism accounting for the observed downregulation of miRNAs in human cancer is CpG island hypermethylation, in a manner similar to that which is now well accepted for classic tumor-suppressor genes. In fact, patterns of histone modifications associated with aberrantly hypermethylated miRNAs are similar to those found in hypermethylated tumor-suppressor genes (Lujambio et al., 2007). Also these data provide examples of epigenetic silencing of miRNAs and their functional consequences on the activity of specific oncogenes (Fig. 9.1). In addition, the function of these miRNAs can be restored by erasing DNA methylation, in a similar scenario shown for tumor-suppressor genes. Thus, the epigenetic dysregulation of miRNAs in human cancer constitutes an emerging scientific field where cascades of regulatory mechanisms may have significant consequences for cancer patients undergoing treatment with DNA-demethylating drugs (Fig. 9.3).

Cascades of epigenetic deregulation also involve the interconnection between genetic mutations and epigenetic changes. Classic examples are represented, for instance, by chromosomal translocations in leukemia that generate fusion proteins that involve transcription factors and histone-modifying enzymes. Altered behavior of fusion factors result in misregulation of their downstream targets. Another example is represented by epigenetic inactivation of DNA repair genes that have an impact on the stability of the genome. This interconnection only reflects a high complexity in the relationship between different types of defects and the need for extraordinary research efforts to identify the hierarchy between all these events.

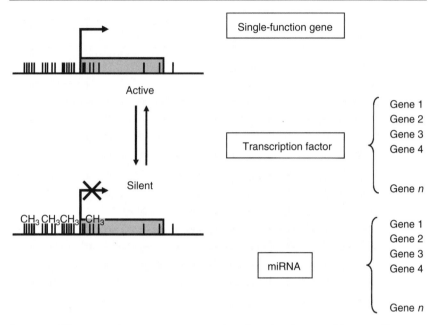

Figure 9.3. The impact of epigenetic inactivation depends on the relevance of the target. Epigenetic inactivation can affect a single-function gene, a transcription factor, or a miRNA. Silencing of both transcription factor and a miRNA can impact several hundreds of genes.

V. WHAT ARE THE MECHANISMS THAT LEAD TO ABERRANT METHYLATION PATTERNS IN CANCER?

One of the fundamental questions to be answered in cancer epigenetics concerns the determinants of tumor-suppressor gene hypermethylation and the mechanisms that define the existence of specific profiles of promoter CpG island hypermethylation in tumor cells (Esteller *et al.*, 2001a; Paz *et al.*, 2003a). Two possible scenarios have been proposed: (a) CpG island hypermethylation occurs in a random manner and the epigenetic inactivation of certain genes confers a selective advantage in a particular tumor type and (b) a set of nuclear factors specifically target hypermethylation to different subsets of genes in a tumor-type-specific fashion.

Evidently, both scenarios are compatible with the observation of the specificity in the profiles of epigenetic alterations, which seems to be the result of the selection of clones with clear advantages in proliferation and dispersion. Several authors have proposed that the mechanisms that prevent CpG islands from methylation are no longer able to maintain this process in cancer cells. However, to date no experimental evidence has been found to support this hypothesis. Alternatively, intense research is currently underway in identifying the mechanisms that determine targeting of methylation of DNA. In this sense, it is important to distinguish between genes that become naturally methylated in normal cells and those that are aberrantly methylated in cancer cells, as the mechanisms that apply for normal cells are probably disrupted in cancer cells.

The first evidence of targeting mechanisms for DNA methylation comes from studies on plant cells, which identified the existence of RNA-based mechanisms (Wassenegger et al., 1994). One link is through RNA-directed DNA methylation, which can involve an RNA made in the nucleus or cytoplasm. A second possible link that remains to be confirmed experimentally is the transcription of aberrant RNAs from methylated DNA templates. These aberrant RNAs have been postulated to trigger RNA turnover in the cytoplasm and methylation of unlinked homologous DNA copies. The ability of RNAs produced in the cytoplasm to feedback and induce epigenetic changes in a sequence-specific manner was most clearly shown in plants, where nuclear transgenes become methylated by an RNA virus engineered with transgene sequences (Jones et al., 1998a). That dsRNAs might be the actual inducers of DNA methylation has been suggested by the ability of viroids, a plant pathogen consisting solely of a noncoding RNA duplex, to trigger methylation of homologous nuclear DNA (Wassenegger et al., 1994). During the past few years, many cardinal insights on the mechanisms that guide DNA methylation in plants through the activity of small RNAs have been uncovered (Chan et al., 2005).

In contrast, the potential role of RNA in directing methylation in mammals is poorly understood, although there is some evidence that antisense RNA molecules may target methylation of regulatory elements (Tufarelli et al., 2003). In a transgenic model of α-thalassemia and in differentiating embryonic stem cells, transcription of antisense RNA mediates silencing and methylation of promoter CpG island of the α-globin gene. Other reports indicate that RNA-mediated methylation might be a mechanism in targeting DNA methylation in mammals; however, the generality of these pathways needs to be determined.

It was also proposed that transcription factors might target methylation. An example is provided by the Myc transcription factor, an essential mediator of cell growth and proliferation through its ability to both positively and negatively regulate transcription. It has been recently shown that Myc binds the corepressor DNMT3a and associates with DNA methyltransferase activity in vivo (Brenner et al., 2005). These authors found that cells with reduced DNMT3a levels exhibit specific reactivation of the Myc-repressed p21Cip1 gene, whereas the expression of Myc-activated E-box genes is unchanged. In addition, selective binding of Myc and targeting of DNMT3a to the promoter of p21Cip1 was observed. Myc is known to be recruited to the p21Cip1 promoter by the DNA-binding factor Miz-1. Consistent with this, Brenner and colleagues observed that Myc and DNMT3a form a ternary complex with Miz-1 and that this complex can corepress the p21Cip1 promoter in a DNA methylation-dependent manner. These findings were among the first suggestions that targeting of DNA methyltransferases by transcription factors could be a wide and general mechanism for the generation of specific DNA methylation patterns within a cell.

Recent data implicate the Polycomb group (PcG) proteins in the targeting of epigenetic inactivation. The PcG proteins are a class of epigenetic regulators that function as transcriptional repressors. This family of proteins plays a key role in development. Specifically, stem cells rely on PcG proteins to reversibly repress genes encoding transcription factors required for differentiation. Recently, Vire et al. (2006) have shown that the PcG protein EZH2 (Enhancer of Zeste homolog 2) interacts—within the context of the Polycomb repressive complexes 2 and 3 (PRC2/3)—with DNMTs and associates with DNMT activity in vivo. EZH2 is required for DNA methylation of EZH2-target promoters (Vire et al., 2006). Several pieces of evidence indicate that PcG target genes in stem cells are significantly more likely to have cancer-specific promoter DNA hypermethylation than nontargets (Widschwendter et al., 2007). This PcG-dependent recruitment of DNMTs and subsequent de novo methylation occurs only in cancer cells, in opposition to normal cells (Schlesinger et al., 2007) These data are relevant not only because they shed some light on the genes that are potential targets for aberrant hypermethylation but also because they support a stem cell origin of cancer (Widschwendter et al., 2007).

Other lines of evidence suggest that at least in certain contexts, target-driven mechanisms account for epigenetic alterations. Acute leukemia, for instance, has been a rich source of information about the targeted disruption of histone modifications in cancer. Leukemias are well characterized for nonrandom chromosomal translocations disrupting genes residing in the breakpoint region of translocation (see, for instance, Pui and Evans, 2004). Genes residing at these breakpoint regions are often master regulators of hematopoietic cell differentiation, apoptosis, or proliferation. Many of the chromosomal translocations associated with acute leukemia disrupt genes that either encode histone-modifying factors or express transcription factors that recruit histone-modifying enzymes. Although many of these enzymes show remarkable specificity, histone-modifying enzymes can also modify nonhistone proteins, including other transcription factors, making it difficult to pinpoint the action of histone-modifying enzymes in leukemogenesis. Nevertheless, these chromosomal translocations indicate how disrupted enzyme functions that control chromatin structure can cause alterations of the histone modification profile in a target-specific fashion, resulting in an altered chromatin structure that has an impact on gene expression at specific loci, eventually causing cellular transformation. A typical example is mixed-lineage leukemia (MLL), a histone H3 K4-specific methyltransferase that is a positive regulator of Hox expression. MLL rearrangements and amplification are common in acute lymphoid and myeloid leukemias and myelodysplastic disorders and are associated with abnormal upregulation of Hox gene expression (Hess, 2004). At any rate, these chromosomal translocations indicate how disruptions of the function of the enzymes that control chromatin structure can cause alterations of the histone pattern and chromatin structure to alter gene expression at specific loci, eventually causing cellular transformation. It is likely that both general deregulation of the physiological processes that direct DNA methyltransferases to their normal targets and direct mistargeting of histone-modifying enzymes—through chromosomal rearrangements—take place in cancer cells.

VI. EPIGENETIC THERAPY FOR CANCER TREATMENT

The evidence that epigenetic alterations play a prominent role in cancer development together with their potential reversibility opens up a number of possibilities in the design of specific drugs for cancer therapy. Several enzymes of the epigenetic machinery constitute attractive targets for this approach. However, only two types of epigenetic drugs, neither of which is very specific, have been demonstrated to be clinically relevant: DNA-demethylating agents and histone deacetylase inhibitors (HDACis) (Esteller, 2005; Villar-Garea and Esteller, 2004). Novel epigenetic compounds with potential interest in a clinical context as therapeutic drugs include histone acetyltransferase inhibitors, such as anacardic acid, curcumin, and peptide CoA conjugates. In addition, histone methyltransferase inhibitors (Greiner et al., 2005) or those HDACis that are specific for SIRT1 (class III HDAC), such as nicotinamide and splitomycin, are now under intense analyses.

Of the class of DNA-demethylating agents, the first drug used to inhibit DNA methylation was 5-azacytidine (Vidaza). This substance causes covalent arrest of DNMTs, resulting in cytotoxicity. 5-Azacytidine was tested for its usefulness as an antileukemic drug before its demethylating activity was known (Esteller, 2005). The analogue 5-aza-2'-deoxycytidine (Decitabine) is one of the most commonly used demethylating drugs in assays with cultured cells. Zebularine is another recently developed cytidine analogue (Yoo et al., 2004). It forms a covalent complex with DNA methyltransferases (Yoo et al., 2004). Furthermore, zebularine has also shown promising antitumoral effects in xenografts (Yoo et al., 2004) and thymic lymphomas (Herranz et al., 2005) in mice. Perhaps the most interesting feature of this DNA-demethylating agent is that it is chemically stable and of low toxicity (Herranz et al., 2005; Yoo et al., 2004), and can be taken orally. It is in the field of hematological malignancies that DNA-demethylating agents have had their greatest success so far, especially in high-risk myelodysplastic syndrome using 5-aza-2-deoxycytidine (Esteller, 2005b). In 2004, the US Food and Drug Administration approved the use of 5-azacytidine (Vidaza) for the treatment of all myelodysplastic syndrome subtypes.

On the contrary, naturally occurring and synthetic HDACis are also the focus of interest because of their great potential utility against cancer. Overall, HDACis manifest a wide range of activities against all HDACs. These compounds can be classified into the following groups according to their chemical nature: hydroxamic acids, such as trichostatin A, suberoylanilide hydroxamic acid (SAHA), PXD101, and NVP-LAQ-824; carboxylic acids, such as sodium valproate and butyrate; benzamides, such as MS-272; and others, including trapoxins and FK228 (Villar-Garea and Esteller, 2004). It is believed that the anticancer effects of HDACis are mediated by the reactivation of the expression of tumor-suppressor genes. However, the treatment of cancer cell lines with HDACis has pleiotropic effects inducing differentiation, cell-cycle arrest, and apoptosis. In this regard, the observation that cancer cells have lost monoacetylated K16 histone H4 (Fraga et al., 2005) implies a new molecular pathway that may explain the beneficial effects of HDAC inhibitors because these compounds may promote the restoration of normal histone H4 acetylation levels in the whole cell, restoring the normal chromatin status of repetitive DNA sequences (Fraga et al., 2005). It is clear from in vitro preclinical studies and ongoing clinical trials that HDACis have enormous potential as anticancer drugs. In this regard, SAHA may soon be approved for the treatment of cutaneous lymphoma.

The success of compounds that target epigenetic enzymes as a successful source of therapy depends on the identification of molecules that are specific enough to revert epigenetic alterations at specific genomic sites. Detailed information on the type, specific genomic location, and mechanisms of epigenetic alteration, as well as the timing by which epigenetic events interact or cooperate with genetic mutations will surely contribute to design of more powerful and specific drugs.

Acknowledgments

The authors declare no conflicts of interest. This work has been supported by grants BFU2004–02073/BMC and CSD2006–49 from the Spanish Ministry of Education and Science (MEC).

References

Aissani, B., and Bernardi, G. (1991). CpG islands: Features and distribution in the genomes of vertebrates. Gene **106**, 173–183.

Akiyama, Y., Watkins, N., Suzuki, H., Jair, K. W., van Engeland, M., Esteller, M., Sakai, H., Ren, C. Y., Yuasa, Y., Herman, J. G., and Baylin, S. B. (2003). GATA-4 and GATA-5 transcription factor genes and potential downstream antitumor target genes are epigenetically silenced in colorectal and gastric cancer. Mol. Cell Biol. **23**, 8429–8439.

Allfrey, V. G. (1966). Structural modifications of histones and their possible role in the regulation of ribonucleic acid synthesis. *Proc. Can. Cancer Conf.* **6**, 313–335.

Ballestar, E., and Wolffe, A. P. (2001). Methyl-CpG-binding proteins: Targeting specific gene repression. *Eur. J. Biochem.* **268**, 1–6.

Ballestar, E., Paz, M. F., Valle, L., Wei, S., Fraga, M. F., Espada, J., Cigudosa, J. C., Huang, T. H.-M., and Esteller, M. (2003). Methyl-CpG binding proteins identify novel sites of epigenetic inactivation in human cancer. *EMBO J.* **22**, 6335–6345.

Bauer, U. M., Daujat, S., Nielsen, S. J., Nightingale, K., and Kouzarides, T. (2002). Methylation at arginine 17 of histone H3 is linked to gene activation. *EMBO Rep.* **3**, 39–44.

Brenner, C., Deplus, R., Didelot, C., Loriot, A., Viré, E., De Smet, C., Gutierrez, A., Danovi, D., Bernard, D., Boon, T., Pelicci, P. G., Amati, B., *et al.* (2005). Myc represses transcription through recruitment of DNA methyltransferase corepressor. *EMBO J.* **24**, 336–346.

Chahal, S. S., Matthews, H. R., and Bradbury, E. M. (1980). Acetylation of histone H4 and its role in chromatin structure and function. *Nature* **287**, 76–79.

Chan, S. W., Henderson, I. R., and Jacobsen, S. E. (2005). Gardening the genome: DNA methylation in Arabidopsis thaliana. *Nat. Rev. Genet.* **6**, 351–360.

Chen, R. Z., Pettersson, U., Beard, C., Jackson-Grusby, L., and Jaenisch, R. (1998). DNA hypomethylation leads to elevated mutation rates. *Nature* **395**, 89–93.

Costello, J. F., Smiraglia, D. J., and Plass, C. (2002). Restriction landmark genome scanning. *Methods* **27**, 144–149.

de la Cruz, X., Lois, S., Sanchez-Molina, S., and Martinez-Balbas, M. A. (2005). Do protein motifs read the histone code? *Bioessays* **27**, 164–175.

Espada, J., Ballestar, E., Fraga, M. F., Villar-Garea, A., Juarranz, A., Stockert, J. C., Robertson, K. D., Fuks, F., and Esteller, M. (2004). Human DNMT1 is essential to maintain the histone H3 modification pattern. *J. Biol. Chem.* **279**, 37175–37184.

Esteller, M. (2002). CpG island hypermethylation and tumor suppressor genes: A booming present, a brighter future. *Oncogene* **21**, 5427–5440.

Esteller, M. (2005). DNA methylation and cancer therapy: New developments and expectations. *Curr. Opin. Oncol.* **17**, 55–60.

Esteller, M., and Almouzni, G. (2005). How epigenetics integrates nuclear functions. *EMBO Rep.* **6**, 624–628.

Esteller, M., Levine, R., Baylin, S. B., Ellenson, L. H., and Herman, J. G. (1998). MLH1 promoter hypermethylation is associated with the microsatellite instability phenotype in sporadic endometrial carcinomas. *Oncogene* **17**, 2413–2417.

Esteller, M., Catasus, L., Matias-Guiu, X., Mutter, G., Baylin, S. B., Prat, J., and Herman, J. G. (1999). hMLH1 Promoter hypermethylation is an early event in human endometrial tumorigenesis. *Am. J. Pathol.* **155**, 1767–1772.

Esteller, M., Tortola, S., Toyota, M., Capella, G., Peinado, M. A., Baylin, S. B., and Herman, J. G. (2000a). Hypermethylation-associated inactivation of $p14^{ARF}$ is independent of $p16^{INK4a}$ methylation and $p53$ mutational status. *Cancer Res.* **60**, 129–133.

Esteller, M., Toyota, M., Sanchez-Cespedes, M., Capella, G., Peinado, M. A., Watkins, D. N., Issa, J. P., Sidransky, D., Baylin, S. B., and Herman, J. G. (2000b). Inactivation of the DNA repair gene O^6-*methylguanine-DNA methyltransferase* by promoter hypermethylation is associated with G to A mutations in *K-ras* in colorectal tumorigenesis. *Cancer Res.* **60**, 2368–2371.

Esteller, M., Sparks, A., Toyota, M., Sanchez-Cespedes, M., Capella, G., Peinado, M. A., Gonzalez, S., Tarafa, G., Sidransky, D., and Meltzer, S. J. (2000c). Analysis of adenomatous polyposis coli promoter hypermethylation in human cancer. *Cancer Res.* **60**, 4366–4371.

Esteller, M., Corn, P. G., Baylin, S. B., and Herman, J. G. (2001a). A gene hypermethylation profile of human cancer. *Cancer Res.* **61**, 3225–3229.

Esteller, M., Fraga, M. F., Guo, M., Garcia-Foncillas, J., Hedelfank, I., Godwin, A. K., Trojan, J., Vaurs-Barrière, C., Bignon, Y.-J., Ramus, S., Benitez, J., Akiyama, Y., et al. (2001b). DNA methylation patterns in hereditary human cancers mimic sporadic tumorigenesis. *Hum. Mol. Genet.* **10,** 3001–3007.

Feinberg, A. P., and Vogelstein, B. (1983). Hypomethylation distinguishes genes of some human cancers from their normal counterparts. *Nature* **301,** 89–92.

Filion, G. J., Zhenilo, S., Salozhin, S., Yamada, D., Prokhortchouk, E., and Defossez, P. A. (2006). A family of human zinc finger proteins that bind methylated DNA and repress transcription. *Mol. Cell Biol.* **26,** 169–181.

Fisher, A. G. (2002). Cellular identity and lineage choice. *Nat. Rev. Immunol.* **2,** 977–982.

Fleisher, A. S., Esteller, M., Wang, S., Tamura, G., Suzuki, H., Yin, J., Zou, T. T., Abraham, J. M., Kong, D., Smolinski, K. N., Shi, Y. Q., Rhyu, M. G., et al. (1999). Hypermethylation of the hMLH1 gene promoter in human gastric cancers with microsatellite instability. *Cancer Res.* **59,** 1090–1095.

Fleisher, A. S., Esteller, M., Tamura, G., Rashid, A., Stine, O. C., Yin, J., Zou, T. T., Abraham, J. M., Kong, D., Nishizuka, S., James, S. P., Wilson, K. T., et al. (2001). Hypermethylation of the hMLH1 gene promoter is associated with microsatellite instability in early human gastric neoplasia. *Oncogene* **20,** 329–335.

Fraga, M. F., Ballestar, E., Villar-Garea, A., Boix-Chornet, M., Espada, J., Schotta, G., Bonaldi, T., Haydon, C., Petrie, K., Ropero, S., Perez-Rosado, A., Calvo, E., et al. (2005). Loss of acetylated lysine 16 and trimethylated lysine 20 of histone H4 is a common hallmark of human cancer. *Nat. Genet.* **37,** 391–400.

Fuks, F., Burgers, W. A., Godin, N., Kasai, M., and Kouzarides, T. (2001). Dnmt3a binds deacetylases and is recruited by a sequence-specific repressor to silence transcription. *EMBO J.* **20,** 2536–2544.

Fuks, F., Hurd, P. J., Deplus, R., and Kouzarides, T. (2003). The DNA methyltransferases associate with HP1 and the SUV39H1 histone methyltransferase. *Nucleic Acids Res.* **31,** 2305–2312.

Fukui, T., Kondo, M., Ito, G., Maeda, O., Sato, N., Yoshioka, H., Yokoi, K., Ueda, Y., Shimokata, K., and Sekido, Y. (2005). Transcriptional silencing of secreted frizzled related protein 1 (SFRP 1) by promoter hypermethylation in non-small-cell lung cancer. *Oncogene* **24,** 6323–6327.

Gama-Sosa, M. A., Wang, R. Y., Kuo, K. C., Gehrke, C. W., and Ehrlich, M. (1983). The 5-methylcytosine content of DNA from human tumors. *Nucleic Acids Res.* **11,** 6883–6894.

Gardiner-Garden, M., and Frommer, M. (1987). CpG islands in vertebrate genomes. *J. Mol. Biol.* **196,** 261–282.

Greiner, D., Bonaldi, T., Eskeland, R., Roemer, E., and Imhof, A. (2005). Identification of a specific inhibitor of the histone methyltransferase SU (VAR) 3–9. *Nat. Chem. Biol.* **1,** 143–145.

Hendrich, B., and Bird, A. (1998). Identification and characterization of a family of mammalian methyl-CpG binding proteins. *Mol. Cell. Biol.* **18,** 6538–6547.

Herman, J. G., Latif, F., Weng, Y., Lerman, M. I., Zbar, B., Liu, S., Samid, D., Duan, D. S., Gnarra, J. R., Linehan, W. M., and Baylin, S. B. (1994). Silencing of the VHL tumorsuppressor gene by DNA methylation in renal carcinoma. *Proc. Natl. Acad. Sci. USA* **91,** 9700–9704.

Herman, J. G., Jen, J., Merlo, A., and Baylin, S. B. (1996). Hypermethylation-associated inactivation indicates a tumor suppressor role for p15INK4B. *Cancer Res.* **56,** 722–727.

Herman, J. G., Umar, A., Polyak, K., Graff, J. R., Ahuja, N., Issa, J. P., Markowitz, S., Willson, J. K., Hamilton, S. R., Kinzler, K. W., Kane, M. F., Kolodner, R. D., et al. (1998). Incidence and functional consequences of hMLH1 promoter hypermethylation in colorectal carcinoma. *Proc. Natl. Acad. Sci. USA* **95,** 6870–6875.

Herranz, M., Martín-Caballero, J., Fraga, M. F., Ruiz-Cabello, J., Flores, J. M., Desco, M., Marquez, V., and Esteller, M. (2005). The novel DNA methylation inhibitor zebularine is effective against the development of murine T-cell lymphoma. *Blood* **107,** 1174–1177.

Hess, J. L. (2004). MLL: A histone methyltransferase disrupted in leukemia. *Trends Mol. Med.* **10**, 500–507.

Jones, A. L., Thomas, C. L., and Maule, A. J. (1998a). *De novo* methylation and cosuppression induced by a cytoplasmically replicating plant RNA virus. *EMBO J.* **17**, 6385–6393.

Jones, P. L., Veenstra, G. J., Wade, P. A., Vermaak, D., Kass, S. U., Landsberger, N., Strouboulis, J., and Wolffe, A. P. (1998b). Methylated DNA and MeCP2 recruit histone deacetylase to repress transcription. *Nat. Genet.* **19**, 187–191.

Keshet, I., Lieman-Hurwitz, J., and Cedar, H. (1986). DNA methylation affects the formation of active chromatin. *Cell* **44**, 535–543.

Kondo, Y., Shen, L., Yan, P. S., Huang, T. H., and Issa, J. P. (2004). Chromatin immunoprecipitation microarrays for identification of genes silenced by histone H3 lysine 9 methylation. *Proc. Natl. Acad. Sci. USA* **101**, 7398–7403.

Lachner, M., O'Carroll, D., Rea, S., Mechtler, K., and Jenuwein, T. (2001). Methylation of histone H3 lysine 9 creates a binding site for HP1 proteins. *Nature* **410**, 116–120.

Lewis, J. D., Meehan, R. R., Henzel, W. J., Maurer-Fogy, I., Jeppesen, P., Klein, F., and Bird, A. (1992). Purification, sequence, and cellular localization of a novel chromosomal protein that binds to methylated DNA. *Cell* **69**, 905–914.

Lujambio, A., Ropero, S., Ballestar, E., Fraga, M. F., Cerrato, C., Setien, F., Casado, S., Suarez-Gauthier, A., Sanchez-Cespedes, M., Gitt, A., Spiteri, I., Das, P. P., *et al.* (2007). Genetic unmasking of an epigenetically silenced microRNA in human cancer cells. *Cancer Res.* **67**, 1424–1429.

Magdinier, F., and Wolffe, A. P. (2001). Selective association of the methyl-CpG binding protein MBD2 with the silent p14/p16 locus in human neoplasia. *Proc. Natl. Acad. Sci. USA* **98**, 4990–4995.

Miller, O. J., Schnedl, W., Allen, J., and Erlanger, B. F. (1974). 5-Methylcytosine localised in mammalian constitutive heterochromatin. *Nature* **251**, 636–637.

Myohanen, S. K., Baylin, S. B., and Herman, J. G. (1998). Hypermethylation can selectively silence individual P16ink4A alleles in neoplasia. *Cancer Res.* **58**, 591–593.

Nakayama, J., Rice, J. C., Strahl, B. D., Allis, C. D., and Grewal, S. I. (2001). Role of histone H3 lysine 9 methylation in epigenetic control of heterochromatin assembly. *Science* **292**, 110–113.

Nan, X., Ng, H. H., Johnson, C. A., Laherty, C. D., Turner, B. M., Eisenman, R. N., and Bird, A. (1998). Transcriptional repression by the methyl-CpG-binding protein MeCP2 involves a histone deacetylase complex. *Nature* **393**, 386–389.

Ng, H. H., Zhang, Y., Hendrich, B., Johnson, C. A., Turner, B. M., Erdjument-Bromage, H., Tempst, P., Reinberg, D., and Bird, A. (1999). MBD2 is a transcriptional repressor belonging to the MeCP1 histone deacetylase complex. *Nat. Genet.* **23**, 58–61.

Paz, M. F., Fraga, M. F., Avila, S., Guo, M., Pollan, M., Herman, J. G., and Esteller, M. (2003a). A systematic profile of DNA methylation in human cancer cell lines. *Cancer Res.* **63**, 1114–1121.

Paz, M. F., Wei, S., Cigudosa, J. C., Rodriguez-Perales, S., Peinado, M. A., Huang, T. H., and Esteller, M. (2003b). Genetic unmasking of epigenetically silenced tumor suppressor genes in colon cancer cells deficient in DNA methyltransferases. *Hum. Mol. Genet.* **12**, 2209–2219.

Prokhortchouk, A., Hendrich, B., Jorgensen, H., Ruzov, A., Wilm, M., Georgiev, G., Bird, A., and Prokhortchouk, E. (2001). The p120 catenin partner Kaiso is a DNA methylation-dependent transcriptional repressor. *Genes Dev.* **15**, 1613–1618.

Pui, C. H., and Evans, W. E. (2004). Treatment of acute lymphoblastic leukemia. *N. Engl. J. Med.* **354**, 166–178.

Reik, W., Collick, A., Norris, M. L., Barton, S. C., and Surani, M. A. (1987). Genomic imprinting determines methylation of parental alleles in transgenic mice. *Nature* **328**, 248–251.

Robertson, K. D., Ait-Si-Ali, S., Yokochi, T., Wade, P. A., Jones, P. L., and Wolffe, A. P. (2000). DNMT1 forms a complex with Rb, E2F1 and HDAC1 and represses transcription from E2F-responsive promoters. *Nat. Genet.* **25**, 338–342.

Ropero, S., Setien, F., Espada, J., Fraga, M. F., Herranz, M., Asp, J., Benassi, M. S., Franchi, A., Patino, A., Ward, L. S., Bovee, J., Cigudosa, J. C., et al. (2004). Epigenetic loss of the familial tumor-suppressor gene exostosin-1 (EXT1) disrupts heparan sulfate synthesis in cancer cells. Hum. Mol. Genet. 13, 2753–2765.

Rundlett, S. E., Carmen, A. A., Suka, N., Turner, B. M., and Grunstein, M. (1998). Transcriptional repression by UME6 involves deacetylation of lysine 5 of histone H4 by RPD3. Nature 392, 831–835.

Saito, Y., Liang, G., Egger, G., Friedman, J. M., Chuang, J. C., Coetzee, G. A., and Jones, P. A. (2006). Specific activation of microRNA-127 with downregulation of the protooncogene BCL6 by chromatin-modifying drugs in human cancer cells. Cancer Cell 9, 435–443.

Santos-Rosa, H., Schneider, R., Bannister, A. J., Sherriff, J., Bernstein, B. E., Emre, N. C., Schreiber, S. L., Mellor, J., and Kouzarides, T. (2002). Active genes are trimethylated at K4 of histone H3. Nature 419, 407–411.

Schlesinger, Y., Straussman, R., Keshet, I., Farkash, S., Hecht, M., Zimmerman, J., Eden, E., Yakhini, Z., Ben-Shushan, E., Reubinoff, B. E., Bergman, Y., Simon, I., et al. (2007). Polycomb-mediated methylation on Lys27 of histone H3 pre-marks genes for de novo methylation in cancer. Nat. Genet. 39, 232–236.

Schotta, G., Lachner, M., Sarma, K., Ebert, A., Sengupta, R., Reuter, G., Reinberg, D., and Jenuwein, T. (2004). A silencing pathway to induce H3-K9 and H4-K20 trimethylation at constitutive heterochromatin. Genes Dev. 18, 1251–1262.

Seligson, D. B., Horvath, S., Shi, T., Yu, H., Tze, S., Grunstein, M., and Kurdistani, S. K. (2005). Global histone modification patterns predict risk of prostate cancer recurrence. Nature 435, 1262–1266.

Shen, C. K., and Maniatis, T. (1980). Tissue-specific DNA methylation in a cluster of rabbit beta-like globin genes. Proc. Natl. Acad. Sci. USA 77, 6634–6638.

Sjoblom, T., Jones, S., Wood, L. D., Parsons, D. W., Lin, J., Barber, T. D., Mandelker, D., Leary, R. J., Ptak, J., Silliman, N., Szabo, S., Buckhaults, P., et al. (2006). The consensus coding sequences of human breast and colorectal cancers. Science 314, 268–274.

Strahl, B. D., and Allis, C. D. (2000). The language of covalent histone modifications. Nature 403, 41–45.

Suzuki, H., Gabrielson, E., Chen, W., Anbazhagan, R., van Engeland, M., Weijenberg, M. P., Herman, J. G., and Baylin, S. B. (2002). A genomic screen for genes upregulated by demethylation and histone deacetylase inhibition in human colorectal cancer. Nat. Genet. 31, 141–149.

Tufarelli, C., Stanley, J. A., Garrick, D., Sharpe, J. A., Ayyub, H., Wood, W. G., and Higgs, D. R. (2003). Transcription of antisense RNA leading to gene silencing and methylation as a novel cause of human genetic disease. Nat. Genet. 34, 157–165.

Villar-Garea, A., and Esteller, M. (2004). Histone deacetylase inhibitors: Understanding a new wave of anticancer agents. Int. J. Cancer 112, 171–178.

Vire, E., Brenner, C., Deplus, R., Blanchon, L., Fraga, M., Didelot, C., Morey, L., Van Eynde, A., Bernard, D., Vanderwinden, J. M., Bollen, M., Esteller, M., et al. (2006). The Polycomb group protein EZH2 directly controls DNA methylation. Nature 439, 871–874.

Wade, P. A., Gegonne, A., Jones, P. L., Ballestar, E., Aubry, F., and Wolffe, A. P. (1999). Mi-2 complex couples DNA methylation to chromatin remodelling and histone deacetylation. Nat. Genet. 23, 62–66.

Wang, Y., Fischle, W., Cheung, W., Jacobs, S., Khorasanizadeh, S., and Allis, C. D. (2004). Beyond the double helix: Writing and reading the histone code. Novartis Found. Symp. 259, 3–17.

Wassenegger, M., Heimes, S., Riedel, L., and Sanger, H. L. (1994). RNA-directed de novo methylation of genomic sequences in plants. Cell 76, 567.

Weber, M., Davies, J. J., Wittig, D., Oakeley, E. J., Haase, M., Lam, W. L., and Schubeler, D. (2005). Chromosome-wide and promoter-specific analyses identify sites of differential DNA methylation in normal and transformed human cells. *Nat. Genet.* **37,** 853–862.

Widschwendter, M., Fiegl, H., Egle, D., Mueller-Holzner, E., Spizzo, G., Marth, C., Weisenberger, D. J., Campan, M., Young, J., Jacobs, I., and Laird, P. W. (2007). Epigenetic stem cell signature in cancer. *Nat. Genet.* **39,** 157–158.

Wolf, S. F., and Migeon, B. R. (1982). Studies of X chromosome DNA methylation in normal human cells. *Nature* **295,** 667–671.

Yeager, T. R., DeVries, S., Jarrard, D. F., Kao, C., Nakada, S. Y., Moon, T. D., Bruskewitz, R., Stadler, W. M., Meisner, L. F., Gilchrist, K. W., Newton, M. A., Waldman, F. M., *et al.* (1998). Overcoming cellular senescence in human cancer pathogenesis. *Genes Dev.* **15,** 163–174.

Yoder, J. A., Walsh, C. P., and Bestor, T. H. (1997). Cytosine methylation and the ecology of intragenomic parasites. *TIG* **13,** 335–340.

Yoo, C. B., Cheng, J. C., and Jones, P. A. (2004). Zebularine: A new drug for epigenetic therapy. *Biochem. Soc. Trans.* **32,** 910–912.

10 Genomic Identification of Regulatory Elements by Evolutionary Sequence Comparison and Functional Analysis

Gabriela G. Loots

Biosciences and Biotechnology Division, Chemistry, Materials and Life Sciences Directorate, Lawrence Livermore National Laboratory, Livermore, California 94550

I. Introduction
II. Genomic Architecture of the Human Genome
    A. Distant regulatory elements controlling transcription
    B. Noncoding mutations causing human disease
III. Computational Methods of Predicting Regulatory Elements
    A. Identifying evolutionarily conserved noncoding sequences
    B. Predicting TFBSs
IV. *In Vivo* Validation and Characterization of Transcriptional Regulatory Elements
    A. Enhancer validation using transient transgenesis
    B. Identifying distant enhancers using large genomic constructs
    C. Mutating candidate regulatory elements in engineered mice
V. Conclusions
    References

ABSTRACT

Despite remarkable recent advances in genomics that have enabled us to identify most of the genes in the human genome, comparable efforts to define transcriptional *cis*-regulatory elements that control gene expression are lagging behind. The difficulty of this task stems from two equally important problems: our knowledge of how regulatory elements are encoded in genomes remains

Advances in Genetics, Vol. 61

0065-2660/08 $35.00
DOI: 10.1016/S0065-2660(07)00010-7

elementary, and there is a vast genomic search space for regulatory elements, since most of mammalian genomes are noncoding. Comparative genomic approaches are having a remarkable impact on the study of transcriptional regulation in eukaryotes and currently represent the most efficient and reliable methods of predicting noncoding sequences likely to control the patterns of gene expression. By subjecting eukaryotic genomic sequences to computational comparisons and subsequent experimentation, we are inching our way toward a more comprehensive catalog of common regulatory motifs that lie behind fundamental biological processes. We are still far from comprehending how the transcriptional regulatory code is encrypted in the human genome and providing an initial global view of regulatory gene networks, but collectively, the continued development of comparative and experimental approaches will rapidly expand our knowledge of the transcriptional regulome. © 2008, Elsevier Inc.

I. INTRODUCTION

In contrast to the genomic landscape of many prokaryotic organisms that are compact and gene rich, most eukaryotic, particularly metazoan, genomes have a small ratio of genes to noncoding DNA since only a minority of the genome is transcribed and translated into proteins. The focus of the initial analysis of both the human (Lander et al., 2001; Venter et al., 2001) and mouse genomes (Waterston et al., 2002) has been to catalog all mammalian protein-coding genes, now estimated to be in the vicinity of ~25,000 unique transcripts (Collins, 2004), and spanning less than 2% of the human genome. An additional 40–45% of the human genome is covered by repetitive DNA elements, while the remaining ~53% is composed of noncoding DNA (Lander et al., 2001; Venter et al., 2001; Waterston et al., 2002). Despite this vast amount of noncoding DNA, little progress has been made in conclusively determining whether it plays any vital functional role. Although some parts of noncoding regions within our genome will eventually reveal no detectable biological function, a growing hypothesis speculates that much of an organism's genetic complexity is due to elaborate transcriptional regulatory signals embedded in our noncoding DNA that determine when, where, and what amount of a gene transcript is expressed.

Natural selection is a major driving force in stabilizing functionally important regions within genomes, preserving the sequences of orthologous coding exons and transcriptional regulatory elements. Evolutionary comparisons have long held the promise for identifying transcriptional response elements in eukaryotic genomes, where initially searches were conducted using consensus sequences and positional weight matrices in a method often termed *phylogenetic footprinting* (Blanchette et al., 2002; Chiang et al., 2003; Fink et al., 1996; Wasserman and Fickett, 1998). A new generation of *ab initio* approaches have increasingly shown great potential for identifying novel functional motifs, but in general the sheer size and complexity of mammalian genomes has precluded the extension of these approaches to study mammalian transcription on a genome scale. This has been primarily because these simple motifs are short and highly degenerate and create overwhelming predictions with high rates of false positives when used in whole-genome analysis. The availability of large amounts of sequence data from numerous organisms and new user-friendly alignment tools have totally altered our contemporary approach to transcriptional regulation, where evolutionary comparisons have become the first tier of analysis routinely performed when searching for regulatory elements. This is reflected by the dramatic increase in the number of studies reporting the identification of functional sequences through the use of comparative genomics, and these emerging studies are providing compelling evidence in support for the use of evolutionary comparisons as a robust strategy for highlighting functional coding (Gilligan et al., 2002; Pennacchio et al., 2001) and noncoding sequences (Gottgens et al., 2000; Loots et al., 2000; Nobrega et al., 2003; Pennacchio et al., 2006; Touchman et al., 2000). In particular, aligning whole genomes and identifying evolutionarily conserved regions (ECR) on a large scale has become a robust approach for discovering transcriptional regulatory elements in noncoding DNA (Pennacchio et al., 2006; Woolfe et al., 2005). Here we will discuss methods of applying comparative genomics to the identification of transcriptional regulatory elements in the human genome, and functional approaches for validating and characterizing computationally predicted elements.

II. GENOMIC ARCHITECTURE OF THE HUMAN GENOME

It is not fully understood how one could precisely define a human gene locus, since all functional elements have yet to be determined for each transcript, but in general, one could view a typical animal gene as a promoter linked to the transcript, both of which are embedded in a sea of positively and negatively regulating elements positioned anywhere in relation to the transcriptional start site (5′, 3′, and intronic), and acting at a distance across large segments of DNA (up to megabases in lengths) (Fig. 10.1). The positively regulating elements or *enhancers* are each responsible for a subset of the total gene expression pattern and usually drive transcription within a specific tissue or subset of cell types. A typical enhancer can range in size from as little as 100 base pairs (bp) (Banet et al., 2000; Catena et al., 2004; Krebsbach et al., 1996) to several kilobases (kb) in length (Chi et al., 2005; Danielian et al., 1997), but on average would be about 500 bp in length (Kamat et al., 1999; Loots et al., 2005; Marshall et al., 2001). Within enhancer elements are docking sites for regulatory proteins or transcription factors (TFs) that physically interact with specific DNA sequences or

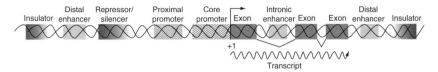

Figure 10.1. Schematic representation of a transcriptional locus in the human genome that consists of a complex arrangement of multiple clustered regulatory modules that may include elements such as enhancers, silencers, and insulators which interact with proximal and core promoter elements to specify transcription. These modules can be located across large stretches of DNA up to megabases in length and can be located either upstream or downstream of the transcriptional start site.

transcription factor binding sites (TFBSs). TFs recognize and bind to short (6–12 bp), highly degenerate sequence motifs that occur very frequently in a genome; therefore, computationally predicting TFBSs that are functionally significant is a great challenge. It is not yet known how many TFBSs are needed to build a functional enhancer, nor how many different TFs need to synergistically cooperate to drive expression. One hypothesis suggests that a typical enhancer contains a minimum of 10 TFBSs for at least 3 different TFs (Levine and Tjian, 2003).

The core promoter serves as a platform for the assembly of transcriptional preinitiation complex (PIC) that includes TFIIA, TFIIB, TFIID, TFIIE, TFIIF, TFIIH, and RNA polymerase II (Pol II), which function collectively to specify the transcription start site. The PIC usually begins with TFIID binding to the TATA box (a TATA box is a DNA sequence found in the promoter region of eukaryotic genes, specified as 5′-TATAA-4-3′ or a variant), initiator, and/or downstream promoter element (DPE) found in most core promoters, followed by the entry of other general transcription factors (GTFs) and Pol II through either a sequential assembly or a preassembled Pol II holoenzyme pathway (Thomas and Chiang, 2006). This promoter-bound complex is sufficient to drive basal level of transcription, but would require additional cofactors to transmit regulatory signals between gene-specific activators and the general transcription machinery. Three classes of general cofactors, including TATA-binding protein (TBP)-associated factors (TAFs), mediator, and upstream stimulatory activity (USA)-derived positive cofactors (PC1/PARP-1, PC2, PC3/DNA topoisomerase I, and PC4) and negative cofactor 1 (NC1/HMGB1), normally function independently or in combination to fine-tune the promoter activity in a gene-specific or cell-type-specific manner. In addition, other cofactors, such as TAF1, BTAF1, and NC2, can also modulate TBP or TFIID binding to the core promoter. Many genes also contain binding TFBSs for proximal regulatory factors located just 5′ of the core promoter. These factors do not always function as classic activators or repressors; instead, they might serve as a connecter between distal enhancers and the core promoter. In addition, a different class of regulatory elements, *insulators* or *boundary elements*, serve as gatekeepers and prevent enhancers from inappropriately regulating neighboring genes.

A. Distant regulatory elements controlling transcription

What makes transcriptional genomics in vertebrates a highly intricate problem stems from two recent observations: (1) all regulatory elements associated with a transcript can be scattered over great distances that can reach megabases (Mb) in length (Nobrega *et al.*, 2003; Sagai *et al.*, 2005), and (2) some regulatory elements are capable of controlling multiple transcripts, skip intercalating genes, or regulate one transcript while being positioned within a different transcript (Loots *et al.*, 2000; Zuniga *et al.*, 2004). In Fig. 10.2 we depict three such examples. For the human interleukin 4 (*IL4*) gene cluster on human chromosome 5, it was long hypothesized that a common regulatory element or locus control region possibly controls the Th2 expression of several cytokine genes. Using comparative genomics one highly conserved element positioned between *IL4* and *IL13* was removed from a human yeast artificial chromosome transgene (Loots *et al.*, 2000) as well as from the mouse genome (Mohrs *et al.*, 2001) to show that by removing this element the expression of three cytokines *IL4*, *IL13* and *IL5* is affected in Th2 cells. What was peculiar about this discovery was the fact that this regulatory element was positioned 120 kb away from the promoter of *IL5* gene, and it was able to exclusively control these three cytokines at the transcriptional level, leaving the intercalating gene, *RAD50*, unaffected when removed from the genome (Fig. 10.2A) (Loots *et al.*, 2000; Mohrs *et al.*, 2001).

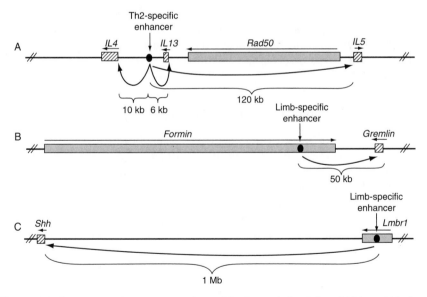

Figure 10.2. Long-range enhancer activity observed for the *interleukin 4, 5,* and *13* locus (A), the *formin-gremlin* locus (B), and the *sonic hedgehog-Lmbr1* locus (C).

In the case of limb deformity mutations originally mapped to the C-terminal region of the *formin* gene almost two decades ago (Mass *et al.*, 1990; Woychik and Alagramam, 1998; Woychik *et al.*, 1985, 1990), it was long hypothesized that *formin* is the gene responsible for the disruption of limb bud morphogenesis. Recently when a null mutation in the neighboring gene *gremlin* was generated, it was shown that by removing this BMP antagonist one recapitulates the limb phenotypes recorded for all limb mutations mapped to the *formin* locus (Khokha *et al.*, 2003). Consequent complementation studies together with *in vivo* enhancer expression assays confirmed that the limb deformity phenotypes were indeed a result of *gremlin*-specific regulatory element mutations, and that this limb-specific enhancer resides in an intron, in the C terminus region of the large *formin* transcript, and that *formin* does not contribute to the limb morphogenic abnormalities due to these mutations (Fig. 10.2B) (Zuniga *et al.*, 2004). In a more dramatic example, a region in the fifth intron of *Lmbr1* gene has long been suggested as the responsible element for preaxial polydactyly recorded in mouse and human mutations (Heutink *et al.*, 1994). In these mice, *Shh* expression is perturbed in the anterior margin of the limb bud mesenchyme (Masuya *et al.*, 1995), a gene positioned 1 Mb away from the *Lmbr1* transcript, recapitulating phenotypic aspects of the *Shh* knockout. These studies were followed by transgenic reporter experiments which revealed that this intronic region from *Lmbr1* is indeed an enhancer that drives expression in the posterior mesenchyme of the developing mouse limb bud (Lettice *et al.*, 2003). Most recently, this enhancer was removed from the mouse genome to conclusively show that this element located 1 Mb away from the *Shh* transcriptional start site is required for the limb-specific expression of *Shh*, and its ablation results in a limb phenotype identical to the limb phenotype observed in *Shh* knockout, and therefore it functions as a limb-specific *Shh* enhancer (Fig. 10.2C) (Sagai *et al.*, 2005).

B. Noncoding mutations causing human disease

In the absence of sequence information, one method that was employed to genetically map congenital abnormalities was to karyotype affected individuals and determine whether chromosomal abnormalities in the form of deletions and translocations segregate with the phenotype in affected families. While generally, detailed mapping and targeted sequencing of affected individuals lead to the discovery of the causative gene and identification of the deleterious mutations, in some instances these chromosomal aberrations do not disrupt any genes or coding regions—several such examples are listed in Table 10.1 (Ahituv *et al.*, 2004). For example, mutations in the coding region of the gene encoding for the developmental transcription factor *SALL1* lead to autosomal dominant Townes–Brocks syndrome, while a thoroughly characterized translocation in one patient 180 kb telomeric to *SALL1* also leads to a similar phenotype

Table 10.1. Human Abnormalities Mapped to Noncoding Regions

Gene	Disease	References
FOXC2	Lymphedema-distichiasis	(Fang et al., 2000)
FOXL2	Blepharophimosis/ptosis/epi-canthus inversus syndrome	(Crisponi et al., 2004)
FSHD	Facioscapulohumeral dystrophy	(van Deutekom et al., 1996)
GLI3	Greig cephalopolysyndactyly	(Vortkamp et al., 1991)
HBB complex	Gb-Thalassemia	(Kioussis et al., 1983)
PAX6	Aniridia	(Fantes et al., 1995)
PITX2	Rieger syndrome	(Flomen et al., 1997)
PLP1	Pelizaeus–Merzbacher disease	(Inoue et al., 2002)
POU3F4	X-linked deafness	(de Kok et al., 1995)
SALL1	Townes–Brocks syndrome	(Marlin et al., 1999)
Shh	Holoprosencephaly	(Roessler et al., 1997)
SIX3	Holoprosencephaly	(Wallis et al., 1999)
SOST	Van Buchem disease	(Balemans et al., 2002)
SOX9	Campomelic displasia	(Wirth et al., 1996)
SRY	Sex reversal TWIST Saethre–Chotzen syndrome	(McElreavy et al., 1992)

(Marlin et al., 1999). One likely explanation for these observations is that noncoding cis-regulatory sequences have been mutated or removed from the genome, affecting the expression pattern or expression level of the gene they normally regulate. Since many recorded diseases have no documented coding mutations, it is likely that disruption in the communication between a vital cis-regulatory sequence and the gene it regulates could potentially result in a disease that resembles hypomorphic or null alleles of the causative gene. However, providing definitive proof that the noncoding sequence change is indeed causing a particular phenotype is a highly complex problem, difficult to address experimentally. Recently, it has been suggested that engineered bacterial artificial chromosomes (BAC) may be used to determine if noncoding deletions deleteriously impact gene expression of disease-causing genes (Loots et al., 2005). The authors investigated whether a homozygous 52-kb noncoding deletion linked to the sclerosteosis disease-causing gene, SOST (Balemans et al., 2001), and homozygous in Van Buchem (VB) patients affects SOST gene expression by expressing a wild-type BAC and a genetically modified BAC mimicking the VB allele. They proceeded to show that a SOST wild-type allele expresses human SOST according to its endogenous expression pattern, primarily in the adult bone, while the VB allele fails to drive SOST expression in the bone. They further proceeded to use comparative sequence analysis and enhancer assays to identify a distant enhancer element that is able to drive transgenic expression in osteocyte-like cell lines, and in the mouse skeletal anlage at E14.5 (Loots et al., 2005).

III. COMPUTATIONAL METHODS OF PREDICTING REGULATORY ELEMENTS

The majority of available computational tools for predicting regulatory elements are based on constructing alignments between orthologous sequences and/or detecting TF DNA binding motifs. Investigators now have the option to deduce phylogenetic relationships among sequences either by generating their own alignments (Bray et al., 2003; Brudno et al., 2003; Mayor et al., 2000; Schwartz et al., 2003) or by using ready-made DNA conservation plots available at various genome browsers (Kuhn et al., 2007; Ovcharenko et al., 2004b; Schwartz et al., 2003). There are several different approaches for scanning sequences for putative regulatory elements using pattern recognition. First, a number of computational tools predict TFBS using a library of known motifs (Heinemeyer et al., 1998, 1999; Loots and Ovcharenko, 2004), or identify conserved sequence blocks in a multiple sequence alignment (Blanchette et al., 2002; Hertz and Stormo, 1999). Clustering of TFBS has been implemented as a second approach for predicting regulatory elements or cis-regulatory modules (CRMs). A few programs analyze homogenous clusters of a single overrepresented DNA motif or heterogenous clusters of multiple different sequence motifs (synergistic motifs)(Berman et al., 2002; Kim et al., 2006; Loots et al., 2002). A third approach for predicting sequences with specific regulatory properties is to identify CRMs shared by multiple functionally related sequences from the same organism (Jegga et al., 2005, 2007; Sharan et al., 2004). Expression profiling experiments have the potential to identify groups of coexpressed genes that respond to similar environmental and metabolic stimuli, and it has been speculated that such gene sets often share similar types of CRMs because their coregulation is mediated by the same set of regulatory proteins. Several new computational approaches use microarray expression data to predict tissue-specific regulatory elements in coexpressed set of genes (Jegga et al., 2005; Ovcharenko and Nobrega, 2005; Pennacchio et al., 2007).

A. Identifying evolutionarily conserved noncoding sequences

Since sequences that mediate gene expression tend to be evolutionarily conserved, one can identify putative enhancers by comparing genomes and determining regions of high homology (Loots et al., 2000, 2005; Nobrega et al., 2003). To identify evolutionarily conserved noncoding sequences (ncECRs), one needs to be able to generate reliable alignments between orthologous noncoding regions from different organisms. Aligning short sequences from closely related organisms is a straightforward process, while determining sequence similarity between larger, highly divergent regions is a more difficult task due to significant DNA rearrangements. Even highly orthologous regions are rich in insertions and deletions as well as many single base-pair mutations which can lack orthologous counterparts and be represented as gaps within alignments. An additional potential difficulty in obtaining accurate syntenic alignments is created by the large numbers of tandem and segmental duplications found in the human genome. Some assembly strategies are unable to differentiate highly homologous duplications from true overlapping sequences, resulting in erroneous genomic assemblies with underrepresented paralogs. The most complex problem is posed by lineage-specific segmental duplications that arose since the separation of two species from their most recent common ancestor. In this situation identifying true orthologous syntenic sequences from paralogous ones is a difficult task since determining true orthology and synteny represents a major challenge in the absence of a one-to-one sequence match.

The majority of pairwise sequence alignment programs utilize dynamic programming of global alignments (Needleman and Wunsch, 1970), local alignments (Altschul et al., 1990), or database searches (Altschul et al., 1997). A small fraction of alignment programs use hidden Markov models (HMM) such as WABA (Kent and Zahler, 2000) or suffix tree such as MUMmer (Majoros et al., 2005) and AVID (Bray et al., 2003). In a very simplistic view, global alignments assume that there is a colinearity of DNA sequences while local alignments focus on detecting short matches between two sequences independent of their location and orientation. Local alignments are very powerful in detecting evolutionary rearrangements resulting in DNA reshuffling and segmental duplications (paralogs) as well as species-specific tandem gene expansions. In addition, local alignment tools are also useful when highly divergent genomes are compared, since gene structure and order is not well preserved over large evolutionary distances.

The first available alignment tools were designed to recognize and align highly homologous protein sequences. The basic local alignment search tool (BLAST) was created to rapidly match a relatively short stretch of DNA with homologous regions from a large collection of sequences stored in the National Center for Biotechnology Information (NCBI) database. Most recently, BLAST has evolved into a family of alignment tools able to detect matches for various types of sequences and evolutionary distances including blastn (nucleotide), blastp (protein), blastx [nucleotide query—protein database (db)], tblastn (protein query—translated db), tblastx (nucleotide query—translated db), and megablast (highly conserved matches). The "Blast2sequences" (bl2seq) tool was created to apply all the BLASTs to both nucleotide and protein pairwise sequence comparisons and is extremely powerful in annotating genomic sequences by comparing large contigs with mRNA sequences (Altschul et al., 1990, 1997). Despite the great versatility of the BLAST, its application becomes limited when trying to align large genomic loci, megabases in length. Processing large alignments require graphical interfaces that allow the compact visualization of genes and repetitive elements along with the evolutionarily conservation profile of the aligned sequences.

A new generation of alignment tools can efficiently process two or more input sequences that can be up to genome size in length, are publicly accessible, and have user-friendly web interfaces. Some examples are listed in Table 10.2. PipMaker (Schwartz *et al.*, 2000), zPicture (Ovcharenko *et al.*, 2004a), and Mulan (Ovcharenko *et al.*, 2005) are based on the BLASTZ (Schwartz *et al.*,

Table 10.2. Comparative Genomic Tools

Alignment tools	Web address	Generated alignments
Alfresco	http://www.sanger.ac.uk/Software/Alfresco/	M
AVID	http://baboon.math.berkeley.edu/mavid/	G, M
BALSA	http://bayesweb.wadsworth.org/balsa/balsa.html	L
BLAST	http://www.ncbi.nlm.nih.gov/BLAST/	L
Blast2Sequences	http://www.ncbi.nlm.nih.gov/blast/bl2seq/bl2.html	L
BLAT	http://genome.ucsc.edu/cgi-bin/hgBlat	L
ClustalW	http://www.ebi.ac.uk/clustalw/	M
EMBOSS	http://www.ebi.ac.uk/emboss/align/	L, G
GLASS	http://glass.lcs.mit.edu/	
LAGAN	http://lagan.stanford.edu/	L, G, M
LALIGN	http://www.ch.embnet.org/software/LALIGN_form.html	L, G
Mulan	http://mulan.dcode.org	L, M
MUMmer	http://www.tigr.org/software/mummer/	
Pfam	http://www.sanger.ac.uk/Software/Pfam/	M
PipMaker	http://bio.cse.psu.edu/pipmaker/	L, M
Poa	http://www.bioinformatics.ucla.edu/poa/	M
SAM	http://www.cse.ucsc.edu/research/compbio/sam.html	HMM, M
SSAHA	http://www.sanger.ac.uk/Software/analysis/SSAHA/	
SSEARCH	http://www.biology.wustl.edu/gcg/ssearch.html	
SynPlot	http://www.sanger.ac.uk/Users/jgrg/SynPlot	
Tcoffee	http://www.ch.embnet.org/software/TCoffee.html	
VISTA	http://www-gsd.lbl.gov/VISTA/index.html	G, M
zPicture	http://zpicture.dcode.org	L
WUBlast2	http://www.ebi.ac.uk/blast2/	L
Genome Browsers		
ECR Browser	http://ecrbrowser.dcode.org/	(Ovcharenko et al., 2004b)
Ensembl	http://www.ensembl.org/	(Stalker et al., 2004)
UCSC	http://www.genome.ucsc.edu/	(Kent et al., 2002; Kuhn et al., 2007)
VISTA	http://pipeline.lbl.gov/	(Brudno et al., 2004)

G, global; M, multiple; L, local alignments; HMM, hidden Markov model. This is not meant to be a comprehensive list but a sampling of available resources.

2003) local alignment program. They combine suffix tree algorithms with dynamic programming techniques and have the capacity to align very long genomic intervals in a very short period of time. VISTA (Frazer *et al.*, 2004; Mayor *et al.*, 2000) is a visualization tool for global alignments generated by AVID (Bray *et al.*, 2003) or LAGAN (Brudno *et al.*, 2003) programs. All these alignment engines provide the user with informative, high-resolution graphical displays of the resulting alignments depicting both the genomic location of the conserved regions in the reference sequence and the degree of similarity for each aligned DNA segment. Individual features of the DNA sequence, such as coding exons, untranslated regions (UTR), and repetitive elements, can be distinguished in the graphical output through the use of different color schemes, allowing the identification of evolutionarily conserved sequences present in noncoding regions.

Comparative sequence alignment data can also be retrieved from genome browsers. Genome browsers are web-based database interfaces designed to allow the navigation across an entire genome by scrolling and zooming through any region of DNA and visualizing all available annotation data. In general, annotations include mRNAs, expressed sequence tags (EST), gene predictions, single nucleotide polymorphisms (SNPs), as well as many other features. Users can enter a region of the genome by searching for a landmark such as the name or acronym of a known gene, the accession number of a DNA sequence, the numeric position within a chromosome, or even through a homology search by providing a piece of sequence. The two main genome browsers are Ensembl (Stalker *et al.*, 2004) and the UCSC Genome Browser database (Kent *et al.*, 2002; Kuhn *et al.*, 2007), both of which were originally designed to support the assembly and annotation needs of the Human Genome Project by creating an efficient, user-friendly data storage and retrieval system with a compact visual presentation. The UCSC Browser has rapidly expanded to provide access to other available genome assemblies and their accompanying annotations, which now include 14 vertebrate species. Recently, UCSC Browser has also incorporated comparative genomic tracks to visualize regions of DNA conservation between two fully sequenced genomes, as well as a regulatory potential track based on normalized log-odds scores calculated using an HMM model that distinguishes known regulatory regions from ancestral repeats. Similar pairwise whole-genome alignments have been generated using a combination of local and global alignment strategies and can be visualized in the ECR Browser (Ovcharenko *et al.*, 2004b) and in the VISTA Genome Browser (Brudno *et al.*, 2004). In addition to ready-made pairwise alignments, the ECR Browser also aligns user-provided sequences to available genomes and incorporates any available annotation into the visual display of the generated alignment. For a comprehensive review of genome browsers and databases, see Ureta-Vidal *et al.* (2003).

B. Predicting TFBSs

In eukaryotes, modulation of gene expression is achieved through the complex interaction of regulatory proteins (*trans*-factors) with specific DNA regions (*cis*-acting regulatory sequences). Intensive efforts over the last decades have identified numerous regulatory proteins or TFs and the DNA sequences they recognize. TRANSFAC database (http://www.biobase.de) represent the most comprehensive collection of TF- binding specificities, summarized as position weight matrices (PWMs) (Heinemeyer et al., 1998, 1999). Pattern-recognition programs such as MATCH or MatInspector (Quandt et al., 1995) use these libraries of TF-PWMs to identify significant matches in DNA sequences. A major confounding factor in the use of PWMs to identify TFBSs is that TFs bind to short (6–12 bp), degenerate sequence motifs that occur very frequently in a genome, and only a small fraction of the predicted TFBSs are functionally significant.

It has been shown that by combining pattern recognition with comparative sequence analysis the number of false positives is dramatically reduced while the number of functional sites is preserved. These results suggest an alternative strategy for sequence-based discovery of biologically relevant regulatory elements. The rVISTA (Loots and Ovcharenko, 2004) and Consite (Sandelin et al., 2004) web-based tools combine TFBS motif searches and cross-species sequence analysis. rVISTA analysis proceeds in four major steps: (1) identify TFBS matches in each individual sequence using PWM from TRANSFAC database, (2) detect and calculate the percent identity of each locally aligned TFBS, (3) select TFBSs present in regions of high DNA conservation, and (4) graphically display individual or clustered TFBSs (Loots and Ovcharenko, 2004).

Phylogenetic footprinting is a method for identifying highly conserved DNA motifs present in a multiple sequence alignment. It is usually performed by computing a global multiple alignment of three or more orthologous sequences, and by identifying regions of high conservation in the alignment. Foot-Printer (Blanchette and Tompa, 2003; Fang and Blanchette, 2006), FOOTER (Corcoran et al., 2005), TRES, and PhyloGibbs (Siddharthan et al., 2005) are some of the algorithms available for generating motif predictions and reporting motif sequences with the lowest parsimony scores, calculated with respect to the phylogenetic tree relating the input sequences. A more successful recent approach to phylogenetic footprinting is to use motif discovery algorithms such as MEME (Bailey et al., 2006). Programs like MEME neither take into account the phylogenetic relationship among the input sequences nor do they rely on precalculated PWM stored in a database; they treat input sequences individually and the patterns are learned through several rounds of ungapped local alignments. The sampled alignments are used to fit a set of weights and the best weights are used to define an alignment, similar to the Gibbs sampling method (Schug and Overton, 1997).

In eukaryotes, transcriptional gene regulation is directed by a cohort of several different TFs that cooperatively bind to clusters of TFBS known as gene CRMs. One of the main objectives of transcriptional genomics is to decode the structure of CRMs and distinguish between the footprints of functional TFBS from genomic intervals devoid of biological significance and determine which CRM structures confer which tissue specificity. Searches for clusters of multiple adjacent binding sites for regulatory proteins have been successful in analyzing regulatory regions involved in mammalian muscle (Wasserman and Fickett, 1998) and liver-specific gene expression (Krivan and Wasserman, 2001). MSCAN (Alkema *et al.*, 2004) and rVISTA (Loots and Ovcharenko, 2004) are two examples of web-based tools that allow users to search for clusters of *cis*-elements, either using precalculated matrices from the TRANSFAC database or using consensus sequences provided by the user. Using these tools one could search for regions of high density of repeated TFBS for a single or multi different TFs. Recently, a new tool has been made available, SYNOR, which allows users to search for any configurations of TFBS across the whole human genome to predict functional regions with a distinct TFBS profile (Ovcharenko and Nobrega, 2005).

A new generation of computational tools aimed at predicting tissue-specific regulatory elements are blending together three elements: (a) comparative sequence analysis, (b) TFBS analysis, and (c) microarray expression analysis. In this approach, coexpressed genes are mapped to a genome and their promoters and surrounding conserved noncoding regions are used to identify CRMs that are overrepresented in the data set when compared with the distribution of the same CRMs across the whole genome. Pilpel and colleagues proposed such a method for modeling transcription regulatory networks in complex eukaryotes by combining microarray expression data with insights from combinatorial structure of promoter regions (Pilpel *et al.*, 2001). They were able to show that it is possible to discover novel functional PWMs by identifying statistically significant synergistic motifs in promoters of coexpressed genes, using a process called transcription factor centric clustering (TFCC). TFCC strategies are designed to create an explicit link between CRMs and the TFs that bind to them (Zhu *et al.*, 2002). These methods permit the detection of enriched TFBSs that are used as a seed to bicluster genes and compare gene expression with TFs' distribution in a two-dimensional space. Sharan *et al.* (2003) built on Pilpel's strategy by analyzing humans–mouse conserved promoter elements of cell cycle and stress response-related genes. Their analyses revealed several clusters of TFBS specific to the coexpressed genes. The significance of such co-occurrences was statistically evaluated and showed direct correlation between the identified CRMs and the biologically validated target genes derived from the microarray expression data. They proceeded to incorporate this method into a publicly available software,

CREME, which can perform combinatorial cluster analysis, statistically evaluate the detected co-occurrences, and graphically display predicted CRMs in a browser (Sharan et al., 2003, 2004).

IV. *IN VIVO* VALIDATION AND CHARACTERIZATION OF TRANSCRIPTIONAL REGULATORY ELEMENTS

In general, computational predictions have strongly correlated with functionally characterized regulatory elements mostly because the training sets used for these analyses have been carefully chosen from biologically validated data sets. On the contrary, the majority of novel predictions have not yet been biologically confirmed, and the limitations of computational tools have not been carefully assessed. Functional characterization of noncoding sequences represents the largest bottleneck which prevents us from expanding the annotation of regulatory elements from small target regions to entire genomes. The field of *in silico* biology is still in its infancy, but is evolving at a fast pace, presenting researchers with new theoretical solutions for the analysis of noncoding sequences. The computationally derived regulatory predictions may not all be functionally significant at this point, but by centering biological focus on a handful of high-priority regions to be tested, computational tools have already surpassed expectations for identifying regulatory elements. In Section IV.A we will review several experimental approaches employed to validate and characterize predicted transcriptional regulatory elements.

A. Enhancer validation using transient transgenesis

Almost two decades ago, Kothary et al. (1988) created a mouse heat shock 68 promoter (*hsp68*) β-galactosidase (*LacZ*) transgene (*hsp68-LacZ*) and generated several independent lines of transgenic mice aiming to study heat shock gene regulation *in vivo*. In six of these transgenic lines, the transgene was consistently silent until subjected to heat shock treatment; however, one line of transgenic mice expressed *LacZ* in a neural-specific pattern independent of heat shock. The transgene integrated into the gene responsible for *dystonia musculorum* (OMIM 113810) (Bressman, 2003), mutated the gene, and acquired its transcriptional profile. This study was able to show that the *hsp68* promoter which is normally silent at physiological temperature is able to activate transcription of a reporter gene in response to positive regulatory elements and therefore can be used to trap enhancers in mammalian genomes (Kothary et al., 1988). It was not until the human and mouse genome projects were well underway that this transgene was fully exploited and transformed into an essential tool for validating tissue-specific enhancer elements in transient transgenic mice.

Unlike stable transgenic lines that are screened for germline transmission of exogenous DNA, transient transgenic mice are transgenic animals that are analyzed in F0, without having to pass the transgene to future generations. In this method, embryos are injected with the reporter construct, transferred to the recipient mom, allowed to develop to a desired embryonic stage (usually between E10.5 and E14.5) when the moms are sacrificed and the embryos are harvested and examined for reporter transgene expression. This method dramatically shortens the experimental time for collecting expression data, since founder mice do not need to be established before carrying out the expression analysis, making this procedure highly efficient for validating a set of putative enhancers at a desired developmental time point. It is also a more cost-efficient approach, since it eliminates the need for breeding mice and establishing founder lines. The use of this method as a validation and characterization tool has dramatically grown over the past years, but in general has remained gene centric, where individual investigators have focused on testing conserved elements in the context of a well-characterized locus to identify tissue-specific enhancers that follow the expression pattern of one gene of interest (Bejerano et al., 2006; Forghani et al., 2001; Loots et al., 2005; Nobrega et al., 2003; Rojas et al., 2005; Wang et al., 2001; Zhu et al., 2004). In an attempt to apply this method to more comprehensive, whole-genome analysis, this approach has proven very useful in validating highly conserved human elements grouped into two major categories: deeply conserved (evolutionarily conserved from human to fish) (Nobrega et al., 2003; Pennacchio et al., 2006) and ultraconserved (elements that have close to 100% identify from human to mouse) (Bejerano et al., 2006; Pennacchio et al., 2006; Poulin et al., 2005).

While this method is currently considered *high throughput* in mice, there are several obstacles that preclude it from effectively being applied on a genome-wide scale. First, since mouse embryos develop *in utero*, collecting transgenic embryos is a terminal procedure that permits a litter of mice to be analyzed at only one given time point. In the absence of gene expression information for the genes putative enhancers are expected to regulate, screening for enhancer function could become a fishing expedition with low probability of success. For example, if an enhancer drives expression only at E17.5 in the medulla oblongata, with no detectable expression at any other time or tissue during development, the investigator would have to assay this particular time point to be able to detect its function. By assaying any other time point, the consistent lack of expression would drive the investigator to erroneously assume that the element has no function. A second drawback is posed by the transgene visualization method for *LacZ*. *LacZ* is a bacterial gene whose gene product, galactosidase, catalyzes the hydrolysis of galactosides or X-gal, and produces a blue color that can be visualized. This procedure requires fixation, and hence is terminal. Other caveats to this experimental approach include position effect, promoter specificity, and restriction to enhancer detection (one cannot detect a repressor or silencer element).

To overcome some of these problems, some alternatives have been proposed that include (1) the use of a transgenic model system that develops *ex utero* [fish (zebra fish) or frogs (*Xenopus*)]; (2) the use of reporter genes that do not require terminal fixation for visualization, such as green fluorescent protein (GFP); and (3) the use of larger transgene and "knocked" in reporters that track the protein expression from the endogenous promoter. Such transgenic systems would allow investigators to more efficiently monitor the transgene expression during development and determine both the temporal and spatial window of enhancer activity an element may pose. Using zebra fish transgenesis, Woolfe and his colleagues tested 25 ECRs from a set of 1400 elements identified by comparisons between the human and the puffer fish *Fugu rubripes* genomes. This study is of great significance because it makes two very important points. First, the authors were able to show that distant comparisons between human and *Fugu* enrich for a special category of regulatory elements which cluster around genes that have vital roles during embryonic development (e.g., TFs). Second, they were able to confirm that 23 of the total 25 tested ncECRs exhibit tissue-specific enhancer activity, suggesting that most of deeply conserved elements do indeed function as transcriptional regulatory elements (Woolfe *et al.*, 2005).

Recently, several reports have emerged that describe transposon-based gene delivery methods in both zebra fish and *Xenopus* embryos. These technologies can potentially evolve into a rapid system for transgenesis and expedite enhancer validation. In this approach, an ECR is cloned upstream of a minimal promoter driving a fluorescent reporter, where the entire transgenic cassette is flanked by transposable elements, such as *Tol2* (Allende *et al.*, 2006; Balciunas *et al.*, 2006; Hamlet *et al.*, 2006), *Sleeping Beauty* (Sinzelle *et al.*, 2006), *piggyBAC* (Wu *et al.*, 2006), or *Frog Prince* (Miskey *et al.*, 2003), and the reporter constructs are assayed for *cis*-regulatory activity. Using the *Tol2* transposable system in both zebra fish and *Xenopus* transgenic experiments, Allende and his colleagues have recently shown that most of the 50 ECRs they tested behave as positive modulators of gene expression and contribute to the specific temporal and spatial expression patterns of the endogenous genes they regulate (Allende *et al.*, 2006). The continuation of studies such as the ones described above will further our understanding of tissue-specific transcriptional regulation and will aid the discovery of all functional regulatory elements in the human genome.

B. Identifying distant enhancers using large genomic constructs

An alternative approach to the use of heterologous minimal promoters to assay for enhancer activity in transient transgenics is to tag a transcript with a reporter gene (*LacZ* or *GFP*) within the context of a larger genomic region by modifying a yeast artificial chromosome (YAC) or BAC. This method has several advantages. First, the reporter gene is driven by the endogenous promoter and responds to all regulatory elements included in the transgenic construct; therefore, by comparing the expression pattern of the reporter transgene to the endogenous expression pattern of the mouse gene one can determine what components of the complete tissue-specific expression profile is controlled by elements residing within the BAC region (Bouchard *et al.*, 2005; Gebhard *et al.*, 2007; Mortlock *et al.*, 2003; Tallini *et al.*, 2006). Second, transgenic animals generated using larger DNA constructs are less likely to be affected by position effects; therefore, most of the time the transgenic expression faithfully resembles the endogenous gene expression (Gebhard *et al.*, 2007). Third, by mutating individual ECRs within a BAC construct, one can identify not only regulatory elements that positively modulate transcription but also *cis*-elements that act as negative regulators, such as repressors, silencers, or boundary elements.

C. Mutating candidate regulatory elements in engineered mice

The methods described in the previous sections are considered "gain-of-function" approaches to determining whether a conserved noncoding sequence possesses biological activity. However, these experiments do not provide any information whether these elements are functionally critical, and whether mutating them can lead to serious congenital abnormalities or whether they contribute to susceptibility to disease. The ultimate functional test that confirms essential physiological activity of a ncECR is to mutate these elements by changing individual base pairs or by removing them from an animal's genome. Loss of function alleles can be generated by two main methods: random mutagenesis or targeted knock-out (KO). In a random mutagenesis experiment, one would subject an animal to a mutagen that either causes large chromosomal abnormalities, such as deletions and translocations, or has smaller effects by mutating single base pairs or removing a few nucleotides. Mutagenesis experiments are feasible for most experimental organisms, but require rigorous screenings to detect individuals that carry a desired mutation. A targeted mutation can only be engineered in animals for which KO technologies have been established, but unfortunately rodents are the only mammals for which targeted deletions can be carried out efficiently, primarily because embryonic stem cell lines have not been derived for other mammals. In this section, we will discuss functionally characterizing putative regulatory elements through loss-of-function experiments *in vivo*, in genetically modified mice by either removing an ECR from a large transgene or removing it from the mouse genome.

Homologous recombination techniques in yeast and bacteria have facilitated the genetic modification of artificial chromosomes [YACs, Pl-based artificial chromosomes (PACs), and BACs] (Imam *et al.*, 2000; Lee *et al.*, 2001; Loots, 2006; Nistala and Sigmund, 2002; Warming *et al.*, 2005) to study gene expression and transcriptional regulation across large genomic loci.

The attraction of using such DNA constructs is primarily because they carry large fragments of genomic DNA (>100 kb) and therefore are likely to contain most of the *cis*-regulatory elements required for the expression of a gene; therefore, even when inserted randomly into the mouse genome, these transgenes are likely to behave similarly to their native environment recapitulating the gene expression pattern of endogenous loci. An extra advantage is our ability to modify these transgenes by inserting sequences prone to recombination such as *loxP* or *FRT* sites in the presence of *recombinases*. By flanking a ncECR with *loxP* sites, one can determine *in vivo* the transcriptional effects an element will have on a transcript when the ncECR is present or absent from the locus (Loots *et al.*, 2000), independent of position effects. In most situations, the integrated *loxP* sites do not affect gene expression and the floxed allele behaves equivalent to the wild-type allele. Upon administration of recombinase protein, the *loxP* sites excise the ncECR element, leaving behind a new deleted allele. Finally the investigator compares expression of the transgene with and without the ncECR to determine if the ncECR has any impact on transcription. This method was used to show that a highly conserved noncoding DNA sequence controls the expression of three cytokine genes, *IL4*, *IL13*, and *IL5*, in the context of a human YAC (Loots *et al.*, 2000). Similarly, one can use *in vitro* recombination to create several variants of a transgene, by either removing a putative regulatory element (Xu *et al.*, 2006) or removing a large noncoding region (Loots *et al.*, 2005). Since each transgene randomly integrates into the mouse genome, to ensure that an observed difference in expression is due to the mutation and not to position effect, several independent transgenic lines have to be analyzed for each allele. Finally, the most informative and reliable method of testing whether a ncECR impacts gene expression and causes a deleterious phenotype is to remove it from the mouse genome through targeted KO strategies. Although this approach remains the ultimate proof of biological significance because it is technically challenging, laborious, expensive, and time-consuming to generate KO animals, to date very few ncECR have been mutated in mice (Mohrs *et al.*, 2001; Sagai *et al.*, 2005).

V. CONCLUSIONS

Comparative genomic approaches are having a remarkable impact on the study of transcriptional regulation in eukaryotes. Many eukaryotic genome sequences are being explored by new computational methods and high-throughput experimental tools. These tools are enabling efficient searches for common regulatory motifs which will eventually lead to elucidating the genome's second code: understanding the building blocks of tissue-specific gene regulation encoded in noncoding DNA. Experimental validation and characterization, however, continues to be a major bottleneck, and hence extending the limits of current techniques will greatly enhance the discovery of transcriptional regulatory elements in mammals, moving us closer to a systematic deciphering of transcriptional regulatory elements and providing the first global insights into gene regulatory networks. In addition to the methods described here, other recent advances in transcriptional regulation approaches that include probing DNA–protein interactions by ChIP-chip or comparing patterns of gene expression are moving the field of transcriptional genomics forward.

References

Ahituv, N., Rubin, E. M., and Nobrega, M. A. (2004). Exploiting human–fish genome comparisons for deciphering gene regulation. *Hum. Mol. Genet.* **13**(Spec. No. 2), R261–R266.

Alkema, W. B., Johansson, O., Lagergren, J., and Wasserman, W. W. (2004). MSCAN: Identification of functional clusters of transcription factor binding sites. *Nucleic Acids Res.* **32**, W195–W198.

Allende, M. L., Manzanares, M., Tena, J. J., Feijoo, C. G., and Gomez-Skarmeta, J. L. (2006). Cracking the genome's second code: Enhancer detection by combined phylogenetic footprinting and transgenic fish and frog embryos. *Methods* **39**, 212–219.

Altschul, S. F., Gish, W., Miller, W., Myers, E. W., and Lipman, D. J. (1990). Basic local alignment search tool. *J. Mol. Biol.* **215**, 403–410.

Altschul, S. F., Madden, T. L., Schaffer, A. A., Zhang, J., Zhang, Z., Miller, W., and Lipman, D. J. (1997). Gapped BLAST and PSI-BLAST: A new generation of protein database search programs. *Nucleic Acids Res.* **25**, 3389–3402.

Bailey, T. L., Williams, N., Misleh, C., and Li, W. W. (2006). MEME: Discovering and analyzing DNA and protein sequence motifs. *Nucleic Acids Res.* **34**, W369–W373.

Balciunas, D., Wangensteen, K. J., Wilber, A., Bell, J., Geurts, A., Sivasubbu, S., Wang, X., Hackett, P. B., Largaespada, D. A., McIvor, R. S., and Ekker, S. C. (2006). Harnessing a high cargo-capacity transposon for genetic applications in vertebrates. *PLoS Genet.* **2**, 1715–1724.

Balemans, W., Ebeling, M., Patel, N., Van Hul, E., Olson, P., Dioszegi, M., Lacza, C., Wuyts, W., Van Den Ende, J., Willems, P., Paes-Alves, A. F., Hill, S., *et al.* (2001). Increased bone density in sclerosteosis is due to the deficiency of a novel secreted protein (SOST). *Hum. Mol. Genet.* **10**, 537–543.

Balemans, W., Patel, N., Ebeling, M., Van Hul, E., Wuyts, W., Lacza, C., Dioszegi, M., Dikkers, F. G., Hildering, P., Willems, P. J., Verheij, J. B., Lindpaintner, K., *et al.* (2002). Identification of a 52 kb deletion downstream of the SOST gene in patients with van Buchem disease. *J. Med. Genet.* **39**, 91–97.

Banet, G., Bibi, O., Matouk, I., Ayesh, S., Laster, M., Kimber, K. M., Tykocinski, M., de Groot, N., Hochberg, A., and Ohana, P. (2000). Characterization of human and mouse H19 regulatory sequences. *Mol. Biol. Rep.* **27**, 157–165.

Bejerano, G., Lowe, C. B., Ahituv, N., King, B., Siepel, A., Salama, S. R., Rubin, E. M., Kent, W. J., and Haussler, D. (2006). A distal enhancer and an ultraconserved exon are derived from a novel retroposon. *Nature* **441**, 87–90.

Berman, B. P., Nibu, Y., Pfeiffer, B. D., Tomancak, P., Celniker, S. E., Levine, M., Rubin, G. M., and Eisen, M. B. (2002). Exploiting transcription factor binding site clustering to identify cis-regulatory modules involved in pattern formation in the Drosophila genome. *Proc. Natl. Acad. Sci. USA* **99**, 757–762.

Blanchette, M., and Tompa, M. (2003). FootPrinter: A program designed for phylogenetic footprinting. *Nucleic Acids Res.* **31**, 3840–3842.

Blanchette, M., Schwikowski, B., and Tompa, M. (2002). Algorithms for phylogenetic footprinting. *J. Comput. Biol.* **9**, 211–223.

Bouchard, M., Grote, D., Craven, S. E., Sun, Q., Steinlein, P., and Busslinger, M. (2005). Identification of Pax2-regulated genes by expression profiling of the mid-hindbrain organizer region. *Development* **132**, 2633–2643.

Bray, N., Dubchak, I., and Pachter, L. (2003). AVID: A global alignment program. *Genome Res.* **13**, 97–102.

Bressman, S. B. (2003). Dystonia: Phenotypes and genotypes. *Rev. Neurol. (Paris)* **159**, 849–856.

Brudno, M., Do, C. B., Cooper, G. M., Kim, M. F., Davydov, E., Green, E. D., Sidow, A., and Batzoglou, S. (2003). LAGAN and Multi-LAGAN: Efficient tools for large-scale multiple alignment of genomic DNA. *Genome Res.* **13**, 721–731.

Brudno, M., Poliakov, A., Salamov, A., Cooper, G. M., Sidow, A., Rubin, E. M., Solovyev, V., Batzoglou, S., and Dubchak, I. (2004). Automated whole-genome multiple alignment of rat, mouse, and human. *Genome Res.* **14**, 685–692.

Catena, R., Tiveron, C., Ronchi, A., Porta, S., Ferri, A., Tatangelo, L., Cavallaro, M., Favaro, R., Ottolenghi, S., Reinbold, R., Schöler, H., and Nicolis, S. K. (2004). Conserved POU binding DNA sites in the Sox2 upstream enhancer regulate gene expression in embryonic and neural stem cells. *J. Biol. Chem.* **279**, 41846–41857.

Chi, X., Chatterjee, P. K., Wilson, W., 3rd, Zhang, S. X., Demayo, F. J., and Schwartz, R. J. (2005). Complex cardiac Nkx2–5 gene expression activated by noggin-sensitive enhancers followed by chamber-specific modules. *Proc. Natl. Acad. Sci. USA* **102**, 13490–13495.

Chiang, D. Y., Moses, A. M., Kellis, M., Lander, E. S., and Eisen, M. B. (2003). Phylogenetically and spatially conserved word pairs associated with gene-expression changes in yeasts. *Genome Biol.* **4**, R43.

Collins, F. (2004). Finishing the euchromatic sequence of the human genome. *Nature* **431**, 931–945.

Corcoran, D. L., Feingold, E., and Benos, P. V. (2005). FOOTER: A web tool for finding mammalian DNA regulatory regions using phylogenetic footprinting. *Nucleic Acids Res.* **33**, W442–W446.

Crisponi, L., Uda, M., Deiana, M., Loi, A., Nagaraja, R., Chiappe, F., Schlessinger, D., Cao, A., and Pilia, G. (2004). FOXL2 inactivation by a translocation 171 kb away: Analysis of 500 kb of chromosome 3 for candidate long-range regulatory sequences. *Genomics* **83**, 757–764.

Danielian, P. S., Echelard, Y., Vassileva, G., and McMahon, A. P. (1997). A 5.5-kb enhancer is both necessary and sufficient for regulation of Wnt-1 transcription in vivo. *Dev. Biol.* **192**, 300–309.

de Kok, Y. J., Merkx, G. F., van der Maarel, S. M., Huber, I., Malcolm, S., Ropers, H. H., and Cremers, F. P. (1995). A duplication/paracentric inversion associated with familial X-linked deafness (DFN3) suggests the presence of a regulatory element more than 400 kb upstream of the POU3F4 gene. *Hum. Mol. Genet.* **4**, 2145–2150.

Fang, F., and Blanchette, M. (2006). FootPrinter3: Phylogenetic footprinting in partially alignable sequences. *Nucleic Acids Res.* **34**, W617–W620.

Fang, J., Dagenais, S. L., Erickson, R. P., Arlt, M. F., Glynn, M. W., Gorski, J. L., Seaver, L. H., and Glover, T. W. (2000). Mutations in FOXC2 (MFH-1), a forkhead family transcription factor, are responsible for the hereditary lymphedema-distichiasis syndrome. *Am. J. Hum. Genet.* **67**, 1382–1388.

Fantes, J., Redeker, B., Breen, M., Boyle, S., Brown, J., Fletcher, J., Jones, S., Bickmore, W., Fukushima, Y., Mannens, M., Danes, S., van Heyningen, V., et al. (1995). Aniridia-associated cytogenetic rearrangements suggest that a position effect may cause the mutant phenotype. *Hum. Mol. Genet.* **4**, 415–422.

Fink, D. L., Chen, R. O., Noller, H. F., and Altman, R. B. (1996). Computational methods for defining the allowed conformational space of 16S rRNA based on chemical footprinting data. *RNA* **2**, 851–866.

Flomen, R. H., Gorman, P. A., Vatcheva, R., Groet, J., Barisic, I., Ligutic, I., Sheer, D., and Nizetic, D. (1997). Rieger syndrome locus: A new reciprocal translocation t(4;12)(q25;q15) and a deletion del(4)(q25q27) both break between markers D4S2945 and D4S193. *J. Med. Genet.* **34**, 191–195.

Forghani, R., Garofalo, L., Foran, D. R., Farhadi, H. F., Lepage, P., Hudson, T. J., Tretjakoff, I., Valera, P., and Peterson, A. (2001). A distal upstream enhancer from the myelin basic protein gene regulates expression in myelin-forming schwann cells. *J. Neurosci.* **21,** 3780–3787.

Frazer, K. A., Pachter, L., Poliakov, A., Rubin, E. M., and Dubchak, I. (2004). VISTA: Computational tools for comparative genomics. *Nucleic Acids Res.* **32,** W273–W279.

Gebhard, S., Hattori, T., Bauer, E., Bosl, M. R., Schlund, B., Poschl, E., Adam, N., de Crombrugghe, B., and von der Mark, K. (2007). BAC constructs in transgenic reporter mouse lines control efficient and specific LacZ expression in hypertrophic chondrocytes under the complete Col10a1 promoter. *Histochem. Cell Biol.* **127,** 183–194.

Gilligan, P., Brenner, S., and Venkatesh, B. (2002). Fugu and human sequence comparison identifies novel human genes and conserved non-coding sequences. *Gene* **294,** 35–44.

Gottgens, B., Barton, L. M., Gilbert, J. G., Bench, A. J., Sanchez, M. J., Bahn, S., Mistry, S., Grafham, D., McMurray, A., Vaudin, M., Amaya, E., Bentley, D. R., et al. (2000). Analysis of vertebrate SCL loci identifies conserved enhancers. *Nat. Biotechnol.* **18,** 181–186.

Hamlet, M. R., Yergeau, D. A., Kuliyev, E., Takeda, M., Taira, M., Kawakami, K., and Mead, P. E. (2006). Tol2 transposon-mediated transgenesis in Xenopus tropicalis. *Genesis* **44,** 438–445.

Heinemeyer, T., Wingender, E., Reuter, I., Hermjakob, H., Kel, A. E., Kel, O. V., Ignatieva, E. V., Ananko, E. A., Podkolodnaya, O. A., Kolpakov, F. A., Podkolodny, N. L., and Kolchanov, N. A. (1998). Databases on transcriptional regulation: TRANSFAC, TRRD and COMPEL. *Nucleic Acids Res.* **26,** 362–367.

Heinemeyer, T., Chen, X., Karas, H., Kel, A. E., Kel, O. V., Liebich, I., Meinhardt, T., Reuter, I., Schacherer, F., and Wingender, E. (1999). Expanding the TRANSFAC database towards an expert system of regulatory molecular mechanisms. *Nucleic Acids Res.* **27,** 318–322.

Hertz, G. Z., and Stormo, G. D. (1999). Identifying DNA and protein patterns with statistically significant alignments of multiple sequences. *Bioinformatics* **15,** 563–577.

Heutink, P., Zguricas, J., van Oosterhout, L., Breedveld, G. J., Testers, L., Sandkuijl, L. A., Snijders, P. J. L. M., Weissenbach, J., Lindhout, D., Hovius, S. E. R., and Oostra, B. A. (1994). The gene for triphalangeal thumb maps to the subtelomeric region of chromosome 7q. *Nat. Genet.* **6,** 287–292.

Imam, A. M., Patrinos, G. P., de Krom, M., Bottardi, S., Janssens, R. J., Katsantoni, E., Wai, A. W., Sherratt, D. J., and Grosveld, F. G. (2000). Modification of human beta-globin locus PAC clones by homologous recombination in Escherichia coli. *Nucleic Acids Res.* **28,** E65.

Inoue, K., Osaka, H., Thurston, V. C., Clarke, J. T., Yoneyama, A., Rosenbarker, L., Bird, T. D., Hodes, M. E., Shaffer, L. G., and Lupski, J. R. (2002). Genomic rearrangements resulting in PLP1 deletion occur by nonhomologous end joining and cause different dysmyelinating phenotypes in males and females. *Am. J. Hum. Genet.* **71,** 838–853.

Jegga, A. G., Gupta, A., Gowrisankar, S., Deshmukh, M. A., Connolly, S., Finley, K., and Aronow, B. J. (2005). CisMols Analyzer: Identification of compositionally similar cis-element clusters in ortholog conserved regions of coordinately expressed genes. *Nucleic Acids Res.* **33,** W408–W411.

Jegga, A. G., Chen, J., Gowrisankar, S., Deshmukh, M. A., Gudivada, R., Kong, S., Kaimal, V., and Aronow, B. J. (2007). GenomeTrafac: A whole genome resource for the detection of transcription factor binding site clusters associated with conventional and microRNA encoding genes conserved between mouse and human gene orthologs. *Nucleic Acids Res.* **35,** D116–D121.

Kamat, A., Graves, K. H., Smith, M. E., Richardson, J. A., and Mendelson, C. R. (1999). A 500-bp region, approximately 40 kb upstream of the human CYP19 (aromatase) gene, mediates placenta-specific expression in transgenic mice. *Proc. Natl. Acad. Sci. USA* **96,** 4575–4580.

Kent, W. J., and Zahler, A. M. (2000). Conservation, regulation, synteny, and introns in a large-scale C. briggsae-C. elegans genomic alignment. *Genome Res.* **10,** 1115–1125.

Kent, W. J., Sugnet, C. W., Furey, T. S., Roskin, K. M., Pringle, T. H., Zahler, A. M., and Haussler, D. (2002). The human genome browser at UCSC. *Genome Res.* **12,** 996–1006.

Khokha, M. K., Hsu, D., Brunet, L. J., Dionne, M. S., and Harland, R. M. (2003). Gremlin is the BMP antagonist required for maintenance of Shh and Fgf signals during limb patterning. *Nat. Genet.* **34,** 303–307.

Kim, J. D., Hinz, A. K., Bergmann, A., Huang, J. M., Ovcharenko, I., Stubbs, L., and Kim, J. (2006). Identification of clustered YY1 binding sites in imprinting control regions. *Genome Res.* **16,** 901–911.

Kioussis, D., Vanin, E., deLange, T., Flavell, R. A., and Grosveld, F. G. (1983). Beta-globin gene inactivation by DNA translocation in gamma beta-thalassaemia. *Nature* **306,** 662–666.

Kothary, R., Clapoff, S., Brown, A., Campbell, R., Peterson, A., and Rossant, J. (1988). A transgene containing lacZ inserted into the dystonia locus is expressed in neural tube. *Nature* **335,** 435–437.

Krebsbach, P. H., Nakata, K., Bernier, S. M., Hatano, O., Miyashita, T., Rhodes, C. S., and Yamada, Y. (1996). Identification of a minimum enhancer sequence for the type II collagen gene reveals several core sequence motifs in common with the link protein gene. *J. Biol. Chem.* **271,** 4298–4303.

Krivan, W., and Wasserman, W. W. (2001). A predictive model for regulatory sequences directing liver-specific transcription. *Genome Res.* **11,** 1559–1566.

Kuhn, R. M., Karolchik, D., Zweig, A. S., Trumbower, H., Thomas, D. J., Thakkapallayil, A., Sugnet, C. W., Stanke, M., Smith, K. E., Siepel, A., Rosenbloom, K. R., Rhead, B., *et al.* (2007). The UCSC genome browser database: Update 2007. *Nucleic Acids Res.* **35,** D668–D673.

Lander, E. S., Linton, L. M., Birren, B., Nusbaum, C., Zody, M. C., Baldwin, J., Devon, K., Dewar, K., Doyle, M., FitzHugh, W., Funke, R., Gage, D., *et al.* (2001). Initial sequencing and analysis of the human genome. *Nature* **409,** 860–921.

Lee, E. C., Yu, D., Martinez de Velasco, J., Tessarollo, L., Swing, D. A., Court, D. L., Jenkins, N. A., and Copeland, N. G. (2001). A highly efficient Escherichia coli-based chromosome engineering system adapted for recombinogenic targeting and subcloning of BAC DNA. *Genomics* **73,** 56–65.

Lettice, L. A., Heaney, S. J., Purdie, L. A., Li, L., de Beer, P., Oostra, B. A., Goode, D., Elgar, G., Hill, R. E., and de Graaff, E. (2003). A long-range Shh enhancer regulates expression in the developing limb and fin and is associated with preaxial polydactyly. *Hum. Mol. Genet.* **12,** 1725–1735.

Levine, M., and Tjian, R. (2003). Transcription regulation and animal diversity. *Nature* **424,** 147–151.

Loots, G. G. (2006). Modifying yeast artificial chromosomes to generate Cre/LoxP and FLP/FRT site-specific deletions and inversions. *Methods Mol. Biol.* **349,** 75–84.

Loots, G. G., and Ovcharenko, I. (2004). rVISTA 2.0: Evolutionary analysis of transcription factor binding sites. *Nucleic Acids Res.* **32,** W217–W221.

Loots, G. G., Locksley, R. M., Blankespoor, C. M., Wang, Z. E., Miller, W., Rubin, E. M., and Frazer, K. A. (2000). Identification of a coordinate regulator of interleukins 4, 13, and 5 by cross-species sequence comparisons. *Science* **288,** 136–140.

Loots, G. G., Ovcharenko, I., Pachter, L., Dubchak, I., and Rubin, E. M. (2002). rVista for comparative sequence-based discovery of functional transcription factor binding sites. *Genome Res.* **12,** 832–839.

Loots, G. G., Kneissel, M., Keller, H., Baptist, M., Chang, J., Collette, N. M., Ovcharenko, D., Plajzer-Frick, I., and Rubin, E. M. (2005). Genomic deletion of a long-range bone enhancer misregulates sclerostin in Van Buchem disease. *Genome Res.* **15,** 928–935.

Majoros, W. H., Pertea, M., Delcher, A. L., and Salzberg, S. L. (2005). Efficient decoding algorithms for generalized hidden Markov model gene finders. *BMC Bioinformatics* **6,** 16.

Marlin, S., Blanchard, S., Slim, R., Lacombe, D., Denoyelle, F., Alessandri, J. L., Calzolari, E., Drouin-Garraud, V., Ferraz, F. G., Fourmaintraux, A., Philip, N., Toublanc, J. E., *et al.* (1999). Townes-Brocks syndrome: Detection of a SALL1 mutation hot spot and evidence for a position effect in one patient. *Hum. Mutat.* **14,** 377–386.

Marshall, P., Chartrand, N., and Worton, R. G. (2001). The mouse dystrophin enhancer is regulated by MyoD, E-box-binding factors, and by the serum response factor. *J. Biol. Chem.* **276,** 20719–20726.

Mass, R. L., Zeller, R., Woychik, R. P., Vogt, T. F., and Leder, P. (1990). Disruption of formin-encoding transcripts in two mutant limb deformity alleles. *Nature* **346,** 853–855.

Masuya, H., Sagai, T., Wakana, S., Moriwaki, K., and Shiroishi, T. (1995). A duplicated zone of polarizing activity in polydactylous mouse mutants. *Genes Dev.* **9,** 1645–1653.

Mayor, C., Brudno, M., Schwartz, J. R., Poliakov, A., Rubin, E. M., Frazer, K. A., Pachter, L. S., and Dubchak, I. (2000). VISTA: Visualizing global DNA sequence alignments of arbitrary length. *Bioinformatics* **16,** 1046–1047.

McElreavy, K., Vilain, E., Abbas, N., Costa, J. M., Souleyreau, N., Kucheria, K., Boucekkine, C., Thibaud, E., Brauner, R., Flamant, F., and Fellous, M. (1992). XY sex reversal associated with a deletion 5' to the SRY "HMG box" in the testis-determining region. *Proc. Natl. Acad. Sci. USA* **89,** 11016–11020.

Miskey, C., Izsvak, Z., Plasterk, R. H., and Ivics, Z. (2003). The Frog Prince: A reconstructed transposon from Rana pipiens with high transpositional activity in vertebrate cells. *Nucleic Acids Res.* **31,** 6873–6881.

Mohrs, M., Blankespoor, C. M., Wang, Z. E., Loots, G. G., Afzal, V., Hadeiba, H., Shinkai, K., Rubin, E. M., and Locksley, R. M. (2001). Deletion of a coordinate regulator of type 2 cytokine expression in mice. *Nat. Immunol.* **2,** 842–847.

Mortlock, D. P., Guenther, C., and Kingsley, D. M. (2003). A general approach for identifying distant regulatory elements applied to the Gdf6 gene. *Genome Res.* **13,** 2069–2081.

Needleman, S. B., and Wunsch, C. D. (1970). A general method applicable to the search for similarities in the amino acid sequence of two proteins. *J. Mol. Biol.* **48,** 443–453.

Nistala, R., and Sigmund, C. D. (2002). A reliable and efficient method for deleting operational sequences in PACs and BACs. *Nucleic Acids Res.* **30,** e41.

Nobrega, M. A., Ovcharenko, I., Afzal, V., and Rubin, E. M. (2003). Scanning human gene deserts for long-range enhancers. *Science* **302,** 413.

Ovcharenko, I., and Nobrega, M. A. (2005). Identifying synonymous regulatory elements in vertebrate genomes. *Nucleic Acids Res.* **33,** W403–W407.

Ovcharenko, I., Loots, G. G., Hardison, R. C., Miller, W., and Stubbs, L. (2004a). zPicture: Dynamic alignment and visualization tool for analyzing conservation profiles. *Genome Res.* **14,** 472–477.

Ovcharenko, I., Nobrega, M. A., Loots, G. G., and Stubbs, L. (2004b). ECR Browser: A tool for visualizing and accessing data from comparisons of multiple vertebrate genomes. *Nucleic Acids Res.* **32,** W280–W286.

Ovcharenko, I., Loots, G. G., Giardine, B. M., Hou, M., Ma, J., Hardison, R. C., Stubbs, L., and Miller, W. (2005). Mulan: Multiple-sequence local alignment and visualization for studying function and evolution. *Genome Res.* **15,** 184–194.

Pennacchio, L. A., Olivier, M., Hubacek, J. A., Cohen, J. C., Cox, D. R., Fruchart, J. C., Krauss, R. M., and Rubin, E. M. (2001). An apolipoprotein influencing triglycerides in humans and mice revealed by comparative sequencing. *Science* **294,** 169–173.

Pennacchio, L. A., Ahituv, N., Moses, A. M., Prabhakar, S., Nobrega, M. A., Shoukry, M., Minovitsky, S., Dubchak, I., Holt, A., Lewis, K. D., Plajzer-Frick, I., Akiyama, J., et al. (2006). In vivo enhancer analysis of human conserved non-coding sequences. *Nature* **444,** 499–502.

Pennacchio, L. A., Loots, G. G., Nobrega, M. A., and Ovcharenko, I. (2007). Predicting tissue-specific enhancers in the human genome. *Genome Res.* **17,** 201–211.

Pilpel, Y., Sudarsanam, P., and Church, G. M. (2001). Identifying regulatory networks by combinatorial analysis of promoter elements. *Nat. Genet.* **29,** 153–159.

Poulin, F., Nobrega, M. A., Plajzer-Frick, I., Holt, A., Afzal, V., Rubin, E. M., and Pennacchio, L. A. (2005). In vivo characterization of a vertebrate ultraconserved enhancer. *Genomics* **85,** 774–781.

Quandt, K., Frech, K., Karas, H., Wingender, E., and Werner, T. (1995). MatInd and MatInspector: New fast and versatile tools for detection of consensus matches in nucleotide sequence data. *Nucleic Acids Res.* **23**, 4878–4884.

Roessler, E., Ward, D. E., Gaudenz, K., Belloni, E., Scherer, S. W., Donnai, D., Siegel-Bartelt, J., Tsui, L. C., and Muenke, M. (1997). Cytogenetic rearrangements involving the loss of the Sonic Hedgehog gene at 7q36 cause holoprosencephaly. *Hum. Genet.* **100**, 172–181.

Rojas, A., De Val, S., Heidt, A. B., Xu, S. M., Bristow, J., and Black, B. L. (2005). Gata4 expression in lateral mesoderm is downstream of BMP4 and is activated directly by Forkhead and GATA transcription factors through a distal enhancer element. *Development* **132**, 3405–3417.

Sagai, T., Hosoya, M., Mizushina, Y., Tamura, M., and Shiroishi, T. (2005). Elimination of a long-range cis-regulatory module causes complete loss of limb-specific Shh expression and truncation of the mouse limb. *Development* **132**, 797–803.

Sandelin, A., Wasserman, W. W., and Lenhard, B. (2004). ConSite: Web-based prediction of regulatory elements using cross-species comparison. *Nucleic Acids Res.* **32**, W249–W252.

Schug, J., and Overton, G. C. (1997). Modeling transcription factor binding sites with Gibbs Sampling and Minimum Description Length encoding. *Proc. Int. Conf. Intell. Syst. Mol. Biol.* **5**, 268–271.

Schwartz, S., Zhang, Z., Frazer, K. A., Smit, A., Riemer, C., Bouck, J., Gibbs, R., Hardison, R., and Miller, W. (2000). PipMaker–a web server for aligning two genomic DNA sequences. *Genome Res.* **10**, 577–586.

Schwartz, S., Kent, W. J., Smit, A., Zhang, Z., Baertsch, R., Hardison, R. C., Haussler, D., and Miller, W. (2003). Human-mouse alignments with BLASTZ. *Genome Res.* **13**, 103–107.

Sharan, R., Ovcharenko, I., Ben-Hur, A., and Karp, R. M. (2003). CREME: A framework for identifying cis-regulatory modules in human-mouse conserved segments. *Bioinformatics* **19** (Suppl. 1), i283–i291.

Sharan, R., Ben-Hur, A., Loots, G. G., and Ovcharenko, I. (2004). CREME: Cis-Regulatory Module Explorer for the human genome. *Nucleic Acids Res.* **32**, W253–W256.

Siddharthan, R., Siggia, E. D., and van Nimwegen, E. (2005). PhyloGibbs: A Gibbs sampling motif finder that incorporates phylogeny. *PLoS Comput. Biol.* **1**, e67.

Sinzelle, L., Vallin, J., Coen, L., Chesneau, A., Pasquier, D. D., Pollet, N., Demeneix, B., and Mazabraud, A. (2006). Generation of transgenic Xenopus laevis using the Sleeping Beauty transposon system. *Transgenic Res.* **15**, 751–760.

Stalker, J., Gibbins, B., Meidl, P., Smith, J., Spooner, W., Hotz, H. R., and Cox, A. V. (2004). The Ensembl Web site: Mechanics of a genome browser. *Genome Res.* **14**, 951–955.

Tallini, Y. N., Shui, B., Greene, K. S., Deng, K. Y., Doran, R., Fisher, P. J., Zipfel, W., and Kotlikoff, M. I. (2006). BAC transgenic mice express enhanced green fluorescent protein in central and peripheral cholinergic neurons. *Physiol. Genomics* **27**, 391–397.

Thomas, M. C., and Chiang, C. M. (2006). The general transcription machinery and general cofactors. *Crit. Rev. Biochem. Mol. Biol.* **41**, 105–178.

Touchman, J. W., Anikster, Y., Dietrich, N. L., Maduro, V. V., McDowell, G., Shotelersuk, V., Bouffard, G. G., Beckstrom-Sternberg, S. M., Gahl, W. A., and Green, E. D. (2000). The genomic region encompassing the nephropathic cystinosis gene (CTNS): Complete sequencing of a 200-kb segment and discovery of a novel gene within the common cystinosis-causing deletion. *Genome Res.* **10**, 165–173.

Ureta-Vidal, A., Ettwiller, L., and Birney, E. (2003). Comparative genomics: Genome-wide analysis in metazoan eukaryotes. *Nat. Rev. Genet.* **4**, 251–262.

van Deutekom, J. C., Lemmers, R. J., Grewal, P. K., van Geel, M., Romberg, S., Dauwerse, H. G., Wright, T. J., Padberg, G. W., Hofker, M. H., Hewitt, J. E., and Frants, R. R. (1996). Identification of the first gene (FRG1) from the FSHD region on human chromosome 4q35. *Hum. Mol. Genet.* **5**, 581–590.

Venter, J. C., Adams, M. D., Myers, E. W., Li, P. W., Mural, R. J., Sutton, G. G., Smith, H. O., Yandell, M., Evans, C. A., Holt, R. A., Gocayne, J. D., Amanatides, P., *et al.* (2001). The sequence of the human genome. *Science* **291**, 1304–1351.

Vortkamp, A., Gessler, M., and Grzeschik, K. H. (1991). GLI3 zinc-finger gene interrupted by translocations in Greig syndrome families. *Nature* **352**, 539–540.

Wallis, D. E., Roessler, E., Hehr, U., Nanni, L., Wiltshire, T., Richieri-Costa, A., Gillessen-Kaesbach, G., Zackai, E. H., Rommens, J., and Muenke, M. (1999). Mutations in the home-odomain of the human SIX3 gene cause holoprosencephaly. *Nat. Genet.* **22**, 196–198.

Wang, D. Z., Valdez, M. R., McAnally, J., Richardson, J., and Olson, E. N. (2001). The Mef2c gene is a direct transcriptional target of myogenic bHLH and MEF2 proteins during skeletal muscle development. *Development* **128**, 4623–4633.

Warming, S., Costantino, N., Court, D. L., Jenkins, N. A., and Copeland, N. G. (2005). Simple and highly efficient BAC recombineering using galK selection. *Nucleic Acids Res.* **33**, e36.

Wasserman, W. W., and Fickett, J. W. (1998). Identification of regulatory regions which confer muscle-specific gene expression. *J. Mol. Biol.* **278**, 167–181.

Waterston, R. H., Lindblad-Toh, K., Birney, E., Rogers, J., Abril, J. F., Agarwal, P., Agarwala, R., Ainscough, R., Alexandersson, M., An, P., Antonarakis, S. E., Attwood, J., *et al.* (2002). Initial sequencing and comparative analysis of the mouse genome. *Nature* **420**, 520–562.

Wirth, J., Wagner, T., Meyer, J., Pfeiffer, R. A., Tietze, H. U., Schempp, W., and Scherer, G. (1996). Translocation breakpoints in three patients with campomelic dysplasia and autosomal sex reversal map more than 130 kb from SOX9. *Hum. Genet.* **97**, 186–193.

Woolfe, A., Goodson, M., Goode, D. K., Snell, P., McEwen, G. K., Vavouri, T., Smith, S. F., North, P., Callaway, H., Kelly, K., Walter, K., Abnizova, I., *et al.* (2005). Highly conserved non-coding sequences are associated with vertebrate development. *PLoS Biol.* **3**, e7.

Woychik, R. P., and Alagramam, K. (1998). Insertional mutagenesis in transgenic mice generated by the pronuclear microinjection procedure. *Int. J. Dev. Biol.* **42**, 1009–1017.

Woychik, R. P., Stewart, T. A., Davis, L. G., D'Eustachio, P., and Leder, P. (1985). An inherited limb deformity created by insertional mutagenesis in a transgenic mouse. *Nature* **318**, 36–40.

Woychik, R. P., Generoso, W. M., Russell, L. B., Cain, K. T., Cacheiro, N. L., Bultman, S. J., Selby, P. B., Dickinson, M. E., Hogan, B. L., and Rutledge, J. C. (1990). Molecular and genetic characterization of a radiation-induced structural rearrangement in mouse chromosome 2 causing mutations at the limb deformity and agouti loci. *Proc. Natl. Acad. Sci. USA* **87**, 2588–2592.

Wu, S. C., Meir, Y. J., Coates, C. J., Handler, A. M., Pelczar, P., Moisyadi, S., and Kaminski, J. M. (2006). piggyBac is a flexible and highly active transposon as compared to sleeping beauty, Tol2, and Mos1 in mammalian cells. *Proc. Natl. Acad. Sci. USA* **103**, 15008–15013.

Xu, X., Scott, M. M., and Deneris, E. S. (2006). Shared long-range regulatory elements coordinate expression of a gene cluster encoding nicotinic receptor heteromeric subtypes. *Mol. Cell. Biol.* **26**, 5636–5649.

Zhu, Z., Pilpel, Y., and Church, G. M. (2002). Computational identification of transcription factor binding sites via a transcription-factor-centric clustering (TFCC) algorithm. *J. Mol. Biol.* **318**, 71–81.

Zhu, L., Lee, H. O., Jordan, C. S., Cantrell, V. A., Southard-Smith, E. M., and Shin, M. K. (2004). Spatiotemporal regulation of endothelin receptor-B by SOX10 in neural crest-derived enteric neuron precursors. *Nat. Genet.* **36**, 732–737.

Zuniga, A., Michos, O., Spitz, F., Haramis, A. P., Panman, L., Galli, A., Vintersten, K., Klasen, C., Mansfield, W., Kuc, S., Duboule, D., Dono, R., *et al.* (2004). Mouse limb deformity mutations disrupt a global control region within the large regulatory landscape required for Gremlin expression. *Genes Dev.* **18**, 1553–1564.

11

Regulatory Variation and Evolution: Implications for Disease

Emmanouil T. Dermitzakis
Wellcome Trust Sanger Institute, Wellcome Trust Genome Campus,
CB10 1SA Cambridge, United Kingdom

I. Introduction
II. Evolution and Variation of Noncoding DNA
III. Natural Selection in Noncoding DNA
IV. Gene Expression Studies
V. Disease Implications
VI. Conclusions
    References

ABSTRACT

In the past few years, it has become apparent that there is a substantial amount of noncoding DNA that contributes to genome function. However, the multidimensionality of noncoding DNA properties does not allow us to readily identify, characterize, and assess the functional impact of mutations, polymorphisms, and interspecific substitutions. In this chapter, we discuss the evolutionary properties of some of the known noncoding genomic elements, namely regulatory regions, and the extensions of this to other potentially functionally important noncoding regions such as conserved noncoding regions. The implications of this analysis for studies looking at molecular phenotypes such as gene expression and whole-organism phenotypes (e.g., disease) are presented in the context of the exploration of noncoding DNA properties. The aim is to take advantage of current and emerging analysis methods for noncoding DNA to elucidate the genetic causes of phenotypic variation. © 2008, Elsevier Inc.

Advances in Genetics, Vol. 61   0065-2660/08 $35.00
Copyright 2008, Elsevier Inc. All rights reserved.   DOI: 10.1016/S0065-2660(07)00011-9

I. INTRODUCTION

Protein-coding genes are the best-understood components of the human genome (IHGSC, 2004). We know the code by which they represent the amino acid components of the proteins they encode, and also the structure of the transcripts that are produced by these genes. Although it is clear that there is still much to learn about protein-coding genes, such as their fine structure, the nature and regulation of alternatively spliced forms, as well as the expected discovery of novel genes, it is safe to say that we have a pretty comprehensive picture of the protein-coding repertoire of the human cell (Hubbard et al., 2005). But protein-coding genes cannot function purely on the basis of the properties of their amino acid sequence. Especially in multicellular organisms, the cellular functions depend highly on a well-coordinated spatial, temporal, and quantitative control of the expression of gene products. In other words, there is a requirement for signals that carry information about when, where, and how much of the protein is produced. This coordination is termed "gene regulation" and describes in a very general way the processes and interactions described above. The sequences conferring regulatory control can be found in the genome, generally in the vicinity of the gene to be regulated, within the transcribed and intronic sequences, as well as outside the gene. Regulation can function at the DNA, at the RNA, or even at the protein sequence level (Wray et al., 2003).

The structure of these sequences is better understood at the DNA level. Most DNA regulatory sequences consist of short 6–12 bp sequences that are known as transcription factor binding sites (TFBSs). The requirements for binding of each transcription factor (TF) are pretty redundant so one cannot readily identify short sequences that are actually bound by specific TFs (Moses, 2002). Many models are currently in use for statistical identification of TFBS. Additional types of DNA regulatory elements are still being identified in multiple projects, the most prestigious and influential being the ENCODE project (Birney et al., 2007).

DNA regulatory and sometimes RNA sequences are the elements that express regulatory signals, which are then transmitted via multiple protein–protein and protein–nucleic acid interactions. These interactions can be intercellular as well as intracellular, and this raises the overall complexity of the system (Wray et al., 2003). In order to study properly the nature and structure of regulatory signals in humans and other mammals, one has to account for all these interactions (Fig. 11.1), a challenging and complex task for the currently available bioinformatics tools and experimental procedures.

One key issue is that all the protein–DNA and protein–protein interactions, and the sequences described above, are not fixed and stable. Regulatory processes evolve over long periods, allowing either conservation of function or alteration of different components where regulatory interactions may be gained or lost, leading to novel structures and functions within cells and the whole body.

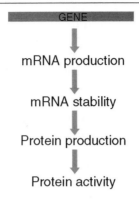

Figure 11.1. The levels at which regulatory variation can be expressed in the cell.

The evolutionary processes are driven by natural selection and also sometimes by genetic drift of mutations that arise in the populations within regulatory components. Some of these mutations may be associated with different, novel phenotypic effects differing between individuals in the population. Perhaps through selection, variants may rise to higher frequencies in the population and achieve polymorphic allele status, associated with distinct clinical outcomes. In this chapter, we will review some of the key features of the evolution and natural variation in regulatory components with our main focus on the DNA regulatory elements. The main characteristics of these regulatory sequences are relatively well characterized with some recognizable features. Finally, the depth and breadth of the regulatory components is so big that a certain degree of focus is required for proper treatment in a chapter like this.

II. EVOLUTION AND VARIATION OF NONCODING DNA

It would be ideal if we could discuss the evolutionary patterns of regulatory sequences and compare them to other functional genomic elements and genes. Unfortunately, this is not possible for a very simple reason: our knowledge of functional regulatory elements is very sparse and possibly biased for specific types of elements that fit the criteria of our experimental assays. So in order to study the patterns of evolution of DNA regulatory elements, we first have to study the patterns of evolution of noncoding DNA as a whole and then attempt to identify the subset that seems to fit the characteristics of regulatory sequences.

One of the most popular and currently most successful methods for identifying regulatory elements is by comparing sequences between species (Blanchette *et al.*, 2004; Dermitzakis *et al.*, 2002; Dubchak *et al.*, 2000; Hardison, 2000). The rationale behind such an approach is that functionally important sequences are maintained in long evolutionary times. Many studies have identified such regulatory sequences, especially those involved in the regulation of developmental genes (see also Greg Elgar's chapter) (Nobrega et al., 2003). We do not discuss this approach in detail since it is the topic of other chapters in this volume, but it is worth noting that a significant fraction of the genome appears to be under selective constraint (>5%), and this is an underestimate since the comparisons involve distant species such as human and mouse, where there has been considerable functional divergence between distant species. It is estimated that at least two third to three quarter of the constrained genome does not code for proteins and that hundreds of thousands of potentially functional noncoding elements are present in the human genome, many of which are far away from protein-coding genes (Waterston et al., 2002).

The comparative studies above provide a first-pass view on noncoding conservation but do not produce a clear picture of the evolutionary patterns of regulatory sequences, since we do not know which of the constrained noncoding elements are regulatory elements, at least in the conventional way we define them (Elnitski et al., 2003). It is useful to consider the evolutionary patterns of known regulatory elements. A few studies have addressed this question, frequently revealing conflicting results with respect to the degree of conservation of regulatory sequences (Dermitzakis and Clark, 2002; Wasserman et al., 2000). The key issue is that TFBSs have highly variable sequence specificity, which suggests that a certain degree of sequence change will not affect protein binding (Berg and von Hippel, 1987). In addition, the same redundant properties make it very likely that TFBSs turn over at an appreciable rate so that different species recruit the same TFs in a given regulatory element but through different and nonorthologous sequences (Dermitzakis and Clark, 2002; Ludwig et al., 1998; Wasserman et al., 2000). Although it has become clear that regulatory sequences demonstrate some degree of selective constraint, it is pretty well supported now that most of the regulatory sequences in the mammalian genomes do not show high degrees of sequence conservation (Birney et al., 2007; Dermitzakis and Clark, 2002). It is worth pointing out that recent results from the ENCODE consortium also demonstrate that we cannot account for the function of a large number of constrained sequences (Consortium, 2007). This suggests that the function of noncoding DNA is much more complicated than we had envisaged and the diversity and abundance of regulatory features will probably reveal new types of regulatory interactions at the DNA level.

Recent studies have attempted to combine TFBS variability characteristics and patterns of nucleotide substitution to reveal new regulatory sequences at the DNA and RNA level (Ettwiller et al., 2005; Moses et al., 2004; Xie et al., 2005). These studies consider sequence motif representation and conservation in confined targeted sequences of the genome, such as promoter sequences 5' to the gene. The reason for this is that such approaches are computationally intensive, so exploration of the full genome sequence of humans or other mammals is not feasible at this stage. Future approaches that account for such properties will likely prove very powerful when the computational limitations of working with the whole genome are surpassed.

Comparative studies are not the only ones that provide useful information for the nature of effects and selective constraints in regulatory sequences. Population genetic variation is also offering clues, particularly with respect to potentially functional variation segregating in human populations. There is not a substantial amount of resequencing data for regulatory regions, so one cannot yet reach firm conclusions about the degree of variation of such sequences. It has been demonstrated pretty convincingly that conserved noncoding sequences are under selective constraint in human populations (Drake *et al.*, 2006). This pattern is stable regardless of the definition or subset of single nucleotide polymorphisms (SNPs) or conserved noncoding sequences. In terms of known functional sequences, regardless of conservation, the most comprehensive study so far has been in the context of the ENCODE project (Birney *et al.*, 2007), which looked at the patterns of SNP frequencies and densities as well as indel densities within functionally active regulatory regions in 1% of the genome (see Fig. 11.2). The patterns were variable depending on the types of elements analyzed, and in some cases power was limited, but it is clear that a large fraction of regulatory elements may not be under strong enough selective constraint in human populations and many potentially small effect polymorphisms may be segregating at high frequencies. This is partly because human populations have

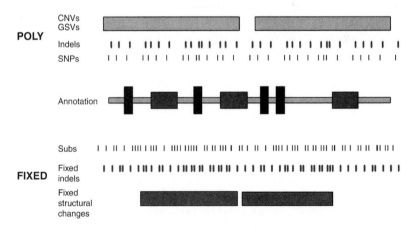

Figure 11.2. Distribution of polymorphic features (single nucleotide polymorphisms, copy number variations, genome structural variants, indels) and fixed features (substitutions, indels, segmental duplications, and genome structural changes) in the context of genome annotation. One has to consider all these features when evaluating the effects of population variation or differences between species.

small effective population size, so selection is not efficient enough to purge small effect deleterious mutations, which may subsequently reach high frequencies within some populations.

III. NATURAL SELECTION IN NONCODING DNA

Elucidating the forces of natural selection in noncoding DNA allows for the proper classification of genomic elements in the human genome. We have discussed briefly the detection and distribution of conserved noncoding sequences. However, noncoding DNA also contains elements that have played a role in recent human evolution. The detection of such elements, and the dissection of their characteristics, is not as straightforward as for genes, because we do not have a well-developed vocabulary to "read" the consequences of evolutionary changes. The methodologies currently used to detect recent evolutionary changes in noncoding DNA rely on contrasting the patterns and levels of sequence divergence in specific lineages (e.g., human) to the patterns and levels of sequence divergence of the same orthologous piece of DNA in other neighboring lineages on the phylogenetic tree (e.g., primates).

Studies focusing on this approach have identified recently accelerated noncoding sequences changes in the human genome, by contrasting their evolutionary rate to that of the chimpanzee lineage, using other species and especially macaque as outgroup (Bird et al., 2007; Pollard et al., 2006; Prabhakar et al., 2006). All these studies have identified hundreds or thousands of such noncoding elements using different methodologies and criteria. Accelerated changes in noncoding elements are close to genes that have specific gene ontology properties, suggesting that specific biological processes and functions are more prone to lineage specific evolution. In one of the studies, natural variation in human populations was considered (Bird et al., 2007). The authors found that the frequency spectrum of SNPs in "accelerated noncoding sequences" is skewed toward excess of high-frequency-derived alleles, a signal consistent with recent positive selection. But the most exciting observation is that SNPs within accelerated noncoding sequence are significantly associated with recent changes in gene expression levels of nearby genes. This suggests that the selective forces driving the fast evolution of these sequences are operating on the gene expression phenotypes.

The fact that change in gene expression is an important driver of evolution has also been demonstrated by various studies looking at divergence of the levels of gene expression in primate species. Such studies were difficult before the wider availability of multiple primate sequences for the design of appropriate probe sequences for microarrays (Enard et al., 2002). However, a recent study has looked at well-designed arrays that have used the specific sequences of genes for each species and have demonstrated that there have been recent changes in gene expression, especially for TFs, further suggesting that gene regulation is an important driver of evolution (Gilad *et al.*, 2006).

IV. GENE EXPRESSION STUDIES

As mentioned previously, gene expression is not only different between species but there is also segregating variation in human populations that drives differential gene expression levels between individuals. The availability of cell lines from multiple individuals in a population, and from multiple populations, has allowed the dissection of genetic effects in gene regulation, in particular in lymphoblastoid cells. In addition, the HapMap project has generated millions of genotypes in 270 individuals (IHMC, 2005) for which there are lymphoblastoid cell lines and these have been used extensively for expression at quantitative trait loci (eQTL) studies.

First, it has been demonstrated that gene expression is a heritable trait when families were studied (Cheung *et al.*, 2003; Morley *et al.*, 2004; Schadt *et al.*, 2003). These served as a baseline for many subsequent studies that attempted to detect regions of the genome carrying sequence variation with either direct (*cis*) or indirect (*trans*) effect on gene expression levels. These studies have revealed the presence of substantial numbers of genes with *cis* regulatory variation, which, in some cases, generate very big differences in gene expression levels between individuals (Cheung *et al.*, 2005; Monks *et al.*, 2004; Stranger *et al.*, 2005, 2007a,b). Most of these *cis* signals are found very close to the gene not only in the promoter regions but also in the introns and other genic regions. Distant elements have not been found much in such studies, possibly because these regulators may be acting in developmental processes not expressed in lymphocytes. *Trans*-regulatory interactions are harder to detect, and in fact very few reliable signals have been found, and mainly in model organisms such as yeast (Brem *et al.*, 2002). This is primarily due to the small sample sizes that have been used in these studies. In addition, the biology of the cell lines out of the context of the blood may not manifest many of the genetic effects that are *trans* deriving from intercellular interactions.

All the eQTL studies discussed above have mainly considered SNP data as the proxy for human genomic variation and this is of course only one part of the picture. Recent studies have looked at the effects of structural variation and in particular copy number variation (CNV) in gene expression either in targeted regions (McCarroll *et al.*, 2006) or genome wide (Stranger *et al.*, 2007a). Although the data on CNV are still limited (Redon *et al.*, 2006), it has become clear that it plays an important role in driving different levels of gene expression between individuals. The effects of CNV on gene expression is not just because of increased or decreased gene copy number that correlate directly with gene expression levels, but additional effects result from alteration or disturbance of the regulatory landscape at the sites of integration or deletion of copies in different genomic regions (Stranger et al., 2007a).

The information produced by all these studies has been very useful to understand the properties of variation in gene regulation, especially in *cis*. However, there are also limitations of these analyses to be considered when making generalizations. Sample sizes were small, and it is certain that larger future studies will reveal many more eQTLs, some of smaller effect. In addition, the vast majority of studies have looked at the most readily available cell type, lymphoblastoid cell lines, representing only one of the many tissues in the human body. This restricts the discovery of variation tissue-specific regulatory elements. Although more tissues and cell lines may become available, it will be hard to eventually obtain a comprehensive set of cell types from a large number of individuals. Finally, one needs to consider all types of variation, such as CNV and other structural variants, and the study of these requires the prior detection of such variants. Whole-genome resequencing technologies will make it possible to correlate all types of human genome variation with gene expression phenotypes and obtain an unbiased distribution of causal regulatory effects.

V. DISEASE IMPLICATIONS

Most types of functional variation, regulatory variation, and chromatin structure variation in noncoding DNA may be responsible for phenotypic differences between individuals and susceptibility to disease (Stranger and Dermitzakis, 2006). In the past, it was hypothesized that protein-coding variation was most likely the main source of disease susceptibility, mainly because most known monogenic disorders are due to such mutations (Botstein and Risch, 2003). However, it is becoming increasingly clear that complex disorders such as type II diabetes and cardiovascular disease may be driven by regulatory polymorphisms with smaller effects. Recent genome-wide association studies have indicated that this is very likely because many of the disease-associated SNPs are not in protein-coding regions, but the validity of such variants as causal needs to be verified when identified in such disease association studies (Saxena et al., 2007; Sladek et al., 2007; Zeggini et al., 2007). Studies that have generated genome-wide eQTLs may assist in this direction because SNPs or genomic regions that carry both a disease signal and an eQTL may demonstrate that the disease effect is regulatory. These studies are currently being performed, and we will soon see the outcome and validity of such an argument (Fig. 11.3).

But the use of genomic information in noncoding regions also helps the identification of mutations for monogenic disorders. There are substantial numbers of disease individuals with clear phenotypic categorization in whom it cannot be shown that they carry any protein-coding or posttranscriptional processing

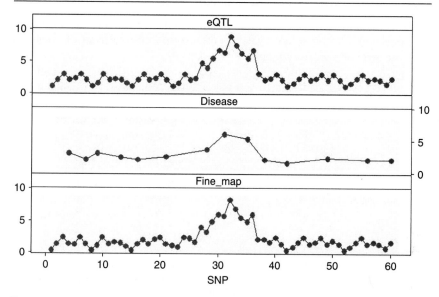

Figure 11.3. Examples of associations of single nucleotide polymorphisms with gene expression variation (expression at quantitative trait loci, eQTL), disease phenotype, and fine of a disease-associated region. Note that the initial disease signal may be weaker than the eQTL but become stronger when more markers are considered. The combination of eQTL data (even from different populations or samples) with disease association data facilitates the interpretation of the primary biological effect behind the disease association.

mutations in the established disease-causing gene mutated in most other individuals. Exploration of the noncoding space for mutations is the obvious next step, but until recently that was not possible (Antonarakis and Beckmann, 2006). In the last few years, the identification of conserved noncoding sequences as well as other functional noncoding elements from coordinated projects such as EN-CODE (Birney *et al.*, 2007) allow proper prioritization of target sequences to explore (see Chapter 13 by Kleinjan and Lettice, this volume).

VI. CONCLUSIONS

For the first time, we see the human genome as an integrated whole instead of merely as a collection of genes. We begin to realize the value of the whole sequence space and start to see the potential functions and interactions of individual elements. The desire to discover new disease and other causal variants leads to an immediate use of all genomic information to explore sequence variation and assist biological interpretation. The main message from recent studies is that lots of new biology will be discovered in noncoding DNA, some of it in the form of regulatory signals.

References

Antonarakis, S. E., and Beckmann, J. S. (2006). Mendelian disorders deserve more attention. *Nat. Rev. Genet.* **7,** 277–282.

Berg, O. G., and von Hippel, P. H. (1987). Selection of DNA binding sites by regulatory proteins. Statistical-mechanical theory and application to operators and promoters. *J. Mol. Biol.* **193,** 723–750.

Bird, C. P., Stranger, B. E., Liu, M., Thomas, D. J., Ingle, C. E., Beazley, C., Miller, W., Hurles, M. E., and Dermitzakis, E. T. (2007). Fast-evolving noncoding sequences in the human genome. *Genome Biol.* **8,** R118.

Birney, E., Stamatoyannopoulos, J. A., Dutta, A., Guigo, R., Gingeras, T. R., Margulies, E. H., Weng, Z., Snyder, M., Dermitzakis, E. T., Thurman, R. E., Kuehn, M. S., Taylor, C. M., *et al.* (2007). Identification and analysis of functional elements in 1% of the human genome by the ENCODE pilot project. *Nature* **447,** 799–816.

Blanchette, M., Kent, W. J., Riemer, C., Elnitski, L., Smit, A. F., Roskin, K. M., Baertsch, R., Rosenbloom, K., Clawson, H., Green, E. D., Haussler, D., and Miller, W. (2004). Aligning multiple genomic sequences with the threaded blockset aligner. *Genome Res.* **14,** 708–715.

Botstein, D., and Risch, N. (2003). Discovering genotypes underlying human phenotypes: Past successes for Mendelian disease, future approaches for complex disease. *Nat. Genet.* **33**(Suppl.), 228–237.

Brem, R. B., Yvert, G., Clinton, R., and Kruglyak, L. (2002). Genetic dissection of transcriptional regulation in budding yeast. *Science* **296,** 752–755.

Cheung, V. G., Conlin, L. K., Weber, T. M., Arcaro, M., Jen, K. Y., Morley, M., and Spielman, R. S. (2003). Natural variation in human gene expression assessed in lymphoblastoid cells. *Nat. Genet.* **33,** 422–425.

Cheung, V. G., Spielman, R. S., Ewens, K. G., Weber, T. M., Morley, M., and Burdick, J. T. (2005). Mapping determinants of human gene expression by regional and genome-wide association. *Nature* **437,** 1365–1369.

Dermitzakis, E. T., and Clark, A. G. (2002). Evolution of transcription factor binding sites in mammalian gene regulatory regions: Conservation and turnover. *Mol. Biol. Evol.* **19,** 1114–1121.

Dermitzakis, E. T., Reymond, A., Lyle, R., Scamuffa, N., Ucla, C., Deutsch, S., Stevenson, B. J., Flegel, V., Bucher, P., Jongeneel, C. V., and Antonarakis, S. E. (2002). Numerous potentially functional but non-genic conserved sequences on human chromosome 21. *Nature* **420,** 578–582.

Drake, J. A., Bird, C., Nemesh, J., Thomas, D. J., Newton-Cheh, C., Reymond, A., Excoffier, L., Attar, H., Antonarakis, S. E., Dermitzakis, E. T., and Hirschhorn, J. N. (2006). Conserved noncoding sequences are selectively constrained and not mutation cold spots. *Nat. Genet.* **38,** 223–227.

Dubchak, I., Brudno, M., Loots, G. G., Pachter, L., Mayor, C., Rubin, E. M., and Frazer, K. A. (2000). Active conservation of noncoding sequences revealed by three-way species comparisons. *Genome Res.* **10,** 1304–1306.

Elnitski, L., Hardison, R. C., Li, J., Yang, S., Kolbe, D., Eswara, P., O'Connor, M. J., Schwartz, S., Miller, W., and Chiaromonte, F. (2003). Distinguishing regulatory DNA from neutral sites. *Genome Res.* **13,** 64–72.

Enard, W., Khaitovich, P., Klose, J., Zollner, S., Heissig, F., Giavalisco, P., Nieselt-Struwe, K., Muchmore, E., Varki, A., Ravid, R., Doxiadis, G. M., Bontrop, R. E., *et al.* (2002). Intra- and interspecific variation in primate gene expression patterns. *Science* **296,** 340–343.

Ettwiller, L., Paten, B., Souren, M., Loosli, F., Wittbrodt, J., and Birney, E. (2005). The discovery, positioning and verification of a set of transcription-associated motifs in vertebrates. *Genome Biol.* **6,** R104.

Gilad, Y., Oshlack, A., Smyth, G. K., Speed, T. P., and White, K. P. (2006). Expression profiling in primates reveals a rapid evolution of human transcription factors. *Nature* **440,** 242–245.

Hardison, R. C. (2000). Conserved noncoding sequences are reliable guides to regulatory elements. *Trends Genet.* **16,** 369–372.

Hubbard, T., Andrews, D., Caccamo, M., Cameron, G., Chen, Y., Clamp, M., Clarke, L., Coates, G., Cox, T., Cunningham, F., Curwen, V., Cutts, T., *et al.* (2005). Ensembl 2005. *Nucleic Acids Res.* **33,** D447–D453.

IHGSC (2004). Finishing the euchromatic sequence of the human genome. *Nature* **431,** 931–945.

IHMC (2005). A haplotype map of the human genome. *Nature* **437,** 1299–1320.

Ludwig, M. Z., Patel, N. H., and Kreitman, M. (1998). Functional analysis of eve stripe 2 enhancer evolution in Drosophila: Rules governing conservation and change. *Development* **125,** 949–958.

McCarroll, S. A., Hadnott, T. N., Perry, G. H., Sabeti, P. C., Zody, M. C., Barrett, J. C., Dallaire, S., Gabriel, S. B., Lee, C., Daly, M. J., and Altshuler, D. M. (2006). Common deletion polymorphisms in the human genome. *Nat. Genet.* **38,** 86–92.

Monks, S. A., Leonardson, A., Zhu, H., Cundiff, P., Pietrusiak, P., Edwards, S., Phillips, J. W., Sachs, A., and Schadt, E. E. (2004). Genetic inheritance of gene expression in human cell lines. *Am. J. Hum. Genet.* **75,** 1094–1105.

Morley, M., Molony, C. M., Weber, T. M., Devlin, J. L., Ewens, K. G., Spielman, R. S., and Cheung, V. G. (2004). Genetic analysis of genome-wide variation in human gene expression. *Nature* **430,** 743–747.

Moses, A. (2002). Site-specific substitution rate. *BMC Evolutionary Biology to be added* .

Moses, A. M., Chiang, D. Y., Pollard, D. A., Iyer, V. N., and Eisen, M. B. (2004). MONKEY: Identifying conserved transcription-factor binding sites in multiple alignments using a binding site-specific evolutionary model. *Genome Biol.* **5,** R98.

Nobrega, M. A., Ovcharenko, I., Afzal, V., and Rubin, E. M. (2003). Scanning human gene deserts for long-range enhancers. *Science* **302,** 413.

Pollard, K. S., Salama, S. R., King, B., Kern, A. D., Dreszer, T., Katzman, S., Siepel, A., Pedersen, J. S., Bejerano, G., Baertsch, R., Rosenbloom, K. R., Kent, J., *et al.* (2006). Forces shaping the fastest evolving regions in the human genome. *PLoS Genet.* **2,** e168.

Prabhakar, S., Noonan, J. P., Paabo, S., and Rubin, E. M. (2006). Accelerated evolution of conserved noncoding sequences in humans. *Science* **314,** 786.

Redon, R., Ishikawa, S., Fitch, K. R., Feuk, L., Perry, G. H., Andrews, T. D., Fiegler, H., Shapero, M. H., Carson, A. R., Chen, W., Cho, E. K., Dallaire, S., *et al.* (2006). Global variation in copy number in the human genome. *Nature* **444,** 444–454.

Saxena, R., Voight, B. F., Lyssenko, V., Burtt, N. P., de Bakker, P. I., Chen, H., Roix, J. J., Kathiresan, S., Hirschorm, J. N., Daly, M. J., Hughes, T. E., Groop, L., *et al.* (2007). Genome-wide association analysis identifies loci for type 2 diabetes and triglyceride levels. *Science* **316,** 1331–1336.

Schadt, E. E., Monks, S. A., Drake, T. A., Lusis, A. J., Che, N., Colinayo, V., Ruff, T. G., Milligan, S. B., Lamb, J. R., Cavet, G., Linsley, P. S., Mao, M., *et al.* (2003). Genetics of gene expression surveyed in maize, mouse and man. *Nature* **422,** 297–302.

Sladek, R., Rocheleau, G., Rung, J., Dina, C., Shen, L., Serre, D., Boutin, P., Vincent, D., Belisle, A., Hadjadj, S., Balkau, B., Heude, B., *et al.* (2007). A genome-wide association study identifies novel risk loci for type 2 diabetes. *Nature* **445,** 881–885.

Stranger, B. E., and Dermitzakis, E. T. (2006). From DNA to RNA to disease and back: The "central dogma" of regulatory disease variation. *Hum. Genomics* **2,** 383–390.

Stranger, B. E., Forrest, M. S., Clark, A. G., Minichiello, M. J., Deutsch, S., Lyle, R., Hunt, S., Kahl, B., Antonarakis, S. E., Tavare, S., Deloukas, P., and Dermitzakis, E. T. (2005). Genome-wide associations of gene expression variation in humans. *PLoS Genet.* **1,** e78.

Stranger, B. E., Forrest, M. S., Dunning, M., Ingle, C. E., Beazley, C., Thorne, N., Redon, R., Bird, C. P., de Grassi, A., Lee, C., Tyler-Smith, C., Carter, N., et al. (2007a). Relative impact of nucleotide and copy number variation on gene expression phenotypes. Science 315, 848–853.

Stranger, B. E., Nica, A. C., Forrest, M. S., Dimas, A., Bird, C. P., Beazley, C., Ingle, C. E., Dunning, M., Flicek, P., Koller, D., Montgomery, S., Tavare, S., et al. (2007b). Population genomics of human gene expression. Nat. Genet. 39, 1217–1224.

Wasserman, W. W., Palumbo, M., Thompson, W., Fickett, J. W., and Lawrence, C. E. (2000). Human-mouse genome comparisons to locate regulatory sites. Nat. Genet. 26, 225–228.

Waterston, R. H., Lindblad-Toh, K., Birney, E., Rogers, J., Abril, J. F., Agarwal, P., Agarwala, R., Ainscough, R., Alexandersson, M., An, P., Antonarakis, S. E., Attwood, J., et al. (2002). Initial sequencing and comparative analysis of the mouse genome. Nature 420, 520–562.

Wray, G. A., Hahn, M. W., Abouheif, E., Balhoff, J. P., Pizer, M., Rockman, M. V., and Romano, L. A. (2003). The evolution of transcriptional regulation in eukaryotes. Mol. Biol. Evol. 20, 1377–1419.

Xie, X., Lu, J., Kulbokas, E. J., Golub, T. R., Mootha, V., Lindblad-Toh, K., Lander, E. S., and Kellis, M. (2005). Systematic discovery of regulatory motifs in human promoters and 3′ UTRs by comparison of several mammals. Nature 434, 338–345.

Zeggini, E., Weedon, M. N., Lindgren, C. M., Frayling, T. M., Elliott, K. S., Lango, H., Timpson, N. J., Perry, J. R., Rayner, N. W., Freathy, R. M., Barrett, J. C., Shields, B., et al. (2007). Replication of genome-wide association signals in U.K. samples reveals risk loci for type 2 diabetes. Science 316, 1336–1341.

12 Organization of Conserved Elements Near Key Developmental Regulators in Vertebrate Genomes

Adam Woolfe[1] and Greg Elgar
School of Biological and Chemical Sciences, Queen Mary, University of London, London E1 4NS, United Kingdom

   I. Introduction
  II. Gene-Regulatory Networks in Development
 III. Identification of Evolutionarily Constrained Sequences Using Phylogenetic Footprinting
  IV. Searches for Regulatory Elements Using Evolutionary Conservation
   V. *Takifugu Rubripes*: A Compact Model Genome
  VI. Identification of Enhancer Elements Through Fish-Mammal Comparisons
 VII. Fish-Mammal Conserved Noncoding Elements Are Associated with Vertebrate Development
VIII. High-Resolution Analysis of the Organization of CNEs Around Key Developmental Regulators
  IX. General Genomic Environment Around CNEs
   X. CNEs Present in Transcripts
  XI. CNEs Located Within UTRs
 XII. CNEs Are Located Large Distances Away from Their Putative Target Gene
       A. Mutation events in and around CNEs over evolution
XIII. Discussion
       References

[1]Current address: Genome Technology Branch, National Human Genome Research Institute, National Institutes of Health, Rockville, Maryland 20870

Advances in Genetics, Vol. 61
Copyright 2008, Elsevier Inc. All rights reserved.

0065-2660/08 $35.00
DOI: 10.1016/S0065-2660(07)00012-0

ABSTRACT

Sequence conservation has traditionally been used as a means to target functional regions of complex genomes. In addition to its use in identifying coding regions of genes, the recent availability of whole genome data for a number of vertebrates has permitted high-resolution analyses of the noncoding "dark matter" of the genome. This has resulted in the identification of a large number of highly conserved sequence elements that appear to be preserved in all bony vertebrates. Further positional analysis of these conserved noncoding elements (CNEs) in the genome demonstrates that they cluster around genes involved in developmental regulation. This chapter describes the identification and characterization of these elements, with particular reference to their composition and organization.   © 2008, Elsevier Inc.

I. INTRODUCTION

Complex multicellular organisms must overcome the fact that whil all constituent cells contain the same genetic instructions, they do not all express the same set of genes. Specific populations or groups of cells must be able to differentiate, specialize, and perform defined roles, so enabling the survival of the whole. Once a differentiated state has become established, further changes in gene expression allow the organism to respond to insult or challenge. The most complex, dynamic, and exquisitely controlled period of an organism's life occurs during development, from single cell to fully differentiated state. In vertebrates, this process results in billions of differentiated cells, communicating and functioning in a particular and coordinated manner, responding and adapting as organized cohorts.

Biologists have been studying how genes are regulated for many decades and despite, or perhaps because of, its complex nature, a great deal of attention has been focused on the control of vertebrate development. This has resulted in the elucidation of many of the regulatory genes and pathways involved, and yet surprisingly, there is scant knowledge of either the molecular mechanisms or the underlying regulatory language involved.

Clearly, one of the limitations in identifying regulatory elements around vertebrate genes was, until a few years ago, the lack of genomic sequence data available. The completion of the human genome in 2001, followed by those of a host of other vertebrates in the proceeding few years, provided an opportunity to mine these vast datasets for specific sequences that might have function. The simplest and most commonly used approach has been to identify regions of noncoding DNA that have been conserved between two or more species, implying that some functional constraint has prevented these sequences from mutating at a neutral rate. However, despite the simplicity of the approach, interpretation of the results has often been confounded by a number of factors: rates of evolution are not homogenous across or between genomes, regulatory sequences are often poorly conserved at the primary sequence level, and our knowledge of the regulatory language of the genome is rather limited. In fact, regulatory sequences, it appears, are often not conserved at all.

However, under certain circumstances, some of these analyses have been insightful, particularly those aimed at identifying sequences with regulatory potential around genes that coordinate early development. Given the critical nature and exquisite complexity of developmental processes, both in terms of time and in terms of 3-dimensional space, one might expect to find an abundance of regulatory control elements encoded in the genomic DNA surrounding these genes. This chapter describes the identification of such a set of sequences, and how these sequences are organized with respect to the genes upon which they act in vertebrate genomes.

II. GENE-REGULATORY NETWORKS IN DEVELOPMENT

Animal development is thought to be controlled by progression through a number of transcriptional states that are transiently positioned in embryonic space (Levine and Davidson, 2005). This is orchestrated by regulated expression of transcription factor genes in a specific spatiotemporal manner. These genes are controlled by multiple *cis*-regulatory modules (CRMs) (often termed "enhancers," "repressors," or "silencers") that act as sites for combinatorial protein binding in order to actuate complex patterns of expression. This complex set of interactions during development can be represented in the form of a gene regulatory network (GRN). GRNs involving specific developmental pathways have so far been dissected in a number of invertebrates. These include endomesoderm specification in the sea urchin (Davidson *et al.*, 2002), dorsal-ventral specification in *Drosophila melanogaster* (Levine and Davidson, 2005; Ochoa-Espinosa *et al.*, 2005), and the general specification of developmental pathways in *Ciona intestinalis* (Shi *et al.*, 2005). Evidence on which these GRNs are based derives from large-scale experimental efforts which involve obtaining spatial and temporal expression data on all genes in the network and the use of large-scale perturbation analysis to see the effect of gene deletion on the expression of other genes in the network (Davidson *et al.*, 2002). Among the more complex vertebrates, *Xenopus laevis* has been proposed as a model system in which to study GRNs, because of the ease with which genes can be manipulated, and a small GRN involving mesoderm specification has already been proposed (Koide *et al.*, 2005; Loose and Patient, 2004). However, the problem of identifying the CRMs controlling these networks is still a major challenge.

III. IDENTIFICATION OF EVOLUTIONARILY CONSTRAINED SEQUENCES USING PHYLOGENETIC FOOTPRINTING

Prior to the availability of whole genome sequences, classical searches for distal cis-regulatory elements typically involved various trial-and-error strategies. Experimental approaches to the identification of regulatory elements include deletion constructs of upstream sequences to determine the minimal sequences necessary for transcription in cell culture-based systems, DNAse I hypersensitivity studies to identify sequences potentially available for transcription factor binding, and DNA footprinting to determine sequences that bind various regulatory proteins (see Pennacchio and Rubin, 2001). Large-scale promoter- and enhancer-trapping studies have also been carried out in mice (Durick et al., 1999) and zebra fish (Ellingsen et al., 2005), but on the whole, laboratory attempts at identification and characterization of distal regulatory elements have been unguided, highly laborious, and time-consuming. Access to the sequences of several vertebrate genomes has presented an unprecedented opportunity for the discovery of functional elements in the human genome through comparative genomics. The discovery of putative functional elements through comparison of sequences from several species, known as phylogenetic footprinting, is based on the assumption that these elements evolve more slowly than surrounding nonfunctional DNA, as they are under negative (purifying) selection. Thus, sequences that are more highly conserved than would be expected under a reasonable model of neutral evolution are likely to be important for function. One of the key decisions inherent in phylogenetic footprinting is the choice of organisms with which the comparison will be made. Given the common and fundamental nature of early embryogenesis across the vertebrate lineage, it might be expected that many of the regulatory instructions hardwired into the genome would be highly conserved across large evolutionary distances.

IV. SEARCHES FOR REGULATORY ELEMENTS USING EVOLUTIONARY CONSERVATION

Genomic comparisons have frequently been used as a method to identify regulatory elements. Early studies using comparisons between distantly related D. melanogaster species proved successful in identifying conserved enhancer elements (Martinez-cruzado et al., 1988), as the rapid rate of evolution in this species helped resolve functional sequences from background conservation. The release of a number of draft vertebrate genomes has in parallel spurred the development of appropriate genomic alignment, visualization, and analytical bioinformatics tools have made large-scale sequence comparisons not only possible but an increasingly popular approach for the discovery of functional elements in these genomes. Several studies have attempted to identify regulatory elements in mammals using human–mouse comparisons (e.g., Frazer et al., 2004; Göttgens et al., 2000; Hardison et al., 1997; Oeltjen et al., 1997; Wasserman et al., 2000). Unfortunately, owing to differences in mutation rates across the genome, relatively small evolutionary divergence between mammals and the slow rate of neutral divergence among vertebrates, many more sequences may be conserved than actually play functional roles (Tautz, 2000). Indeed, although ~40% of the human and mouse genomes are alignable, only ~5% is estimated to be under evolutionary constraint (Waterston et al., 2002). Consequently, specific conservation criteria were proposed to distinguish functional conservation from background. An arbitrary criterion of 70% identity over at least 100 bp of ungapped alignment (which is above the average rate of neutral conservation) between human and mouse sequences has been used to successfully identify a number of regulatory elements (Göttgens et al., 2000; Loots et al., 2000). Using this criterion across whole-genome human–mouse alignments identified ~327,000 conserved elements, making up around 1% of the human genome, which were located in noncoding regions and had little or no evidence of transcription (Dermitzakis et al., 2003, 2005). These sequences appear to be distributed uniformly across the genome and are negatively correlated with the distribution of genes, suggesting roles which are distance-independent or that are not directly involved in gene expression. An alternative approach for the identification of noncoding constrained elements was proposed by Margulies et al. (2003), who devised two strategies based on parsimony and binomial-based models applied to multispecies alignments. Unlike simple percent-identity-based approaches, these models take into account the derived local neutral mutation rate as well as the divergence times between sequences based on a phylogenetic tree. Combining these approaches, on a 1.7 Mb region around the CFTR locus, they were able to successfully distinguish between neutrally evolving sequence such as known ancestral repeats and constrained elements such as exons, and identified a large number of conserved elements, ~70% of which were located in noncoding regions. Many more constrained sequences were identified using this multiple-alignment approach than could be identified using human–mouse pairwise alignments alone, demonstrating the power of multispecies alignments (Margulies et al., 2006).

Nevertheless, it is currently unclear what proportion of these constrained noncoding elements is regulatory or even functional at all. This was highlighted by Nobrega et al. (2004) who deleted two large noncoding regions of roughly 1 Mb in size on mouse chromosomes 3 and 19. These regions contain a total of 1243 sequences that are >70% identical across at least 100 bp between human and mouse (a small number of which were also conserved in chicken and frog) and yet their deletion caused no loss of viability or any other overt phenotypic changes. Quantitative polymerase chain reaction (PCR) analysis showed that expression levels of the genes flanking the deleted sections were unaffected in all but 2 of the 108 tissues assayed. In addition, enhancer assays in transgenic mice of 15 of the most highly conserved elements in these regions (10 of which were conserved in chicken and 5 in frog) found only one that upregulated the reporter gene in a tissue-specific manner. Although it is possible that these deletions may cause abnormalities undetected in this time setting or environmental context, this study highlights the possibility that many sequences that have remained conserved across large evolutionary distances may not play critical functional roles. Furthermore, the ability to use comparative genome alignments to discover thousands of potentially functional regulatory elements in mammalian genomes overwhelms our current ability to test their function *in vivo*.

Commonly used techniques for testing putative distal regulatory sequences for enhancer activity, such as transgenic reporter gene assays in developing mouse embryos, are still very expensive, time-consuming, and laborious. Cheaper and faster *in vivo* reporter gene assays have been developed for use in other model organisms such as frog, zebra fish, and medaka (Müller *et al.*, 2002), although it would still take many decades to test even a fraction of the elements thought to be functional in mammals. In light of the inability of current technologies to test large numbers of mammalian-conserved elements with "regulatory potential," it is important to be able to prioritize a smaller set of elements with high regulatory potential for more focused studies. To address this, Bejerano *et al.* (2004) searched for sequences that were identical over at least 200 bp between human, mouse, and rat (termed "ultraconserved" elements) and identified 481 such sequences, of which over half were located in noncoding regions. By contrast with other mammalian-conserved sequences, these elements often clustered in the vicinity of genes involved in transcriptional regulation and/or development and overlapped a number of known enhancers, suggesting that they are likely to function as CRMs. Another highly successful approach for prioritizing CRMs is the identification of sequences that are conserved across extreme evolutionary distances such as fish and mammals. Teleost fish are well suited for comparisons with mammals, as they last shared a common ancestor 400–450 million years ago and therefore it is assumed that only critical functional sequences would be conserved between genomes that are otherwise diverged. One teleost genome, in particular, that of the puffer fish *Takifugu rubripes*, has been used extensively for the discovery of CRMs.

V. *TAKIFUGU RUBRIPES*: A COMPACT MODEL GENOME

The sequencing of the Japanese puffer fish *Takifugu rubripes* (commonly known as *Fugu*) was first proposed by Sydney Brenner and colleagues in 1993 as a compact model vertebrate genome (Brenner *et al.*, 1993). They showed *Fugu* has one of the smallest genomes of any known vertebrate at around 390 Mb in length (around one eighth the size of the human genome), but as a vertebrate, it has a similar complement of genes to that of mammals (Elgar *et al.*, 1996). In addition, given the enormous cost of sequencing large genomes at that point, *Fugu* represented a genome that could be completed within a few years at a fraction of the cost of the human genome whil providing a resource for its annotation. The publication of the draft version of the genome in 2002 (Aparicio *et al.*, 2002) hailed the release of only the second vertebrate genome to be sequenced after human. The justification for sequencing was demonstrated immediately, with the first comparison of the *Fugu* and human genome revealing more than 1000 genes that had previously remained unidentified (Aparicio *et al.*, 2002). The compact nature of the *Fugu* genome is, principally, due to the low abundance of repeat sequences and a reduction in the size of intronic and intergenic sequences, with a resultant increase in gene density. The remaining genomic sequence is, consequently, enriched for regulatory and other functional elements, helping reduce the search space for such elements in comparative analyses with other genomes. Several other interesting aspects of fish in general have spurred the sequencing of a number of other teleosts. Teleosts are a highly diverse group, making up half of all extant vertebrate species (Nelson, 1994). Teleosts underwent a whole genome-duplication event around 330 million years ago that coincided with this huge burst in diversification (Hoegg *et al.*, 2004) leading many to believe that the two events were linked. Some teleosts are used as experimental developmental models, in particular zebra fish and medaka. Fish are therefore excellent models in the study of speciation, genome evolution, gene duplication, and development.

VI. IDENTIFICATION OF ENHANCER ELEMENTS THROUGH FISH-MAMMAL COMPARISONS

Although initial interest in the *Fugu* genome centered on its application for gene identification, several studies preceding the release of the genome sequence indicated that fish–mammal comparisons may also be useful in the identification of some regulatory enhancers. The first example of conserved noncoding sequences (CNSs) identified between *Fugu* and mammals was found around the *Hoxb-1* gene and were shown to be able to recapitulate *Hoxb-1* neuroectoderm expression in developing mouse embryos (Marshall *et al.*, 1994). This was followed up by a similar survey around the *Hoxb4* gene which found a number of other conserved enhancer elements through transgenic testing in mouse (Aparicio *et al.*, 1995). More recently, comparative analyses with *Fugu* identified a global control region responsible for tissue-specific expression of the HoxD cluster by pinpointing a core sequence within a 40-kb region known to harbor this element (Spitz *et al.*, 2003). Several other studies utilizing *Fugu*–mammal comparisons have identified CNSs indicative of important regulatory elements and in each case the genes under study are implicated in developmental control. Such studies have identified enhancers responsible for the regulation of *PAX6* (Griffin *et al.*, 2002; Kammandel *et al.*, 1999; Miles *et al.*, 1998), *SOX9* (Bagheri-Fam *et al.*, 2001), the Dlx genes (Ghanem *et al.*, 2003), *Wnt-1* (Rowitch *et al.*, 1998), *PAX9/Nkx2-9* (Santagati *et al.*, 2001, 2003), and the Iroquois clusters (De la Calle-Mustienes *et al.*, 2005). In a pioneering study aimed at characterizing intriguing regions of low gene density in the human genome, known as gene deserts, Nobrega *et al.* (2003) identified over 1000 conserved elements between human and mouse around and within the introns of *DACH1*, a gene involved in development of the brain, limbs, and sensory organs. To narrow the search for sequences likely to be functional, they looked for sequences also conserved in Frog and *Fugu*, shortening the list to 32 candidates. Nine of these were tested in reporter gene assays in mouse embryos and seven were found to reproducibly drive β-galactosidase expression in a manner that recapitulated several aspects of *DACH1* endogenous expression. These studies demonstrate the power of using highly divergent sequences to prioritize sequences likely to be *cis*-regulatory in function. A growing number of acronym-based names, such as CNSs and multispecies conserved sequences (MCS), have been used to describe noncoding sequences under evolutionary constraint (Table 12.1), reflective of the varied nature of the identification processes (i.e., different neutral models and across different evolutionary distances) as well as current ambiguity in relation to their function. For ease of reference, I have chosen the acronym conserved noncoding elements (CNEs) to refer specifically to a noncoding sequence which is conserved between fish and mammals and therefore has a high regulatory potential. Despite the identification of a number of *Fugu*–mammal CNSs within a limited

Table 12.1. Acronyms Used for Conserved Noncoding Regions in Vertebrate Genomes

Acronym	Meaning	Reference
ANCOR	Ancestral noncoding conserved region	Aloni and Lancet, 2005
CNC	Conserved noncoding	Couronne *et al.*, 2003
CNE	**Conserved noncoding element**	**Woolfe *et al.*, 2005**
CNG	Conserved nongenic	Dermitzakis *et al.*, 2003
CNS	Conserved noncoding sequence	Dubchak *et al.*, 2000
CST	Conserved sequence tag	Mignone *et al.*, 2003
ECR	Evolutionary conserved region	Ovcharenko *et al.*, 2004
HCR	Highly conserved region	Duret and Bucher, 1997
MCS	Multispecies conserved sequence	Thomas *et al.*, 2003
UCE	Ultraconserved element	Bejerano *et al.*, 2004

Notes: References refer to the chapter in which the acronym was first used. In this study, the acronym CNE (highlighted in bold) is used to refer specifically to DNA sequences conserved from mammals to fish. Table adapted from Aloni and Lancet, 2005.

number of well-studied developmental genes, it was not clear whether this was indicative of a more extensive genome-wide trend and whether similar CNEs were located around other types of genes. To answer this question, a pairwise comparison of the *Fugu* and human genomes was performed to identify genome-wide conservation of noncoding sequences (Woolfe *et al.*, 2005).

VII. FISH-MAMMAL CONSERVED NONCODING ELEMENTS ARE ASSOCIATED WITH VERTEBRATE DEVELOPMENT

The *Fugu* genome was masked for the majority of coding and tRNA content and the remaining regions compared to the human genome using basic local alignment search tool (MegaBLAST) with a stringent "seeding" word size of 20 bp. All resulting matches were filtered to include alignments of over 100 bp in length and exclude any repetitive elements, protein coding or ncRNA sequences, that had been missed. One thousand three hunhdred and seventy-three CNEs were identified in this way, with little or no evidence of transcription [except for ∼6% located within untranslated regions (UTRs) of known mRNA molecules]. Unsurprisingly, the majority of the CNEs are also conserved in other mammals such as the mouse and rat as well as in the chicken and zebra fish genomes, indicating these sequences are likely to be common to all bony vertebrates. By contrast, no significant similarity with these CNEs can be found within invertebrate chordate or other invertebrate genomes, suggesting these sequences are a vertebrate innovation. In human, CNEs are found in all chromosomes except chromosome 21 and Y, but their distribution is highly clustered. One hundred and sixty-five clusters were defined, with over 85% of clusters containing five or more CNEs. A statistical analysis of the gene ontology (GO) assignments and InterPro structural domains for all genes located within these clusters found 12 of the 13 most overrepresented GO terms relate to transcriptional regulation and development (Table 12.2) as well as enrichment for DNA-binding motifs (such as the C_2H_2 zinc finger domain), in particular the homeobox domain. Indeed, over 96% of clusters are located in the vicinity of one or more genes with such functional assignments. Initial observations of tight association of CNEs with genes involved in transcriptional regulation and/or development (which for ease are referred to as *trans-dev* genes) were therefore confirmed on a genome-wide scale.

A number of the *trans-dev* genes identified have previously been shown to have highly conserved *cis*-regulatory elements associated with them. The association of noncoding sequences conserved deep in the vertebrate lineage with *trans-dev* genes has since been confirmed by a number of other studies (Ahituv *et al.*, 2005; Ovcharenko *et al.*, 2005; Sandelin *et al.*, 2004; Sironi *et al.*, 2005). CNEs tend to be located in regions of low gene density, also referred to as "gene deserts" (Nobrega *et al.*, 2003; Ovcharenko *et al.*, 2005), confirming the likelihood that these regions harbor large numbers of distal CRMs. Sensitive

Table 12.2. Table of Overrepresented GO Terms in 170 Annotated Genes with *Fugu*–Human Conservation in the UTR of both Orthologues

GO Id	GO term	Group Count	Total Count	P-value
GO: 0003700	Transcription factor activity	40	1292	4.92×10^{-26}
GO: 0045449	Regulation of transcription	60	2922	4.71×10^{-21}
GO: 0019219	Nucleobase, nucleoside, nucleotide, and nucleic acid metabolism	60	2953	8.14×10^{-21}
GO: 0031323	Regulation of cellular metabolism	60	2991	1.9×10^{-20}
GO: 0006355	Regulation of transcription, DNA-dependent	57	2799	6.07×10^{-20}
GO:0051244	Regulation of cellular physiological process	70	3907	6.07×10^{-20}
GO: 0019222	Regulation of metabolism	61	3134	6.18×10^{-20}
GO: 0006350	Transcription	60	3061	7.16×10^{-20}
GO: 0006351	Transcription, DNA-dependent	57	2875	3.58×10^{-19}
GO: 0005634	Nucleus	77	5243	8.45×10^{-15}
GO: 0007399	Neurogenesis	17	448	2.62×10^{-07}
GO: 0043227	Membrane-bound organelle	80	7195	3.71×10^{-07}
GO: 0043231	Intracellular membrane-bound organelle	80	7195	3.71×10^{-07}
GO:0007492	Endoderm development	2	2	0.00157
GO:0007420	Brain development	4	37	0.00332
GO:0048523	Negative regulation of cellular process	13	592	0.00401

Notes: Group count indicates the number of genes from within the 170 that posesss a particular GO term, whereas Total count refers to the number of genes in the human genome with that GO term. The *P*-value is an indicator as to how enriched the group count is compared to the total count.

multiple alignments using additional mammalian orthologous sequence around a number of the CNE cluster regions identified more than double the number of CNEs of over 100 bp in length, demonstrating the power of this approach. To confirm their regulatory potential, 25 CNEs, derived from both whole genome comparison and multiple-alignment approaches, located in clusters surrounding four different *trans-dev* genes (*SOX21*, *HLXB9*, *SHH*, and *PAX6*), were tested for enhancer activity using a transient enhancer green fluorescent protein (GFP) reporter-gene assay in zebrafish (*Danio rerio*) embryos. Twenty-three of the 25 CNEs tested were shown to have reproducible enhancer activity, inducing expression of a GFP reporter gene in a temporal and tissue-specific manner that frequently coincides with the endogenous expression domains of their nearby *trans-dev* gene. As this assay measures only upregulation of gene expression, those CNEs that showed no enhancer activity may represent negative regulatory elements such as silencers or insulators, or may be involved in indirect processes such as chromatin remodeling. These results demonstrate the frequent functional nature of CNEs identified through fish–mammal comparisons. The extreme conservation of these CNEs across the vertebrate lineage together with their tight association with intricately regulated early developmental genes implies that they are enriched for putative CRMs.

VIII. HIGH-RESOLUTION ANALYSIS OF THE ORGANIZATION OF CNEs AROUND KEY DEVELOPMENTAL REGULATORS

In order to examine the conserved regulatory architecture around key developmental genes in greater detail, multiple alignments have been performed using MLAGAN (Brudno *et al.*, 2003a,b), and the resulting data stored in a publicly accessible relational database, conserved non-coding orthologous regions (CONDOR), that also includes functional data on selected CNEs [Woolfe *et al.* (submitted for publication)] and http://condor.fugu.biology.qmul.ac.uk/). Multiple alignments allow greater sensitivity, as the probability that a CNE will occur in more than two genomes around the same gene by chance is greatly reduced compared with a pairwise alignment. Despite evidence that there is a degree of shuffling of conserved regulatory signatures (Sanges *et al.*, 2006), the vast majority of CNEs identified associated with *trans-dev* genes appear to be co-linear (Sun *et al.*, 2006) and are thus well suited to this form of analysis, which uses "chained anchors" along the sequence, that is, small blocks of well-conserved sequence along the alignment. Another aligner, shuffle LAGAN (SLAGAN) (Brudno *et al.*, 2003c), can identify shuffled CNEs.

 CONDOR contains detailed data on 7000 CNEs spanning nearly 100 gene regions and provides an unprecedented opportunity to look at the distribution of CNEs on a large scale.

IX. GENERAL GENOMIC ENVIRONMENT AROUND CNEs

CNEs are by definition located in noncoding sequence, although the genic environment surrounding the noncoding sequence can be very different. The distribution of CNEs around *trans-dev* genes is of particular note because it contradicts the traditional view that *cis*-regulatory sequences are located just upstream of the promoter. Figure 12.1 shows the proportion of CNEs from CONDOR that are located in introns (of either the *trans-dev* gene upon which the CNE acts, or of neighboring genes), between genes, or in UTRs. Around 4% of CNEs appear to be located in UTRs, and this is discussed in Section X. Interestingly, nearly a third of CNEs are found in introns, both of the *trans-dev* gene upon which they act (18%) and in the introns of neighboring genes (12%). When CNEs are located in neighboring genes, these genes are invariably found in conserved synteny in all vertebrates, presumably as a result of having to maintain the regulatory repertoire of the *trans-dev* gene intact. By examining the position of 4950 CNEs that are in the vicinity of just a single *trans-dev* gene, we can determine whether there is any positional bias toward maintaining either upstream or downstream CNEs. Figure 12.2 demonstrates that there are almost as

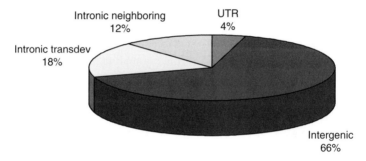

Figure 12.1. Distribution of CNEs across *trans-dev* genes. CNEs are classified as being within untranslated regions (UTR), between genes (Intergenic), within the introns of a *trans-dev* gene (Intronic *trans-dev*), or within the introns of a neighboring (non-*trans-dev*) gene (Intronic neighboring).

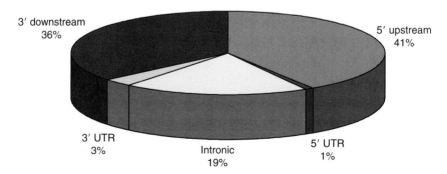

Figure 12.2. Position of CNEs relative to the *trans-dev* gene they are associated with.

many CNEs located downstream of *trans-dev* genes as there are upstream. Indeed, if we include intronic CNEs as well as those located in the 3′ UTR, over 58% of CNEs are located 3′ of the transcription start site (TSS).

Not surprisingly, the distribution of CNEs across any one specific *trans-dev* gene varies considerably, but positional bias of CNEs either 5′ or 3′ generally correlates with a similar bias in the amount of flanking noncoding sequence at that position. For example, the largest bias in both CNE position and flanking DNA is seen around the human *BCL11A* gene, which has virtually no flanking DNA 5′ of the gene (and no CNEs), whil 80% of CNEs are located within a large (2.21 Mb) gene desert located 3′ of the gene. Conversely, 92% of the CNEs around the *FOXD3* gene are found in a 450-kb gene desert 5′ of the gene, whereas the region 3′ is gene rich and has no CNEs (Fig. 12.3). The variable distribution of CNEs therefore suggests that they can function irrespective of position around their associated gene unlike other position-specific *cis*-regulatory elements such as promoters.

X. CNEs PRESENT IN TRANSCRIPTS

Although all CNEs that are located within the introns of genes are by definition transcribed, only ~5% of CNEs are located in UTRs and therefore constitute part of the mature mRNA molecule. These are dealt with in more detail below. To investigate whether the remaining 95% "nongenic" CNEs are likely to form part of mature transcripts, sequences were BLAST searched against expressed sequence tags (ESTs) databases encompassing all currently deposited ESTs from vertebrates. Seventy-seven percent of these CNEs have no significant hits to ESTs, 11.1% have one hit, 7.2% have between two and four hits to ESTs, and 4.7% have five or more hits. In contrast, BLAST searches of 1000 randomly conserved elements overlapping known coding exons from CONDOR regions identified just 4.5% with two to four hits to ESTs, and 95.5% with five or more, demonstrating that transcribed sequences are characterized by hits to large numbers of ESTs. The vast majority of CNEs therefore have little or no evidence of transcription, indicating likely roles as CRMs. The small proportion of CNEs with larger hits to ESTs may represent unannotated coding sequences or noncoding RNAs, but equally likely, these sequences may represent good candidates for CRMs. A number of these sequences (e.g., CRCNE00007308 located in the gene desert between *IRX3* and *IRX5* and CRCNE00006651 located in the gene desert between *IRX1* and *IRX2*, both of which have five hits to ESTs) have been shown to drive reporter constructs in zebra fish embryos in a spatial- and temporal-specific manner (De la Calle-Mustienes *et al.*, 2005), suggesting at least some of these sequences are regulatory in function. This observation also indicates the possibility that some regulatory elements are part of sequences that

A

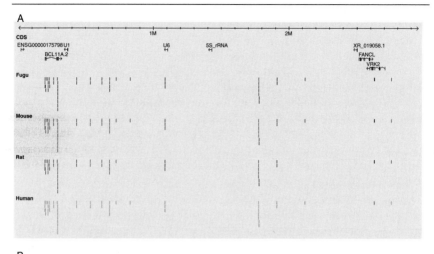

B

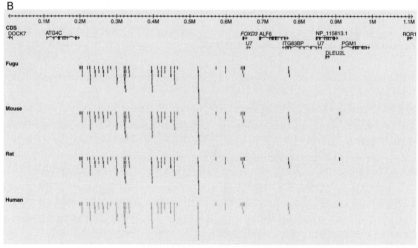

Figure 12.3. CONDOR view of (A) *BCL11A* and (B) *FOXD3* gene regions showing positions of CNEs in human, rat, mouse, and Fugu. Note that the majority of the CNEs are 3' of the first exon of the *BCL11A* gene, and 5' of the *FOXD3* gene, correlating with the location of gene deserts in each region.

play a dual function as coding sequences or work through an RNA intermediate (e.g., Jones and Flavell, 2005) in addition to or as part of their regulatory function. There is also a possibility that the matching ESTs derive from genomic contamination and are not transcribed at all or they are the product of illegitimate transcription (Sorek and Safer, 2003).

XI. CNEs LOCATED WITHIN UTRs

A small number of CNEs are located within the UTRs of some mammalian *trans-dev*, or neighboring genes (presence within a UTR of a gene in *Fugu* is not currently discernable due to lack of full-length cDNA data). Within the entire CNE dataset (around all *trans-dev* genes in CONDOR), 279 sequences (4.2%) are located in a UTR. Of these, 85% are located in the UTR of a gene annotated as *trans-dev* and the rest are located in UTRs of neighboring (non-*trans-dev*) genes. Over 74% of CNEs located in *trans-dev* UTRs are located in the 3' UTR, reflecting the increased size of 3' UTRs compared to 5' UTRs in vertebrate genomes (the average length of 5' UTRs in human Ensembl is only 255 bp, compared with 989 bp for 3' UTRs). By contrast, equal numbers of CNEs are located in the 5' and 3' UTRs of genes neighboring *trans-dev* genes. As these sequences, unlike other CNEs, are transcribed, it is not known whether they function at the pre-transcriptional (i.e. *cis*-regulatory) or post-transcriptional level (e.g., directing mRNA stability).

Nevertheless, a genome-wide analysis between all human UTRs and the *Fugu* genome indicates that sequence conservation is rare, and in most cases coincides with a *trans-dev* gene region that already has catalogued CNEs. Furthermore, there are at least eight examples where a CNE is located in the UTR of a human gene for which there is either no (gene) orthologue in Fugu, or the orthologue is in a different location to the CNE in Fugu. This sheds light on the possible function of CNEs in UTRs and the utility of using highly diverged genomes in such cases. Five examples have no orthologue in the *Fugu* genome, with three of these being primate-specific genes, indicating that the CNE has been coincidentally incorporated into the UTR of a "new" gene sometime in primate evolution. The other two are located in UTRs of the *HOXA7* and *HOXB7* genes in human, genes that have been lost sometime during the evolution of Fugu (Aparicio *et al.*, 1997), yet the CNEs have been retained in the same place, located between *Hoxa/b6* and *Hoxa/b8*, indicating a putative *cis*-regulatory role for the surrounding *Hox* genes rather than post-transcriptional regulation of *Hoxa7/b7*. The final three examples also suggest that CNEs are unlikely to be involved in post-transcriptional regulation of the genes in which they are located. These are located in *Fugu* regions where gene synteny around the CNE has been retained, but the gene in which they are located in human is present elsewhere in the *Fugu* genome. An example of this can be seen in Fig. 12.4. A CNE (corresponding to CNE CRCNE00007244 in CONDOR) is located within the UTR of an ancient, relatively uncharacterized gene Q9Y2K8, which is conserved from mammals through to *Caenorhabditis elegans* (Ensembl Compara v.36). It is located upstream of the Iroquois B cluster (made up of Irx3, -5, and -6), a region characterized by large numbers of CNEs which extends through the neighboring genes (including *Q9Y2K8)* across a region encompassing almost 4.9 Mb on human chromosome 16. CRCNE00007244 and Q9Y2K8

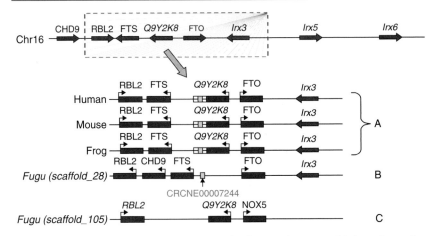

Figure 12.4. Location of CNEs in the UTR of nondevelopmental genes is unlikely to play a role in their function. CNE CRCNE00007244 (shaded box) is located in *IRX3, -5, and -6* region and is found in the UTR of a relatively uncharacterized gene *Q9Y2K8* (*KIAA1005*). This gene is conserved back to *C. elegans* and the gene order is preserved in this region in tetrapods from mammals down to frog (A), indicating the likely ancestral state. In *Fugu*, this gene is no longer present in the *Irx3, -5, and -6* region (scaffold_28) (B), but the CNE remains, indicating it plays no role in this gene and is associated with neighboring genes, or the IRX cluster. The synteny of the rest of the genes in this region has remained the same in *Fugu*, although RBL2 and CHD9 have undergone an inversion event, so they are in the opposite orientation relative to their orthologues in tetrapods. The orthologue of *Q9Y2K8* is present in *Fugu* but is located on scaffold_105 (C). Presence of a second *RBL2* gene downstream of *Q9Y2K8* on this scaffold suggests these genes underwent a fish-specific duplication and the copy on scaffold_28 was lost through nonfunctionalizsation over evolution.

are conserved in the same relative position in all currently available tetrapod genomes as is the gene order across the rest of the IrxB cluster. In *Fugu*, the CNE is located in the same relative position (on scaffold_28, ensembl) containing the IrxB cluster, but *Q9Y2K8* is located on another scaffold. Interestingly, in *Fugu*, *Q9Y2K8* is positioned next to another gene paralogous to the *RBL2* transcriptional cell cycle regulator in the IrxB region (see Fig. 12.4). This suggests that these two genes are likely to be remnants of an extra whole genome duplication thought to have occurred early in the teleost-lineage (Vandepoele et al., 2004), after which, probably due to nonfunctionalization, the copy of *Q9Y2K8* on scaffold_28 was lost. These examples suggest that at least in these cases, and most probably for all non-*trans-dev* genes, the location of CNEs in their UTRs is merely incidental and plays no role in their regulation but rather regulates other genes in the vicinity. Like CNEs in general, there is a tight correlation of UTR CNEs with *trans-dev* genes, but there are no known post-transcriptional regulatory elements in UTRs that are as large or as well conserved as CNEs. Therefore, the location here is also likely to be incidental to their function, which in these cases appears to be *cis*-regulation.

XII. CNEs ARE LOCATED LARGE DISTANCES AWAY FROM THEIR PUTATIVE TARGET GENE

Distal CRMs are known to act at large distances from their target gene. Traditionally, prior to the availability of whole genomes, searches for such elements have occurred relatively close to the gene because of limitations in data as well as the concern of mis-assigning elements from further distances. The current distance limit between a target developmental gene and experimentally verified enhancers is ~1 Mb, for example, *SHH* (Lettice *et al.*, 2003), *SOX9* (Bishop *et al.*, 2000), and *MAF* (Jamieson *et al.*, 2002). CNEs represent excellent candidates for CRMs since many are located in the vicinity of a single *trans-dev* target gene, which provides an opportunity to gauge the distance limits over which such elements may act. Using only those elements conserved in *Fugu*, mouse, rat, and human, we examined the distribution of distances between CNEs and the gene TSS for all four species (Fig. 12.5). In the mammalian genomes, over a quarter of elements are within 100 kb of their target gene and over 88% are within 1 Mb. Incredibly, over 11% of elements were located between 1 Mb and 2 Mb from the target gene and ~1% are over 2 Mb, exceeding current limits for any known CRMs. A relative shift toward longer distance bins is seen in *Fugu* when compared to mammals, possibly reflecting the decreased level of compaction seen in a number of regions, although many regions are more compact than expected. This suggests that despite the tendency for compaction in the *Fugu* genome, it may be constrained by spatial requirements for elements interspersed among the CNEs identified here, including those that may be fish or species-specific CRMs. The furthest distance between a CNE and its likely target gene is 2.54–2.58 Mb in the mammalian genomes and 358.7 kb in *Fugu*. This is a 98 bp element (CRCNE00002604) located 3′ of the transcription factor *BCL11A*. The *bcl11a* gene is duplicated in *Fugu*, but the CNE is conserved only in one copy downstream of *bcl11a.1* (data from CONDOR). *BCL11A* is characterized by a large gene desert downstream containing over 100 CNEs with 13 CNEs located further than 2 Mb away from the gene. Regions containing *BCL11B*, *SOX11*, *TFAP2A*, and *NR2F1* also have CNEs located in excess of 2 Mb from these genes in mammals, although the distances vary in *Fugu* from 113.8 kb to 278.2 kb, reflecting asymmetrical rates of compaction previously noted. In addition, 42% (28/67) of single *trans-dev* genes have CNEs more than 1 Mb away in mammals. These data reflect those of Vavouri *et al.* (2006), who carried out a similar analysis but focused on duplicated CNEs that can be uniquely associated with paralogous genes in the human genome. One of the consequences

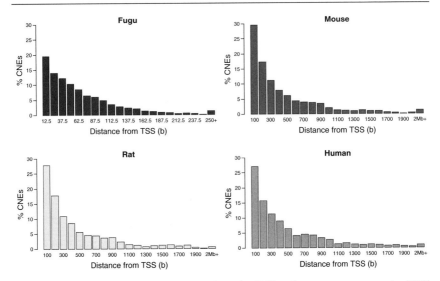

Figure 12.5. Proportion of CNEs at distances from the CNE to the transcription start site (TSS) of their target *trans-dev* gene in *Fugu*, mouse, rat, and human. Distance in mammals is measured at increments of 100 kb (i.e., 100 = 0–100, 200 = >100–200, etc.). Distance bins in *Fugu* are measured in increments of 12.5 kb (i.e., reflecting an average compaction rate of one-eighth in *Fugu* compared to mammalian genomes). The last bin in each graph contains all elements over 2 Mb away in mammals and 250 kb in *Fugu*.

of the very large distances across which CNEs act is that they often encompass neighboring genes. Invariably, this results in conservation not only of synteny but also of gene order in these regions (Ahituv *et al.*, 2005; Goode *et al.*, 2005; Kikuta *et al.*, 2007; Spitz *et al.*, 2003).

A. Mutation events in and around CNEs over evolution

Mutation events such as insertions/deletions (*indels*), inversions, and transloca-tions are a major feature of genome evolution. Insertions and deletions involve changes in the sequence of the DNA, whil inversions and translocations act on a larger scale and involve predominantly structural rather than sequence changes. Given the distances over which CNEs are dispersed in the genome, it is likely, unless there is strong evolutionary selection against it, for CNEs to be involved in, or undergo, one or more of these events. Two approaches, BLAST and SLAGAN, have been used to identify CNEs that have undergone two forms of rearrangement, inversion and transposition.

1. Inversions

Mutational events involving inversions occur when a subsegment of DNA is removed from the sequence and then inserted back in the same location but in opposite orientation, often through repeat-mediated homologous recombination (Shaw and Lupski, 2004). CRMs, such as enhancers, have been shown to act both irrespective of orientation (e.g., Hill-Kapturczak *et al.*, 2003) and in an orientation-dependent manner (e.g., Swamynathan and Piatigorsky, 2002). It is therefore of interest to ascertain the proportion of elements which have undergone changes in orientation and whether this is a common feature of putative *cis*-regulatory elements. CNE orientation can be assessed relative to the transcriptional direction of the closest *trans-dev* gene in different species to identify inversions. We looked at orthologous CNEs in nine vertebrate genomes. Two hundred and twenty-four CNEs (~3.5% of this CNE set) were found to have changed orientation in one or more of these genomes, and are located around 50 of the 90 *trans-dev* regions in CONDOR. These changes are due to 114 individual inversion events, 40 of which involve two or more CNEs (Woolfe, 2006). Using orientation information from all genomes, it was possible to date these inversion events. A summary of this can be seen in Fig. 12.6.

The majority of the elements (62%) that have undergone changes in orientation are organism-specific (i.e., they have undergone inversion in only one genome), with the largest number occurring in frog and rat. In contrast to frog for which most of the changes in orientation are due to individual inversion events, the 43 CNEs showing changes in orientation in rat are the result of just three large-scale inversions. Two of these (containing all but one of the elements) are located in regions in the vicinity of *Barhl2* (Chr14) and *Sall3* (Chr18) genes. These regions are characterized by large gaps in the sequence that surround the inversion, suggesting that they are possibly due to errors in sequence assembly rather than a true inversion. Some 33% of inversion events appear to predate the mammalian lineage. Because of the incomplete nature of many vertebrate genomes, it is not always possible to identify the CNE and orthologous reference gene, or to locate both these features on the same scaffold, in order to confirm orthology. This was specially true of the genome of *Xenopus tropicalis*, which, like the *Fugu* genome, has no chromosomal mapping data and is highly fragmentary. Therefore, in many cases, dating of the inversion event could only be derived as the earliest point given the data available. Of elements with data available in all species, 10 derived from an inversion that occurred prior to the mammalian lineage but after the fish-tetrapod split. CNEs for which the change in orientation was seen between fish and tetrapods were placed at the root of the tree as in most cases it is not possible to know whether the inversion occurred within the fish lineage or early in the tetrapod lineage prior to the split with amphibians. A small number of these CNEs are duplicated in the *Fugu* genome, both from ancient vertebrate-specific

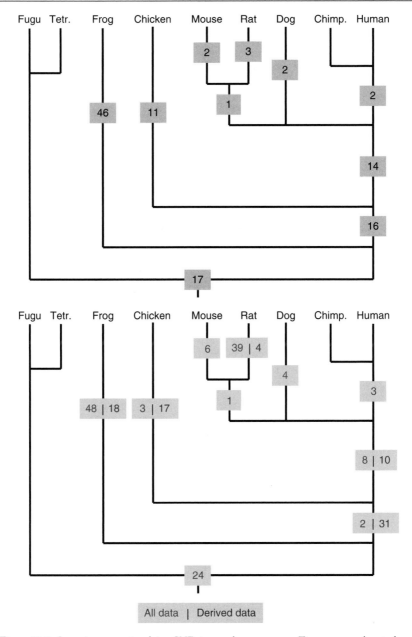

Figure 12.6. Inversion events involving CNEs in vertebrate genomes. Top tree—numbers in boxes represent individual inversion events involving one or more CNEs occurring at the evolutionary time point represented in the phylogenetic tree. Bottom tree—numbers

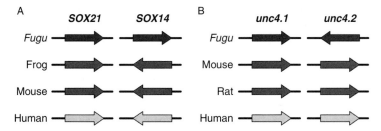

Figure 12.7. Changes in orientation between duplicated elements suggest both tetrapod-specific and *Fugu*-specific inversions. CNEs and their orientation are represented by block arrows. The relative orientation of the CNE is measured in relation to a reference *trans-dev* gene which is indicated above each set of arrows. (A) Two paralogous CNEs located upstream of *SOX21* and *SOX14* (CRCNE00000571 and CRCNE00000750) derive from whole genome duplication events at the origin of vertebrates (see Chapter 4, this volume). While the sequence of the CNE upstream of *SOX21* is found in the same relative orientation in all genomes, the same paralogous sequence in *SOX14* is orientated in opposite orientation in tetrapods whil remaining in the same orientation in fish, suggesting the inversion occurred early in tetrapod evolution. (B) *unc4.1* is a single copy gene in tetrapods but is duplicated in the *Fugu* genome and each gene region has a CNE [CRCNE00009841 (*unc4.1.2*) and CRCNE00009909 (*unc4.1.2*)] conserved to the same single-copy sequence in mammals. The CNE in *unc4.1.2* has undergone an inversion, indicating this is a fish- or *Fugu*-specific inversion event.

duplication events or duplications occurring early in teleost evolution providing a rare opportunity, through comparison, to date these inversions to the fish or tetrapod lineage. Two of these are highlighted in more detail in Fig. 12.7 and provide evidence that both fish-specific and tetrapod-specific inversion events have occurred around these elements and that the change in orientation has been conserved across evolution. This is strong evidence, at least for these elements, that CNEs may act as orientation-independent *cis*-regulatory elements.

2. Transpositions

In addition to inversions, transpositions are known to occur within genomes where a subsegment of DNA is removed from the sequence and then inserted back elsewhere in the genome in the same orientation and is thought to occur via mobile-repeat elements such as transposons. Changes in the order of a CNE

in boxes represent individual CNEs involved in inversion events. Where two numbers are side by side, the left hand figure represents the numbers where data is available at all time points in the tree. The figure on the right represents the best approximation to the inversion event in the absence of complete data.

between the *Fugu* and mouse, rat, dog (when utilized in the alignment), and human genomes were identified for all CNEs in CONDOR (Woolfe, 2006). Of the 6232 CNEs (224 elements that had undergone inversion were excluded from this analysis) representing 18,028 individual elements in these four mammalian genomes, just 7 CNEs were found to have undergone transposition events. Three derived from a single transposition event specific to rat within the gene desert upstream of the gene *ZFHX4*. Two are specific to mouse and occur upstream of *Fidgetin* gene and *Znf503*. One element appears to have undergone two separate transposition events in the mouse and rat genomes, which given the extremely low rate of transposition in other CNEs is highly unusual. There are no transpositions common to rodents or mammals, suggesting these events occurred after these organisms diverged. The transposed CNEs have a range of sizes between 45 bp and 180 bp and are significantly conserved in single copy in orthologous regions throughout the tetrapod lineage, indicating that they are not sequences identified by chance. Assuming the genomes in these regions have been assembled correctly, the extremely low rate of transposition involving CNEs is possibly due to deleterious effects of moving CNEs away from their target gene(s). It could also indicate that the order of elements along the sequence is of functional importance. If the isolated cases of transposition described above involve CNEs that represent true functional *cis*-regulatory elements, they are likely to have been fixed due to their fortuitous transposition to the same region in proximity to their target gene. Interestingly, more recent transpositions of conserved sequences have been identified, although these are not associated with *trans-dev* genes and are not conserved in fish genomes (Xie *et al.*, 2006; Bejerano *et al.*, 2006).

XIII. DISCUSSION

The identification and characterization of sequences responsible for the tight and precise regulation of vertebrate development remain one of the main challenges of the post-genomic era. In contrast to other functional elements such as coding exons, which have been the subject of intense study for many decades, our knowledge of the language, mode of action, and evolution of *cis*-regulatory elements are scant, in part because they are difficult to identify. The availability of ~6500 CNEs conserved between *Fugu* and mammals and located in the vicinity of key developmental regulator genes across a range of genomic environments provides the opportunity to study a large but focused noncoding cohort. These elements overlap no known coding transcripts or ncRNAs and are therefore likely to be highly enriched for putative *cis*-regulatory elements, providing the first opportunity to study their sequence and evolutionary character on a large scale.

The majority of CNEs are conserved across all currently available vertebrate genomes with the likelihood that missing CNEs are due to incompleteness of sequence coverage. Despite deep sequence conservation in the highly diverged genomes of fish and tetrapods, no similarity to invertebrate sequences can be found. A significant proportion of the *trans-dev* genes in this study has an ancient origin and is homologous to those identified in nonvertebrate chordates such as the sea squirt and in invertebrates such as flies and worms. Nevertheless, pairwise local alignments and more sensitive multiple alignments of orthologous noncoding DNA surrounding these genes do not identify any similarity outside the coding regions, even for the most highly conserved CNEs such as those found around *SOX21*. Thus it appears unlikely that the same set of sequences that appear to regulate vertebrate *trans-dev* genes are found in invertebrates, although intriguingly parallel sets of noncoding regulatory sequences have been identified around similar gene sets in nematodes (Vavouri *et al.*, 2007). A number of CNEs can be identified within a limited set of sequence read from the genome of the cartilaginous elephant fish, which, like other members of chondrichthyes such as sharks and rays, diverged prior to the split between Actinopterygii (ray-finned fishes) and Sarcoptergii (lobe-finned fishes) 480–570 million years ago (Venkatesh *et al.*, 2005). Thus far, it is possible to date the origin and fixation of these sequences in an ancestral genome sometime after invertebrate chordates but prior to the evolution of chondrichthyes 500–650 million years ago.

CNEs are extraordinarily constrained within the tetrapod lineage (with most CNEs ranging between 90% and 98% identity compared to the human sequence) despite being separated by up to 350 million years of evolution. In contrast, CNEs within teleost genomes appear to have evolved relatively rapidly both in comparison to their tetrapod orthologues and within teleost species. A faster rate of evolution in teleost genomes in comparison to mammalian genomes has been observed previously in relation to the neutral rate, which was found to be up to four times faster in the puffer fish than in mammals (Aparicio *et al.*, 2002, Jaillon *et al.*, 2004). This is highlighted most clearly by the significantly higher level of conservation observed in CNEs between elephant fish CNEs and human and teleost CNEs, despite the greater phylogenetic divergence (Venkatesh *et al.*, 2006). Teleosts, unlike sharks, have undergone an additional whole genome duplication event (Taylor *et al.*, 2003), which may be a reason for their elevated rate of evolution.

The CNEs discussed in this chapter cover a wide range of sizes (30–869 bp) and conservation (58–97%) between the *Fugu* and mammalian sequences. These attributes appear to be directly associated with their function as size and conservation of CNEs is not influenced by their genomic environment. For most length categories, CNEs have lower or equal levels of conservation than surrounding exons. This may not necessarily be true when compared against exons across the whole genome, as many of the exons within CNE regions derive from transcription factors and/or developmental genes which are known to be more highly conserved than other types of genes (Chuang and Li, 2004). CNEs of longer length (311–800 bp) generally exhibit higher sequence identity than exons of the same size, many CNEs have long stretches of complete identity between *Fugu* and mammalian sequences. Whil sequence conservation implies function, the precise function of the vast majority of CNEs remains unknown. A comparison of CNEs with protein coding exons through several analyses uncovers a number of revealing characteristics that shed light on their possible function. In the case of coding exons, the mechanism for conservation is known to be due to the constraint of the genetic code in specifying sequence coding for proteins. That most CNEs have little or no evidence of transcription suggests they do not form parts of mature transcripts, and this is consistent with a *cis*-regulatory role. A small proportion do match a large numbers of ESTs, similar to that seen in coding exons, indicating they may be missed, unannotated exons. However, some of these have proven roles as enhancers [e.g., CNEs in the Iroquois clusters (De la Calle-Mustienes *et al.*, 2005)], indicating that either the ESTs they match derive from genomic contamination or illegitimate transcription, or that some CNEs may act through transcribed intermediaries. Indeed, experiments on putative enhancers, located adjacent to the cytokine IL10 (Jones and Flavell, 2005) and the dlx-5/6 region (Feng *et al.*, 2006), demonstrate that a number of them transcribe intergenic RNA, leading to speculation that these regulatory elements function via an intermediate regulatory RNA. There is also intriguing evidence from work on the α-globin cluster that intergenic transcription from specific elements is used to modify chromatin structure leading to gene activation (Gribnau *et al.*, 2000). There is also recent evidence that some conserved elements have been distributed in the genome by ancient retrotransposons and can play roles as both enhancers and alternatively spliced exons depending on their genomic context (Bejerano *et al.*, 2006). Finally, there is also a possibility that some CNEs may function as novel ncRNAs as ~10% overlap predictions of significant RNA secondary structure. Again a number of these have proven *cis*-regulatory roles (Woolfe *et al.*, 2005), and it is not clear what proportion of these predictions represents real ncRNAs. It is therefore possible that CNEs may function through a number of different mechanisms.

CNEs have a number of evolutionary characteristics which are also distinct from those of coding exons. Regions of highly constrained blocks in multiple alignments are often signals of functional requirements. Coding exons show clear periodicity in the size of conserved block sizes indicative of the genetic code and the specificity of the first two bases of a triplet codon and the degeneracy of the third base. In contrast, CNEs show no periodicity in block sizes. However, CNEs do show a significant enrichment for blocks of size 7–13 bp in length, which correspond to the size of one to two transcription factor binding sites, commonly 5–10 bp in length (Woolfe, 2006). Intriguingly, CNEs also have a significant proportion of conserved block sizes of extreme length (>25 bp), which are, unsurprisingly, not as prevalent within block sizes for coding exons. This extreme conservation is analogous to that seen in "ultraconserved elements" defined as at least 200 bp of complete identity between human, mouse, and rat (Bejerano et al., 2004), and indeed many of these sequences overlap CNEs, although this level of conservation is all the more surprising given the vast evolutionary distance involved. This degree of conservation cannot be fully explained by current models of CRMs, even for large numbers of transcription factor binding sites, as these binding sites are generally rather short and normally exhibit a level of redundancy. A number of ideas on the evolutionary mechanisms responsible for "ultraconservation" have been proposed, including increased DNA repair, multiple overlapping transcription factor binding sites, and decreased mutation (Bejerano et al., 2004; Boffelli et al., 2004). The last possibility is unlikely given the relatively rapid rate of evolution seen for many CNEs within teleosts, as well as a recent study that used single nucleotide polymorphism (SNP) data from the HapMap project and showed CNSs in the human genome are not located in mutation cold-spots but rather are selectively constrained (Drake et al., 2006). It is also possible that if CNEs do bind transcription factors, these factors may have larger and more constrained binding site specificities than those currently known. It is clear therefore that until more is known about the mode of action of CNEs, the reasons behind these extreme patterns of conservation will remain both enigmatic and highly speculative.

Patterns of sequence insertion/deletion (indels) between *Fugu* and mammals are also distinct in CNEs compared to those in exons. Only a third of exons have any indels at all and those that do occur show a clear third base periodicity. This is, of course, due to the constraints of the genetic code which does not allow insertions of sequence lengths which would cause frameshift mutations. In contrast, almost 80% of CNEs have at least one indel, but these are mostly small and they follow a negative exponential distribution (i.e., larger indels are increasingly rare) (Woolfe, 2006). Nevertheless, many of the indels occur within some of the most highly constrained sections of the CNEs where we might predict changes would be most likely to have critical effects on function. It therefore appears that CNEs may, in some cases, be more accommodating to sequence indels than their highly constrained nature suggests. This was recently demonstrated in a functional study of an ultraconserved enhancer located within the introns of the *DACH1* gene (corresponding to CNE *CRCNE00000232* in CONDOR) (Poulin et al., 2005). The central core of the enhancer has a region of 144 bp with 100% identity between *Fugu* and human and is critical to the function of the enhancer. However, the insertions of two nonfunctional 16 bp linkers into this core region caused no detectable modification of its *in vivo* activity, at least as far as the resolution of their transgenic mouse assay.

If we assume that CNEs represent modules for a number of protein binding sites, it is possible that sequence insertions can occur between binding sites without disrupting function and/or that some large CNEs represent a number of closely positioned but independent CRMs and that many of the large species-specific insertions lie at their boundaries. A whole range of further mutagenesis studies are required to see exactly which parts of a CNE can and cannot tolerate sequence insertions without disrupting their action and whether there is a limit to the distance at which parts of a CNE may be separated without affecting the function. This would indicate whether parts of a CNE are functionally dependent in some way or are completely independent.

CNEs are located within intergenic regions, the introns of trans-dev genes as well as the introns of neighboring non-trans-dev genes. There is no evidence that CNEs have any consistent bias in their position around their likely target gene consistent with the fact that many enhancers can act in a position-independent way. CNEs also appear to act, within limits, in a distance-independent manner. This is revealed through the simple fact that CNEs located many hundreds of kilobases away from their likely target gene in mammals are located at much shorter distances from the orthologous gene in the Fugu genome, and yet we assume that they function in the same way. This is consistent with current models of enhancer action which see the enhancer interact directly with the transcriptional machinery by the looping out of intervening DNA. Over 11% of human CNEs are located at distances from their likely target gene that exceed the distance limits over which currently experimentally verified enhancers are known to act ($\sim$1 Mb). This has major implications in the search for regulatory elements implicated in human disease. An increasing number of conditions are now being associated with SNPs or chromosomal break points/deletions that occur within noncoding regions at significant distances from a disease-linked gene (see other chapters in this volume, Kleinjan and van Heyningen, 2005). The knowledge that many critical regulatory elements may lie beyond these genomic disruptions will help in identifying disease-candidates. For example, Townes-Brock syndrome (TBS) is an autosomal dominant developmental disorder characterized by anal and thumb malformations and by loss of hearing. While most mutations for the disease map to the coding sequence of the transcription factor SALL1, a translocation breakpoint 180 kb upstream of SALL1 on human chromosome 16 has also been linked to the TBS (Marlin et al., 1999). Using CONDOR, 43 out of a total of 71 CNEs identified around SALL1 map beyond this breakpoint, suggesting the removal of many critical regulatory elements away from the SALL1 transcriptional unit is the cause of TBS in these cases.

The likelihood that CNEs work in a distance-independent manner does not preclude the possibility that the order of CNEs across the genome has important functional implications. Over 96% of all the CNEs in this study are conserved not only in sequence but have remained identical in order along the

DNA strand across vertebrate evolution. This suggests that some CNEs might play indirect regulatory roles that are position-dependent, such as structuring the genomic architecture around *trans-dev* genes through control of chromatin conformation or by acting as insulators that shield surrounding genes from the effects of enhancer action. The extremely low rate of CNE transposition suggests evolutionary constraint against the insertion and movement of transposons in these regions. A study of rearrangement events in the human and mouse genomes since their divergence showed that inversions and transpositions are a common occurrence (Kent *et al.*, 2003). There are a number of regions in the genome, though, which are lacking in any transposons [known as transposon free regions (TFRs)] which are enriched for gene regions containing genes with *trans-dev* ontologies (Simons *et al.*, 2006). It is therefore likely that constraint against the insertion and proliferation of transposons derives from the presence of large numbers of critical *cis*-regulatory elements which could become disrupted if transposed away from their target gene(s). Localized inversion events are also relatively rare within this set of CNEs, although they are more common than transpositions, possibly due to the orientation-independent nature of some enhancer elements.

In conclusion, the identification of thousands of potentially critical regulatory elements in the human genome has created an invaluable data resource. The organization of these elements throughout the genome has major implications in how future studies will now approach gene regulation and the characterization of regulatory disorders. Furthermore, knowledge of the function and distribution of CNEs around key developmental regulators will lead to a better understanding of developmental processes in vertebrates and lead to new paradigms of genetic disease.

References

Ahituv, N., Prabhakar, S., Poulin, F., Rubin, E. M., and Couronne, O. (2005). Mapping *cis*-regulatory domains in the human genome using multi-species conservation of synteny. *Hum. Mol. Genet.* **14,** 3057–3063.

Aloni, R., and Lancet, D. (2005). Conservation anchors in the vertebrate genome. *Genome Biol.* **6**(7), 115.

Aparicio, S., Morrison, A., Gould, A., Gilthorpe, J., Chaudhuri, C., Rigby, P., Krumlauf, R., and Brenner, S. (1995). Detecting conserved regulatory elements with the model genome of the Japanese puffer fish, Fugu rubripes. *Proc. Natl. Acad. Sci. USA* **92,** 1684–1688.

Aparicio, S., Hawker, K., Cottage, A., Mikawa, Y., Zuo, L., Venkatesh, B., Chen, E., Krumlauf, R., and Brenner, S. (1997). Organization of the Fugu rubripes Hox clusters: Evidence for continuing evolution of vertebrate Hox complexes. *Nat. Genet.* **16,** 79–83.

Aparicio, S., Chapman, J., Stupka, E., Putnam, N., Chia, J. M., Dehal, P., Christoffels, A., Rash, S., Hoon, S., Smit, A., Gelpke, M. D., Roach, J., *et al.* (2002). Whole-genome shotgun assembly and analysis of the genome of *Fugu rubripes*. *Science* **297,** 1301–1310.

Bagheri-Fam, S., Ferraz, C., Demaille, J., Scherer, G., and Pfeifer, D. (2001). Comparative genomics of the SOX9 region in human and *Fugu rubripes*: Conservation of short regulatory sequence elements within large intergenic regions. *Genomics* **78,** 73–82.

Bejerano, G., Pheasant, M., Makunin, I., Stephen, S., Kent, W. J., Mattick, J. S., and Haussler, D. (2004). Ultraconserved elements in the human genome. *Science* **304,** 1321–1325.

Bejerano, G., Lowe, C. B., Ahituv, N., King, B., Siepel, A., Salama, S. R., Rubin, E. M., Kent, W. J., and Haussler, D. (2006). A distal enhancer and an ultraconserved exon are derived from a novel retroposon. *Nature* **441**(7089), 87–90.

Bishop, C. E., Whitworth, D. J., Qin, Y., Agoulnik, A. I., Agoulnik, I. U., Harrison, W. R., Behringer, R. R., and Overbeek, P. A. (2000). A transgenic insertion upstream of sox9 is associated with dominant XX sex reversal in the mouse. *Nat. Genet.* **26,** 490–494.

Boffelli, D., Nobrega, M. A., and Rubin, E. M. (2004). Comparative genomics at the vertebrate extremes. *Nat. Rev. Genet.* **5,** 456–465.

Brenner, S., Elgar, G., Sandford, R., Macrae, A., Venkatesh, B., and Aparicio, S. (1993). Characterization of the pufferfish (Fugu) genome as a compact model vertebrate genome. *Nature* **366,** 265–268.

Brudno, M., Chapman, M., Gottgens, B., Batzoglou, S., and Morgenstern, B. (2003a). Fast and sensitive multiple alignment of large genomic sequences. *BMC Bioinformatics* **4,** 66–77.

Brudno, M., Do, C. B., Cooper, G. M., Kim, M. F., Davydov, E., Green, E. D., Sidow, A., and Batzoglou, S. (2003b). LAGAN and Multi-LAGAN: Efficient tools for large-scale multiple alignment of genomic DNA. *Genome Res.* **13,** 721–731.

Brudno, M., Malde, S., Poliakov, A., Do, C. B., Couronne, O., Dubchak, I., and Batzoglou, S. (2003c). Glocal alignment: Finding rearrangements during alignment. *Bioinformatics* **19,** i54–i62.

Chuang, J. H., and Li, H. (2004). Functional bias and spatial organization of genes in mutational hot and cold regions in the human genome. *PLoS Biol.* **2,** E29.

Couronne, O., Poliakov, A., Bray, N., Ishkhanov, T., Ryaboy, D., Rubin, E., Pachter, L., and Dubchak, I. (2003). Free Full Text Strategies and tools for whole-genome alignments. *Genome Res.* **13**(1), 73–80.

De la Calle-Mustienes, E., Feijoo, C. G., Manzanares, M., Tena, J. J., Rodriguez-Seguel, E., Letizia, A., Allende, M. L., and Gomez-Skarmeta, J. L. (2005). A functional survey of the enhancer activity of conserved non-coding sequences from vertebrate Iroquois cluster gene deserts. *Genome Res.* **15,** 1061–1072.

Davidson, E. H., Rast, J. P., Oliveri, P., Ransick, A., and Calestani, C. (2002). A genomic gene regulatory network for development. *Science* **295,** 1669–1678.

Dermitzakis, E. T., Reymond, A., Scamuffa, N., Ucla, C., Kirkness, E., Rossier, C., and Antonarakis, S. E. (2003). Evolutionary discrimination of mammalian conserved nongenic sequences (CNGs). *Science* **302,** 1033–1035.

Dermitzakis, E. T., Reymond, A., and Antonarakis, S. E. (2005). Conserved non-genic sequences—an unexpected feature of mammalian genomes. *Nat. Rev. Genet.* **6,** 151–157.

Drake, J. A., Bird, C., Nemesh, J., Thomas, D. J., Newton-Cheh, C., Reymond, A., Excoffier, L., Attar, H., Antonarakis, S. E., and Dermitzakis, E. T. (2006). Conserved noncoding sequences are selectively constrained and not mutation cold spots. *Nat. Genet.* **38,** 223–227.

Dubchak, I., Brudno, M., Loots, G. G., Pachter, L., Mayor, C., Rubin, E. M., and Frazer, K. A. (2000). Active conservation of noncoding sequences revealed by three-way species comparisons. *Genome Res.* **10,** 1304–1306.

Duret, L., and Bucher, P. (1997). Searching for regulatory elements in human noncoding sequences. *Curr. Opin. Struct. Biol.* **7,** 399–406.

Durick, K., Mendlein, J., and Xanthopoulos, K. G. (1999). Hunting with traps: Genome-wide strategies for gene discovery and functional analysis. *Genome Res.* **9,** 1019–1025.

Elgar, G., Sandford, R., Aparicio, S., Macrae, A., Venkatesh, B., and Brenner, S. (1996). Small is beautiful: Comparative genomics with the pufferfish (Fugu rubripes). *Trends Genet.* **12**, 145–150.

Ellingsen, S., Laplante, M. A., Konig, M., Kikuta, H., Furmanek, T., Hoivik, E. A., and Becker, T. S. (2005). Large-scale enhancer detection in the zebrafish genome. *Development* **132**, 3799–3811.

Feng, J., Bi, C., Clark, B. S., Mady, R., Shah, P., and Kohtz, J. D. (2006). The Evf-2 noncoding RNA is transcribed from the Dlx-5/6 ultraconserved region and functions as a Dlx-2 transcriptional coactivator. *Genes Dev.* **20**, 1470–1484.

Frazer, K. A., Tao, H., Osoegawa, K., de Jong, P. J., Chen, X., Doherty, M. F., and Cox, D. R. (2004). Noncoding sequences conserved in a limited number of mammals in the SIM2 interval are frequently functional. *Genome Res.* **14**, 367–372.

Ghanem, N., Jarinova, O., Amores, A., Long, Q., Hatch, G., Park, B. K., Rubenstein, J. L., and Ekker, M. (2003). Regulatory roles of conserved intergenic domains in vertebrate Dlx bigene clusters. *Genome Res.* **13**, 533–543.

Goode, D. K., Snell, P., Smith, S. F., Cooke, J. E., and Elgar, G. (2005). Highly conserved regulatory elements around the SHH gene may contribute to the maintenance of conserved synteny across human chromosome 7q36.3. *Genomics* **86**, 172–181.

Göttgens, B., Barton, L. M., Gilbert, J. G., Bench, A. J., Sanchez, M. J., Bahn, S., Mistry, S., Grafham, D., McMurray, A., Vaudin, M., Amaya, E., and Bentley, D. R. (2000). Analysis of vertebrate SCL loci identifies conserved enhancers. *Nat. Biotechnol.* **18**, 181–186.

Gribnau, J., Diderich, K., Pruzina, S., Calzolari, R., and Fraser, P. (2000). Intergenic transcription and developmental remodeling of chromatin subdomains in the human beta-globin locus. *Mol Cell* **5**, 377–386.

Griffin, C., Kleinjan, D. A., Doe, B., and van Heyningen, V. (2002). New 3′ elements control Pax6 expression in the developing pretectum, neural retina and olfactory region. *Mech. Dev.* **112**, 89–100.

Hardison, R. C., Oeltjen, J., and Miller, W. (1997). Long human-mouse sequence alignments reveal novel regulatory elements: A reason to sequence the mouse genome. *Genome Res.* **7**, 959–966.

Hill-Kapturczak, N., Sikorski, E., Voakes, C., Garcia, J., Nick, H. S., and Agarwal, A. (2003). An internal enhancer regulates heme- and cadmium-mediated induction of human heme oxygenase-1. *Am. J. Physiol. Renal Physiol.* **285**, F515–F523.

Hoegg, S., Brinkmann, H., Taylor, J. S., and Meyer, A. (2004). Phylogenetic timing of the fish specific genome duplication correlates with the diversification of teleost fish. *J. Mol. Evol.* **59**, 190–203.

Jaillon, O., Aury, J. M., Brunet, F., Petit, J. L., Stange-Thomann, N., Mauceli, E., Bouneau, L., Fischer, C., Ozouf-Costaz, C., Bernot, A., Nicaud, S., and Jaffe, D. (2004). Genome duplication in the teleost fish Tetraodon nigroviridis reveals the early vertebrate proto-karyotype. *Nature* **431**, 946–957.

Jamieson, R. V., Perveen, R., Kerr, B., Carette, M., Yardley, J., Heon, E., Wirth, M. G., van Heyningen, V., Donnai, D., Munier, F., and Black, G. C. (2002). Domain disruption and mutation of the bZIP transcription factor, MAF, associated with cataract, ocular anterior segment dysgenesis and coloboma. *Hum. Mol. Genet.* **11**, 33–42.

Jones, E. A., and Flavell, R. A. (2005). Distal enhancer elements transcribe intergenic RNA in the IL-10 family gene cluster. *J. Immunol.* **175**, 7437–7446.

Kammandel, B., Chowdhury, K., Stoykova, A., Aparicio, S., Brenner, S., and Gruss, P. (1999). Distinct cis-essential modules direct the time-space pattern of the Pax6 gene activity. *Dev. Biol.* **205**, 79–97.

Kent, W. J., Baertsch, R., Hinrichs, A., Miller, W., and Haussler, D. (2003). Evolution's cauldron: Duplication, deletion, and rearrangement in the mouse and human genomes. *Proc. Natl. Acad. Sci. USA* **100**, 11484–11489.

Kikuta, H., Laplante, M., Navratilova, P., Komisarczuk, A. Z., Engstrom, P. G., Fredman, D., Akalin, A., Caccamo, M., Sealy, I., Howe, K., Ghislain, J., Pezeron, G., et al. (2007). Genomic regulatory blocks encompass multiple neighboring genes and maintain conserved synteny in vertebrates. *Genome Res.* **17,** 545–555.

Kleinjan, D. A., and van Heyningen, V. (2005). Long-range control of gene expression: Emerging mechanisms and disruption in disease. *Am. J. Hum. Genet.* **76,** 8–32.

Koide, T., Hayata, T., and Cho, K. W. (2005). Xenopus as a model system to study transcriptional regulatory networks. *Proc. Natl. Acad. Sci. USA* **102,** 4943–4948.

Lettice, L. A., Heaney, S. J., Purdie, L. A., Li, L., de Beer, P., Oostra, B. A., Goode, D., Elgar, G., Hill, R. E., and de Graaff, E. (2003). A long-range Shh enhancer regulates expression in the developing limb and fin and is associated with preaxial polydactyly. *Hum. Mol. Genet.* **12,** 1725–1735.

Levine, R., and Davidson, E. H. (2005). Gene regulatory networks for development. *Proc. Natl. Acad. Sci. USA* **102,** 4936–4942.

Loose, M., and Patient, R. (2004). A genetic regulatory network for Xenopus mesendoderm formation. *Dev. Biol.* **271,** 467–478.

Loots, G. G., Locksley, R. M., Blankespoor, C. M., Wang, Z. E., Miller, W., Rubin, E. M., and Frazer, K. A. (2000). Identification of a coordinate regulator of interleukins 4, 13, and 5 by cross-species sequence comparisons. *Science* **288,** 136–140.

Marlin, S., Blanchard, S., Slim, R., Lacombe, D., Denoyelle, F., Alessandri, J. L., Calzolari, E., Drouin-Garraud, V., Ferraz, F. G., Fourmaintraux, A., Philip, N., and Toublanc, J. E. (1999). Townes-Brocks syndrome: Detection of a SALL1 mutation hot spot and evidence for a position effect in one patient. *Hum. Mutat.* **14,** 377–386.

Margulies, E. H., Blanchette, M., Haussler, D., and Green, E. D. (2003). Identification and characterization of multi-species conserved sequences. *Genome Res.* **13,** 2507–2518.

Margulies, E. H., Chen, C. W., and Green, E. D. (2006). Differences between pair-wise and multisequence alignment methods affect vertebrate genome comparisons. *Trends Genet.* **22,** 187–193.

Marshall, H., Studer, M., Popperl, H., Aparicio, S., Kuroiwa, A., Brenner, S., and Krumlauf, R. (1994). A conserved retinoic acid response element required for early expression of the homeobox gene Hoxb-1. *Nature* **370,** 567–571.

Martinez-Cruzado, J. C., Swimmer, C., Fenerjian, M. G., and Kafatos, F. C. (1988). Evolution of the autosomal chorion locus in *Drosophila*. I. General organization of the locus and sequence comparisons of genes s15 and s19 in evolutionary distant species. *Genetics* **119,** 663–677.

Mignone, F., Grillo, G., Liuni, S., and Pesole, G. (2003). Computational identification of protein-coding potential of conserved sequence tags through cross-species evolutionary analysis. *Nucleic Acids Res.* **31,** 4639–4645.

Miles, C., Elgar, G., Coles, E., Kleinjan, D. J., van Heyningen, V., and Hastie, N. (1998). Complete sequencing of the Fugu WAGR region from WT1 to PAX6: Dramatic compaction and conservation of synteny with human chromosome 11p13. *Proc. Natl. Acad. Sci. USA* **95,** 13068–13072.

Müller, F., Blader, P., and Strahle, U. (2002). Search for enhancers: Teleost models in comparative genomic and transgenic analysis of *cis* regulatory elements. *Bioessays* **24,** 564–572.

Nelson, J. S. (1994). "Fishes of the World" 3rd edn., John Wiley and Sons, New York, USA.

Nobrega, M. A., Ovcharenko, I., Afzal, V., and Rubin, E. M. (2003). Scanning human gene deserts for long-range enhancers. *Science* **302,** 413.

Nobrega, M. A., Zhu, Y., Plajzer-Frick, I., Afzal, V., and Rubin, E. M. (2004). Megabase deletions of gene deserts result in viable mice. *Nature* **431**(7011), 988–993.

Ochoa-Espinosa, A., Yucel, G., Kaplan, L., Pare, A., Pura, N., Oberstein, A., Papatsenko, D., and Small, S. (2005). The role of binding site cluster strength in Bicoid-dependent patterning in Drosophila. *Proc. Natl. Acad. Sci. USA* **102,** 4960–4965.

Oeltjen, J. C., Malley, T. M., Muzny, D. M., Miller, W., Gibbs, R. A., and Belmont, J. W. (1997). Large-scale comparative sequence analysis of the human and murine Bruton's tyrosine kinase loci reveals conserved regulatory domains. *Genome Res.* **7,** 315–329.

Ovcharenko, I., Nobrega, M. A., Loots, G. G., and Stubbs, L. (2004). ECR Browser: A tool for visualizing and accessing data from comparisons of multiple vertebrate genomes. *Nucleic Acids Res.* **32**(suppl 2): W280–W286.

Ovcharenko, I., Loots, G. G., Nobrega, M. A., Hardison, R. C., Miller, W., and Stubbs, L. (2005). Evolution and functional classification of vertebrate gene deserts. *Genome Res.* **15,** 137–145.

Pennacchio, L. A., and Rubin, E. M. (2001). Genomic strategies to identify mammalian regulatory sequences. *Nat. Rev. Genet.* **2,** 100–109.

Poulin, F., Nobrega, M. A., Plajzer-Frick, I., Holt, A., Afzal, V., Rubin, E. M., and Pennacchio, L. A. (2005). *In vivo* characterization of a vertebrate ultraconserved enhancer. *Genomics* **85,** 774–781.

Rowitch, D. H., Echelard, Y., Danielian, P. S., Gellner, K., Brenner, S., and McMahon, A. P. (1998). Identification of an evolutionarily conserved 110 base-pair cis-acting regulatory sequence that governs Wnt-1 expression in the murine neural plate. *Development* **125,** 2735–2746.

Sandelin, A., Bailey, P., Bruce, S., Engstrom, P. G., Klos, J. M., Wasserman, W. W., Ericson, J., and Lenhard, B. (2004). Arrays of ultraconserved noncoding regions span the loci of key developmental genes in vertebrate genomes. *BMC Genomics* **5,** 99.

Sanges, R., Kalmar, E., Claudiani, P., D'Amato, M., Muller, F., and Stupka, E. (2006). Shuffling of cis-regulatory elements is a pervasive feature of the vertebrate lineage. *Genome Biol.* **7,** R56.

Santagati, F., Gerber, J. K., Blusch, J. H., Kokubu, C., Peters, H., Adamski, J., Werner, T., Balling, R., and Imai, K. (2001). Comparative analysis of the genomic organization of Pax9 and its conserved physical association with Nkx2–9 in the human, mouse, and pufferfish genomes. *Mamm. Genome* **12,** 232–237.

Santagati, F., Abe, K., Schmidt, V., Schmitt-John, T., Suzuki, M., Yamamura, K., and Imai, K. (2003). Identification of cis-regulatory elements in the mouse Pax9/Nkx2–9 genomic region: Implication for evolutionary conserved synteny. *Genetics* **165,** 235–242.

Shaw, C. J., and Lupski, J. R. (2004). Implications of human genome architecture for rearrangement-based disorders: The genomic basis of disease. *Hum. Mol. Genet.* **13,** R57–R64.

Shi, W., Levine, M., and Davidson, B. (2005). Unraveling genomic regulatory networks in the simple chordate, Ciona intestinalis. *Genome Res.* **15,** 1668–1674.

Simons, C., Pheasant, M., Makunin, I. V., and Mattick, J. S. (2006). Transposon-free regions in mammalian genomes. *Genome Res.* **16,** 164–172.

Sironi, M., Menozzi, G., Comi, G. P., Cagliani, R., Bresolin, N., and Pozzoli, U. (2005). Analysis of intronic conserved elements indicates that functional complexity might represent a major source of negative selection on non-coding sequences. *Hum. Mol. Genet.* **14,** 2533–2546.

Sorek, R., and Safer, H. M. (2003). A novel algorithm for computational identification of contaminated EST libraries. *Nucleic Acids Res.* **31,** 1067–1074.

Spitz, F., Gonzalez, F., and Duboule, D. (2003). A global control region defines a chromosomal regulatory landscape containing the HoxD cluster. *Cell* **113,** 405–417.

Sun, H., Skogerbo, G., and Chen, R. (2006). Conserved distances between vertebrate highly conserved elements. *Hum. Mol. Genet.* **15,** 2911–2922.

Swamynathan, S. K., and Piatigorsky, J. (2002). Orientation-dependent influence of an intergenic enhancer on the promoter activity of the divergently transcribed mouse Shsp/alpha B-crystallin and Mkbp/HspB2 genes. *J. Biol. Chem.* **277,** 49700–49706.

Tautz, D. (2000). Evolution of transcriptional regulation. *Curr. Opin. Genet. Dev.* **10,** 575–579.

Taylor, J. S., Braasch, I., Frickey, T., Meyer, A., and van de Peer, Y. (2003). Genome duplication, a trait shared by 22,000 species of ray-finned fish. *Genome Res.* **13,** 382–390.

Thomas, J. W., Touchman, J. W., Blakesley, R. W., Bouffard, G. G., Beckstrom-Sternberg, S. M., Margulies, E. H., Blanchette, M., Siepel, A. C., Thomas, P. J., McDowell, J. C., Maskeri, B., Hansen, N. F., et al. (2003). Comparative analyses of multi-species sequences from targeted genomic regions. Nature **424**(6950), 788–793.

Vandepoele, K., de Vos, W., Taylor, J. S., Meyer, A., and van de Peer, Y. (2004). Major events in the genome evolution of vertebrates: Paranome age and size differ considerably between ray-finned fishes and land vertebrates. Proc. Natl. Acad. Sci. **101**, 1638–1643.

Vavouri, T., McEwen, G. K., Woolfe, A., Gilks, W. R., and Elgar, G. (2006). Defining a genomic radius for long-range enhancer action: Duplicated conserved non-coding elements hold the key. Trends Genet. **22**, 5–10.

Vavouri, T., Walter, K., Gilks, W. R., Lehner, B., and Elgar, G. (2007). Parallel evolution of conserved non-coding elements that target a common set of developmental regulatory genes from worms to humans. Genome Biol. **8**, R15.

Venkatesh, B., Tay, A., Dandona, N., Patil, J. G., and Brenner, S. (2005). A compact cartilaginous fish model genome. Curr. Biol. **15**, R82–R83.

Venkatesh, B., Kirkness, E. F., Loh, Y. H., Halpern, A. L., Lee, A. P., Johnson, J., Dandona, N., Viswanathan, L. D., Tay, A., Venter, J. C., Strausberg, R. L., and Brenner, S. (2006). Ancient noncoding elements conserved in the human genome. Science **314**, 1892.

Wasserman, W. W., Palumbo, M., Thompson, W., Fickett, J. W., and Lawrence, C. E. (2000). Human: Mouse genome comparisons to locate regulatory sites. Nat. Genet. **26**, 225–228.

Waterston, R. H., Lindblad-Toh, K., Birney, E., Rogers, J., Abril, J. F., Agarwal, P., Agarwala, R., Ainscough, R., Alexandersson, M., An, P., Antonarakis, S. E., and Attwood, J. (2002). Initial sequencing and comparative analysis of the mouse genome. Nature **420**, 520–562.

Woolfe, A. (2006). Computational detection and analysis of putative cis-regulatory elements in vertebrate genomes. PhD Thesis, University of Cambridge.

Woolfe, A., Goodson, M., Goode, K., Snell, P., McEwen, G. K., Vavouri, T., Smith, S. F., North, P., Callaway, H., Kelly, K., Walter, K., and Abnizova, I. (2005). Highly conserved noncoding sequences are associated with vertebrate development. PLoS Biol. **3**, e7.

Woolfe, A., Goode, D., Cooke, J., Callaway, H., Smith, S., Snell, P., McEwen, G., and Elgar, G. (2007). CONDOR: A database resource of developmentally associated conserved non-coding elements. BMC Dev. Biol. **7**, 100.

Xie, X., Kamal, M., and Lander, E. S. (2006). A family of conserved noncoding elements derived from an ancient transposable element. Proc. Natl. Acad. Sci. USA **103**, 11659–11664.

13

Long-Range Gene Control and Genetic Disease

Dirk A. Kleinjan and Laura A. Lettice
MRC Human Genetics Unit, Western General Hospital,
Edinburgh EH4 2XU, United Kingdom

   I. From Genetic Disease to Long-Range Gene Regulation
  II. Position Effect Revisited
      A. Thalassemias and the α- and β-globin loci
 III. Loss of a Positive Regulator
      A. Van Buchem disease
      B. Leri-Weill dyschondrosteosis
  IV. TWIST, POU3F4, PITX2, SOX3, GLI3, and FOXP2
   V. The "Bystander" Effect
      A. Aniridia and PAX6
  VI. MAF, SDC2, TGFB2, REEP3, and PLP1
 VII. Two Position Effects—Different Outcomes
      A. Sonic hedgehog, holoprosencephaly, and
         preaxial polydactyly
VIII. Phenotypes Resulting from Position Effects on
      More than One Gene
      A. Split hand foot malformation locus 1
      B. Combination of two position effects in
         SHH and RUNX2
  IX. Global Control Regions; HOXD, Gremlin, and
      Limb Malformations
   X. FOX Genes and Position Effects
      A. FOX genes in eye anomalies
  XI. SOX9 and Campomelic Displasia

Advances in Genetics, Vol. 61
Copyright 2008, Elsevier Inc. All rights reserved.

0065-2660/08 $35.00
DOI: 10.1016/S0065-2660(07)00013-2

XII. Facioscapulohumeral Dystrophy
XIII. Aberrant Creation of an Illegitimate
       siRNA Target Site
XIV. Genetic Disease Due to Aberrant Gene Transcription Can Be
       Caused by Many Different Mechanisms
       A. The problem: How to find and assess regulatory mutations?
       B. Long-range control and genome organization
       C. Implications for (common) genetic disease
XV. Concluding Remarks
       References

ABSTRACT

The past two decades have seen great progress in the elucidation of the genetic basis of human genetic disease. Many clinical phenotypes have been linked with mutations or deletions in specific causative genes. However, it is often less recognized that in addition to the integrity of the protein-coding sequences, human health critically also depends on the spatially, temporally, and quantitatively correct expression of those genes. Genetic disease can therefore equally be caused by disruption of the regulatory mechanisms that ensure proper gene expression. The term "position effect" is used in those situations where the expression level of a gene is deleteriously affected by an alteration in its chromosomal environment, while maintaining an intact transcription unit. Here, we review recent advances in our understanding of the possible mechanisms of a number of "position effect" disease cases and discuss the findings with respect to current models for genome organization and long-range control of gene expression. © 2008, Elsevier Inc.

I. FROM GENETIC DISEASE TO LONG-RANGE GENE REGULATION

Large advances have been made over the past two decades in determining the genetic basis of human genetic disease. Mutations in an impressive number of genes have been linked with specific clinical phenotypes (Valle, 2004). The completion of the human genome project has undoubtedly been invaluable in speeding up the search for new genes responsible for specific medical disorders. It has, however, also contributed to the recognition that in a small but significant number of patients the molecular lesions do not disrupt the transcribed region of the gene directly, but rather interfere with its transcriptional regulation. In many instances, individual patient cases provided the first indications that long-range gene regulation might be involved in the control of expression of the genes underlying congenital malformations.

The spatially, temporally, and quantitatively correct activity of a gene requires the presence of not only intact coding sequence but also properly functioning regulatory control. There may be little difference between a coding region mutation that reduces the protein activity by half and a regulatory defect that causes a twofold reduction in expression. Thus, it comes as no surprise that apart from deleterious mutations in the protein encoding part of the gene, genetic disease can equally be caused by disruption of the regulatory mechanisms that ensure its proper expression. Due in part to the generally greater flexibility and robustness in the regulatory control sequences than in protein-coding sequences and in part to ascertainment bias, by far the majority of reported cases of disease-associated mutations alter the protein-coding sequence of the gene in some way. However, a number of different mechanisms can interfere with normal gene function by disrupting the proper regulatory controls of gene expression and lead to pathological states. In this chapter, we catalogue the various known mechanisms with examples of genetic diseases thought to be caused in those ways.

Understanding the mechanisms of gene expression control is one of the great challenges of the post-genome era. It is generally accepted that for most genes, transcriptional regulation is the main level at which control of gene expression takes place, although it is certainly true that in specific cases post-transcriptional events can also play an important role. Transcriptional control itself is influenced at two levels: (1) through modification of the chromatin structure of the gene locus and (2) through the action of *trans*-acting factors on *cis*-regulatory sequences. While often described as separate events, the two are intimately linked, with chromatin structure influencing accessibility of DNA to transcription factors, and the binding of *trans*-acting factors to the DNA involved in triggering chromatin modifications. It will be clear that when either the *cis-trans* regulatory system of a gene or the normal context of its chromatin structure is disrupted, the expression of the gene may be adversely affected, in some cases leading to disease. Conversely, in cases of genetic disease where the disease-causing gene product is known, but no mutations are found in the transcribed portion of the gene, disruption of its normal expression control can be suspected. Such cases have been termed "position effect" cases in reference to the phenomenon of position-effect variegation whereby expression of a gene is variably, but in a clonally heritable manner, inhibited because of juxtaposition with a region of heterochromatin (see Karpen, 1994). In the case of human genetic disease, we define the term "position effect" to refer to situations where the level of expression of a gene is deleteriously affected by an alteration in its chromosomal environment, while maintaining an intact transcription unit. Over the past years, an increasing number of such disease-related position effect cases have come to light. The study of these cases has been valuable in high-lighting the dependence on long-range gene control for the underlying gene, and in some cases has been instrumental in pinpointing the *cis*-regulatory elements.

An important aspect to come out of these studies is the observation that *cis*-regulatory sequences can be located at very large distances from their linked gene, and can even be found in locations beyond or within the introns of neighboring genes. As described below, severe genetic defects can be the result of even single nucleotide substitutions in such a long-range enhancer. Further-more, it is important to be aware that a translocation or deletion could very well coincidentally disrupt the coding region of one gene as well as remove a distant enhancer from a second gene, and that the resultant alteration of expression of this second gene rather than the physical disruption of the first is the cause of the disease. As will be discussed later, regulatory mutations may disrupt only a spatiotemporally specific subset of the normal expression pattern, and hence the clinical phenotype observed in a patient with a position effect can be different from that due to a complete loss of the underlying gene. Thus, analysis of such patients may uncover novel functions for that gene.

While juxtaposition with heterochromatin is a potential mechanism in a small number of genetic diseases, it has become clear that in the majority of "position effect" cases studied to date the genetic defect appears to be caused by a disruption of the normal *cis*-regulatory architecture of the gene locus. Even so, a number of different mechanisms leading to such *cis*-regulatory disruption have been encountered as will be described below. We review recent advances in our understanding of the possible mechanisms of a number of "position effect" disease cases and try to incorporate the findings into a model for genome organization and long-range control of gene expression.

II. POSITION EFFECT REVISITED

A. Thalassemias and the α- and β-globin loci

Some of the earliest cases where it was recognized that disruption of normal gene regulation can lead to genetic disease stem from research on the blood disorder thalassemia. Thalassemias result from an imbalance in the levels of the α- and β-globin chains that make up the oxygen carrying hemoglobin in our red blood cells. Mostly this imbalance is caused by mutation or deletion of one or more of the globin genes, but, as in the case of the Spanish and Dutch thalassemias, it can also occur through translocations that remove the locus control region (LCR), a major upstream control region for the β-globin locus. The LCR consists of a set of *cis*-regulatory sites (the hypersensitive sites) with the ability to drive high-level expression of the β-globin genes in erythroid cells (Grosveld *et al.*, 1987, and reviewed in the Palstra, de Laat, and Grosveld chapter in this volume). $\gamma\beta$-Thalassemias caused by translocations in the β-globin locus thus represent one of the earliest recognized position effect cases (Driscoll *et al.*, 1989; Kioussis *et al.*, 1983), and control of β-globin expression continues to be the subject of intense study and many ideas about mechanisms of gene regulation stem from these analyses. Recently, detailed studies of the murine β-globin locus using two novel techniques, 3C (Tolhuis *et al.*, 2002) and RNA-TRAP (Carter *et al.*, 2002), have shown evidence for a mechanism of long-range interaction that involves close contact between the enhancer and the promoter. The more versatile 3C technique has since been applied to a number of other gene loci and most of the published findings provide support for a looping model of enhancer–promoter interactions, whereby distal *cis*-elements and promoters come together in nuclear space with the intervening sequences looping out (Fraser, 2006).

The α-globin gene cluster also depends on long-range control elements for its expression, and deletions removing these elements from the locus are a cause of α-thalassemia (Hatton *et al.*, 1990; Viprakasit *et al.*, 2003). In contrast to the β-globin locus, the α-globin locus resides in a gene-rich region of open chromatin surrounded by a number of housekeeping genes (Vyas *et al.*, 1995). Continued analysis of new patient cases with small deletions outside the α-globin genes has homed in on the most highly conserved element, HS-40, as the dominant α-globin regulatory element in humans (Viprakasit *et al.*, 2006), even though deletion of the equivalent element in the mouse, HS-26, has a remarkably mild phenotype (Anguita *et al.*, 2002).

A novel type of position effect mechanism was revealed in a recently reported case of α-thalassemia (Tufarelli *et al.*, 2003). In this particular α-thalassemia case, an 18-kb deletion, encompassing the HBA1 and HBQ1 genes, was identified in a Polish family. In normal individuals, α-globin is transcribed from four major genes (HBA1 and HBA2 on both copies of chromosome 16) in the α-globin cluster, under the control of four regulatory regions (R1–R4) of which the 5' HS-40 region (R2) plays the major role. The patients' HBA2 gene and control elements remained completely intact. However, while one copy of the HBA1 gene was indeed deleted in this family, the severity of their phenotype suggested that expression from the intact HBA2 gene on the deleted chromosome could also be affected. The absence of HBA2 expression from the abnormal chromosome was confirmed in an experiment using permissive cell hybrids. Furthermore, a 2-kb region including the HBA2 CpG island was found to be densely methylated in all tissues, while under normal circumstances the α-globin CpG islands always remain unmethylated, even in nonexpressing tissues. Further analysis of this case showed that rather than the initially suspected disruption of a control element, silencing and methylation of HBA2 were strongly correlated with the presence of antisense RNA transcripts derived from the truncated neighboring LUC7L gene on the opposite strand. In addition to the HBA1 and HBQ genes, the 18-kb deletion had removed the final three exons of the LUC7L gene, including its polyadenylation signal, causing RNA polymerase to read through into the HBA2 promoter and coding region. The antisense-induced silencing and methylation of the promoter was confined to the deleted chromosome, indicating it occurred through a purely cis-acting mechanism (Tufarelli et al., 2003).

This case presents a novel and currently unusual mechanism by which a "position effect" can occur, because it does not involve the disruption of normal long-range gene control required for normal expression. However, it can be envisaged that such a juxtaposition of a truncated, (highly) expressed gene, lacking a polyA signal, in close proximity to a disease gene on the opposite strand as a result of a deletion or translocation is a mechanism that could easily cause problems in many other gene loci, in particular in gene-dense regions of the genome.

Detailed study of the α-globin locus in a set of α-thalassemia patients from Melanesia has led to the discovery of yet another novel mechanism of human genetic disease. Analysis of these patients failed to find any of the previously described molecular defects known to cause α-thalassemia, such as mutations or deletions in the globin-coding regions or regulatory elements. The inheritance pattern suggested a codominant defect, and as linkage to the α-globin locus was confirmed it seemed the disease would most likely be due to a gain-of-function mutation that downregulates expression of the α-globin genes. Resequencing of a bacterial artificial chromosome (BAC) containing the globin locus derived from patient DNA identified a large number of single nucleotide polymorphisms (SNPs) compared to the database sequence. This raised the problem of discerning which of these SNPs might be functionally relevant. When the RNA expression profiles of the locus in normal and patient erythroblasts were compared, a major new peak of transcription was identified in the patient sample in the genomic region between the regulatory elements and the globin genes. The peaks of transcription of the globin genes were concomitantly reduced. Of the seven SNPs located underlying the new transcription peak, only one was unique to the patient and involved a T to C change that creates a new GATA factor binding site. GATA-1 is a key regulator of erythroid cell differentiation and binding of GATA-1 as well as other transcription factors, SCL, E2A, LMO2, and Ldb-1, which are often found in association with GATA-1 at erythroid-specific enhancers, was confirmed by chromatin immunoprecipitation. Thus, it appears that this single base pair mutation has created a new promoter, located between the upstream regulatory elements and their cognate promoters, that interferes with the normal regulatory mechanism, thereby causing downregulation of the α-globin genes.

III. LOSS OF A POSITIVE REGULATOR

Arguably the simplest way by which a position effect can occur is a situation where loss of a regulatory element leads to misexpression of the gene. Such situations have been identified for a number of human conditions and form the majority of currently recognized position effect cases. Mostly they entail the loss of one or more positive acting cis-elements, but they could also result from loss of a repressor or the gain of an inappropriate regulatory element. The latter mechanisms are known to play a part in certain cancers, but these will not be discussed here. The loss of cis-elements can lead to disease in both homozygous (recessive) and heterozygous [(semi-)dominant] states, though because of selection and ascertainment bias most cases are the result of haploinsufficiency of the target gene.

A. Van Buchem disease

The homozygous recessive disorder Van Buchem disease, a severe sclerosing bone dysplasia, has been mapped to chromosome 17p21 (Balemans et al., 2002; Van Hul et al., 1998). It is characterized by a progressive increase in bone density which results in facial distortion, head and mandible enlargement, entrapment of cranial nerves, and a general increase in bone strength and weight (Van Hul et al., 1998). The closely related, but phenotypically more severe bone malformation sclerosteosis (MIM 269500) has also been mapped to the same region. A gene encoding a negative regulator of bone formation, sclerostin (SOST), was found in this region (Balemans et al., 2001; Brunkow et al., 2001) and its expression was shown to be affected in both sclerosteosis and Van Buchem disease. While sclerosteosis patients were shown to carry homozygous null SOST mutations, Van Buchem patients have an intact SOST-coding region but carry a homozygous deletion of 52 kb starting 35 kb downstream of the SOST polyadenylation signal and 10 kb upstream of the neighboring MEOX1 gene (Balemans et al., 2001; Staehling-Hampton et al., 2002). The similarities between sclerosteosis and Van Buchem disease strongly suggested that expression of the SOST gene was affected by the deletion, implicating sequences within the deletion in the regulation of SOST. A BAC carrying the human SOST locus faithfully reproduced the endogenous SOST expression pattern in transgenic mice, but mice carrying a BAC engineered to contain the 52-kb deletion had dramatically reduced levels of human SOST mRNA expression (Loots et al., 2005). Using human/mouse sequence comparisons of the 52-kb region, seven highly conserved elements were identified and tested by luciferase assay in osteosarcoma cells. Only one, ECR5, was able to activate the human SOST promoter and this element was subsequently shown to drive reporter gene expression in the skeleton of E14.5 transgenic mice. Thus, it is highly likely that the cause of Van Buchem disease is the removal of the 250 bp ECR5 element from the SOST locus (Loots *et al.*, 2005).

B. Leri-Weill dyschondrosteosis

The growth of bones in our body is attained by proliferation and differentiation of chondrocytes in the growth plates of the bones, a process that requires tight genetic (and environmental) control and misregulation can lead to skeletal dysplasias or short stature syndromes. One such syndrome, Leri-Weill dyschondrosteosis (OMIM 127300), is caused by haploinsufficiency of the short stature homeobox gene (SHOX) in about 50–70% of cases. Homozygous loss of SHOX results in the more severe Langer dysplasia (OMIM 249700). Both are characterized by mesomelic, disproportionate short stature and a characteristic curving of the radius (Madelung deformity). The SHOX gene is located on the pseudoautosomal region of the sex chromosomes and has therefore also been implicated in the skeletal deformities of Turner syndrome. While exonic mutations have been found in a large proportion of LWS cases, a number of studies have found patients with an intact SHOX intragenic region but carrying microdeletions in the region downstream from SHOX. These studies highlight a common deleted region of ~40 kb, (Fukami *et al.*, 2005), 30 kb (Benito-Sanz *et al.*, 2005), or as little as 10 kb (Huber *et al.*, 2006). As no other known genes are present in the 750 kb between SHOX and the next gene CSF2RA, a recent study conducted an *in vivo* analysis of putative enhancer activity in this area. Data from four patients with deletions of between 220 and 360 kb were used to define a 200 kb common deletion interval (Sabherwal *et al.*, 2007). Scanning of the region for the presence of multispecies conserved sequences revealed eight evolutionary conserved regions (ECRs) in comparison to dog, chicken, and frog. SHOX like other genes in the Xp22 region is absent in mouse and rat. Enhancer activity of the eight ECRs was therefore tested by electroporation of reporter constructs into the chicken limb bud *in ovo* and analyzed for expression at the appropriate stages of development. Three of the ECRs showed reporter gene expression in the proximal part of the limb, consistent with endogenous chicken SHOX expression. The most distal of those three ECRs lies within the small deletion intervals identified in two of the other LWS position effect studies and thus seems to represent a frequently deleted ECR in patients with LWS and downstream deletions.

IV. TWIST, POU3F4, PITX2, SOX3, GLI3, AND FOXP2

There are now several other cases of human genetic disease where the removal of a positive acting regulatory element is strongly suspected, but for which the exact *cis*-element has not yet been identified and characterized. Examples of these are found in patients with Saethre-Chotzen syndrome (TWIST), X-linked deafness (POU3F4), Rieger syndrome (PITX2), hypoparathyroidism (SOX3), and Greig cephalopolysyndactyly (GLI3) (see Table 13.1).

Saethre-Chotzen syndrome is a common autosomal dominant form of craniosynostosis, the premature fusion of the sutures of the calvarial bones of the skull. Haploinsufficiency of TWIST, a basic helix-loop-helix transcription factor, is implicated as the pathogenic mechanism by virtue of the identification of mutations and deletions in the coding region (Chun et al., 2002). Using real-time polymerase chain reaction (PCR) to analyze allele dosage by "walking" across the critical region, translocation or inversion breakpoints were found in two patients located at least 260 kb downstream from the TWIST gene (Cai et al., 2003).

The POU3F4 (Brn-4) gene is involved in the pathogenesis of X-linked deafness type 3 (DFN3), the most common form of X-linked inherited deafness. Clinical features include fixation of the stapes and a widening of the internal auditory canal, allowing entry of cerebrospinal fluid into the inner ear. In addition to a spectrum of missense and truncating mutations, a cohort of genomic deletion cases has been studied (de Kok et al., 1995). The observed deletions either remove the POU3F4 gene itself or overlap in a small region ~900 kb upstream of the gene (de Kok et al., 1995, 1996). The smallest of these deletions comprises an 8 kb small fragment containing a 2-kb sequence that is 80% conserved between mouse and human (Cremers and Cremers, 2004). Interestingly, a mouse mutant generated in a random mutagenesis screen with developmental malformations of the inner ear resulting in hearing loss and vestibular dysfunction, the sex-linked fidget (slf) mutant, was shown to be a regulatory mutant of Pou3f4/brn-4 (Phippard et al., 2000). The mutant carries a large X-linked inversion with one breakpoint near but not in the Pou3f4/brn-4 transcription unit, with no gross structural rearrangements being detected within 6–10 kb of the coding region. Expression of Pou3f4/brn-4 was abolished in the embryonic inner ear of the mutant; however, the gene continued to be expressed normally within the neural tube, consistent with a study that mapped two neural tube enhancers within 6 kb upstream of the gene and which would be unaffected by the inversion (Heydemann et al., 2001).

Rieger syndrome (RIEG) type 1 is an autosomal dominant disorder characterized by dental hypoplasia and malformation of the umbilicus and anterior segment of the eye. The main RIEG locus is mapped to chromosome 4q25–27, and mutations in PITX2, a paired-related homeobox gene with multiple isoforms, have been shown to cause Rieger syndrome (Alward et al., 1998). In addition to deletions and mutations in the gene itself, translocation breakpoints 15 and 90 kb upstream of PITX2 have been identified in three separate patients (Flomen et al., 1998; Trembath et al., 2004).

Hypoparathyroidism (HPT) is an endocrine disorder due to deficiency of parathyroid hormone (PTH) leading to hypocalcemia and hyperphosphatemia. It can exist as part of larger syndromes or as an isolated endocrinopathy for

Table 13.1. Position Effect Genes in Human Genetic Disease

Gene	Gene function	Domains/Motifs	Disease	OMIM	Distance of furthest breakpoint	3' or 5' side	References
PAX6	TF	Paired box and homeodomain	Aniridia	106210	125 kb	3'	Fantes et al., 1995; Kleinjan et al., 2001
TWIST	TF		Saethre-Chotzen syndrome	101400	260 kb	3'	Cai et al., 2003
POU3F4	TF	POU homeodomain	X-linked deafness	304400	900 kb	5'	de Kok et al., 1996
PITX2	TF	Homeodomain	Rieger syndrome	180500	90 kb	5'	Trembath et al., 2004
GLI3	TF	Zinc finger	Greig cephalopolysyndactyly	175700	10 kb	3'	Wild et al., 1997
MAF	TF	bZIP	Cataract, ocular anterior segment dysgenesis, and coloboma	610202	1000 kb	5'	Jamieson et al., 2002
FOXC1	TF	Forkhead	Glaucoma/autosomal-dominant iridogoniodysgenesis	601631	25 kb	5'	Davies et al., 1999
FOXC2	TF	Forkhead	Lymphedema-distichiasis syndrome	153400	120 kb	3'	Fang et al., 2000
FOXL2	TF	Forkhead	BPES (blepharophimosis-ptosis-epicanthus inversus syndrome)	110100	170 kb	5'	Crisponi et al., 2004; Beysen et al., 2005
FOXP2	TF	Forkhead	Speech-language disorder	602081	>680 kb	5'	Scherer et al., 2003

(*Continues*)

Table 13.1. (*Continued*)

Gene	Gene function	Domains/Motifs	Disease	OMIM	Distance of furthest breakpoint	3' or 5' side	References
TGFβ2	Signaling		Peters anomaly	114290	500 kb	3'	David et al., 2003
SOX9	TF	HMG box	Campomelic dysplasia		932 kb	5'	Leipoldt et al., 2007; Velagaleti et al., 2005
					1300 kb	3'	
SIX3	TF	Homeodomain	Holoprosencephaly (HPE2)	157170	<200 kb	5'	Wallis et al., 1999
SHH	Signaling		Holoprosencephaly (HPE3)	142945	265 kb	5'	Roessler et al., 1997; Fernandez et al., 2005
SHH	Signaling		Preaxial polydactyly	174500	1000 kb	5'	Lettice et al., 2003
RUNX2	TF		Cleidocranial dysplasia	119600	700 kb	5'	Fernandez et al., 2005
SHFM1	TF	DLX5/DLX6?	Split hand split foot malformation	183600	~450 kb	5'/3'	Crackower et al., 1996
ALX4	TF	Homeobox	Potocki-Schaffer syndrome	601224	>15 kb	3'	Wakui et al., 2005
REEP3	TH receptor regulator		Autism		43 kb		Castermans et al., 2007
PLP1	Proteolipid protein		Spastic paraplegia type 2 with axonal neuropathy	312920	136 kb	3'	Lee et al., 2006;

(*Continues*)

Table 13.1. (*Continued*)

Gene	Gene function	Domains/Motifs	Disease	OMIM	Distance of furthest breakpoint	3' or 5' side	References
			Pelizaeus–Merzbacher disease	312080	70 kb	5'	Muncke et al., 2004
FSHD	Unknown		Facioscapulo humeral dystrophy	158900	Unknown	3'	Gabellini et al., 2002 Jiang et al., 2003 Masny et al., 2004
HBB	Oxygen carrier	Globin	gamma;β-Thalassemia	141900	50 kb	5'	Kioussis et al., 1983 Driscoll et al., 1989
HBA	Oxygen carrier	Globin	α-Thalassemia	141800	30 kb	5'	Viprakasit et al., 2003
				141850	18 kb	3'	Tufarelli et al., 2003
HOX D complex	TF	Homeodomain	Mesomelic dysplasia and vertebral defects		56 kb	3' 950 kb	Spitz et al., 2002 Dlugaszewska et al., 2006
RET	Receptor Tyr kinase		Increased risk of Hirschsprung disease	142623	Intronic enhancer	Intronic	Emison et al., 2005
SHOX	TF		Langer mesomelic dysplasia	249700	Micro-deletion 200 kb	3'	Sabherwal et al., 2007; Fukami et al., 2005
			Leri-Weill dyschondrosteosis	127300			
SOST	Sclerostin	BMP antagonists	Van Buchem disease	239100	35 kb		Loots et al., 2005
SALL1	TF		Townes-Brocks syndrome	107480	>180 kb	5'	Marlin et al., 1999

(*Continues*)

Table 13.1. (*Continued*)

Gene	Gene function	Domains/Motifs	Disease	OMIM	Distance of furthest breakpoint	3' or 5' side	References
SOX3	TF	HMG box	X-linked recessive hypoparathyroidism	307700	67 kb	3'	Bowl et al., 2005
LCT	Enzyme	Lactase	Lactase persistence	223100	15 kb	5'	Enattah et al., 2002; Tishkoff et al., 2007

In case of 3' breakpoints the distance refers to the distance from the breakpoint to the 3' end of the gene or complex.

which there are several etiologies including autosomal dominant and recessive forms as well as an X-linked recessive form (OMIM 307700), mapped to a 900-kb region on Xq27. A recent study of a multigeneration family with recessive X-linked HPT revealed a deletion of a 25-kb fragment located around 67 kb downstream from the SOX3 gene, which was replaced by a 340-kb insertion derived from chromosome 2p25. Although the insertion contains a portion of a gene, SNTG2, the inserted fragment lacks an open reading frame and trisomy of the segment is unlikely to be the cause of the disease as patients with 2p trisomy do not suffer from HPT. Instead, it is thought that the 25 kb deleted fragment contains cis-regulatory sequences for SOX3 or that cis-elements beyond the deletion fragment have been moved beyond the reach of the promoter by the 340-kb insertion (Bowl et al., 2005).

The zinc finger gene GLI3 has a role in the embryonic development of limbs and the skull, and is an important effector of the hedgehog signaling network. Greig cephalo-polysyndactyly syndrome (GCPS), characterized by PPD, syndactyly, and craniofacial abnormalities, is caused by haploinsufficiency of GLI3 on chromosome 7p13 (Wild et al., 1997). All identified mutations leading to GCPS are thought to be loss of function while truncating mutations within Gli3 lead to Pallister Hall syndrome, which is characterized by postaxial/central polydactyly, hypothalamic hamartoma, and bifid epiglottis. A probable position effect translocation has been described in a GCPS patient with a breakpoint 10 kb downstream from the last exon of GLI3 (Vortkamp et al., 1991). In the mouse, two dominant mutant alleles, the extra toes (Xt and Xt^J) phenotypes, are caused by mutations within the murine Gli3 gene (Vortkamp et al., 1992). However, the weak recessive Xt allele, anterior digit deformity (add), is caused by a transgene integration combined with the deletion of an 80-kb region at ~40 kb upstream of Gli3 (van der Hoeven et al., 1993). Another mouse mutant, the polydactyly/arhinencephaly mouse (Polydactyly Nagoya-Pdn/Pdn), shows phenotypic similarity to GCPS. It is caused by insertion of a transposon into intron 3 of the Gli3 gene, resulting in suppression of Gli3 expression (Ueta et al., 2002).

Two translocation cases with a diagnosis of autism have been implicated to be due to position effects. In the first case, the translocation breakpoint mapped near the FOXP2 gene, which has previously been shown to cause a form of speech and language disorder, suggesting that FOXP2 might also be involved in autism (Scherer et al., 2003). The child inherited the translocation from the mother, who had speech delay. Seven isoforms of FOXP2 (spanning 545 kb) were characterized, but all mapped at least 680 kb 3' of the translocation breakpoint in a gene-poor region, raising the possibility of a position-effect mutation. In the other case, the breakpoint was nearest to the neuronal pentraxin 2 (NPTX2) gene, which is thought to be involved in excitatory synaptogenesis and could therefore be considered a functional candidate for autism (Scherer et al., 2003).

V. THE "BYSTANDER" EFFECT

The situation is slightly more complicated in a second group of position effect cases, where the gene disrupted by the chromosomal rearrangements is not directly involved in causing the disease, but rather appears as a bystander, implicated by virtue of harboring regulatory elements for an adjacent gene located some distance away.

A. Aniridia and PAX6

Among the first human genetic diseases where a disruption of the transcriptional control of the gene was recognized as a cause of disease is the congenital eye malformation aniridia. Aniridia (OMIM 106210) is characterized by severe hypoplasia of the iris, usually accompanied by foveal hypoplasia, cataracts, and corneal opacification (Fig. 13.1). It is caused by haploinsufficiency of the PAX6 gene at chromosome 11p13 as shown by deletion cases (Ton *et al.*, 1991), and through loss-of-function point mutations (van Heyningen and Williamson, 2002). In the mouse, heterozygosity for a number of mutation and deletion alleles of Pax6 is the cause of the small eye (Sey) phenotype. A number of aniridia patients with translocation breakpoints mapping downstream of the PAX6 gene have been described (Crolla and van Heyningen, 2002; Fantes *et al.*, 1995; Lauderdale *et al.*, 2000). Detailed mapping of these breakpoints placed them at various positions downstream from an intact PAX6-coding region, with the furthest located 125 kb beyond the final exon (Fantes *et al.*, 1995). Analysis of the sequence downstream of PAX6 revealed the presence of a ubiquitously expressed gene, ELP4 (a subunit of Elongator, a protein complex associated with the elongating RNA polymerase II). Even though all breakpoints map within the final intron of this gene, presumably interfering with its activity, it was shown through transgenic yeast artificial chromosome (YAC) rescue of a Sey deletion mutant that heterozygosity of Elp4 was unlikely to be the cause of the eye phenotype (Kleinjan *et al.*, 2002). A naturally occurring Pax6 mutant carrying a truncating mutation (Hill *et al.*, 1991), the Sey[Ed] mutant, was employed to further study the PAX6 position effect. A 420-kb human *PAX6* YAC-containing sequence extending a further 80 kb beyond the position of the most distal patient breakpoint was shown to rescue the Sey lethality and give full phenotypic correction in both heterozygous and homozygous Sey mice (Schedl *et al.*, 1996). In contrast, a shorter YAC extending only as far as the breakpoint failed to rescue or correct the mutant phenotype (Kleinjan *et al.*, 2001). Essential regulatory elements were identified in the sequence between the ends of the shorter and longer YACs using DNaseI hypersensitive site mapping and evolutionary sequence comparison. Reporter studies in transgenic mice revealed that some of these elements direct eye- and brain-specific expression, providing strong

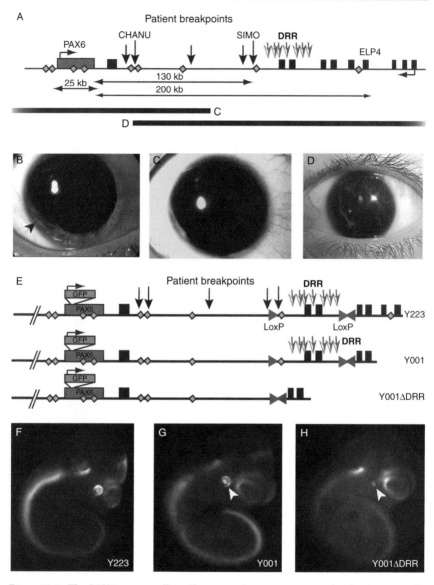

Figure 13.1. The PAX6 position effect. Chromosomal rearrangements with a breakpoint in the region downstream of PAX6 give rise to the congenital eye malformation aniridia (absence of iris). The eye phenotype of the position effect patients is indistinguishable from aniridia due to deletion or mutation of the PAX6 coding region itself. (A) Schematic representation of the PAX6 locus on chromosome 11p13. The 13 exon PAX6 gene is shown as a single box. The exons of the adjacent ubiquitously expressed ELP4 gene are shown as black boxes. Downward solid arrows show the

evidence that the PAX6 position effect is due to the removal of these elements, collectively termed the downstream regulatory region (DRR), from the PAX6 locus (Kleinjan *et al.*, 2001). To obtain further evidence that removal of the DRR would lead to aniridia, YAC transgenic mice were engineered in which a green fluorescent protein (GFP) reporter gene was inserted into the PAX6-coding region and LoxP sites were placed around a 30-kb region containing the DRR. Strong GFP signal was observed in a proper PAX6 expression pattern in full-length YAC transgenics, but after Cre mediated deletion of the DRR, expression was no longer found in the neuroretina, iris, and ciliary body (Fig. 13.1). As multiple enhancers for these expression sites have been identified, not only within the DRR but also in upstream and intronic regions of the PAX6 locus, this study not only demonstrates the essential nature of enhancers in the DRR but also highlights the interdependence of multiple *cis*-elements spread throughout the gene locus.

VI. MAF, SDC2, TGFB2, REEP3, AND PLP1

Awareness of the possibility that an adjacent gene rather than the directly disrupted gene could be the true cause of the disease was helpful in the case of a family with lens and ocular anterior segment anomalies with a t(5;16) translocation (Jamieson *et al.*, 2002). Breakpoint analysis indicated that the translocation occurred in an intron of the WWOX gene (Paige *et al.*, 2000). However, since WWOX is a widely expressed putative tumor suppressor gene and the phenotype was observed in both balanced and unbalanced forms of the translocation, WWOX seemed an unlikely candidate to cause an eye phenotype. Based on its expression pattern and known involvement in eye development, a more likely candidate was the bZIP transcription factor gene MAF, located ~1 Mb

position of a number of patient breakpoints. Grey/open arrows indicate a region beyond the furthest patient breakpoint termed the downstream regulatory region (DRR), containing a number of hypersensitive sites, some of which have been shown to have enhancer activity in transgenic mice. Diamond shapes indicate positions of several tissue-specific enhancers characterized in transgenic mice. (B) Eye of an aniridia patient with a point mutation in PAX6 (G36R). Arrow indicates a bit of residual iris tissue (C) Eye of a patient carrying a large deletion encompassing PAX6. (D) Eye of a patient (CHANU) with a PAX6 position effect due to an inversion of a large chromosome 11 segment (E) Reporter YAC transgenic experiment. GFP was inserted into the PAX6-coding region and LoxP sites were introduced surrounding the DRR. The full-length YAC largely reproduces the endogenous PAX6 expression pattern (F), while a slightly shorter YAC insertion lacks part of the expression in diencephalon. (G) Consistent with the position effect cases, Cre-mediated deletion of the DRR from this line results in loss of expression in the retina (white arrowheads) as well as iris and ciliary body at later embryonic stages.

telomeric of the breakpoint. Subsequent identification of a missense mutation within MAF in another family with lens and iris anomalies confirmed its role in lens development (Jamieson *et al.*, 2002). The identification of affected individuals carrying both balanced and unbalanced forms of the t(5,16) translocation suggests the phenotype may be caused by the dominant misregulation of MAF due to sequences on the der(5) chromosome.

A similar situation occurred with the breakpoint mapping of an X;8 reciprocal translocation in a patient manifesting multiple exostoses and autism with mental retardation and epilepsy (Ishikawa-Brush *et al.*, 1997). The X-chromosomal breakpoint was shown to locate within the first intron of the gastrin-releasing peptide receptor (GRPR), while the breakpoint on chromosome 8 occurred near the syndecan-2 (SDC2) gene, located 30 kb from the breakpoint, but left intact by the translocation. Gastrin-releasing peptide (GRP) has various neurobiological activities in the brain, lungs, and gastrointestinal tract, and while it appears to escape X-inactivation, a lower dosage of GRPR could well be a cause of autism. It is, however, unlikely to be a cause for exostoses. Awareness that breakpoints near but not directly disrupting the coding region should also be considered led to closer examination of the chromosome 8 breakpoint region and the SDC2 gene. SDC2 is a member of a family of cell surface heparan sulphate proteoglycans with a role in bone formation, and a position effect on SDC2 is therefore the most likely reason for the exostosis phenotype (Ishikawa-Brush *et al.*, 1997).

A third example is found in a case of Peters anomaly, which affects the anterior chamber of the eye. The translocation between chromosomes 1 and 7 disrupts the HDAC9 gene at 7p21.1, while the breakpoint on chromosome 1 is located 500 kb from TGFβ2. Nevertheless, since the knockout of TGFβ2 in the mouse has a very similar eye phenotype to Peters anomaly, TGFβ2 must be considered to be the causal gene (David *et al.*, 2003).

A further example of such a situation occurs in a patient with autism and a *de novo* balanced paracentric inversion 46,XY,inv(10)(q11.1;q21.3). The distal breakpoint directly disrupts the TRIP8 gene, a transcriptional regulator associated with nuclear thyroid hormone receptors, but no link between thyroid gland and autism has ever been reported. A nearby gene, receptor expression-enhancing protein (REEP)3, located 43 kb distal to the breakpoint with a probable role in regulating cellular vesicle trafficking between the ER and Golgi, would seem a much more likely candidate. Using an SNP in the 3'UTR of this gene expression analysis in patient- and control-derived cell lines indicated a normal biallelic expression in the controls, but a distinctly monoallelic expression of REEP3 in the patient cell line (Castermans *et al.*, 2007).

Finally, position effects have been found in two clinically distinct allelic disorders Pelizaeus-Merzbacher disease (PMD) and spastic paraplegia type 2 (SPG2), both caused by mutations in the proteolipid protein gene (PLP1), which codes for the major myelin component in the central nervous system (CNS). PMD (OMIM 312080) is an X-linked disorder of varying severity characterized by dysmyelination of the CNS. SPG2 (OMIM 312920) is generally less severe and presents as progressive weakness and spasticity of the lower extremities. PMD-causing mutations fall into two categories, duplications of PLP1, which account for the majority of cases, and sequence variations within the gene. A male PMD patient carrying an inversion on the X-chromosome (inv (X) (p22.3; q22) displayed a subset of PMD symptoms. His Xq22 breakpoint was mapped about 70 kb upstream from the PLP1 gene, within GLRA4, a putative glycine receptor family pseudogene. Because of the established link between PMD and PLP1 and failure to identify intragenic mutations in GLRA4 in a cohort of comparable patients, a position effect on PLP1 is considered the most likely cause of the disease. The PMD phenotype of the patient most closely resembles that of patients carrying PLP1 duplications. As loss of PLP1 is expected to give a different phenotype, the most likely result of this position effect is the upregulation of PLP1 (Muncke et al., 2004). In a separate study, semiquantitative PCR was used to detect a 150-kb duplication located 136 kb downstream from PLP1 in a patient with SPG2, at the milder end of the PMD/SPG2 spectrum. In this case, the duplication is therefore thought to silence PLP1 repression (Lee et al., 2006).

VII. TWO POSITION EFFECTS—DIFFERENT OUTCOMES

A. Sonic hedgehog, holoprosencephaly, and preaxial polydactyly

In the case of the Sonic hedgehog (SHH) gene on chromosome 7q36, transcriptional misregulation through disruption of long-range control of the same gene, but via different mechanisms, is the cause of two very different human genetic disorders (Fig. 13.2). The first, holoprosencephaly type 3 (HPE3; OMIM 142945), is caused by deletions and point mutations in the SHH gene itself, but can also be caused by translocations up to 265 kb upstream of the gene (Belloni et al., 1996; Roessler et al., 1996, 1997). Phenotypic expressivity is variable, ranging from a single cerebral ventricle and cyclopia to clinically unaffected carriers in some familial HPE cases. In humans, HPE3 is caused by haploinsufficiency due to the loss of one allele of SHH, whereas in the mouse both alleles need to be inactivated to produce a similar phenotype (Chiang et al., 1996), indicating a more critical role for correct SHH dosage in humans. The clinical variability is especially prominent in two familial position effect cases with translocation breakpoints at 235 and 265 kb upstream of SHH, where some family members are phenotypically normal or mildly affected, while others are much more severely affected (Roessler et al., 1997). It is thought that the removal

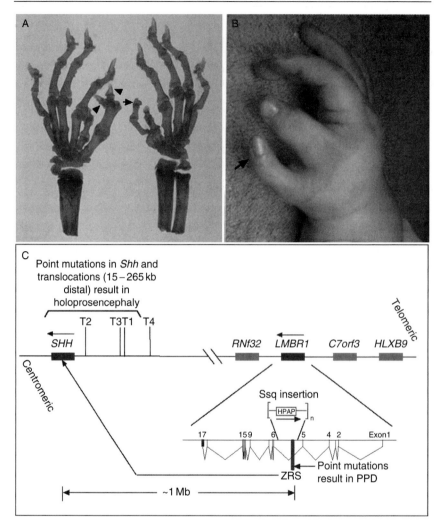

Figure 13.2. Limb-specific malformation by a long-range position effect on the Sonic Hedgehog (SHH) gene a chromosome 7q36. (A) Alizarin red stained bones from the fore limbs of an adult *Ssq* homozygous mouse. Left paw has two sets of terminal phalanges (arrowheads) from a single metacarpal, while the right paw exhibits complete duplication of all the bones of digit one (arrow). (B) Photograph of the hand of a patient with preaxial polydactyly (PPD) showing duplication of the thumb (arrow). (C) Schematic depiction of the genomic organization of the mouse *Shh* locus from *Hlxb9* to *Shh* itself. Positions of the translocations which result in holoprosencephaly (Belloni et al., 1996; Roessler et al., 1996, 1997) are indicated (T1-T4). Also marked is the position of the zone of polarizing activity regulatory sequence (ZRS), located 1 Mb upstream of SHH, beyond RNF32 and within an intron of the LMBR1 gene.

of a cis-regulatory element located more than 265 kb upstream leads to a reduced SHH expression level that fluctuates around the critical level. A further translocation, located 315 kb upstream, has not produced a detectable HPE phenotype in the seven individuals examined, thus limiting the region in which to look for the putative control element(s) (Roessler et al., 1997). A functional screen, conducted in mouse (Jeong et al., 2006), has identified a number of Shh enhancers which drive expression in the developing forebrain. Some lie at distances of over 400 kb from the promoter and would predictably be disrupted by translocations such as these.

The second disorder linked to misregulation of SHH expression is the limb developmental malformation preaxial polydactyly (PPD OMIM 174500). Much progress has been made in recent years on long-range gene regulation of SHH in the limb through the analysis of a mouse limb mutant, the polydactylous sasquatch mouse (Ssq) (Sharpe et al., 1999). The Ssq mutation arose as a result of a random insertion of a reporter cassette during transgenic studies analyzing HoxB1 regulatory sequences. Mice carrying the Ssq insertion display PPD, with homozygotes being more affected than heterozygotes. Genetic analysis of the mutant showed that the reporter transgene segregates with the limb phenotype and that the insertion site was physically linked to the Shh gene but situated almost 1 Mb away. The effect could not be attributed to the reporter cassette itself as none of the other transgenic lines with the same construct showed a similar phenotype (Sharpe et al., 1999). The linkage to Shh was interesting because the nature of the Ssq phenotype was consistent with the limb phenotypes of a group of mouse mutants called the hemimelia-luxate group which all exhibit misexpression of Shh during limb development (Hill et al., 2003). Analysis of Shh expression in the limb buds of mutant embryos indeed showed an additional region of ectopic expression in the anterior of the limb bud. During normal limb development, Shh is believed to act as a morphogen, emanating from a region (called the zone of polarizing activity, ZPA) at the posterior end of the limb bud to set up a concentration gradient across the anterior/posterior axis of the limb, with high levels of SHH resulting in digits with a posterior identity and lower levels more anterior digits. Ectopic anterior expression will clearly disrupt such gradient, providing an explanation for the observed phenotype. Interestingly, when the insertion site was cloned it proved to lie 1 Mb upstream of Shh, beyond the adjacent testis- and ovary-specific Rnf32 gene, within intron 5 of a gene identified as Lmbr1 (Limb region 1) (Lettice et al., 2002). The human LMBR1 gene was originally identified as a candidate gene within the critical region for PPD,

---

The ZRS drives Shh expression within the limb and point mutations in this Shh limb-specific regulatory element are associated with human PPD and with extra digits in mouse polydactyly models.

one of the most common human congenital limb malformations (Heus et al., 1999). Analysis of a de novo chromosomal translocation in a PPD patient also indicated a breakpoint in intron 5 of LMBR1, close to the Ssq insertion site (Lettice et al., 2002). Furthermore, Lmbr1 was shown to be located in the critical region for the closely linked mouse limb mutants, the polydactylous Hemimelia extra-toes (Hx) and the syndactylous Hammertoe (Hm) (Clark et al., 2000), and its expression was found to be downregulated in Hx embryos at the time when the phenotype is first observed (Clark et al., 2000). However, examination of the LMBR1 and Lmbr1 structural genes in the various patients and mouse mutants failed to uncover any coding region mutations. The observed Shh misexpression and known importance of Shh in limb development suggested that the Ssq phenotype could be due to disruption of long-range regulation of Shh and that the linkage with Lmbr1 is merely coincidental. A genetic cis-trans experiment was set up to generate by recombination a chromosome in which the Ssq insertion was located in cis to an Shh null allele. Analysis of mice carrying the recombinant chromosome showed no abnormal limb phenotype (as is the case for the heterozygous Shh null by itself), indicating that Ssq is a dominant mutation that interferes with limb-specific expression of Shh (Lettice et al., 2002). A similar cis-trans test was subsequently done for the Hx mutant with the same result (Sagai et al., 2004). Using phylogenetic sequence comparisons, a small, highly conserved region, termed the ZRS (for Zone of polarising activity Regulatory Sequence), was identified in LMBR1 intron 5 that was shown to act as a limb-specific regulatory element in transgenic reporter mice (Lettice et al., 2003). To assess whether mutations in the ZRS could account for the PPD phenotype in families displaying a normal karyotype, the element was sequenced in a large number of affected and unaffected family members as well as controls. Amazingly single-point mutations in the ZRS were found in four PPD families, which were observed in all affected and none of the unaffected individuals. Analysis of the ZRS in the Hx mouse also identified a single base pair change segregating with the phenotype (Lettice et al., 2003). Subsequent analyses in a number of ethylnitrosourea (ENU)-induced mouse mutants (Sagai et al., 2004, Masuya et al., 2007) and in other PPD families have identified more mutations (Gurnett et al., 2007). The ZRS is thought to have a dual function in the regulation of SHH, (1) driving the initiation of expression in the limb bud and (2) restricting this expression to the posterior margin. Expression studies using ZRS sequences to drive reporter expression (Maas and Fallon, 2005; Masuya et al., 2007) demonstrate that the mutations affect the latter but not the former activity of the ZRS, and lead to ectopic SHH expression in the anterior. All of the mutations lie in different parts of the ZRS, and are scattered throughout its length. The lack of clustering of the point mutations suggests that the phenotype cannot simply arise from the inactivation of a single transcription factor binding site. In an interesting aside, it was shown that the ZRS enhancer is conserved in species with limbs, wings, and fins (Lettice *et al.*, 2003), but no conservation is seen in several limbless species, such as snakes and a limbless newt (Sagai *et al.*, 2004).

Several more congenital limb abnormalities map to the same 7q36 region. Patients for one of these, acheiropodia, an autosomal recessive disease in which all bones of the hand and feet are missing and the tibia are truncated, were shown to carry a small deletion on both alleles of LMBR1 (Ianakiev *et al.*, 2001). The deletion, which includes LMBR1 exon 4 and about 6 kb of surrounding sequence, is slightly more distal than the site of the ZRS, but could similarly be hypothesized to disrupt long-range regulation of SHH. In that case however, rather than result in ectopic misexpression, the deletion would be predicted to disrupt a positive regulatory activity driving the normal ZPA-specific expression of SHH. In accordance, the phenotype observed in the acheiropodia families very closely resembles the limb truncations observed in the *Shh* loss-of-function mice (Chiang *et al.*, 1996). A similar phenotype has been identified in the chick mutant *oligozeugodactyly* (*ozd*) (Ros *et al.*, 2003) also hypothesized to result from a defect in a regulatory element that controls limb-specific expression of *Shh*.

Two further mouse mutations, both carrying radiation-induced large-scale inversions of chromosome 5, implicate the long-range disregulation of *Shh* in a wide range of developmental defects. Replicated anterior zeugopod (raz) (Krebs *et al.*, 2003) mice exhibit two anterior skeletal elements in the zeugopod and symmetrical central polydactyly. It results from a huge inversion of much of proximal chromosome 5, covering both *Shh* and ZRS, which when homozygous downregulates expression of Shh mRNA and protein in the limb bud to about 20% of wild-type levels (expression in other tissues is unaffected). *Short-digits* (*Dsh*) mice (Niedermaier *et al.*, 2005) carry a much smaller deletion of 11.7 Mb with one breakpoint lying just over 13 kb upstream of *Shh* (between it and a number of putative enhancers including ZRS). Homozygous mice exhibit a number of defects and are phenotypically nearly identical to the *Shh* loss-of-function mice; the two mutations also fail to complement each other, demonstrating that *Dsh* is a regulatory mutation of *Shh* resulting in almost complete downregulation of Shh during E9.5–12.5. Heterozygous mice, however, show a phenotype similar to human brachydactyly type 1A with a fusion and shortening of the proximal and middle phalanges in all digits. Analysis shows that *Shh* is ectopically expressed in a skeletogenic domain typically occupied by a closely related gene, Indian hedgehog (*Ihh*), within the developing digits at E13.5–14.5. *Ihh* is normally downregulated in regions that will become the joint space, but in *Dsh*/ + mice, *Shh* bypasses this regulatory control and persists; accordingly, cells maintain their chondrogenic fate and the developed digits are shorter than normal. This suggests that perhaps the role of one of the endogenous *Shh* regulatory elements is to act as repressor segregating the activity of *Shh* from

Ihh (de la Fuente and Helms, 2005). Direct misregulation of *Ihh* itself has been implicated in the Doublefoot mutant that shows PPD and craniofacial abnormalities (Hayes *et al.*, 2001; Yang *et al.*, 1998).

VIII. PHENOTYPES RESULTING FROM POSITION EFFECTS ON MORE THAN ONE GENE

In a small but growing number of cases the observed phenotypes cannot easily be explained by hypothesizing the involvement of a single gene, but are more likely to be the result of a combination of effects on two or more genes, either neighboring genes under common control or artificially juxtaposed by a translocation.

A. Split hand foot malformation locus 1

Split hand foot malformation type I (SHFM1) (OMIM 183600), a limb developmental disorder characterized by missing digits, fusion of remaining digits, and a deep median cleft in the hands and feet, is a genetically heterogeneous human disorder also referred to as ectrodactyly or lobster claw deformity. The SHFM1 critical region has been mapped on the basis of chromosomal rearrangements to ~1.5-Mb interval on 7q21.3 (Crackower *et al.*, 1996; Ignatius *et al.*, 1996; Scherer *et al.*, 1994). Despite a decade-long search, no disease gene has so far been identified. Three candidate genes, DSS1 and the distalless homeobox genes DLX5 and DLX6, are located in this region. Chromosomal breakpoints from 12 unrelated ectrodactyly patients were found to be scattered throughout the critical region and do not collectively interrupt any single gene. No missense mutations within the genes have been identified in patients. In mouse, inactivation of either Dlx5 or Dlx6 alone does not produce a limb phenotype (Robledo *et al.*, 2002). However, the targeted double inactivation of Dlx5 and Dlx6, by a targeted 17-kb deletion covering both coding regions and the intervening DNA, causes bilateral ectrodactyly with a severe defect of the central ray of the hindlimbs, a malformation typical of SHFM1. Dss1 continues to be normally expressed in these double knockout mice. Even though there are differences in the two phenotypes—the human condition is believed to be inherited in an autosomal dominant manner, while affected mice must be homozygous for the loss-of-function mutation and in man both hands and feet are affected, while the phenotype is observed only on murine hindlimbs—the overall similarities in phenotype suggest the possibility that SHFM1 is due to a position effect on the two DLX genes, and possibly DSS1 (Merlo *et al.*, 2002). The simplest mechanism would involve the chromosomal breakpoints located up to 1 Mb centromeric of human *DLX5* and *DLX6* separating critical shared regulatory elements from the genes and lead to their misregulation during development.

B. Combination of two position effects in SHH and RUNX2

Interestingly, there is one report in the literature of a translocation resulting in two proposed position effects on two different genes which both contribute to the observed phenotype. The patient carries a *de novo* reciprocal translocation with breakpoints at 6p21.1 and 7q36 and displays premaxillary agenesis, skeletal abnormalities, and impacted teeth (Fernandez *et al.*, 2005). The 7q36 breakpoint has been mapped 15 kb telomeric to the 5′ of the SHH gene (Belloni *et al.*, 1996) and is thought to contribute to the premaxillary agenesis (part of the HPE spectrum). However, fine mapping of the chromosome 6 breakpoint positions it further than 700 kb of the 5′ end of the osteoblast-specific transcription factor CBFA/RUNX2 (with one gene, SUPT3H, a probable transcriptional activator, lying in the interval). Mutations in CBFA/RUNX2 result in cleidocranial dysplasia, a dominant disorder characterized by dental, pelvic, and clavicular disorders; some of which show distinct similarities to the phenotype displayed by this patient.

IX. GLOBAL CONTROL REGIONS; HOXD, GREMLIN, AND LIMB MALFORMATIONS

A particularly interesting locus depending on long-range control is the HOXD gene locus on human chromosome 2q31 (see also the Spitz and Duboule chapter in this volume). The locus contains a number of genes, including the HOXD cluster and the EVX2 and lunapark (LNP) genes, involved in patterning of the axial skeleton and in limb development where they are required for correct digit formation (Spitz *et al.*, 2003). The HoxD cluster, with its characteristic collinearity in spatiotemporal expression of the genes correlating with gene order along the chromosome, has been extensively studied in mouse models. These studies have led to the notion that the ancestral role of the cluster was the specification of morphogenesis along the main body axis, while later in evolution the HoxD genes were co-opted to function in the development of novel structures such as the limbs (Spitz *et al.*, 2001). Consistent with this scenario, it was shown that the regulatory controls for the ancestral and collinear expression are located within the HoxD cluster, while the co-opted expression domains depend on enhancers located at remote positions outside the cluster. During limb development, HoxD10–13, Evx2, and Lnp are coexpressed in the presumptive digits with very similar profiles, suggesting the possibility that a single enhancer would control digit expression for all genes (van der Hoeven *et al.*, 1996). A search for such enhancer using BAC transgenics and sequence comparisons subsequently led to the identification of a region far upstream of the locus that controls tissue-specific expression in multiple tissues of a contiguous set of genes in the locus. This region, termed the global control region (GCR), is proposed to create a widespread regulatory landscape, sharing its enhancing activity over a defined number of genes in a tissue-specific manner (Spitz et al., 2003). Digit activity of the GCR spreads over six genes and at least 240 kb, while CNS activity is limited to the Lnp and Evx2 genes. A direct demonstration of the role of the GCR was provided by analysis of a semidominant mouse limb mutant, Ulnaless. Ulnaless carries a paracentric inversion with one breakpoint in the Lnp gene and the other 770 kb more telomeric, thus resulting in an inversion of Evx2 and the HoxD cluster. When bred against a targeted allele with a deletion of the Evx2 to HoxD11 region, expression of Evx2 and HoxD13 in distal limb was shown to be lost from the Ulnaless allele (Herault et al., 1997; Spitz et al., 2003). This suggests that the inversion either has moved the genes out of reach of the GCR (about 700 kb) or has introduced some insulating activity between the genes and the GCR. Interestingly, while no human cases with chromosomal disruptions between the GCR and the HoxD genes have been described, one human translocation patient has been identified with a translocation centromeric to the GCR (Dlugaszewska et al., 2006) The patient, who exhibits severe brachy-dactyly and syndactyly among other abnormalities, carries an apparently balanced t(2;10)(q31.1;q26.3) translocation. The chromosome 2 breakpoint lies ~390 kb from the end of HOXD13 and no known genes have been identified on a breakpoint spanning BAC, implicating a role of misregulation of the HoxD genes in the phenotype.

An elegant series of deletion and duplication experiments in mouse (Tarchini and Duboule, 2006) has demonstrated that an early limb bud-activating element (ELCR) lies telomeric to the HoxD cluster, an idea first hypothesized by Zakany et al. (2004). The element controls Hox gene expression such that the closer a gene is positioned to the telomeric end of the cluster, either naturally or by artificial means, the earlier it is activated during development and the more anterior within the limb bud it is expressed. This early expression pattern appears to correlate with subsequent expression in the developing forearm and while the element responsible has not been located, it is interesting to note that a patient suffering from mesomelic dysplasia (shortening of the forearms and forelegs) with vertebral defects was found to have a balanced translocation 56 kb telomeric from the HoxD1 gene (Spitz et al., 2002). Two other patients with abnormalities restricted to the limbs have been identified carrying balanced de novo trans-locations with breakpoints 450 and 950 kb telomeric to the HoxD complex (Dlugaszewska et al., 2006).

Over the years, much work has gone into studying the mouse limb deformity (ld) mutations. Mice display disrupted epithelial–mesenchymal inter-actions between the polarizing region and the apical ectodermal ridge, which alters the pattern of the distal limb and results in fusion and loss of digits (see Zeller *et al.*, 1999). A number of ld alleles have been described and as these disrupt the C terminal domain of formin, this was proposed as the causative mutation (Maas *et al.*, 1990; Woychik *et al.*, 1990). However, this was rather at odds with the discovery that the remaining two alleles (ld^{OR} and ld^J) disrupt the coding region of the neighboring gene Gremlin (the intragenic distance is 38 kb) and are allelic to loss-of-function mutations within it. This discrepancy has been resolved by the discovery that a GCR, required for correct expression of gremlin (and formin) within the posterior limb bud mesenchyme, lies within the region of exons 19–24 of formin and is deleted or disturbed in the relevant ld alleles (Zuniga *et al.*, 2004). Taken together, the GCRs from the HoxD complex and the Gremlin locus reveal a mechanism whereby expression of a number of genes within a region can be controlled in concert.

X. FOX GENES AND POSITION EFFECTS

A. FOX genes in eye anomalies

A number of position effect cases are associated with genes of the forkhead/winged helix family of transcription factors, indicating the importance of gene dosage for members of this family. Forkhead genes are involved in a diverse range of developmental pathways, but a large number of them appear to be involved in eye development (Lehmann *et al.*, 2003). FOXC1 (previously named FKHL7) lies in a cluster of forkhead genes on chromosome 6p25. Mutations cause glaucoma-associated ocular developmental anomalies with varying degrees of iris and extraocular abnormality, in addition to abnormalities in a range of other organ systems. Segmental duplication or deletion of the 6p25 region results in developmental defects of the anterior segment of the eye, indicating that precise gene dosage is critical for normal eye development. In addition to mutations and deletions in the FOXC1 gene itself in patients with Axenfeld Rieger anomaly and iris hypoplasia, a balanced translocation mapping 25 kb from the gene was found in a patient with primary congenital glaucoma (Nishimura *et al.*, 1998). A further patient, with glaucoma- and autosomal-dominant iridogoniodysgenesis, was shown to carry an interstitial deletion of 6p24-p25 with the proximal breakpoint estimated to lie at least 1200 kb proximal to the FOXC1 locus (Davies *et al.*, 1999). The case for a position effect in this patient is possible but speculative considering the large distance and the presence of a number of other possible candidate genes, most notably AP2α, in the deletion interval.

A typical case of a position effect is provided by a closely related forkhead member, the FOXC2 gene on 16q24 (formerly MFH-1). FOXC2-inactivating mutations have been found in patients with lymphedema-distichiasis (LD; OMIM 153400), an autosomal-dominant disorder that classically presents as lymphedema of the limbs and double rows of eyelashes (distichiasis). Other complications may include cardiac defects, cleft palate, extradural cysts, and photophobia, highlighting the pleiotrophic effects of FOXC2 during development. A t(Y;16) (q12;q24.3) translocation with the breakpoint mapping 120 kb 3′ of the FOXC2 gene was found in a patient with neonatal lymphedema. The translocation did not appear to interrupt a gene on chromosome 16, nor were any candidate genes found on the Y chromosome, making a clear case for a position effect on the FOXC2 gene. Interestingly, the FOXL1 gene maps in between FOXC2 and the breakpoint, and thus could also be inactivated and have phenotypic effects in this patient (Fang et al., 2000).

A third forkhead family member implicated in a position effect scenario is the FOXL2 gene on chromosome 3q23. Coding region mutations in FOXL2 have been shown to cause blepharophimosis-ptosis-epicanthus inversus syndrome (BPES; OMIM 110100), an eyelid and forehead dysmorphology in both sexes, often associated with gonadal dysgenesis and premature ovarian failure (POF) in women (Crisponi et al., 2001). Consistent with the disease phenotypes, FOXL2 is selectively expressed in the mesenchyme of developing mouse eyelids and in adult ovarian follicles. Three reported translocations that cause BPES, mapping 170 kb from FOXL2, all fall within intron 6 of MRPS22, a ubiquitously expressed gene located upstream of FOXL2 (Crisponi et al., 2004). Sequence comparisons between human and mouse reveal the presence of three highly conserved segments beyond the furthest breakpoint, in introns 6, 11, and 12 of MRPS22. In a more recent study, MLPA was used to specifically detect deletions in the FOXL2 region. In addition to cases removing the FOXL2 gene itself, five deletion cases with a phenotype identical to intragenic mutations were found mapping both upstream (four cases) and downstream (one case) of the intact FOXL2 gene (Beysen et al., 2005). The smallest of the upstream deletion cases involves a 126-kb region that includes the human orthologous region to the goat polled intersex syndrome (PIS) locus. Polled intersex syndrome (PIS) is a genetic syndrome in goats, which combines a craniofacial defect resulting in polledness (an absence of horns), female infertility, and XX sex reversal, and is considered an animal model for BPES. It is caused by an 11.7-kb deletion, located 280 kb upstream of goat FOXL2, distal of the three translocation breakpoints (Pailhoux et al., 2001). It encompasses a block of several conserved sequences in intron 11 of MRPS22, making these strong candidates to be distant cis-regulatory elements affecting FOXL2 expression.

XI. SOX9 AND CAMPOMELIC DISPLASIA

A locus with strong evidence for long-range regulatory elements located at extremely large distances from its cognate gene is the SOX9 locus on chromosome 7q24.3. SOX9 mutation has been identified as the cause of campomelic dysplasia (CD; OMIM 114290), an autosomal-dominant osteochondrodysplasia, characterized by congenital shortening and bowing of the long bones (campomelia) in combination with other skeletal anomalies, such as hypoplasia of the scapular and pelvic bones, lack of mineralization of thoracic pedicles, a missing pair of ribs, and clubbed feet. Lung hypoplasia and a malformed thorax, resulting in severe respiratory distress, may lead to the neonatal death of some patients. XY sex reversal is found in about two thirds of karyotypically male CD patients. In most CD cases, heterozygous loss-of-function mutations are found within the coding region of SOX9 on chromosome 17q, implying that CD results from haploinsufficiency for SOX9. However, as with other position effect cases already described, CD patients have been identified who carry two intact copies of the compact 5.4 kb SOX9 transcription unit, with chromosomal rearrangements in the vicinity of one SOX9 allele. In all but one case, the rearrangements are found upstream of SOX9, and the breakpoints have been mapped at distances from 50 kb up to 950 kb (Hill-Harfe et al., 2005; Pfeifer et al., 1999). In a single case with a complex, balanced translocation, the 17q breakpoint was mapped 1.3 Mb downstream from Sox9 (Velagaleti et al., 2005). Examination of a large genomic interval around SOX9 has failed to find other genes, giving rise to the notion that SOX9 resides in a so-called "gene desert." The lack of any protein-coding gene in the region combined with the similarity, though generally less severe, of the phenotype of these position effect patients to SOX9 loss-of-function cases strongly suggests that the chromosomal rearrangements remove one or more cis-regulatory elements. Detailed analysis of the breakpoints of 15 CD patients suggests that they are found in two clusters: ten translocation breakpoints and one inversion fall in a proximal cluster between 50 and 375 kb upstream, while four more breakpoints lie in a distal cluster between 789 and 932 kb away (Leipoldt et al., 2007). A further patient with CD presented with a large deletion from 380 to 1860 kb upstream (Pop et al., 2004). In accordance with the human phenotype caused by SOX9 mutations, studies in the mouse have shown that Sox9 functions as an essential developmental regulator at various steps of chondrogenesis and during the initial phase of testis determination and differentiation. Furthermore, heterozygous Sox9 knockout mice recapitulate essentially all the symptoms seen in CD patients except for the sex reversal (Kist et al., 2002). The involvement of long-range gene control is supported by the fact that mice transgenic for human SOX9 spanning YACs showed transgene expression patterns (except in gonads) that were similar to endogenous Sox9 only when the YAC transgene contained 350 kb of sequence upstream of SOX9, but not with a truncated YAC containing only 75 kb of SOX9 5′ flanking sequence (Wunderle et al., 1998). Comparative sequence analysis of the SOX9 genomic region between human, mouse, chicken, and fugu has revealed the presence of many conserved elements, some of which have been shown to drive reporter expression in a SOX9 sub-pattern (Bagheri-Fam et al., 2001, 2006; Qin et al., 2004).

In the mouse, a complex position effect has been reported as the cause of the dominant insertional mutation *Odd sex* (ocular degeneration with sex reversal, *Ods*) (Bishop *et al.*, 2000). *Ods* is the result of a transgenic insertion of a tyrosinase minigene driven by the dopachrome tautomerase (Dct) promoter accompanied by a 134-kb deletion at ~980 kb upstream of Sox9. In contrast to the male to female sex reversal found in many human CD patients, *Ods* mice show female to male sex reversal, as well as microphthalmia with pigmentation defects and cataracts. The XX sex reversal phenotype of these mice is thought to be the result of *Sox9* misexpression. Normally *Sox9* expression is repressed in the XX fetal gonad at the time of sex determination, and this repression is counteracted by the presence of *Sry* in the male gonads. Initially, it was proposed that the *Ods* deletion had removed gonad-specific long-range regulatory element(s) that normally mediate the female-specific repression of *Sox9*, resulting in upregulation of *Sox9* in the absence of *Sry* and the consequent male development (Bishop *et al.*, 2000). However, more recent experiments have shown that the 134-kb deletion in itself is insufficient to cause the sex reversal (Qin *et al.*, 2004). When a double gene targeting strategy was used to recreate the *Ods* deletion combined with the introduction of a tyrosinase minigene driven by its own promoter, no eye or sex reversal phenotype was observed, suggesting instead a mechanism of long-range interaction between the *Dct* promoter and *Sox9* in the *Ods* mutant. The *Ods* eye phenotype was also generated in transgenic mice with a *Dct-Sox9* minigene cassette, and a temporal misexpression of *Sox9* under the control of the *Dct* promoter was demonstrated. This suggests that in the eyes of *Ods* mice, the *Dct* promoter acts as a long-range activator on the *Sox9* promoter over a distance of 980 kb. The mechanism behind the sex reversal phenotype is more complex. No sex-reversal was seen in *Dct-Sox9* transgenics, suggesting that in the gonads the *Dct* promoter interacts with *Sox9* via an indirect mechanism possibly involving endogenous gonad-specific *Sox9* enhancers and chromosomal conformation changes due to the deletion (Qin *et al.*, 2004).

XII. FACIOSCAPULOHUMERAL DYSTROPHY

Facioscapulohumeral dystrophy (FSHD) is a neuromuscular disorder affecting predominantly the facial and shoulder girdle muscles. It is inherited as an autosomal dominant trait, but the onset and severity of the disease can be highly variable even within patient families carrying the same genetic lesion, and about 10–30% of patients carry *de novo* mutations, some of them as somatic mosaics. At present the mechanism underlying the disease, although still not very well understood, appears to have a unique epigenetic etiology. The characteristic molecular event is the deletion of an integral number of 3.3 kb tandem repeats from the subtelomeric region of the long arm of chromosome 4 (Wijmenga *et al.*, 1992). Unaffected individuals carry between 11 and 150 copies of this repeat, named D4Z4, while patients have 10 or fewer on one of their chromosomes (van Deutekom et al., 1993). In general, the severity and onset of the disease correlates with the number of repeats, with a lower repeat number associating with more severe disease manifestations. Interestingly, no FSHD patients have been reported completely lacking all D4Z4 repeats, and individuals carrying an unbalanced translocation where the entire 4q35 region is lost do not suffer from FSHD. This suggests that the loss of repeats results in a gain-of-function mutation (Tupler et al., 1996), but much debate continues over the mechanism. The D4Z4 repeat contains internal VNTR-type repeats, has CpG island characteristics, and contains a putative open reading frame (DUX4), but no protein-coding transcripts have been identified from the repeat sequence despite intense efforts. The short D4Z4 repeat at 4q35 appears to cause FSHD indirectly by some unknown cis-interaction, because an almost identical repeat array (plus further homology on both sides of the repeat) is present on chromosome 10q26, but does not cause the disease when shortened (Bakker et al., 1995). Thus, FSHD is the result of a mutation outside the gene or genes underlying the disease, suggesting some kind of position effect mechanism. Furthermore, in addition to the D4Z4 contraction at least one other, so far unidentified cis-element is required to develop the disease. The 4q subtelomeric region was found to exist in two allelic variants, 4qA and 4qB, which are almost equally common in the population. FSHD alleles are always of the 4qA subtype, suggesting additional elements on 4qA are necessary to cause the disease or elements on 4qB can act preventively. Some FSHD families have been reported with healthy individuals carrying FSHD-sized repeat contractions on a 4qB allele (Lemmers et al., 2004). A number of genes or putative genes have been identified in the region centromeric to the D4Z4 repeats, ANT1, FRG1, DUX4C, and FRG2. ANT1, though located at 3.5 Mb distance, was considered the most likely candidate because it encodes an adenine nucleotide translocator that has been implicated in myopathy and is predominantly expressed in heart and skeletal muscle (Gabellini et al., 2002). A multiprotein repressor complex consisting of HMG2B, YY1, and nucleolin has been identified that binds to the D4Z4 repeat (Gabellini et al., 2002). This complex was hypothesized to act as a repressor, negatively regulating gene expression of the 4q35 region, possibly through promoting the spreading of heterochromatin in cis throughout the region. D4Z4 deletions remove binding sites for the repressor complex to below a certain threshold, thus allowing local decondensation of chromatin and the consequent derepression of the genes in the region (Gabellini et al., 2002). In support of the hypothesis, all the genes in the region were found to be upregulated in FSHD. However, several follow-up studies have failed to reproduce the upregulation of ANT1 and FRG1 in muscle tissue of FSHD patients compared to unaffected individuals. Furthermore, histone H4 acetylation levels over the various gene promoters suggested that the region adopts a nonexpressed euchromatin-like structure both in control individuals and in FSHD patients (Jiang et al., 2003). Instead, a model was proposed in which a short array of D4Z4 repeats forms a long distance loop to interact directly with an as yet unknown gene or genes on 4q35, while longer arrays of D4Z4 repeats form intra-array loops, thus sequestering the array (Jiang et al., 2003).

The case for involvement of several genes in the region, notably ANT1, PDLIM3, and FRG2, has weakened by recent overexpression studies in transgenic mice where none of the genes appeared to cause muscular dystrophy. Mice overexpressing FRG1, a putative spliceosomal protein gene located 120 kb from the D4Z4 repeats, on the other hand develop muscular dystrophy with a severity proportional to the level of overexpression (Gabellini et al., 2006). Nonetheless, the role of FRG1 remains speculative. Another avenue of exploration that is pursued relates to the observation that the 4q telomere uniquely always localizes at the nuclear periphery, requiring Lamin A/C to do so, unlike all other chromosome ends studied including the highly homologous 10qter (Masny et al., 2004). How this links with alterations in gene expression in FSHD patients is not clear, but disturbed myogenic differentiation detectable by gene expression profiling suggests possible chromatin effects.

XIII. ABERRANT CREATION OF AN ILLEGITIMATE siRNA TARGET SITE

In recent years, small nuclear RNAs have become recognized as important modulators of transcriptional and translational mechanisms of gene expression. A study to identify quantitative trait loci underlying the economically important feature of meatiness of Texel sheep pointed toward involvement of the myostatin (GDF8) gene locus. Detailed analysis has led to the identification of a single base pair substitution in the 3'UTR of the gene (Clop et al., 2006). This mutation creates a target site for mir1 and mir206, two miRNAs that are highly expressed in skeletal muscle, and causes translational inhibition of the myostatin gene leading to muscular hypertrophy. As miRNAs are undoubtedly also involved in regulatory control in humans, such a mechanism could equally well be involved in the etiology of human genetic disease.

One of the first examples of such a disease mechanism is the aberrant strengthening of a weak miRNA site in the 3'UTR of a gene called SLITRK1, as a possible cause of Tourette syndrome (TS). TS is a developmental neuropsychiatric disorder characterized by chronic vocal and motor tics. In this case the gene involved is the Slit and Trk like family member 1 (SLITRK1) gene. Interestingly, SLITRK1 was first identified as a potential TS syndrome candidate from a position effect case in a boy with a de novo chromosome 13 inversion (inv (13)(q31.1;q33.1)). The breakpoint mapped well outside the coding region of the gene, but subsequent sequencing of the single coding exon of SLITRK1 in a TS patient cohort identified a truncating mutation in one of them. More interesting was the finding of an identical sequence variant in the 3'UTR of two further, unrelated patients. The single base pair change mapped to a highly conserved nucleotide within a predicted target site for human miRNA hs-miR-189, and would be predicted to strengthen the binding site. SLITRK1 and miR-189 show substantial overlap in expression pattern in the brain, and luciferase assays suggest that repression of the mutant 3'UTR variant of SLITRK1 by the miRNA is significantly increased over the wild-type version (Abelson et al., 2005).

XIV. GENETIC DISEASE DUE TO ABERRANT GENE TRANSCRIPTION CAN BE CAUSED BY MANY DIFFERENT MECHANISMS

As highlighted by the cases described above, transcriptional control can be disrupted by many different mechanisms. So far the following mechanisms have been encountered: (1) separation of cis-regulatory elements from the gene promoters through chromosomal translocations or inversions, (2) deletion of long range cis-elements, (3) deleterious mutations in cis-elements, (4) disturbing the normal interactions of promoters and cis-acting enhancers through appearance of a new promoter, (5) alteration of local chromatin structure through interference by an antisense transcript, and (6) disturbance of more global chromatin structure through loss of microsatellite repeats.

A further, but at this stage rather speculative, mechanism, with a potential involvement in genetic disease is transvection. Transvection refers to the effect on a gene's expression from one allele through the influence of regulatory elements on the opposite allele through the pairing of homologous chromosomes. In some disease cases, particularly involving dosage-sensitive genes, differences in severity have been observed between inactivating mutations and larger gene deletions. One possible mechanism to explain that phenomenon could be an ameliorating influence through regulatory effects of distal control elements on the remaining intact copy of the gene on the other chromosome. Such a model has been proposed for the observed differences in penetrance of craniofacial anomalies in mouse models of Smith-Magenis syndrome (SMS OMIM 182290) (Yan et al., 2007). Craniofacial abnormality is one of the major clinical manifestations of SMS, and includes midface hypoplasia, broad nasal bridge, and prognathia. SMS is a multiple congenital anomaly and mental retardation condition due to a heterozygous 3.7-Mb deletion on chromosome 17p11.2 in the majority of cases. Frameshift and nonsense mutations have been identified in RAI1, a retinoic acid inducible gene, in some patients with SMS, suggesting it is the major gene involved through haploinsufficiency. Whereas penetrance of the craniofacial phenotype in isogenic strains of mice carrying various sized deletion around the Rai1 gene was more or less complete, it was much lower in a Rai1 insertional inactivation strain which has retained the rest of the locus. Considering the dosage-sensitivity of Rai1, an upregulation of the wild-type copy of the gene through a *trans*-regulatory mechanism could be a plausible mechanism (Yan et al., 2007). Evidence, though mainly anecdotal, exists for similar differences in severity and penetrance between locus-deletion- and gene-mutation-derived disease cases and deeper investigation of such situations should bring new insights into the phenomenon of transvection.

A. The problem: How to find and assess regulatory mutations?

Determining that a genetic disease diagnosed in a patient is caused by a regulatory mutation or deletion is difficult for a number of reasons: Not only is the possibility of regulatory impairment rather than protein product malfunction not always considered, but for most genes too little is known about their regulatory sequences and mechanisms to easily further investigate the possibility. At present there are no simple ways to discern regulatory elements from nonfunctional sequence (though interspecies sequence comparison is proving very useful in this respect) or to know what regulatory and/or evolutionary conserved elements may be relevant to the disease etiology. Furthermore, while it is usually clear when a mutation is found in a protein-coding sequence whether that mutation causes a faulty protein, this is much more problematic for mutations in regulatory elements and requires extensive functional analysis. Finally, it is likely that many regulatory element mutations will be associated with phenotypes distinct from those identified for coding region mutations. For instance, if the regulatory mutation occurs in a tissue-specific element the phenotype may affect just a subset of the full spectrum seen with inactivation of the protein. The SHH gene forms an example where different diseases can be associated with the same gene: Disruption of the protein or loss of expression through proximal rearrangements causes HPE, while mutation of the distal, limb-specific ZRS enhancer leads to limb malformation (Fig. 13.2) (Lettice et al., 2003). Similarly, if a mutation affects a later acting *cis*-element the regulatory mutant phenotype could be completely different as it would normally be masked by earlier effects from complete knockout of the gene. In general phenotypes due to regulatory mutations will be milder, because genes, in particular developmental control genes, are often regulated by multiple enhancers with partly overlapping activities. While in some cases these enhancers will act additively or synergistically, they can also act hierarchically. Both these options could be seen in a YAC transgenic experiment where a PAX6 DRR was deleted via a Cre/LoxP strategy (Fig. 13.1). The DRR contains a number of *cis*-elements including ones for expression of Pax6 in the developing lens and retina (Kleinjan et al, 2001). However, several further enhancers for these expression sites are also found outside the DRR and these are not removed by the deletion. While after DRR deletion reporter expression was maintained in the lens (at least at early embryonic stages and possibly at lower level), expression in the retina, iris, and ciliary body was completely abolished (Kleinjan et al., 2006), indicating that elements within the DRR are critically important for PAX6 expression in these sites.

The cases described above all highlight genes where long-range transcriptional control has been identified through the analysis of patients with genetic malformations. There are of course many more genes for which long-range gene regulation undoubtedly plays an important role, but which have so far not been implicated in a position effect-type genetic disease (e.g., DiLeone et al., 2000; Hadchouel et al., 2003; Kimura-Yoshida et al., 2004; Uchikawa et al., 2003). Many of the disease cases described in this chapter are caused by semi-dominant de novo chromosomal abnormalities. Because expression levels are often critical for many developmental regulators, haploinsufficiency acts as a selection mechanism for the link between these genes and their involvement in genetic disease. For other genes, however, dosage may not be critical; therefore, disruption of long-range control on one allele would not lead to a recognizable phenotype. In other cases, the disruption may have only a subtle effect or give rise to a phenotype that is quite different from the phenotype shown upon complete gene inactivation. Therefore one can safely assume that more genetic diseases caused by regulatory mutations will exist that have thus far not been associated with the correct causative gene.

As is clear from the many cases described above, long-range enhancers can be essential for correct expression of many genes. Mechanisms through which these distal enhancers interact with the promoters of their target genes have been discussed in other chapters. It is tempting to speculate that mutations in factors involved in such long-range enhancer–promoter interactions should also be a cause of genetic disease. One disorder where this may indeed be the case is Cornelia de Lange syndrome, (CDLS, OMIM 122470) (Krantz et al., 2004; Tonkin et al., 2004). CDLS is a multiple malformation disorder, characterized by facial dysmorphology, mental retardation, growth delay, and limb reduction. NIPBL, the human homologue of Drosophila Nipped-B, was identified as the gene mutated in individuals with CDLS. Nipped-B was first identified as a facilitator of enhancer promoter communication in Drosophila in a screen for long-range interactions with the Drosophila cut gene wing margin enhancer located 85 kb upstream of its cognate promoter (Dorsett, 1999; Rollins et al., 1999). In keeping with the putative general role of Nipped-B in facilitating long-range regulation of multiple genes, compromised long-range regulation of several developmental control genes would fit with the diverse phenotypic anomalies seen in CDLS. Based on homology with yeast Scc2 protein, with a role in sister chromatid cohesion, Nipped-B has been proposed to have a dual role, where a more ancient function facilitating *trans*-interactions between sequences on sister chromatids has been adapted to include an additional role in long-range transcriptional regulation (Rollins *et al.*, 2004). The recent findings that two other forms of the disorder, CDLS2 (OMIM 300590) and CDLS3 (OMIM 610759), are caused by mutations in SMC1A and SMC3, both of which encode other components of the cohesin complex, points to a role of the cohesin complex in the regulation of multiple genes.

B. Long-range control and genome organization

The position effect cases presented above highlight a number of intriguing points: First, among the genes thus far shown to depend on long-range regulatory control, both through human disease-related position effects and work in experimental organisms, a large proportion consists of genes encoding developmental regulators. Similarly, it has been noted when using bioinformatic tools to map multispecies conserved elements in a genome-wide manner, that many of these conserved elements cluster near genes with key developmental functions (e.g., Woolfe *et al.*, 2005). We argue that this is not coincidental, but stems from the complex and strictly critical expression patterns these proteins require.

Developmental regulator genes need to be active in specific tissues at defined time-points in development, often at critically defined levels, and have to be strictly inactive in all other tissues and time-points. To achieve such sophisticated expression profiles, these genes require multiple enhancer elements and these regulatory elements all need to be fitted in *cis* in the region surrounding the gene. Moreover, despite the common use of terminology that suggests some sort of engineered design in the structure of gene loci with respect to transcription and other biological processes, the acquisition and loss of regulatory elements has not occurred through conscious design, but rather as a result of evolutionary tinkering (Carroll, 2001; Duboule and Wilkins, 1998). The redeployment of developmental regulatory genes in the development of new and other tissues and pathways has become a recognized feature in the evolution of greater complexity in higher organisms. To a large extent, this depends on the chance appearance of a new combination of sequences with regulatory activity within the vicinity of the appropriate promoter. As long as the activity can "reach" the promoter, does not interfere with the existing regulatory control in a disadvantageous manner, and presents some kind of evolutionary advantage itself, the new *cis*-element might become fixed. The further appearance of elements that synergize with existing ones to more reliably achieve the optimal expression level will similarly be of evolutionary advantage. One example where there is evidence for the redeployment of regulatory factors in more recent evolutionary adaptations is the HoxD cluster, and interestingly a correlation has been noted between the more proximal location of enhancers involved in the ancestral function of the gene and the more distal location of regulatory elements required for expression in tissues in which the gene has been co-opted. As described above, the HoxD cluster is essential for axial patterning along the main body axis as well as for limb development. Control elements for expression in the trunk, the ancestral site of HoxD function, are located within the HoxD cluster, while control elements for limb expression are located in more distal positions (Spitz *et al.*, 2001). However, the separation between proximal, more ancient and distal, more recent *cis*-elements in the HoxD cluster is probably a somewhat special case where the appearance of new *cis*-elements is strongly influenced by the clustered, coregulated nature of the genes in the complex, and the formation of a new element within the cluster would be likely to interfere adversely with the regulatory phenomenon of collinearity that is the hallmark of the Hox complexes (Tarchini *et al.*, 2006) However, when the distant positions of the limb-specific control elements in the HoxD complex as well as in the Shh and Gremlin loci are considered together, it is plausible that their separation from the coding regions represents a more recent add-on to the ancestral axial pattern in concert with the evolution of the limb.

Second, the analysis of several "position effect" cases has highlighted the fact that *cis*-regulatory elements that drive expression of one gene may be located within introns of another, neighboring gene, or can in some cases even be located beyond the adjacent gene. This not only presents a problem for investigators trying to find the causative gene for a disease of interest through mapping of patient breakpoints, but has also initiated new ideas about the concept of genes and gene domains. Eukaryotic genomes are made up of a large number of genes and gene loci that are regulated independently, and often genes/ loci with quite different patterns of expression are located in close proximity to one another on the chromosome. Traditionally the concept of "structural gene domains" was widely established to explain how adjacent gene loci maintain their independence. In the structural domain model, genes enjoy functional autonomy through the physical separation from neighboring domains that carry a different local chromatin structure sometimes in combination with or brought about by specific functional sequences such as boundary or insulator elements (Dillon and Sabbattini, 2000). The chicken β-globin locus is a strong example to support this model (Felsenfeld *et al.*, 2004). However, the fact that genes can overlap and their *cis*-regulatory elements can be found within or beyond neighboring unrelated genes puts the universal applicability of the structural domain model into question. Detailed analysis of some of the disease genes described has highlighted that the structural domain model cannot apply to all genes. For instance, the main critical regulatory element in the human α-globin locus, the −40-kb enhancer, is located within an intron of a neighboring, ubiquitously expressed housekeeping gene located in a large region of open chromatin (Vyas et al., 1995). In the case of PAX6, a number of regulatory elements have been found spread throughout at least three of the introns of the neighboring ELP4 gene (Kleinjan et al., 2001). In the human growth hormone cluster, elements required for tissue-specific expression of the pituitary-specific GHN gene and the placenta-specific CSL, CSA, GHV, and CSB genes are located beyond a B-cell-specific gene, within introns of the muscle-specific SCN4A gene (Bennani-Baiti et al., 1998). Many more examples are now known. These observations provide a compelling argument against a fundamental requirement for the physical isolation of a gene and its regulatory sequences. Rather it has been proposed that for many loci, specificity of enhancer–promoter interactions is the key in maintaining the functional autonomy of adjacent genes (Dillon and Sabbattini, 2000). Thus, while the concept of a gene as a physical entity with a distinct map position on the chromosome is still useful in many instances, a new concept has been proposed, based on the old definition of the gene as a "unit of inheritance" (de Laat and Grosveld, 2003; Dillon, 2003). In that concept genes are defined in functional terms as "functional expression modules," encompassing both the transcribed regions and their cis-regulatory control systems comprising of the specific cis-acting sequences in conjunction with the local chromatin structure. This model works independent of the physical structure of the gene locus, so that a functional expression module can derive equally well from a gene and its cis-elements that are interdigitated with adjacent genes, or located in its own separate domain or in a gene desert.

The completion of the human (and other) genome sequence has confirmed the highly uneven distribution of genes across the genome into gene-rich and gene-poor areas. It has been estimated that around 25% of the human genome consists of long regions lacking any protein-coding sequences or obvious biological function, and these regions have been termed gene deserts. (Venter et al., 2001). Based on comparative conservation levels with the chicken genome, gene deserts can be classed into two categories, stable and variable (Ovcharenko et al., 2005). Some of the gene deserts have been shown to contain long-range cis-regulatory sequences acting on neighboring genes (Kimura-Yoshida et al., 2004; Nobrega et al., 2003; Uchikawa et al., 2003) and these are overrepresented in the stable class of gene deserts. Gene ontology categories of genes neighboring stable gene deserts show a strong bias toward transcriptional and developmental regulators (Ovcharenko et al., 2005), suggesting the presence of long-range enhancers plays a role in the evolutionary maintenance of the stable gene deserts. In contrast, gene deserts of the variable class may be nonessential to genome function, since they can be deleted without obvious phenotypic consequence (Nobrega et al., 2004). The location of long-distance enhancers within or beyond adjacent genes is also thought to be a major force in the conservation of blocks of synteny over large evolutionary timespans (Mackenzie et al., 2004).

In conclusion, the fact that many developmental regulators in higher organisms have extended regulatory landscapes is a consequence of their acquisition of multiple roles in various spatiotemporal defined sites in the embryo. A direct and obvious consequence of having large gene domains is an increased risk of its disruption by chromosomal rearrangements such as translocations, deletions, and inversions.

C. Implications for (common) genetic disease

It is clear from the "position effect" cases reviewed above that long-range gene regulation plays a critical role in many developmental processes and that its disruption can lead to severe congenital disease. While the occurrence of distal control elements may be more common to developmental regulatory proteins due to their more complex spatiotemporal expression requirements, long-range gene control undoubtedly plays a role at many more gene loci. The effects of disruption/mutation of distal regulatory elements in those genes will be more subtle, but may be a significant factor in common disease and quantative traits.

It is now well accepted that a large percentage of quantitative traits are caused by variants that affect gene expression, so-called expression Quantitative trait loci (QTL). In some well-studied cases, the QT is shown to be the result of single nucleotide changes in distal regulatory elements. A genetic trait where such a single base pair substitution has had a dramatic effect is lactase persistence. While worldwide in most humans the ability to digest the milk sugar lactose declines rapidly after weaning because of decreasing levels of the enzyme lactase phlorizin hydrolase, this ability is maintained with high frequency in certain populations, particularly those whose ancestors have traditionally practiced cattle domestication. Lactase persistence in Europeans has been strongly associated with an SNP (C/T-13910) located 13.9 kb upstream of the lactase gene promoter, with a second correlative SNP found 22 kb upstream (Enattah et al., 2002). The lactose tolerance SNP lies in a cis-element that enhances promoter activity (Olds and Sibley, 2003; Lewinsky et al., 2005; Troelsen et al., 2003). In a further study of this trait, three more SNPs in the same putative cis-element, SNPs (G/C-14010, T/G-13915, C/G-13907), were found to correlate with lactase persistence in pastoral African populations (Tishkoff et al., 2007). Association studies demonstrate that the different SNP alleles have arisen independently and have spread rapidly to high frequency because of the strong selective force of adult milk consumption (Tishkoff et al., 2007).

Quantitative traits caused by regulatory SNPs affecting levels of gene expression are likely to lie at the heart of many common diseases with a genetic component, such as diabetes, heart disease, hypertension, and obesity. In a recent study (Herbert et al., 2006), a link was found between a common genetic variant located 10 kb upstream of the insulin-induced gene 2 (INSIG2) gene and risk of obesity. INSIG2 protein can inhibit cholesterol and fatty acid synthesis, and thus makes a plausible QTL candidate gene affecting body mass index. The predisposing genotype was present in 10% of the test population, suggesting it is an ancient allele that has only recently become deleterious.

Quantitative traits can obviously be caused by variants in all kinds of cis-regulatory elements including distal enhancers, but also in promoters and intronic cis-elements. An example of the former is the association of a genetic predisposition to autism with an SNP in the promoter of the pleiotropic MET receptor tyrosine kinase gene on chromosome 7q31 (Campbell et al., 2006). MET signaling has a role in neocortical and cerebellar growth and maturation, immune function, and gastrointestinal repair, consistent with reported medical complications in some children with autism. Functional assays showed that the autism-associated allele results in a twofold decrease in MET promoter activity.

A mutation in an intronic enhancer for the RET proto-oncogene is associated with an increased risk of Hirschsprung disease (HSCR). HSCR, or congenital aganglionosis with megacolon, is a multifactorial, non-mendelian disorder. The RET gene is the main HSCR gene implicated, but mutations also occur in many other genes involved in enteric development. Resequencing of conserved elements in the RET locus of HSCR families with demonstrated RET linkage, but no identified coding sequence mutations, led to the identification of an SNP (RET + 3) in a conserved element in intron 1 of the gene (Emison et al., 2005). This common variant SNP was shown to decrease enhancer activity of the element in an in vitro enhancer assay. This case shows that not all mutations for rare diseases need to be rare themselves or fully penetrant, but that in complex genetic disorders noncoding mutations can conspire with mutations at additional sites or genes for disease to occur. Thus, the customary practice of validating a potential mutation by applying the criterion that it must be absent from nonaffected controls may not always be appropriate (Emison et al., 2005).

XV. CONCLUDING REMARKS

Over the last 40 years, it has become well established that the genes within our genome contain the code to transform genetic information into the amino acid sequence of our proteins. In this post genome era, we are beginning to understand that the genome contains many more layers of biological information, not least of which are the codes that tell the transcriptional machinery where, when, and how much of those proteins to produce. While over the past decades, we have learned much about these latter codes, we are still a long way off fully understanding the mechanisms of gene regulation. Nevertheless, detailed analysis of human genetic malformations has provided invaluable information, first in highlighting the occurrence and scope of long-distance transcriptional regulation in the genome, and second in the elucidation of a number of different mechanisms of long-range gene control. It seems likely that long-range transcriptional control is essential for a relatively small but substantial subset of our genes, and the genes and diseases discussed in this chapter form just a small example of that group. They do, however, serve as paradigms in our efforts to unravel the regulatory codes embedded in our genome and understand the (long-range) regulatory mechanisms that underlie our healthy development and homeostasis and which, when disrupted, sadly lead to genetic disorders.

References

Abelson, J. F., Kwan, K. Y., O'Roak, B. J., Baek, D. Y., Stillman, A. A., Morgan, T. M., Mathews, C. A., Pauls, D. L., Rasin, M. R., Gunel, M., Davis, N. R., Ercan-Sencicek, A. G., et al. (2005). Sequence variants in SLITRK1 are associated with Tourette's syndrome. *Science* **310** (5746), 317–320.

Alward, W. L., Semina, E. V., Kalenak, J. W., Heon, E., Sheth, B. P., Stone, E. M., and Murray, J. C. (1998). Autosomal dominant iris hypoplasia is caused by a mutation in the Rieger syndrome (RIEG/PITX2) gene. *Am. J. Ophthalmol.* **125,** 98–100.

Anguita, E., Sharpe, J. A., Sloane-Stanley, J. A., Tufarelli, C., Higgs, D. R., and Wood, W. G. (2002). Deletion of the mouse alpha-globin regulatory element (HS-26) has an unexpectedly mild phenotype. *Blood* **100**(10), 3450–3456.

Bagheri-Fam, S., Ferraz, C., Demaille, J., Scherer, G., and Pfeifer, D. (2001). Comparative genomics of the SOX9 region in human and Fugu rubripes: Conservation of short regulatory sequence elements within large intergenic regions. *Genomics* **78**, 73–82.

Bagheri-Fam, S., Barrionuevo, F., Dohrmann, U., Gunther, T., Schule, R., Kemler, R., Mallo, M., Kanzler, B., and Scherer, G. (2006). Long-range upstream and downstream enhancers control distinct subsets of the complex spatiotemporal Sox9 expression pattern. *Dev. Biol.* **291**(2), 382–397.

Bakker, E., Wijmenga, C., Vossen, R. H., Padberg, G. W., Hewitt, J., van der, W. M., Rasmussen, K., and Frants, R. R. (1995). The FSHD-linked locus D4F104S1 (p13E-11) on 4q35 has a homologue on 10qter. *Muscle Nerve* **2**, 39–44.

Balemans, W., Ebeling, M., Patel, N., Van Hul, E., Olson, P., Dioszegi, M., Lacza, C., Wuyts, W., Van Den Ende, J., Willems, P., Paes-Alves, A. F., Hill, S., et al. (2001). Increased bone density in sclerosteosis is due to the deficiency of a novel secreted protein (SOST). *Hum. Mol. Genet.* **10**, 537–543.

Balemans, W., Patel, N., Ebeling, M., Van Hul, E., Wuyts, W., Lacza, C., Dioszegi, M., Dikkers, F. G., Hildering, P., Willems, P. J., Verheij, J. B., Lindpaintner, K., et al. (2002). Identification of a 52 kb deletion downstream of the SOST gene in patients with van Buchem disease. *J. Med. Genet.* **39**, 91–97.

Belloni, E., Muenke, M., Roessler, E., Traverso, G., Siegel-Bartelt, J., Frumkin, A., Mitchell, H. F., Donis-Keller, H., Helms, C., Hing, A. V., Heng, H. H., Koop, B., et al. (1996). Identification of Sonic hedgehog as a candidate gene responsible for holoprosencephaly. *Nat. Genet.* **14**, 353–356.

Benito-Sanz, S., Thomas, N. S., Huber, C., Gorbenko del Blanco, D., Aza-Carmona, M., Crolla, J. A., Maloney, V., Rappold, G., Argente, J., Campos-Barros, A., Cormier-Daire, V., and Heath, K. E. (2005). A novel class of Pseudoautosomal region 1 deletions downstream of SHOX is associated with Leri-Weill dyschondrosteosis. *Am. J. Hum. Genet.* **77**, 533–544.

Bennani-Baiti, I. M., Cooke, N. E., and Liebhaber, S. A. (1998). Physical linkage of the human growth hormone gene cluster and the CD79b (Ig beta/B29) gene. *Genomics* **48**, 258–264.

Beysen, D., Raes, J., Leroy, B. P., Lucassen, A., Yates, J. R., Clayton-Smith, J., Ilyina, H., Brooks, S. S., Christin-Maitre, S., Fellous, M., Fryns, J. P., Kim, J. R., *et al.* (2005). Deletions involving long-range conserved nongenic sequences upstream and downstream of FOXL2 as a novel disease-causing mechanism in blepharophimosis syndrome. *Am. J. Hum. Genet.* **77**, 205–218.

Bishop, C. E., Whitworth, D. J., Qin, Y., Agoulnik, A. I., Agoulnik, I. U., Harrison, W. R., Behringer, R. R., and Overbeek, P. A. (2000). A transgenic insertion upstream of sox9 is associated with dominant XX sex reversal in the mouse. *Nat. Genet.* **26**, 490–494.

Bowl, M. R., Nesbit, M. A., Harding, B., Levy, E., Jefferson, A., Volpi, E., Rizzoti, K., Lovell-Badge, R., Schlessinger, D., Whyte, M. P., and Thakker, R. V. (2005). An interstitial deletion-insertion involving chromosomes 2p25.3 and Xq27.1, near SOX3, causes X-linked recessive hypoparathyroidism. *J. Clin. Invest.* **115**(10), 2822–2831.

Brunkow, M. E., Gardner, J. C., Van Ness, J., Paeper, B. W., Kovacevich, B. R., Proll, S., Skonier, J. E., Zhao, L., Sabo, P. J., Fu, Y., Alisch, R. S., Gillett, L., *et al.* (2001). Bone dysplasia sclerosteosis results from loss of the SOST gene product, a novel cystine knot-containing protein. *Am. J. Hum. Genet.* **68**(3), 577–589.

Cai, J., Goodman, B. K., Patel, A. S., Mulliken, J. B., Van Maldergem, L., Hoganson, G. E., Paznekas, W. A., Ben Neriah, Z., Sheffer, R., Cunningham, M. L., Daentl, D. L., and Jabs, E. W. (2001). Increased risk for developmental delay in Saethre-Chotzen syndrome is associated with TWIST deletions: An improved strategy for TWIST mutation screening. *Hum. Genet.* **114**, 68–76.

Campbell, D. B., Sutcliffe, J. S., Ebert, P. J., Militerni, R., Bravaccio, C., Trillo, S., Elia, M., Schneider, C., Melmed, R., Sacco, R., Persico, A. M., and Levitt, P. (2006). A genetic variant that disrupts MET transcription is associated with autism. *Proc. Natl. Acad. Sci. USA* **103**(45), 16834–16839.

Carroll, S. B. (2001). Chance and necessity: The evolution of morphological complexity and diversity. *Nature* **409**, 1102–1109.

Carter, D., Chakalova, L., Osborne, C. S., Dai, Y. F., and Fraser, P. (2002). Long-range chromatin regulatory interactions *in vivo*. *Nat. Genet.* **32**, 623–626.

Castermans, D., Vermeesch, J. R., Fryns, J. P., Steyaert, J. G., Van de Ven, W. J., Creemers, J. W., and Devriendt, K. (2007). Identification and characterization of the TRIP8 and REEP3 genes on chromosome 10q21.3 as novel candidate genes for autism. *Eur. J. Hum. Genet.* **15**(4), 422–431.

Chiang, C., Litingtung, Y., Lee, E., Young, K. E., Corden, J. L., Westphal, H., and Beachy, P. A. (1996). Cyclopia and defective axial patterning in mice lacking Sonic hedgehog gene function. *Nature* **383**, 407–413.

Chun, K., Teebi, A. S., Jung, J. H., Kennedy, S., Laframboise, R., Meschino, W. S., Nakabayashi, K., Scherer, S. W., Ray, P. N., and Teshima, I. (2002). Genetic analysis of patients with the Saethre-Chotzen phenotype. *Am. J. Med. Genet.* **110**, 136–143.

Clark, R. M., Marker, P. C., and Kingsley, D. M. (2000). A novel candidate gene for mouse and human preaxial polydactyly with altered expression in limbs of Hemimelic extra-toes mutant mice. *Genomics* **67**, 19–27.

Clop, A., Marcq, F., Takeda, H., Pirottin, D., Tordoir, X., Bibe, B., Bouix, J., Caiment, F., Elsen, J. M., Eychenne, F., Larzul, C., Laville, E., *et al.* (2006). A mutation creating a potential illegitimate microRNA target site in the myostatin gene affects muscularity in sheep. *Nat. Genet.* **38**, 813–818.

Crackower, M. A., Scherer, S. W., Rommens, J. M., Hui, C. C., Poorkaj, P., Soder, S., Cobben, J. M., Hudgins, L., Evans, J. P., and Tsui, L. C. (1996). Characterization of the split hand/split foot malformation locus SHFM1 at 7q21.3-q22.1 and analysis of a candidate gene for its expression during limb development. *Hum. Mol. Genet.* **5**, 571–579.

Cremers, F. P., and Cremers, C. W. (2004). *In* "Inborn Errors of Development: The Molecular Basis of Clinical Disorders of Morphogenesis" (C. J. Epstein, R. P. Erickson, and A. Wynshaw-Boris, eds.). Oxford University Press, New York.

Crisponi, L., Deiana, M., Loi, A., Chiappe, F., Uda, M., Amati, P., Bisceglia, L., Zelante, L., Nagaraja, R., Porcu, S., Ristaldi, M. S., Marzella, R., *et al.* (2001). The putative forkhead transcription factor FOXL2 is mutated in blepharophimosis/ptosis/ epicanthus inversus syndrome. *Nat. Genet.* **27**, 159–166.

Crisponi, L., Uda, M., Deiana, M., Loi, A., Nagaraja, R., Chiappe, F., Schlessinger, D., Cao, A., and Pilia, G. (2004). FOXL2 inactivation by a translocation 171 kb away: Analysis of 500 kb of chromosome 3 for candidate long-range regulatory sequences. *Genomics* **83,** 757–764.

Crolla, J. A., and van Heyningen, V. (2002). Frequent chromosome aberrations revealed by molecular cytogenetic studies in patients with aniridia. *Am. J. Hum. Genet.* **71,** 1138–1149.

David, D., Cardoso, J., Marques, B., Marques, R., Silva, E. D., Santos, H., and Boavida, M. G. (2003). Molecular characterization of a familial translocation implicates disruption of HDAC9 and possible position effect on TGFbeta2 in the pathogenesis of Peters' anomaly. *Genomics* **81,** 489–503.

Davies, A. F., Mirza, G., Flinter, F., and Ragoussis, J. (1999). An interstitial deletion of 6p24-p25 proximal to the FKHL7 locus and including AP-2alpha that affects anterior eye chamber development. *J. Med. Genet.* **36,** 708–710.

de Kok, Y., Merkx, G. F., van der Maarel, S. M., Huber, I., Malcolm, S., Ropers, H. H., and Cremers, F. P. (1995). A duplication/paracentric inversion associated with familial X-linked deafness (DFN3) suggests the presence of a regulatory element more than 400 kb upstream of the POU3F4 gene. *Hum. Mol. Genet.* **4,** 2145–2150.

de Kok, Y., Vossenaar, E. R., Cremers, C. W., Dahl, N., Laporte, J., Hu, L. J., Lacombe, D., Fischel-Ghodsian, N., Friedman, R. A., Parnes, L. S., Thorpe, P., Bitner-Glindicz, M., *et al.* (1996). Identification of a hot spot for microdeletions in patients with X-linked deafness type 3 (DFN3) 900 kb proximal to the DFN3 gene POU3F4. *Hum. Mol. Genet.* **5,** 1229–1235.

de Laat, W., and Grosveld, F. (2003). Spatial organization of gene expression: The active chromatin hub. *Chromosome Res.* **11**(5), 447–459.

de la Fuente, L., and Helms, J. A. (2005). The fickle finger of fate. *J. Clin. Invest.* **15,** 833–836.

DiLeone, R. J., Marcus, G. A., Johnson, M. D., and Kingsley, D. M. (2000). Efficient studies of long-distance Bmp5 gene regulation using bacterial artificial chromosomes. *Proc. Natl. Acad. Sci. USA* **97,** 1612–1617.

Dillon, N. (2003). Gene autonomy: Positions, please. *Nature* **425,** 457.

Dillon, N., and Sabbattini, P. (2000). Functional gene expression domains: Defining the functional unit of eukaryotic gene regulation. *Bioessays* **22,** 657–665.

Dlugaszewska, B., Silahtaroglu, A., Menzel, C., Kubart, S., Cohen, M., Mundlos, S., Tumer, Z., Kjaer, K., Friedrich, U., Ropers, H. H., Tommerup, N., Neitzel, H., *et al.* (2006). Breakpoints around the HOXD cluster result in various limb malformations. *J. Med. Genet.* **43,** 111–118.

Dorsett, D. (1999). Distant liaisons: Long-range enhancer-promoter interactions in *Drosophila. Curr. Opin. Genet. Dev.* **9,** 505–514.

Driscoll, M. C., Dobkin, C. S., and Alter, B. P. (1989). Gamma delta beta-thalassemia due to a de novo mutation deleting the 5′ beta-globin gene activation-region hypersensitive sites. *Proc. Natl. Acad. Sci. USA* **86,** 7470–7474.

Duboule, D., and Wilkins, A. S. (1998). The evolution of 'bricolage'. *Trends Genet.* **14,** 54–59.

Emison, E. S., McCallion, A. S., Kashuk, C. S., Bush, R. T., Grice, E., Lin, S., Portnoy, M. E., Cutler, D. J., Green, E. D., and Chakravarti, A. (2005). A common sex-dependent mutation in a RET enhancer underlies Hirschsprung disease risk. *Nature* **434**(7035), 857–863.

Enattah, N. S., Sahi, T., Savilahti, E., Terwilliger, J. D., Peltonen, L., and Jarvela, I. (2002). Identification of a variant associated with adult-type hypolactasia. *Nat. Genet.* **30,** 233–237.

Fang, J., Dagenais, S. L., Erickson, R. P., Arlt, M. F., Glynn, M. W., Gorski, J. L., Seaver, L. H., and Glover, T. W. (2000). Mutations in FOXC2 (MFH-1), a forkhead family transcription factor, are responsible for the hereditary lymphedema-distichiasis syndrome. *Am. J. Hum. Genet.* **67,** 1382–1388.

Fantes, J., Redeker, B., Breen, M., Boyle, S., Brown, J., Fletcher, J., Jones, S., Bickmore, W., Fukushima, Y., Mannens, M., Danes, S., van Heyningen, V., and Hanson, I. (1995). Aniridia-associated cytogenetic rearrangements suggest that a position effect may cause the mutant phenotype. *Hum. Mol. Genet.* **4,** 415–422.

Felsenfeld, G., Burgess-Beusse, B., Farrell, C., Gaszner, M., Ghirlando, R., Huang, S., Jin, C., Litt, M., Magdinier, F., Mutskov, V., Nakatani, Y., Tagami, H., et al. (2004). Chromatin boundaries and chromatin domains. Cold Spring Harb. Symp. Quant. Biol. **69,** 245–250.

Fernandez, B. A., Siegel-Bartelt, J., Herbrick, J. A. S., Teshima, I., and Scherer, S. W. (2005). Holoprosencephaly and cleidocranial dysplasia in a patient due to two position-effect mutations: Case report and review of the literature. Clin. Genet. **68**(4), 349–359.

Flomen, R. H., Vatcheva, R., Gorman, P. A., Baptista, P. R., Groet, J., Barisic, I., Ligutic, I., and Nizetic, D. (1998). Construction and analysis of a sequence-ready map in 4q25: Rieger syndrome can be caused by haploinsufficiency of RIEG, but also by chromosome breaks approximately 90 kb upstream of this gene. Genomics **47,** 409–413.

Fraser, P. (2006). Transcriptional control thrown for a loop. Curr. Opin. Genet. Dev. **16**(5), 490–495.

Fukami, M., Okuyama, T., Yamamori, S., Nishimura, G., and Ogata, T. (2005). Microdeletion in the SHOX 3′ region associated with skeletal phenotypes of Langer mesomelic dysplasia in a 45,X/46, X,r(X) infant and Leri-Weill dyschondrosteosis in her 46,XX mother: Implication for the SHOX enhancer. Am. J. Med. Genet. A **137,** 72–76.

Gabellini, D., Green, M. R., and Tupler, R. (2002). Inappropriate gene activation in FSHD: A repressor complex binds a chromosomal repeat deleted in dystrophic muscle. Cell **110,** 339–348.

Gabellini, D., D'Antona, G., Moggio, M., Prelle, A., Zecca, C., Adami, R., Angeletti, B., Ciscato, P., Pellegrino, M. A., Bottinelli, R., Green, M. R., and Tupler, R. (2006). Facioscapulohumeral muscular dystrophy in mice overexpressing FRG1. Nature **439**(7079), 973–977.

Grosveld, F., van Assendelft, G. B., Greaves, D. R., and Kollias, G. (1987). Position-independent, high-level expression of the human beta-globin gene in transgenic mice. Cell **51,** 975–985.

Gurnett, C. A., Bowcock, A. M., Dietz, F. R., Morcuende, J. A., Murray, J. C., and Dobbs, M. B. (2007). Two novel point mutations in the long-range SHH enhancer in three families with triphalangeal thumb and preaxial polydactyly. Am. J. Med. Genet. A **143,** 27–32.

Hadchouel, J., Carvajal, J. J., Daubas, P., Bajard, L., Chang, T., Rocancourt, D., Cox, D., Summerbell, D., Tajbakhsh, S., Rigby, P. W., and Buckingham, M. (2003). Analysis of a key regulatory region upstream of the Myf5 gene reveals multiple phases of myogenesis, orchestrated at each site by a combination of elements dispersed throughout the locus. Development **130,** 3415–3426.

Hatton, C. S., Wilkie, A. O., Drysdale, H. C., Wood, W. G., Vickers, M. A., Sharpe, J., Ayyub, H., Pretorius, I. M., Buckle, V. J., and Higgs, D. R. (1990). Alpha-thalassemia caused by a large (62 kb) deletion upstream of the human alpha globin gene cluster. Blood **76**(1), 221–227.

Hayes, C., Rump, A., Cadman, M. R., Harrison, M., Evans, E. P., Lyon, M. F., Morriss-Kay, G. M., Rosenthal, A., and Brown, S. D. (2001). A high-resolution genetic, physical, and comparative gene map of the doublefoot (Dbf) region of mouse chromosome 1 and the region of conserved synteny on human chromosome 2q35. Genomics **78,** 197–205.

Herault, Y., Fraudeau, N., Zakany, J., and Duboule, D. (1997). Ulnaless (Ul), a regulatory mutation inducing both loss-of-function and gain-of-function of posterior Hoxd genes. Development **124,** 3493–3500.

Herbert, A., Gerry, N. P., McQueen, M. B., Heid, I. M., Pfeufer, A., Illig, T., Wichmann, H. E., Meitinger, T., Hunter, D., Hu, F. B., Colditz, G., Hinney, A., et al. (2006). A common genetic variant is associated with adult and childhood obesity. Science **312**(5771), 279–283.

Heus, H. C., Hing, A., van Baren, M. J., Joosse, M., Breedveld, G. J., Wang, J. C., Burgess, A., Donnis-Keller, H., Berglund, C., Zguricas, J., Scherer, S. W., Rommens, J. M., et al. (1999). A physical and transcriptional map of the preaxial polydactyly locus on chromosome 7q36. Genomics **57,** 342–351.

Heydemann, A., Nguyen, L. C., and Crenshaw, E. B., III (2001). Regulatory regions from the Brn4 promoter direct LACZ expression to the developing forebrain and neural tube. Brain Res. Dev. Brain Res. **128,** 83–90.

Hill, R. E., Favor, J., Hogan, B. M., Ton, C. C. T., Saunders, G. F., Hanson, I. M., Prosser, J., Jordan, T., Hastie, N. D., and van Heyningen, V. (1991). Mouse small eye results from mutations in a paired-like homeobox-containing gene. *Nature* **354,** 522–525.

Hill, R. E., Heaney, S. J., and Lettice, L. A. (2003). Sonic hedgehog: Restricted expression and limb dysmorphologies. *J. Anat.* **202,** 13–20.

Hill-Harfe, K. L., Kaplan, L., Stalker, H. J., Zori, R. T., Pop, R., Scherer, G., and Wallace, M. R. (2005). Fine mapping of chromosome 17 translocation breakpoints > or = 900 Kb upstream of SOX9 in acampomelic campomelic dysplasia and a mild, familial skeletal dysplasia. *Am. J. Hum. Genet.* **76**(4), 663–671.

Huber, C., Rosilio, M., Munnich, A., and Cormier-Daire, V. (2006). High incidence of SHOX anomalies in individuals with short stature. *J. Med. Genet.* **43**(9), 735–739.

Ianakiev, P., van Baren, M. J., Daly, M. J., Toledo, S. P., Cavalcanti, M. G., Neto, J. C., Silveira, E. L., Freire-Maia, A., Heutink, P., Kilpatrick, M. W., and Tsipouras, P. (2001). Acheiropodia is caused by a genomic deletion in C7orf2, the human orthologue of the Lmbr1 gene. *Am. J. Hum. Genet.* **68,** 38–45.

Ignatius, J., Knuutila, S., Scherer, S. W., Trask, B., and Kere, J. (1996). Split hand/split foot malformation, deafness, and mental retardation with a complex cytogenetic rearrangement involving 7q21.3. *J. Med. Genet.* **33,** 507–510.

Ishikawa-Brush, Y., Powell, J. F., Bolton, P., Miller, A. P., Francis, F., Willard, H. F., Lehrach, H., and Monaco, A. P. (1997). Autism and multiple exostoses associated with an X;8 translocation occurring within the GRPR gene and 3′ to the SDC2 gene. *Hum. Mol. Genet.* **6,** 1241–1250.

Jamieson, R. V., Perveen, R., Kerr, B., Carette, M., Yardley, J., Heon, E., Wirth, M. G., van Heyningen, V., Donnai, D., Munier, F., and Black, G. C. (2002). Domain disruption and mutation of the bZIP transcription factor, MAF, associated with cataract, ocular anterior segment dysgenesis and coloboma. *Hum. Mol. Genet.* **11,** 33–42.

Jeong, Y., El-Jaick, K., Roessler, E., Muenke, M., and Epstein, D. J. (2006). A functional screen for sonic hedgehog regulatory elements across a 1 Mb interval identifies long-range ventral forebrain enhancers. *Development* **133,** 761–772.

Jiang, G., Yang, F., Van Overveld, P. G., Vedanarayanan, V., van der, M. S., and Ehrlich, M. (2003). Testing the position-effect variegation hypothesis for facioscapulohumeral muscular dystrophy by analysis of histone modification and gene expression in subtelomeric 4q. *Hum. Mol. Genet.* **12,** 2909–2921.

Karpen, G. H. (1994). Position-effect variegation and the new biology of heterochromatin. *Curr. Opin. Genet. Dev.* **4,** 281–291.

Kimura-Yoshida, C., Kitajima, K., Oda-Ishii, I., Tian, E., Suzuki, M., Yamamoto, M., Suzuki, T., Kobayashi, M., Aizawa, S., and Matsuo, I. (2004). Characterization of the pufferfish Otx2 cis-regulators reveals evolutionarily conserved genetic mechanisms for vertebrate head specification. *Development* **131,** 57–71.

Kioussis, D., Vanin, E., deLange, T., Flavell, R. A., and Grosveld, F. G. (1983). Beta-globin gene inactivation by DNA translocation in gamma beta-thalassaemia. *Nature* **306,** 662–666.

Kist, R., Schrewe, H., Balling, R., and Scherer, G. (2002). Conditional inactivation of Sox9: A mouse model for campomelic dysplasia. *Genesis* **32,** 121–123.

Kleinjan, D. A., Seawright, A., Elgar, G., and van Heyningen, V. (2002). Characterization of a novel gene adjacent to PAX6, revealing synteny conservation with functional significance. *Mamm. Genome* **13,** 102–107.

Kleinjan, D. A., Seawright, A., Schedl, A., Quinlan, R. A., Danes, S., and van Heyningen, V. (2001). Aniridia-associated translocations, DNase hypersensitivity, sequence comparison and transgenic analysis redefine the functional domain of PAX6. *Hum. Mol. Genet.* **10,** 2049–2059.

Kleinjan, D. A., Seawright, A., Mella, S., Carr, C. B., Tyas, D. A., Simpson, T. I., Mason, J. O., Price, D. J., and van Heyningen, V. (2006). Long-range downstream enhancers are essential for Pax6 expression. *Dev. Biol.* **299**(2), 563–581.

Krantz, I. D., McCallum, J., DeScipio, C., Kaur, M., Gillis, L. A., Yaeger, D., Jukofsky, L., Wasserman, N., Bottani, A., Morris, C. A., Nowaczyk, M. J., Toriello, H., *et al.* (2004). Cornelia de Lange syndrome is caused by mutations in NIPBL, the human homolog of *Drosophila* melanogaster Nipped-B. *Nat. Genet.* **36**, 631–635.

Krebs, O., Schreiner, C. M., Scott, W. J., Jr, Bell, S. M., Robbins, D. J., Goetz, J. A., Alt, H., Hawes, N., Wolf, E., and Favor, J. (2003). Replicated anterior zeugopod (raz): A polydactylous mouse mutant with lowered Shh signaling in the limb bud. *Development* **130**, 6037–6047.

Lauderdale, J. D., Wilensky, J. S., Oliver, E. R., Walton, D. S., and Glaser, T. (2000). 3′ deletions cause aniridia by preventing PAX6 gene expression. *Proc. Natl. Acad. Sci. USA* **97**, 13755–13759.

Lee, J. A., Madrid, R. E., Sperle, K., Ritterson, C. M., Hobson, G. M., Garbern, J., Lupski, J. R., and Inoue, K. (2006). Spastic paraplegia type 2 associated with axonal neuropathy and apparent PLP1 position effect. *Ann. Neurol.* **59**(2), 398–403.

Lehmann, O. J., Sowden, J. C., Carlsson, P., Jordan, T., and Bhattacharya, S. S. (2003). Fox's in development and disease. *Trends Genet.* **19**, 339–344.

Leipoldt, M., Erdel, M., Bien-Willner, G. A., Smyk, M., Theurl, M., Yatsenko, S. A., Lupski, J. R., Lane, A. H., Shanske, A. L., Stankiewicz, P., and Scherer, G. (2007). Two novel translocation breakpoints upstream of SOX9 define borders of the proximal and distal breakpoint cluster region in campomelic dysplasia. *Clin. Genet.* **71**(1), 67–75.

Lemmers, R. J., Wohlgemuth, M., Frants, R. R., Padberg, G. W., Morava, E., and van der Maarel, S. M. (2004). Contractions of D4Z4 on 4qB subtelomeres do not cause facioscapulohumeral muscular dystrophy. *Am. J. Hum. Genet.* **75**(6), 1124–1130.

Lettice, L. A., Heaney, S. J., Purdie, L. A., Li, L., de Beer, P., Oostra, B. A., Goode, D., Elgar, G., Hill, R. E., and de Graaff, E. (2003). A long-range Shh enhancer regulates expression in the developing limb and fin and is associated with preaxial polydactyly. *Hum. Mol. Genet.* **12**, 1725–1735.

Lettice, L. A., Horikoshi, T., Heaney, S. J., van Baren, M. J., van der Linde, H. C., Breedveld, G. J., Joosse, M., Akarsu, N., Oostra, B. A., Endo, N., *et al.* (2002). Disruption of a long-range cis-acting regulator for Shh causes preaxial polydactyly. *Proc. Natl. Acad. Sci. USA* **99**, 7548–7553.

Lewinsky, R. H., Jensen, T. G., Moller, J., Stensballe, A., Olsen, J., and Troelsen, J. T. (2005). T-13910 DNA variant associated with lactase persistence interacts with Oct-1 and stimulates lactase promoter activity *in vitro*. *Hum. Mol. Genet.* **14**(24), 3945–3953.

Loots, G. G., Kneissel, M., Keller, H., Baptist, M., Chang, J., Collette, N. M., Ovcharenko, D., Plajzer-Frick, I., and Rubin, E. M. (2005). Genomic deletion of a long-range bone enhancer misregulates sclerostin in Van Buchem disease. *Genome Res.* **15**(7), 928–935.

Maas, R. L., Zeller, R., Woychik, R. P., Vogt, T. F., and Leder, P. (1990). Disruption of formin-encoding transcripts in two mutant limb deformity alleles. *Nature* **346**, 853–855.

Maas, S. A., and Fallon, J. F. (2005). Single base pair change in the long-range Sonic hedgehog limb-specific enhancer is a genetic basis for preaxial polydactyly. *Dev. Dyn.* **232**, 345–348.

Mackenzie, A., Miller, K. A., and Collinson, J. M. (2004). Is there a functional link between gene interdigitation and multi-species conservation of synteny blocks? *Bioessays* **26**(11), 1217–1224.

Marlin, S., Blanchard, S., Slim, R., Lacombe, D., Denoyelle, F., Alessandri, J. L., Calzolari, E., Drouin-Garraud, V., Ferraz, F. G., Fourmaintraux, A., Philip, N., Toublanc, J. E., and Petit, C. (1999). Townes-Brocks syndrome: detection of a SALL1 mutation hot spot and evidence for a position effect in one patient. *Hum. Mutat.* **14**(5), 377–386.

Masny, P. S., Bengtsson, U., Chung, S. A., Martin, J. H., van Engelen, B., van der Maarel, S. M., and Winokur, S. T. (2004). Localization of 4q35.2 to the nuclear periphery: is FSHD a nuclear envelope disease? *Hum. Mol. Genet.* **13**(17), 1857–1871.

Masuya, H., Sezutsu, H., Sakuraba, Y., Sagai, T., Hosoya, M., Kaneda, H., Miura, I., Kobayashi, K., Sumiyama, K., Shimizu, A., Nagano, J., Yokoyama, H., *et al.* (2007). A series of ENU-induced single-base substitutions in a long-range cis-element altering Sonic hedgehog expression in the developing mouse limb bud. *Genomics* **89**, 207–214.

Merlo, G. R., Paleari, L., Mantero, S., Genova, F., Beverdam, A., Palmisano, G. L., Barbieri, O., and Levi, G. (2002). Mouse model of split hand/foot malformation type I. *Genesis* **33**, 97–101.

Muncke, N., Wogatzky, B. S., Breuning, M., Sistermans, E. A., Endris, V., Ross, M., Vetrie, D., Catsman-Berrevoets, C. E., and Rappold, G. (2004). Position effect on PLP1 may cause a subset of Pelizaeus-Merzbacher disease symptoms. *J. Med. Genet.* **41**(12), e121.

Niedermaier, M., Schwabe, G. C., Fees, S., Helmrich, A., Brieske, N., Seemann, P., Hecht, J., Seitz, V., Stricker, S., Leschik, G., Schrock, E., Selby, P. B., *et al.* (2005). An inversion involving the mouse Shh locus results in brachydactyly through dysregulation of Shh expression. *J. Clin. Invest.* **115**, 900–909.

Nishimura, D. Y., Swiderski, R. E., Alward, W. L., Searby, C. C., Patil, S. R., Bennet, S. R., Kanis, A. B., Gastier, J. M., Stone, E. M., and Sheffield, V. C. (1998). The forkhead transcription factor gene FKHL7 is responsible for glaucoma phenotypes which map to 6p25. *Nat. Genet.* **19**, 140–147.

Nobrega, M. A., Ovcharenko, I., Afzal, V., and Rubin, E. M. (2003). Scanning human gene deserts for long-range enhancers. *Science* **302**(5644), 413.

Nobrega, M. A., Zhu, Y., Plajzer-Frick, I., Afzal, V., and Rubin, E. M. (2004). Megabase deletions of gene deserts result in viable mice. *Nature* **431**(7011), 988–993.

Olds, L. C., and Sibley, E. (2003). Lactase persistence DNA variant enhances lactase promoter activity *in vitro*: Functional role as a cis regulatory element. *Hum. Mol. Genet.* **12**, 2333–2340.

Ovcharenko, I., Loots, G. G., Nobrega, M. A., Hardison, R. C., Miller, W., and Stubbs, L. (2005). Evolution and functional classification of vertebrate gene deserts. *Genome Res.* **15**(1), 137–145.

Paige, A. J., Taylor, K. J., Stewart, A., Sgouros, J. G., Gabra, H., Sellar, G. C., Smyth, J. F., Porteous, D. J., and Watson, J. E. (2000). A 700-kb physical map of a region of 16q23.2 homozygously deleted in multiple cancers and spanning the common fragile site FRA16D. *Cancer Res.* **60**, 1690–1697.

Pailhoux, E., Vigier, B., Chaffaux, S., Servel, N., Taourit, S., Furet, J. P., Fellous, M., Grosclaude, F., Cribiu, E. P., Cotinot, C., and Vaiman, D. (2001). A 11.7-kb deletion triggers intersexuality and polledness in goats. *Nat. Genet.* **29**, 453–458.

Pfeifer, D., Kist, R., Dewar, K., Devon, K., Lander, E. S., Birren, B., Korniszewski, L., Back, E., and Scherer, G. (1999). Campomelic dysplasia translocation breakpoints are scattered over 1 Mb proximal to SOX9: Evidence for an extended control region. *Am. J. Hum. Genet.* **65**, 111–124.

Phippard, D., Boyd, Y., Reed, V., Fisher, G., Masson, W. K., Evans, E. P., Saunders, J. C., and Crenshaw, E. B., III (2000). The sex-linked fidget mutation abolishes Brn4/Pou3f4 gene expression in the embryonic inner ear. *Hum. Mol. Genet.* **9**, 79–85.

Pop, R., Conz, C., Lindenberg, K. S., Blesson, S., Schmalenberger, B., Briault, S., Pfeifer, D., and Scherer, G. (2004). Screening of the 1 Mb SOX9 5' control region by array CGH identifies a large deletion in a case of campomelic dysplasia with XY sex reversal. *J. Med. Genet.* **41**, e47.

Qin, Y., Kong, L. K., Poirier, C., Truong, C., Overbeek, P. A., and Bishop, C. E. (2004). Long-range activation of Sox9 in Odd Sex (Ods) mice. *Hum. Mol. Genet.* **13**, 1213–1218.

Robledo, R. F., Rajan, L., Li, X., and Lufkin, T. (2002). The Dlx5 and Dlx6 homeobox genes are essential for craniofacial, axial, and appendicular skeletal development. *Genes Dev.* **16**, 1089–1101.

Roessler, E., Belloni, E., Gaudenz, K., Jay, P., Berta, P., Scherer, S. W., Tsui, L. C., and Muenke, M. (1996). Mutations in the human Sonic Hedgehog gene cause holoprosencephaly. *Nat. Genet.* **14**, 357–360.

Roessler, E., Ward, D. E., Gaudenz, K., Belloni, E., Scherer, S. W., Donnai, D., Siegel-Bartelt, J., Tsui, L. C., and Muenke, M. (1997). Cytogenetic rearrangements involving the loss of the Sonic Hedgehog gene at 7q36 cause holoprosencephaly. *Hum. Genet.* **100**, 172–181.

Rollins, R. A., Korom, M., Aulner, N., Martens, A., and Dorsett, D. (2004). Drosophila nipped-B protein supports sister chromatid cohesion and opposes the stromalin/Scc3 cohesion factor to facilitate long-range activation of the cut gene. *Mol. Cell. Biol.* **24,** 3100–3111.

Rollins, R. A., Morcillo, P., and Dorsett, D. (1999). Nipped-B, a *Drosophila* homologue of chromosomal adherins, participates in activation by remote enhancers in the cut and Ultrabithorax genes. *Genetics* **152,** 577–593.

Ros, M. A., Dahn, R. D., Fernandez-Teran, M., Rashka, K., Caruccio, N. C., Hasso, S. M., Bitgood, J. J., Lancman, J. J., and Fallon, J. F. (2003). The chick oligozeugodactyly (ozd) mutant lacks sonic hedgehog function in the limb. *Development* **130,** 527–537.

Sabherwal, N., Bangs, F., Roth, R., Weiss, B., Jantz, K., Tiecke, E., Hinkel, G. K., Spaich, C., Hauffa, B. P., van der Kamp, H., Kapeller, J., Tickle, C., and Rappold, G. (2007). Long-range conserved non-coding SHOX sequences regulate expression in developing chicken limb and are associated with short stature phenotypes in human patients. *Hum. Mol. Genet.* **16,** 210–222.

Sagai, T., Masuya, H., Tamura, M., Shimizu, K., Yada, Y., Wakana, S., Gondo, Y., Noda, T., and Shiroishi, T. (2004). Phylogenetic conservation of a limb-specific, cis-acting regulator of Sonic hedgehog (Shh). *Mamm. Genome* **15,** 23–34.

Schedl, A., Ross, A., Lee, M., Engelkamp, D., Rashbass, P., van Heyningen, V., and Hastie, N. D. (1996). Influence of PAX6 gene dosage on development: Overexpression causes severe eye abnormalities. *Cell* **86,** 71–82.

Scherer, S. W., Poorkaj, P., Massa, H., Soder, S., Allen, T., Nunes, M., Geshuri, D., Wong, E., Belloni, E., and Little, S. (1994). Physical mapping of the split hand/split foot locus on chromosome 7 and implication in syndromic ectrodactyly. *Hum. Mol. Genet.* **3,** 1345–1354.

Scherer, S. W., Cheung, J., MacDonald, J. R., Osborne, L. R., Nakabayashi, K., Herbrick, J. A., Carson, A. R., Parker-Katiraee, L., Skaug, J., Khaja, R., Zhang, J., *et al.* (2003). Human chromosome 7: DNA sequence and biology. *Science* **300**(5620), 767–772.

Sharpe, J., Lettice, L., Hecksher-Sorensen, J., Fox, M., Hill, R., and Krumlauf, R. (1999). Identification of sonic hedgehog as a candidate gene responsible for the polydactylous mouse mutant Sasquatch. *Curr. Biol.* **9,** 97–100.

Spitz, F., Gonzalez, F., and Duboule, D. (2003). A global control region defines a chromosomal regulatory landscape containing the HoxD cluster. *Cell* **113,** 405–417.

Spitz, F., Gonzalez, F., Peichel, C., Vogt, T. F., Duboule, D., and Zakany, J. (2001). Large scale transgenic and cluster deletion analysis of the HoxD complex separate an ancestral regulatory module from evolutionary innovations. *Genes Dev.* **15,** 2209–2214.

Spitz, F., Montavon, T., Monso-Hinard, C., Morris, M., Ventruto, M. L., Antonarakis, S., Ventruto, V., and Duboule, D. (2002). A t(2;8) balanced translocation with breakpoints near the human HOXD complex causes mesomelic dysplasia and vertebral defects. *Genomics* **79,** 493–498.

Staehling-Hampton, K., Proll, S., Paeper, B. W., Zhao, L., Charmley, P., Brown, A., Gardner, J. C., Galas, D., Schatzman, R. C., Beighton, P., Papapoulos, S., Hamersma, H., *et al.* (2002). A 52-kb deletion in the SOST-MEOX1 intergenic region on 17q12-q21 is associated with van Buchem disease in the Dutch population. *Am. J. Med. Genet.* **110,** 144–152.

Tarchini, B., and Duboule, D. (2006). Control of Hoxd genes' collinearity during early limb development. *Dev. Cell* **10,** 93–103.

Tarchini, B., Duboule, D., and Kmita, M. (2006). Regulatory constraints in the evolution of the tetrapod limb anterior-posterior polarity. *Nature* **443,** 985–988.

Tishkoff, S. A., Reed, F. A., Ranciaro, A., Voight, B. F., Babbitt, C. C., Silverman, J. S., Powell, K., Mortensen, H. M., Hirbo, J. B., Osman, M., Ibrahim, M., Omar, S. A., *et al.* (2007). Convergent adaptation of human lactase persistence in Africa and Europe. *Nat. Genet.* **39**(1), 31–40.

Tolhuis, B., Palstra, R. J., Splinter, E., Grosveld, F., and de Laat, W. (2002). Looping and interaction between hypersensitive sites in the active beta-globin locus. *Mol. Cell* **10,** 1453–1465.

Ton, C. C., Hirvonen, H., Miwa, H., Weil, M. M., Monaghan, P., Jordan, T., van Heyningen, V., Hastie, N. D., Meijers-Heijboer, H., Drechsler, M., Royer-Pokora, B., Collins, F., et al. (1991). Positional cloning and characterization of a paired box- and homeobox-containing gene from the aniridia region. Cell 67, 1059–1074.

Tonkin, E. T., Wang, T. J., Lisgo, S., Bamshad, M. J., and Strachan, T. (2004). NIPBL, encoding a homolog of fungal Scc2-type sister chromatid cohesion proteins and fly Nipped-B, is mutated in Cornelia de Lange syndrome. Nat. Genet. 36, 636–641.

Trembath, D. G., Semina, E. V., Jones, D. H., Patil, S. R., Qian, Q., Amendt, B. A., Russo, A. F., and Murray, J. C. (2004). Analysis of two translocation breakpoints and identification of a negative regulatory element in patients with Rieger's syndrome. Birth Defects Res. Part A Clin. Mol. Teratol. 70, 82–91.

Troelsen, J. T., Olsen, J., Moller, J., and Sjostrom, H. (2003). An upstream polymorphism associated with lactase persistence has increased enhancer activity. Gastroenterology 125(6), 1686–1694.

Tufarelli, C., Stanley, J. A., Garrick, D., Sharpe, J. A., Ayyub, H., Wood, W. G., and Higgs, D. R. (2003). Transcription of antisense RNA leading to gene silencing and methylation as a novel cause of human genetic disease. Nat. Genet. 34, 157–165.

Tupler, R., Berardinelli, A., Barbierato, L., Frants, R., Hewitt, J. E., Lanzi, G., Maraschio, P., and Tiepolo, L. (1996). Monosomy of distal 4q does not cause facioscapulohumeral muscular dystrophy. J. Med. Genet. 33, 366–370.

Uchikawa, M., Ishida, Y., Takemoto, T., Kamachi, Y., and Kondoh, H. (2003). Functional analysis of chicken Sox2 enhancers highlights an array of diverse regulatory elements that are conserved in mammals. Dev. Cell 4, 509–519.

Ueta, E., Nanba, E., and Naruse, I. (2002). Integration of a transposon into the Gli3 gene in the Pdn mouse. Congenit Anom (Kyoto) 42(4), 318–322.

Valle, D. (2004). Genetics, individuality, and medicine in the 21st century. Am. J. Hum. Genet. 74, 374–381.

van der Hoeven, F., Schimmang, T., Vortkamp, A., and Ruther, U. (1993). Molecular linkage of the morphogenetic mutation add and the zinc finger gene Gli3. Mamm. Genome 4, 276–277.

van der Hoeven, F., Zakany, J., and Duboule, D. (1996). Gene transpositions in the HoxD complex reveal a hierarchy of regulatory controls. Cell 85, 1025–1035.

van Deutekom, J. C., Wijmenga, C., van Tienhoven, E. A., Gruter, A. M., Hewitt, J. E., Padberg, G. W., van Ommen, G. J., Hofker, M. H., and Frants, R. R. (1993). FSHD associated DNA rearrangements are due to deletions of integral copies of a 3.2 kb tandemly repeated unit. Hum. Mol. Genet. 2, 2037–2042.

van Heyningen, V., and Williamson, K. A. (2002). PAX6 in sensory development. Hum. Mol. Genet. 11, 1161–1167.

Van Hul, W., Balemans, W., Van Hul, E., Dikkers, F. G., Obee, H., Stokroos, R. J., Hildering, P., Vanhoenacker, F., Van Camp, G., and Willems, P. J. (1998). Van Buchem disease (hyperostosis corticalis generalisata) maps to chromosome 17q12-q21. Am. J. Hum. Genet. 62, 391–399.

Velagaleti, G. V., Bien-Willner, G. A., Northup, J. K., Lockhart, L. H., Hawkins, J. C., Jalal, S. M., Withers, M., Lupski, J. R., and Stankiewicz, P. (2005). Position effects due to chromosome breakpoints that map approximately 900 Kb upstream and approximately 1.3 Mb downstream of SOX9 in two patients with campomelic dysplasia. Am. J. Hum. Genet. 76(4), 652–662.

Venter, J. C., Adams, M. D., Myers, E. W., Li, P. W., Mural, R. J., Sutton, G. G., Smith, H. O., Yandell, M., Evans, C. A., and Holt, R. A. (2001). The sequence of the human genome. Science 291(5507), 1304–1351.

Viprakasit, V., Kidd, A. M., Ayyub, H., Horsley, S., Hughes, J., and Higgs, D. R. (2003). De novo deletion within the telomeric region flanking the human alpha globin locus as a cause of alpha thalassaemia. Br. J. Haematol. 20, 867–875.

Viprakasit, V., Harteveld, C. L., Ayyub, H., Stanley, J. S., Giordano, P. C., Wood, W. G., and Higgs, D. R. (2006). A novel deletion causing alpha thalassemia clarifies the importance of the major human alpha globin regulatory element. *Blood* **107**, 3811–3812.

Vortkamp, A., Gessler, M., and Grzeschik, K. H. (1991). GLI3 zinc-finger gene interrupted by translocations in Greig syndrome families. *Nature* **352**, 539–540.

Vortkamp, A., Franz, T., Gessler, M., and Grzeschik, K. H. (1992). Deletion of GLI3 supports the homology of the human Greig cephalopolysyndactyly syndrome (GCPS) and the mouse mutant extra toes (Xt). *Mamm. Genome* **3**, 461–463.

Vyas, P., Vickers, M. A., Picketts, D. J., and Higgs, D. R. (1995). Conservation of position and sequence of a novel, widely expressed gene containing the major human alpha-globin regulatory element. *Genomics* **29**, 679–689.

Wakui, K., Gregato, G., Ballif, B. C., Glotzbach, C. D., Bailey, K. A., Kuo, P. L., Sue, W. C., Sheffield, L. J., Irons, M., Gomez, E. G., Hecht, J. T., Potocki, L., *et al.* (2005). Construction of a natural panel of 11p11.2 deletions and further delineation of the critical region involved in Potocki-Shaffer syndrome. *Eur. J. Hum. Genet.* **13**(5), 528–540.

Wallis, D. E., Roessler, E., Hehr, U., Nanni, L., Wiltshire, T., Richieri-Costa, A., Gillessen-Kaesbach, G., Zackai, E. H., Rommens, J., and Muenke, M. (1999). Mutations in the homeo-domain of the human SIX3 gene cause holoprosencephaly. *Nat. Genet.* **22**, 196–198.

Wijmenga, C., Hewitt, J. E., Sandkuijl, L. A., Clark, L. N., Wright, T. J., Dauwerse, H. G., Gruter, A. M., Hofker, M. H., Moerer, P., and Williamson, R. (1992). Chromosome 4q DNA rearrangements associated with facioscapulohumeral muscular dystrophy. *Nat. Genet.* **2**, 26–30.

Wild, A., Kalff-Suske, M., Vortkamp, A., Bornholdt, D., Konig, R., and Grzeschik, K. H. (1997). Point mutations in human GLI3 cause Greig syndrome. *Hum. Mol. Genet.* **6**, 1979–1984.

Woolfe, A., Goodson, M., Goode, D. K., Snell, P., McEwen, G. K., Vavouri, T., Smith, S. F., North, P., Callaway, H., Kelly, K., Walter, K., Abnizova, I., *et al.* (2005). Highly conserved non-coding sequences are associated with vertebrate development. *PLoS Biol.* **3**(1), e7.

Woychik, R. P., Generoso, W. M., Russell, L. B., Cain, K. T., Cacheiro, N. L., Bultman, S. J., Selby, P. B., Dickinson, M. E., Hogan, B. L., and Rutledge, J. C. (1990). Molecular and genetic characterization of a radiation-induced structural rearrangement in mouse chromosome 2 causing mutations at the limb deformity and agouti loci. *Proc. Natl. Acad. Sci. USA* **87**, 2588–2592.

Wunderle, V. M., Critcher, R., Hastie, N., Goodfellow, P. N., and Schedl, A. (1998). Deletion of long-range regulatory elements upstream of SOX9 causes campomelic dysplasia. *Proc. Natl. Acad. Sci. USA* **95**, 10649–10654.

Yan, J., Bi, W., and Lupski, J. R. (2007). Penetrance of craniofacial anomalies in mouse models of Smith-Magenis syndrome is modified by genomic sequence surrounding Rai1: Not all null alleles are alike. *Am. J. Hum. Genet.* **80**, 518–525.

Yang, Y., Guillot, P., Boyd, Y., Lyon, M. F., and McMahon, A. P. (1998). Evidence that preaxial polydactyly in the Doublefoot mutant is due to ectopic Indian Hedgehog signaling. *Development* **125**, 3123–3132.

Zakany, J., Kmita, M., and Duboule, D. (2004). A dual role for Hox genes in limb anterior-posterior asymmetry. *Science* **304**, 1669–1672.

Zeller, R., Haramis, A. G., Zuniga, A., McGuigan, C., Dono, R., Davidson, G., Chabanis, S., and Gibson, T. (1999). Formin defines a large family of morphoregulatory genes and functions in establishment of the polarising region. *Cell Tissue Res.* **296**, 85–93.

Zuniga, A., Michos, O., Spitz, F., Haramis, A. P., Panman, L., Galli, A., Vintersten, K., Klasen, C., Mansfield, W., Kuc, S., Duboule, D., Dono, R., *et al.* (2004). Mouse limb deformity mutations disrupt a global control region within the large regulatory landscape required for Gremlin expression. *Genes Dev.* **18**, 1553–1564.

Index

A

Abdominal-B (Abd-B), 53
  regulatory interactions in *Drosophila*, 85
Aberrant imprinted chromosome, 237. *See also* Imprinting defects
Aberrant methylation, maternal CTCF-binding sites and H19 promoter, 241
Accelerated noncoding sequences, 300
Acetylation levels, 112, 113, 128. *See also* β–globin
Acetylation of Lys9 of H3 (H3K9), 228
achaete (ac)-sc complex (AS-C), *Diptera*, 96
achaete-scute complex, 81
Acheiropodia, 361
Active chromatin hub (ACH), 118–121, 131–132
Acute leukemia, 260
Adenomatous polyposis coli (APC), 252–253
AgASH. *See Anopheles gambiae* Ac-Sc-Homolog
Alpha thalassemia, 147, 168
Amine oxidase AOF2, 228
Amy2 allele, 9
Amy4,6 allele, 9
Anacardic acid, 261
Angelman syndrome (AS), 230
Anopheles gambiae Ac-Sc-Homolog, 96. *See also* New regulatory modules, evolution
Antisense transcript (ATS), 236
APC. *See* Adenomatous polyposis coli
Argonaute RNAi pathway genes, mutation, 28
AS-SRO element
  and binding of trans-factors, 236
  establish maternal imprint with female germ line, 238
  establish maternal methylation at PWS-SRO, 242
  upstream exons of *SNURF-SNRPN* transcript, 235

ATP-dependent chromatin remodeling complexes, 160
Axenfeld Rieger anomaly, 365
5-Azacytidine, role in DNA methylation inhibition, 261–262

B

Bacterial artificial chromosomes (BACs), 181, 275, 344
Bacteriophage T4, 116
Bar eye mutation, 3
Bar gene, 3
Barhl2 gene, 325
Bar locus *(B)*, 3
Base pairs (bp), 271
Basic local alignment search tool, 277
B cell, 193
bcl11a gene, 323
Beckwith-Wiedemann syndrome (BWS), 239, 240
Benzamides, 262
Bicistronic systems, 177
Bicoid (Bcd) homeodomain protein, 73
Bithorax complex (BX-C) of Drosophila, 53
BLAST. *See* Basic local alignment search tool
Blast2sequences, 277
Blepharophimosis-ptosis-epicanthus inversus syndrome, 366
bl2seq. *See* Blast2sequences
BMP-antagonist Gremlin, 187
BPES. *See* Blepharophimosis-ptosis-epicanthus inversus syndrome
BRCA1 hypermethylation, 251
bric a brac (bab), regulatory interactions in *Drosophila*, 85
Bromodomain proteins, 20
Brown allele (bw^D), 18
Brown (bw) gene, 18
Butyrate, 262

BWS A–F, deletions, 241
Bystander effects, 200, 192, 353–355. *See also*
    Global regulation, mechanism

C

Caenorhabditis elegans, 176, 321
Callipyge phenotype, 218
Campomelic displasia, 366–368
Cancer
    disruption of DNA methylation, 250–253
    DNA-demethylating drugs treatment, 257
    DNA hypomethylation, 250
    DNA methylation and histone
        modification, 254
    epigenetic deregulation, 256–258
    genetic and epigenetic deregulation, 251
    histone modification, 253–256
    treatment, 261–262
    tumor-suppressor gene hypermethylation
        mechanisms, 258–261
    tumor-suppressor genes role, 251–252
Candidate genes, 359, 362, 365–366, 378
CCAAT box, 152, 160
CDLS. *See* Cornelia de Lange syndrome
C/D-type snoRNAs, 217
Cell, regulatory variation levels, 297
Central nervous system (CNS), 185, 357
ChIP. *See* Chromatin immunoprecipitation
Chip protein, 116
β-Chromatin, 8
Chromatin hub (CH), 120
Chromatin immunoprecipitation, 152, 153, 158,
    160, 251
Chromatin insulators, 56
Chromatin-probing model, 55
Chromosome conformation capture (3C), 118,
    162
Cis- and trans-acting factors, 238–239. *See also*
    Imprinting defects
cis gene expression, 301
cis-regulatory modules, 276, 281, 309
Cis-regulatory sequences, in *Drosophila*
    experimental approaches to, 71–72
        direct genetic analyses, 71
        gene expression patterns, comparisons of,
            71
        green fluorescent protein, encoding, 72
        modularity, 71
        promoter sequences, evolution of, 69–70

Cis-trans coevolution, 91
    bicoid and hunchback P2 promoter,
        interaction, 91–95
    bicoid binding sites in promoters of *hunchback*
        and *knirps*, 94
    bicoid morphogen and target gene
        activation, 92
Clusters, imprinting regulation in, 207
    DLK1/GTL 2 imprinted cluster, 215–218
    insulator model of regulation, 209–212
    NCRNA model of regulation, 212–215
CNEs. *See* Conserved noncoding elements
CNSs. *See* Conserved noncoding sequences
CNV. *See* Copy number variation
Co-expression territories, 187–188
CONDOR, 317–318, 331–332
Congenital glaucoma, 365
Conserved noncoding elements, 308, 314
    comparison with protein coding exons, 330
    conservation, 331
    genomic environment, 317–319
    inversion events, 325–326
    location, 319–320
        in target gene, 323–328
        in UTRs, 321–323
    mutation, 324–328
    proportion to TSS, 324
Conserved noncoding sequences, 313
Consite web tools, 280
Copy number variation, 301–302
Cornelia de Lange syndrome, 373
CpG dinucleotides, cytosine, 248–249
CpG island hypermethylation, 251–253,
    257–258
CpG methylation, 211
Craniofacial abnormality, 362, 371
CREME, 282
CRMs. *See* Cis-regulatory modules
CTCF targets sites (CTS), 241
CTCF, transcription factor, 119, 120, 122
3C-technology. *See* Chromosome conformation
    capture (3C)
CTS cluster organization, 242
Cubitus interruptus (ci), 12
Curcumin, 261

D

DACH1 gene, 314, 331
DamID technique, 55

Danio rerio, 317
Dct. *See* Dopachrome tautomerase
DEAD-box helicase, 28
Differentially methylated region (DMR), 211
Distal CRMs, 323
Dlk1/Gtl2 locus, 215
DLX5 and *DLX6* gene, 362
Dlx genes, 191, 314
DMD methylation, 212
DNA-binding protein, 116
DNA footprinting, 310
DNA hypomethylation, 250
DNA looping, 115, 121
DNA methylation, 215, 218, 226, 227, 230, 232
  and histone modifications, 253–254
  inhibition, 261
  mechanisms, 259
DNA methyltransferase (Dam), 55
DNA methyltransferases (DNMTs), 51,
  227–228, 238, 253, 259–261
DNAse 1 hypersensitive sites (DHSs), 109, 146
DNA-tracking protein, 116
DNMTs. *See* DNA methyltransferases
Dopachrome tautomerase, 368
Dorsocentral enhancer (DCE), 79–83
  achaete-scute complex, 81
  overlapping domains in transgenic
    D. melanogaster flies, expression in, 83
  species of Drosophila, comparison from, 81
double-Bar mutation, 3
doublesex (dsx), regulatory interactions in
  Drosophila, 85
Downstream promoter element, 272
DPE. *See* Downstream promoter element
Drosophila (Iro-C), 179
Drosophila melanogaster, 8, 47, 310
Drosophila Nipped-B, 373
Drosophila simulans, 19
Dsh. *See* Short-digits
DSS1 gene, 362
Dystonia musculorum, 282

E

Early limb budactivating element, 364
Ecdysozoans, 178
Ectopic gene expression, 176
ELCR. *See* Early limb budactivating element
Ellis-van Creveld syndrome, 192
Embryonic gene (ζ), 146

ENCODE project, 298–299
Endogenous Shh regulatory elements,
  role, 361–362
Enhancer-bound proteins, C/EBPα and
  HNF-3β, 123
Enhancer evolution and loss or gain of traits.
  See also Cis-regulatory sequences, in
    Drosophila
  scute and variation in bristle patterns,
    evolution of, 87–90
  yellow and variation in pigment patterns,
    evolution of, 85–87
Ensembl genome browser database, 279
ENU. *See* Ethylnitrosourea
Epigenetic modifications, 248–250
Epithelial–mesenchymal interaction, 364–365
eQTL. *See* Expression at quantitative
  trait loci
EryP cells, 112
Erythroid cell-specific transcription
  factors, 122
Erythroid Kruppel-like factor (EKLF), 110,
  114, 120
Erythroid progenitors, 120
Erythroid-specific transcription factors,
  110, 120
Erythropoiesis, 113, 146
  transcription factors involved in, 148–150
Escherichia coli, 176, 181
ESTs. *See* Expressed sequence tags
Ethylnitrosourea, 360
E3 ubiquitin ligase activity, 48
Eukaryotes, functionally related genes, 177
E(var) phenotype, 28
even-skipped (eve) gene, 75
Eversporting displacement phenotype, 5
eve stripe two enhancer, eve S2E
  Bcd, Kr, Gt, and Hb, binding sites, 76
  of *D. melanogaster*, 78
  of *D. yakuba* and *D. pseudoobscura*, 77, 78
Evolutionarily conserved noncoding
    sequences, 276–277, 286, 346
Evx2 genes, 186
Expressed sequence tags, 279, 319
Expression at quantitative trait loci, 301–302
Extrachromosomal circles (ecc) of rDNA, 29
Eye Disease
  Aniridia 353–355
  Peters anomaly 356
EZH2-repressed genes, 51

F

Facioscapulohumeral dystrophy, 368–370
Factor-dependent cell Patterson (FDCP), 158
Fetal (γ) globin gene expression, 114
Fidgetin gene, 328
Fish-mammal
    comparisons, 313–315
    noncoding elements, 315–317
FK230 gene, 262
Flippase (FLP) Au2 recombinase system, 17
Fluorescent *in situ* hybridization (FISH), 19
fly genome, 53
FOOTER algorithm for motif prediction, 280
Forkhead gene, 365
formin gene, 187, 193
    role in limb bud morphogenesis, 274
FOXC1 gene, 365
FOXC2 gene, 365–366
    on chromosome 16q24, 365–366
    pleiotrophic effects, 366
FOXD3 gene, 319
FOX genes, 365–366
FOXL2 gene, 366
FSHD. *See* Facioscapulohumeral dystrophy
Fugu. See Takifugu rubripes-model genome
Functional expression modules, 376

G

gap genes, Giant (Gt) and Kruppel (Kr), 76
Gastrin-releasing peptide receptor, 356
GATA factor Pannier (Pnr), 79
GCPS. *See* Greig cephalo-polysyndactyly
    syndrome
GCR. *See* Global control region
Gene competition, 114
Gene expression by enhancers. *See also* β−globin
    long-range control, models of
        linking model, 116–117
        looping model, 115–116
        relocation models, 117
        tracking model, 116
Gene ontology, 315
General transcription factors, 144, 154–155,
    159–160, 162, 164–165, 272
Gene regulatory network, 309
Gene(s)
    CRMs, 281
    desert, 367, 376

expression, 301–302
inactivation in cancer mechanism, 250
regulation, 296
transcriptional control disruption
    mechanisms, 371
    common genetic disease, 377–378
    genetic disease diagnosis
        determination, 372–374
    long-range control and genome
        organization, 374–377
Genetic diseases, 341
Genome annotation, polymorphic
    characteristics, 299
Genome-wide analysis, of PcG binding in
    Drosophila, 59
Genomic imprinting, 207, 226
    erasure, establishment, and maintenance
        of, 229–230
    nature of, 226–228
Genomic imprints, 243
    erasure, establishment, and maintenance
        of, 229
    nature of, 225–228
Genomic tools, 278
GFP. *See* Green fluorescent protein
Glaucoma, 365
GLI3 gene, 352
Global Control Region, 185–186, 364
Global controls, vertebrate development and
    evolution
    co-expressed developmental gene
        clusters, 178–181
    Hox cluster, global regulation
        controlling enhancer activity, 184
    shared remote enhancers, 183–184
    shared regulatory elements, 181–182
Global gene control, evolutionary
    implications, 195–197
Global regulation
    cis-acting sequences, 186
    element downstream of *Hoxd12*, 184
    Hox clusters, 181
    mechanism, 188–192
        active chromatin hub at β-globin locus, 200
        bidirectional promoters, 191
        co-expression of adjacent genes, 189
        posttranslational modifications of
            histones, 200
        repressive chromatin, propagation of, 200
        transcription factories, region, 200

phenotype–genotype relationships, complex
mutations, 197
regulatory landscape, 186
α-Globin cluster, 146, 147, 152, 160
α-Globin gene, 151, 152, 153, 157, 158, 160–162
ε-Globin gene, 110
α−Globin gene expression, long-range
regulation
α−globin cluster, structure and
evolution, 146–147
α-globin regulatory domain, 147–152
hematopoiesis, key stages of, 157–158
remote regulatory elements, 161–162
RNA polymerase and GTFS, 160–161
sequential activation, 165–166
transcription factor, 158–160
upstream elements interaction, 162–165
upstream regulatory elements and
promoters, 152–153
α−Globin locus, 343–344, 375
α-Globin promoter, 153, 160–165, 168
α-Globin regulatory domain, 147
α-Heterochromatin, 8
β−Globin
locus, 109–110, 342–344
chromatin structure of, 112–113
cis-regulatory elements and
interaction, 119, 121
expression, developmental
regulation, 113–115
LCR, 110–112
mouse and human loci, 110
regulation and long-range interactions, 108
long-range activation by β−globin
LCR, 117–127
long-range control of gene expression by
enhancers, 115–117
transcription, enhancement of, 127–131
β-Globin LCR, 110, 117, 124, 127. See also
β−globin
enhancement of transcription
promoter escape and elongation, 130–131
promoter remodeling, 127–128
transcription initiation, 128–130
long-range activation by
LCR, in close proximity to
promoter, 118–121
LCR-promoter contacts
establishment, 121–124
transcription efficiency, 124–127

β-like globin, 109
β-Heterochromatin, 8, 11
Globin promoters, 110
GO. See Gene ontology
GO terms, 316
Green fluorescent protein, 317
Greig cephalo-polysyndactyly syndrome, 352
gremlin gene, 187, 274
role in genetic disease, 365
GRN. See Gene regulatory network
GRP. See Gastrin-releasing peptide
GRPR. See Gastrin-releasing peptide receptor
GTFs. See General transcription factors
Gtl2/Dlk1 locus, 217
gypsy insulator sequence, 28

H

Haematopoietic stem cells (HSCs), 153,
157, 158
hairy gene, 9
H2AK121 ubiquitin ligase activity, 51
Hammertoe, 360
hb P2 promoters, in species distantly related,
73
HDAC. See Histone deacetylases
HDACis. See Histone deacetylase inhibitors
Hematopoiesis, stages, 157
Hematopoietic progenitor cells, 113
Hematopoietic-specific activators, 112
Hematopoietic transcription factors, 114
Hemimelia extra-toes, 360
Heparan sulfate synthesis gene, 256
Heterochromatin, 2, 6, 7. See also Position-effect
variegation (PEV)
initiation of formation, 23–24
methylation and spread of, 22–23
terminating heterochromatic
spreading, 24–26
Hidden Markov models, 277
H19/Igf2 locus, 215
Hirschsprung disease, 378
Hispanic deletion, 111
Histone acetylases (HATs), 228
Histone acetylation, 166
Histone deacetylase inhibitors, 261–262
Histone deacetylases, 228
Histone demethylases (HDMs), 228
Histone H3K4 methylation, 166
Histone methyltransferases (HMTs), 228

Histones
  acetyltransferase inhibitors, 261
  code, 249
  deacetylases, 254
  methyltransferase inhibitors, 261
  modifications and transcription factor
      binding, 156, 228
  posttranslational modification, 253
  sumoylation, 228
  trimethylation, 255
  ubiquitylation, 228
Histones H3 and H4, hyperacetylation, 20
H3K27 methyltransferase activity, 53
Hm. See Hammertoe
HMM. *See* Hidden Markov models
Holocomplex, 111, 121. *See also* β−globin
Holoprosencephaly, 357–362
Homeobox genes, 177
HOM/Hox gene, 96
Homologous recombination technique, 285–286
Hoxb1 and *Hoxb4* gene, 313
HoxB cluster, 182
HOX-B gene cluster, 59
Hoxb4 genes, 182
Hoxb4, *Hoxb5*, and *Hoxb6* region, 182
Hox cluster in cephalochordate, 182
Hoxd10 and *Hoxd11* expressions, 182
HoxD cluster, 182
HoxD cluster gene, role in limb
      development, 363
Hoxd genes, 186
HoxD/Lnp regulatory landscape, 188
Hox gene, 181–182
  expression, ELCR control, 364
HOX genes, 47
HP1, and H3K9, methylation, 29
HPE3, 357
HPT. *See* Hypoparathyroidism
HSCR. *See* Hirschsprung disease
Human β-globin locus, 109
Human chromosome, global control
      region, 363–365
Human genetic disease, genes effect, 348–351
Human genome, genomic architecture,
      271–272
  distant regulatory elements, 273–274
  noncoding mutations, 274–275
Hunchback (hb)
  in Drosophila, 73
  P2 promoter of, 73–75

promoter in higher *Diptera*, conservation, 74
tailless promoter, conservation, 75
Hx. See Hemimelia extra-toes
Hydroxamic acids, 262
Hyperacetylation
  of histones at β-major and β-minor
      promoter, 125
  lysine 4 of histone H3,112
  and open chromatin structure of locus, 126
Hypermethylation
  of CTSs and biallelic expression of IGF2, 241
  CTSs and BWS, 242
Hyperphagia, 230
Hypogonadism, 230
Hypoparathyroidism, 347–352

I

ICF. *See* Instability, and facial anomalies
      syndrome
IG-DMR deletion allele, 216
IGF2 gene, 241
Igf2r/Air locus, 215
Igf2r cluster, 212
Igf2r gene, 212
Ihh. See Indian hedgehog
Imprinted Dlk1/Gtl2 locus, structural
      organization, 216. *See also* Clusters,
      imprinting regulation in
Imprinting center deletion, 232–239. *See also*
      Imprinting defects
  in 11p15, 241
  in 15q11-q13, 233
  segregation of, 234
Imprinting centers (IC), 232, 234–239, 241, 242
Imprinting control region (ICR), 208, 213, 218
Imprinting defects. *See also* Genomic imprints
  in 11p15, 239–242
  in 15q11-q13, 230–239
Indian hedgehog, 361
Initiation codon (ATG), 153
INSIG2. *See* Insulin-induced gene 2
Instability, and facial abnormalities
      syndrome, 250
Insulin-induced gene 2, 377–378
Interleukin 4, 5, and 13 locus, enhancer
      activity, 273
Inversion, orientation changes, 327
In(1)w^{m4} chromosome, 18
Iridogoniodysgenesis, 365

Iris hypoplasia, 365
Irx/Iroquois gene clusters, 178

K

Kaiso protein, 253
Kcnq1 gene, 214
Kcnq1ot1 ncRNA, 215
Keratins, 177
Key developmental genes, high-resolution
analysis, 317
Kilobases (kb), 271
Knock-out (KO) technology, 285
Krppel associated box (KRAB) gene family, 181
Kynurenine production, 11

L

Lac genes, 176
lacI gene, 176
LacZ gene, role in transgene visualization
method, 283
LAGAN, 279
Lamins, 27, 28
Langer dysplasia, 346
Late erythroid progenitors, 157
LCR. *See* Locus control region
ld. *See* Limb deformity
ld gene, 187
ld mutation, 186
Leri-Weill dyschondrosteosis, 346
Leukemias, 260
Limb bud morphogenesis, *formin* gene, 274
Limb deformity, 364
limb deformity (ld) locus, 186
Limb morphogenesis, 193
LIM-domain binding protein 1 (Ldb1), 122
LMBR1 gene in PPD, 359
Lmx1b gene, 193
LNP. *See* Lunapark gene
Lnp–Evx2–Hoxd genes, coregulation, 191
Lnp gene, 185
Locus control region (LCR), 108–109, 119,
342
Long-range gene silencing, 51
Lophotrochozoans, 178
Lunapark gene, 363
Lysine acetylation, 20
Lysine residue, homeodomain, 73
Lysine-specific demethylase 1 (LSD1),228

M

MAF gene, 355–357
MatInspector, 280
MBD. *See* Methyl-CpG-binding domain
MCS. *See* Multispecies conserved sequences
MeCP2 proteins, role in DNA methylation, 254
MEME algorithm for motif modification, 280
Mental retardation, 230
Methylation, 111–112, 120. *See also* β–globin
of H3K4,20
of Lys4 of H3 (H3K4),228
Methyl-CpG-binding domain, 253
5,10-methylenetetrahydrofolate reductase
(MTHFR) gene, 239
Microcephalus ataxia, 230
microRNAs (miRNAs), 256–257
miniwhite transgene, 57
miRNA, epigenetic inactivation affect, 258
Mixed-lineage leukemia, 260–261
MLAGAN, 317
MLH1 hypermethylation, 251
MLL. *See* Mixed-lineage leukemia
mod/mdg4 mutation, 28
Monte-Carlo simulations, 62
Mosaic hypermethylation, 241
Mottled phenotype, 3
Mouse erythroid leukemia (MEL), 113, 157
Multigenic complexes, 178
Multipotential progenitors, 157
Multispecies conserved sequences (MCS), 146,
314
Murine *Evx* genes, 185
Muscular hypertrophy, miRNA role, 370
Muscular hypotonia, 234
Mutation in *Ligase4*, 29
Myelodysplastic disorders, 261
Myeloid leukemias, 261

N

N-β-alanyl dopamine synthetase (NBAD), 85
N^{266-54} chromosome, 10
ncRNA Kcnq1ot1,215
ncRNAs, 216
Neonatal muscular hypotonia, 230
New regulatory modules, evolution, 95–98
NF-E2 heterodimer, 158
Nicotinamide, 261
Nipped-B, 373–374

Nkx2–9 gene, 314
Noncoding DNA
    evolution and variation, 297–300
    natural selection, 300–301
Noncoding region, human
        abnormalities, 274–275
Non-erythroid cells, 162
Normal and aberrant imprints, in 11p15, 240.
    See also Imprinting defects
Notch-mutant phenotype, 5
Notch mutations, 5
N-terminal tails of histones, modifications,
    166
Nuclear envelope protein LAP2β, 27
Nuclear PEV modifiers of w^{m4}, 21–22. *See also*
    Position-effect variegation (PEV)
Nucleolar organizer region (NOR), 29
Nucleosome free region, 153–154, 160
NVPLAQ-824 gene, 262
N$^{266-54}$ X chromosome, 6, 7

O

Obesity, 230
Ocular degeneration with sex reversal, 368
Ods. See Ocular degeneration with sex reversal
Olfactory receptor (OR), 109
oligozeugodactyly, 361
Ontogeny, 113
Operons, 176–177
Orthologous even-skipped stripe 2 elements and
    eve protein, 78
Osteochondrodysplasia, 367
ozd. See oligozeugodactyly

P

PACs. *See* Pl-based artificial chromosomes
Pairing-sensitive silencing (PSS), 57
Pairwise sequence alignment, 277
Parasitic DNA and cancer correlation, 250
Parathyroid hormone, 347–348
Parental alleles, epigenetic differences, 227.
    See also Genomic imprints
Preaxial polydactyly, 357–362
PAX6 and *PAX9* gene, 314
PAX6 gene, 353–355
PC. *See* Positive cofactors
p302/CBP, protein complexes, 122
PcG. *See* Polycomb group

PcG domains (PcDs), 50
PcG-mediated silencing, 56
PcG–PRE interaction, 50–51
PcG protein, 260
PcG-regulated genes, 60
P element transposition, 8
Pelizaeus-Merzbacher disease, 356
Peptide CoA conjugates, 261
Peptidylarginin-deiminase 4 (PADI4), 228
Pheromone receptors, 178
Phylogenetic footprinting, 270, 280, 310
PhyloGibbs algorithm for motif
        modification, 280
PIC. *See* Preinitiation complex
p16INK4a hypermethylation in carcinoma
        cell, 252
PIS. *See* Polled intersex syndrome
PITX2 gene, 347
Pl-based artificial chromosomes, 285
Pleiohomeotic (Pho) repressive complex
        (PhoRC), 48
Pleiotropy, 178
PLP1. *See* Proteolipid protein gene
PLP1 gene, 355–357
Pluripotent ES cells, 162, 165
PMD. *See* Pelizaeus-Merzbacher disease
POF. *See* Premature ovarian failure
Point mutations, 69, 181
Pol II. *See* Polymerase II
Polled intersex syndrome, 366
Polyadenylation, 176
Poly-ADP-ribosylation, 228
Polycistronic pre-mRNAs, 176
Polycomb group (PcG), 247
        components, in Drosophila and human
            orthologues, 49
Polycomb group protein (PcG proteins),
        46–48, 56
        genetic and biochemical characterization,
            48
        and long-range gene silencing, 53–57
        mechanisms of action, 49–53
        and very long-range gene silencing, 57–61
Polycomb (PC), 53
Polycomb repressive complex 1 and 2 (PRC1 and
        PRC2),48
Polycomb response elements (PREs), 49, 190
Polyhomeotic (PH), 53
Polymerase II, 272
Polytene chromosomes, 2, 57

Position-effect variegation (PEV), 1
  effect of T(1;4)w$^{m260-21}$, 10
  genome organization and, 19
    Chromatin structure, 20–26
    Nuclear organization, 26–29
  heterochromatin and euchromatin, 7–8
  of inversion chromosome, 4
  modifiers of PEV phenotype, 12–13
  phenotype and heterochromatin, 5, 6
  Transposon insertion PEV, 16–18
  types of, 8
    chromosomal rearrangement, 8–15
    pairing-dependent dominant and
      trans-inactivation, 18–19
    transposon insertion, 16–18
  X chromosome rearrangements, 6
Position weight matrices, 280
Positive cofactors, 272
POU3F4 gene, 347
POZ -domain proteins, 28
PPD. *See* Patient with preaxial polydactyly
Prader-Willi syndrome (PWS), 230–232
PRE–gene interactions, 51
Preinitiation complex (PIC), 145, 159–162,
  164, 272
Premature ovarian failure, 366
Proerythroblasts, 153, 155, 157, 163, 165
Protein-coding genes, 296
Protein–DNA complexes, 164
Protein–protein interactions, 70, 162, 167
Proteins polymerization, 116
Proteolipid protein gene, 356
Protocadherins, 177
PTH. *See* Parathyroid hormone
Putative distal regulatory sequences, 312
Putative regulatory elements, scanning
  sequences, 276
PWMs. *See* Position weight matrices
PWS-SRO
  AS-SRO element interaction, 238
  to establish maternal imprint, 235
  to establish maternal methylation at, 242
  regions of deletion overlap in PWS, 233
PXD103 gene, 262

Q

15q11-q13, normal and aberrant imprints, 231.
  See also Imprinting defects
QTLs expression, 377

Quantitative polymerase chain reaction
  (qPCR), 61, 311
Q9Y2K8 gene, 322

R

RAI1. *See* Retinoic acid inducible gene
Random mutagenesis method for alleles loss, 285
raz. *See* Replicated anterior zeugopod
Receptor expression enhancing protein, 356
REEP. *See* Receptor expression enhancing
  protein
REEP3 gene, 355–357
Regulatory elements, 310–312
  computational tools
    evolutionarily conserved noncoding
      sequences, 276–279
    TFBSs prediction, 279–282
Regulatory landscapes, vertebrate development
  and evolution, 185, 192
Remote regulatory elements, 161–162
Replicated anterior zeugopod, 361
Retinoic acid, 176
Retinoic acid inducible gene, 371–372
Retrotransposons, HeT-A and TART, 8
Rian/Meg8 locus, 216
Ribosomal RNA synthesis in female ovaries, 11
Rieger syndrome, 347
RNA polymerase II, 51, 116–117, 123, 160, 166
RUNX2 gene, position effect, 363
rVISTA web tools and steps, 280–281

S

S-adenosyl-methionine, 228
Saethre-Chotzen syndrome, 347
SAHA, 262
SALL1 gene, 332
Sall3 gene, 325
Sasquatch (Ssq) mutation, 359
Sclerosteosis, 345
Sclerostin, 345
scute (sc) gene, 79. *See also* Dorsocentral
  enhancer (DCE)
SDC2 gene, 355–357
Secreted frizzled-related protein 1, 256
Sequence turnover, 69–70, 72, 89, 98
Sex-linked mutation, 3
Sexual dimorphism, 85

SFRP1. *See* Secreted frizzled-related protein 1
SHFM1. *See* Split hand foot malformation type I
SHH. *See* Sonic hedgehog
SHH expression, misregulation, 359
SHH gene
  limb malformation effect, 358
  position effect, 363
  ZRS function, 360
Shh gene, 359–360
Shh null allele, 360
Short-digits, 361
SHOX. *See* Stature homeobox gene
Silver-Russell syndrome (SRS), 239
Single gene Su(var) and E(var) mutations,
  13–15. *See also* Position-effect variegation
  (PEV)
Single nucleotide polymorphisms, 279, 299, 344
siRNA, role in gene expression, 370–371
Sleep apnea, 230
Slippage, 69
Slit and Trk like family member 1, 370–371
SLITRK1. *See* Slit and Trk like family member 1
small nucleolar (sno) RNA genes, 230
Smith-Magenis syndrome, 371
SMS. *See* Smith-Magenis syndrome
SNPs. *See* Single nucleotide polymorphisms
SNRPN gene, 230
SNURF-SNRPN promoter/exon 1 region, 232
SNURF-SNRPN sense-*UBE3A* antisense
  transcription unit, 236
Sodium valproate, 262
Somatic mosaicism, 237–238. *See also* Imprinting
  defects
Sonic hedgehog, 357–359
SOST. *See* Sclerostin
SOST gene expression, 275
SOX9 gene, 314, 366–368
Spastic paraplegia type 2, 356–357
SPG2. *See* Spastic paraplegia type 2
Split hand foot malformation type I, 362
Splitomycin, 261
Spreading effect, 10–11. *See also* Position-effect
  variegation (PEV)
Sp-XKLF proteins, 160
Sp-XKLF transcription factors, 161
Stabilizing selection, 70, 72, 79, 84
Stable position effect (S-type), 3
Stature homeobox gene, 346
Stripe 2 enhancer, of even-skipped, 75–79
Su(Hw) protein, 28
Su(var) allele, 14

SWI/SNF, chromatin-remodeling complex, 115
SYNOR web tool, 281
Synteny, loci in conservation, 193

T

TAFs. *See* TATA-associated factors
tailless (tll) promoter, 74. *See also* Hunchback (hb)
Takifugu rubripes-model genome, 312–313
TATA-associated factors (TAFs), 153–154, 272
TATA-binding protein, 272
TATA box, 110, 152, 159
TBP. *See* TATA-binding protein
TBP-associated-factors (TAFs), 159
TBS. *See* Townes-Brock syndrome
Teleosts, 313
Teleregulation of gene expression, 57–61. *See
  also* Polycomb group protein (PcG proteins)
TFBSs. *See* Transcription factor binding sites
TFBSs prediction, 280–282
TFCC. *See* Transcription factor centric
  clustering
TFRs. *See* Transposon free regions
TGFB2 gene, 355–357
Thalassemia, 111, 112, 342–344. *See also* Alpha
  thalassemia
Tissue-specific regulatory elements,
  computational tool, 281
Tn7-based transposon, 186
Topoisomerase-I-interacting protein, 28
Tourette syndrome, 370
Townes-Brock syndrome, 274–275, 332
Trans-association/inactivation, euchromatic
  regions, 19
Transcriptional genomics, objectives, 281
Transcriptional regulatory elements, *in vivo*
  validation and characterization
  candidate regulatory elements, 285–286
  distant enhancer, 284–285
  enhancer validation, 282–284
Transcriptional repression domain, 254
Transcription factor binding sites, 272, 296, 298
Transcription factor centric clustering, 178,
  271–272, 281
Transcription start site, 319
trans-dev genes, 315, 317–318
TRANSFAC database, 280
trans gene expression, 301
Transgene visualization method for *LacZ*
  gene, 283
Transpositions in genomes, 327–328

Transposon free regions, 333
Transposon silencing, 28
Transvection, 115, 371
Trapoxins, 262
TRD. *See* Transcriptional repression domain
TRES algorithm for motif modification, 280
Trichostatin A, 262
TRIP8 gene, 356
Trithorax group (trxG), 47, 190
TS. *See* Tourette syndrome
TSS. *See* Transcription start site
Tumor-suppressor genes
    APC, 252
    CpG island hypermethylation, 250–251
    CpG island of, 250–251, 254
    determinants of, 258
    expression of, 262
    hMLH1 and BRCA1, 252
TWIST gene, 347

U

UBE3A gene, 230
Ubiquitin ligase dTopors, 28
Ubiquitous transcription factors, 115
UCSC genome browser database, 279
Ulnaless allele, 364
Ulnaless (Ul) mutation, 186
ultraBar mutation, 3
Unequal recombination, 69
Unilineage erythroid progenitor, 165
Unit of inheritance, 376
Untranslated regions, 279, 315
Upstream stimulatory activity, 272
USA. *See* Upstream stimulatory activity
UTRs. *See* Untranslated regions
U-Ube3a-ATS transcripts, 236, 237

V

Van Buchem disease, 275, 345–346
Variegated position effect (V-type), 3
VB. *See* Van Buchem
Vertebrate genomes, conserved noncoding
    regions, 314
Vertebrate *Irx* genes, 179
Vertebrates transcriptional genomics, problems
    in, 273

VHL. *See* Von Hippel-Lindau
Vidaza. *See* 5-azacytidine
VISTA genome browser, 279
Von Hippel-Lindau, 250

W

w^{258-18} chromosome, 10
white gene, 3, 29
White-mottled mutations, 5
White mutation, 11
Wilms tumors, 239
Wing spot of pigmentation
    yellow expression and regulation, 88 (*see also*
        Enhancer evolution and loss or gain of
        traits)
    yellow protein prefigures pigmentation,
        expression, 86 (*see also* Enhancer
        evolution and loss or gain of traits)
w^{m4} inversion, 28
w^{m11} translocation, 9
Wnt-1 gene, 314
WWOX gene, 355

X

X chromosome, 3, 5, 12
Xenopus embryos, transposon gene delivery
    method, 284
X-linked disorder, 347, 357
X-ray treatment, 5

Y

YAC. *See* Yeast artificial chromosome
Y chromosome, 7, 9, 12–13
Yeast artificial chromosome, 284, 353
Yet gene, 176

Z

Zebra fish, transposon gene delivery method,
    284
Zebularine, 261
Znf503 gene, 328
Zone of polarizing activity regulatory
    sequence, 358
ZRS. *See* Zone of polarizing activity regulatory
    sequence